Teaching Biology

György Kriska

Teaching Biology

Photosynthetic Organisms

Springer

György Kriska
Biological Institute, Group for Methodology in Biology Teaching
Eötvös Loránd University
Budapest, Hungary

Photographs were taken by the author except for the following: Béla Lovas
OPTICAL DEVICES AND THEIR USE: 16, 19
ALGAE: 1-18, 20-23, 26, 28, 29-32
MOSS, LICHENS, FERNS: 10
Drawings are by the author.

ISBN 978-3-031-84084-5 ISBN 978-3-031-84085-2 (eBook)
https://doi.org/10.1007/978-3-031-84085-2

Image caption: Cross section of reed (Phragmites australis) rhizome stained with toluidine blue.

This Springer imprint is published by the registered company Springer Nature Switzerland AG
The registered company address is: Gewerbestrasse 11, 6330 Cham, Switzerland

Introduction of the Book

The primary goal of the publication titled "Teaching Biology - Photosynthetic Organisms" is to support the practice-oriented teaching of biology; therefore, in addition to basic theoretical knowledge, it contains detailed descriptions of examinations and experiments. These are supplemented with digital material: film excerpts and interactive tasks. The mobile phone compatible supplements can be accessed from the printed publication via QR code, while they can be accessed from the electronic version with a click. Thus, the book not only facilitates the execution of observations and experiments, but can also be a useful tool for classroom activities and independent learning as well.

The wide range of exercises can provide a good basis for competition preparation and the implementation of activities that put students' independent research in the foreground.

In addition to the simple observations in the book, the laboratory practices are not too time-consuming, and there is no need for special materials and tools. The investigations can be carried out in an average equipped school as well.

The first chapter of the volume presents the traditional optical tools of experiments (e.g. microscopes). In addition to describing the techniques that can be used during their use, I also touch on their integrated operation with ICT tools (e.g. microscopic photography, filming, and projection).

The second major thematic part of the publication summarises practical knowledge related to the morphology and physiology of photosynthesising organisms.

Like the other chapters, this is also richly illustrated with detailed captioned figures, so in addition to supporting classroom and laboratory activities, it can also be a useful tool for field practices.

The next part, the largest unit of the book, deals with plant histology.

In addition to providing very spectacular results, plant histological examinations are among the simplest activities to carry out. This is confirmed by the fact that I prepared the hundreds of tissue sections in the book by hand with the help of a razor blade. To create the colour cavalcade that faithfully reflects the tissue structure of the specimens, I only had to dip the sections into a water solution of toluidine blue for half a minute. Therefore, it only takes a few minutes to prepare and stain the histological sections presented in the book, while their microscopic examination can provide us with an exceptionally rich body of knowledge. The question inevitably arises as to why these examinations have not really spread in schools if all this is true. The reason is well illustrated by the following picture.

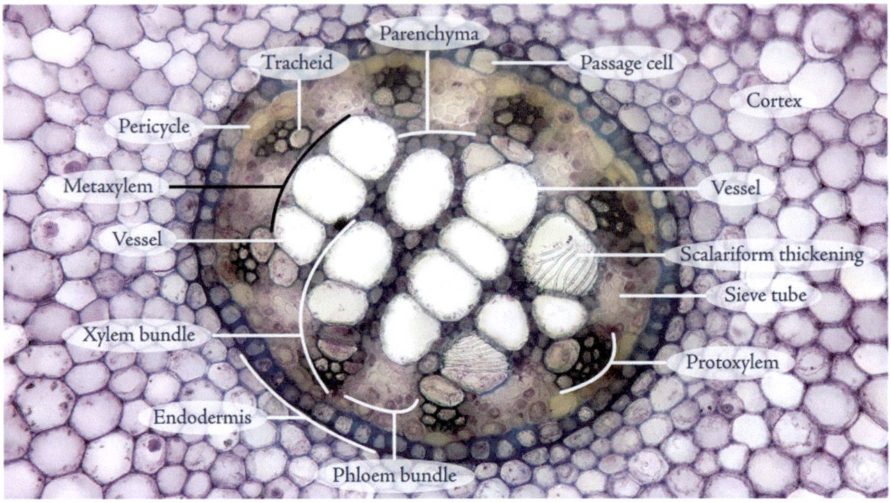

Cross-sectional image of the central cylinder of a lily root

During the light microscopic examination of a root cross-section stained with toluidine blue in a few minutes, an incredibly complex, yet wonderful structure unfolds before our eyes. It would be a challenge for any biology teacher to navigate such a specimen without appropriate educational aids. I hope that this book can help somewhat with the above problem.

It is important to emphasise that the labelling of histological images in the book is often much more detailed than what can be expected in public education.

Nevertheless, I considered it essential to present this knowledge because based on this, biology teachers can professionally carry out the necessary simplifications for their students.

A teacher does not make a professional mistake, for example, when he/she refers to the epidermis, rhizodermis, periderm, and rhytidome as dermal tissue, but as a biology teacher, he/she must know the differences between them, as these are clearly visible on the microscopic sections.

The plant histology section is followed by collections of cell, biochemical, and physiological examinations. The main criteria for their selection were the simplicity of implementation and the achievement of spectacular results. In the descriptions of the experiments tested multiple times, I always provided the simplest implementation methods. As an extreme example, I can mention that a professional book gives the following ideal mineral salt composition for preparing a pollen-germinating sucrose solution:

"To prepare 1 L, 0.417 g of calcium nitrate, 0.200 g of boric acid, 0.101 g of potassium nitrate, 0.217 g of magnesium sulphate and 3.5 cm^3 of 1.0 M ammonium hydroxide solution should be dissolved in water. When mixed with the sucrose solution, the pH of the mineral salt solution should be 8.8". In contrast, a well-functioning pollen-germinating solution can be prepared by dosing three substances (sucrose, gelatine, and water) with a teaspoon.

Finally, adapted to the publication of this book, I conclude my lines with the thoughts of the late professors Sándor Sárkány and István Szalai, published 60 years ago and still valid today:

The successful teaching of biology to students largely depends on how much we can arouse and increase their interest in the basic phenomena of living nature during their studies, how much we can develop their ability to observe and experiment, to deepen biological thinking. The ultimate goal is to educate for a modern biological perspective, for independent problem-solving. In addition to a thorough knowledge of the theoretical material, it is essential for students to have all the knowledge and skills they have gained through the annual subject-specific practices, study trips, gradually independent work, and have permanently recorded. Therefore, it is desirable that future professionals in the field of biology—whether they participate in teaching and education, or work in scientific research—should be in close contact with the primary source, the infinitely varied yet unified world of nature, and strive to see the objects and phenomena of nature in their reality and depth, and thus proceed on the surest path to the knowledge of truth.

I would like to express my gratitude to Dr. Eszter Saláta-Falusi and Prof. Dr. Károly Penksza, an excellent experts and researchers in the field, for her thorough professional review of the book.

Institute of Aquatic Ecology, Lendület Fluvial Ecology Research Group Centre for Ecological Research, Budapest, Hungary

György Kriska

Contents

1 Optical Devices and Their Use

1.1 Triple Bug Viewer

The **Triple Bug Viewer** allows for different viewing modes (Fig. 1.1). The upper part is a transparent, waterproof cup with a lid that can be fitted onto the closed cylindrical part at the bottom. The removable lid of the plastic cup contains two magnifying lenses. One is built-in, while the other is a further magnifying lens that can be folded over it.

To try the Triple Bug Viewer, pull off the magnifying lid of the plastic cup and place a half-centimetre thick lemon wedge or tomato slice at the bottom of the cup! After replacing the lid, start the test!

First, fold up the magnifying lens of the cover and examine the object through the built-in lens only. To get a sharp image, you then need to look into the Triple Bug Viewer from a distance (Fig. 1.2, left image), which will result in two eyes looking at the object. This is important because it gives us a three-dimensional view. Because the magnification is not too high, the depth of field of the image is excellent, which means that every detail of the object is sharply visible. A sharp and three-dimensional image gives a real aesthetic experience during the examination. Next, place the other magnifying lens on the cover and examine the object. With the new lens, you can achieve a higher magnification, but then you have to look closely into the Triple Bug Viewer to see a sharp image (Fig. 1.2, right image). The single-eye examination no longer allows you to see the stereoscopic image, but the higher magnification can compensate for this.

The lower mirror and magnifying glass part of the Triple Bug Viewer allow you to observe the object from below without having to turn it around (Figs. 1.3 and 1.4).

The cup is fully transparent, allowing not only viewing the subject from the top and bottom, but also from the side. The illumination of the object under study does not require a separate light source, so the device can be used in both artificial and natural lighting. The magnifying device, which allows observation from below, consists of two parts, similar to the one on the cup cover. It allows the object to be examined in two dimensions (one eye) at high magnification, and in

G. Kriska, *Teaching Biology*, https://doi.org/10.1007/978-3-031-84085-2_1

Fig. 1.1 Triple Bug Viewer and its parts

three-dimensions at low magnification, as already described. The lower part of the binoculars also allows objects to be examined in transmitted light. This is done by shining a torch into the lower aperture while examining the object from above through the magnifying lenses of the cup cover (Fig. 1.5).

The use of transmitted light illumination is most useful when examining translucent objects such as fleshy fruits (Fig. 1.6).

It is not necessary to make a very thin section of the fruit, because the strong light from the lamp can shine through even 1–2 cm slices. In particular, two-eye scans are a fantastic way of looking at this type of inspection, because they provide a high depth of field and a three-dimensional image, which gives an insight into the spatial structure of the section. The magnifying lenses, which can be placed in the lid of the Triple Bug Viewer or folded out, can also be used on their own, allowing large objects to be examined (Figs. 1.7 and 1.8).

1.2 Light Microscope

Most people use a conventional transmission, or **light microscope**, which uses light rays that pass through the object under examination to form a magnified image (Fig. 1.9).

Fig. 1.2 Two ways to observe an object from above

Fig. 1.3 Viewing the object from below, higher magnification image. Removing the cup reveals the mirror

Fig. 1.4 Viewing the object from below, three-dimensional and low magnification image

Fig. 1.5 Using transmitted light with a flashlight

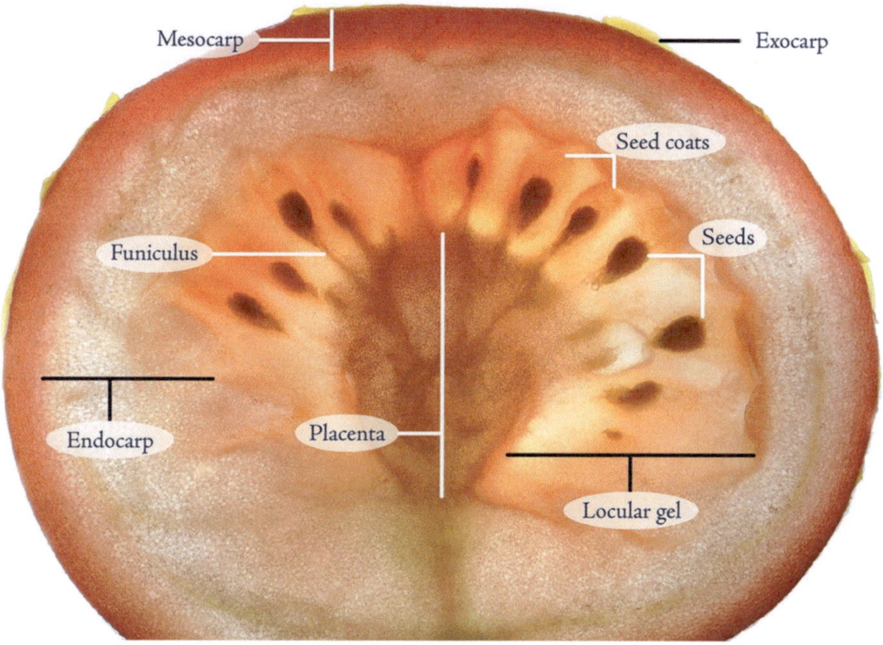

Fig. 1.6 Section of a tomato in the Triple Bug Viewer

Fig. 1.7 Using the side fold-out magnifying lens when examining large objects

Fig. 1.8 The lid of the Triple Bug Viewer as a magnifying device

Accordingly, the light source is located under the stage, or a movable mirror can be used to project the light onto the object from below. The basic function of the light microscope is to examine transparent, translucent objects that are very thin.

Thinness is important not only because of light transmission, but also because the depth of field in microphotography is extremely small.

The eyepiece can be inserted into the tube. The total magnification is the product of the magnification of the eyepiece and the objective lens. The tube continues in the widening head (turret), to the lower part of which is attached the nosepiece into which the objective lenses can be screwed. In most cases, the head is fixed to the microscope stand by screws. By unscrewing it, the head/tube assembly can be removed and a camera can be attached in its place (Figs. 1.10 and 1.11).

The quality of a microscopic image is determined by its resolving power, which is the minimum distance between two pixels at which they are perceived as separate points. The main optical element that determines the resolving power of a microscope is the objective lens, which is a lens system consisting of several lenses. Most of the refraction errors that occur during imaging can be corrected here.

The test object is placed on a microscope slide, usually covered with a thin coverslip, on the stage. Stage clips are used to secure the slide. During the test, the slide can be moved by hand or by means of screws that move the slide in a horizontal plane.

The light source is located under the object table. The narrower the light aperture, the less light enters the eye and the greater the depth of field. Too large a depth of field is not always beneficial, because in the case of multi-cellular sections,

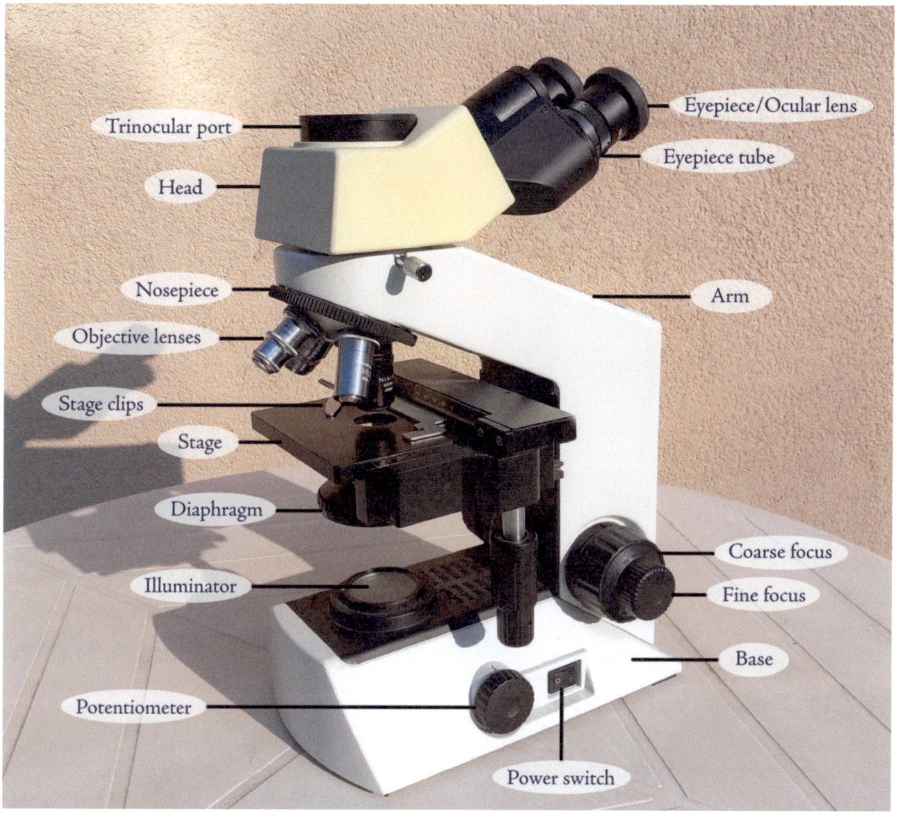

Fig. 1.9 Construction of the light microscope

deeper, lifeless areas may appear, which degrade the quality of the image (Figs. 1.12 and 1.13).

In more serious microscopes, the iris diaphragm is built into the condenser (condenser aperture diaphragm). The condenser comprises several lens elements, is responsible for uniform illumination of the field of view.

The light microscope image can be sharpened with the coarse focus for coarser adjustment and the fine focus for fine adjustment.

If used carelessly, higher magnification lenses can hit or even break the slide, causing damage. Therefore, when adjusting, it is advisable to first bring the objective lens close to the object while observing from the side and then adjust the focus by looking into the microscope.

Newer school light microscopes also incorporate upper illumination. This is a very useful tool because it allows us to examine thicker specimens that we cannot shine through (e.g. branch sections). We can also use microscopic staining techniques on these objects.

Some light microscopes are trinocular, which means that in addition to the two eyepieces, they also have an upright tube to which a camera can be attached.

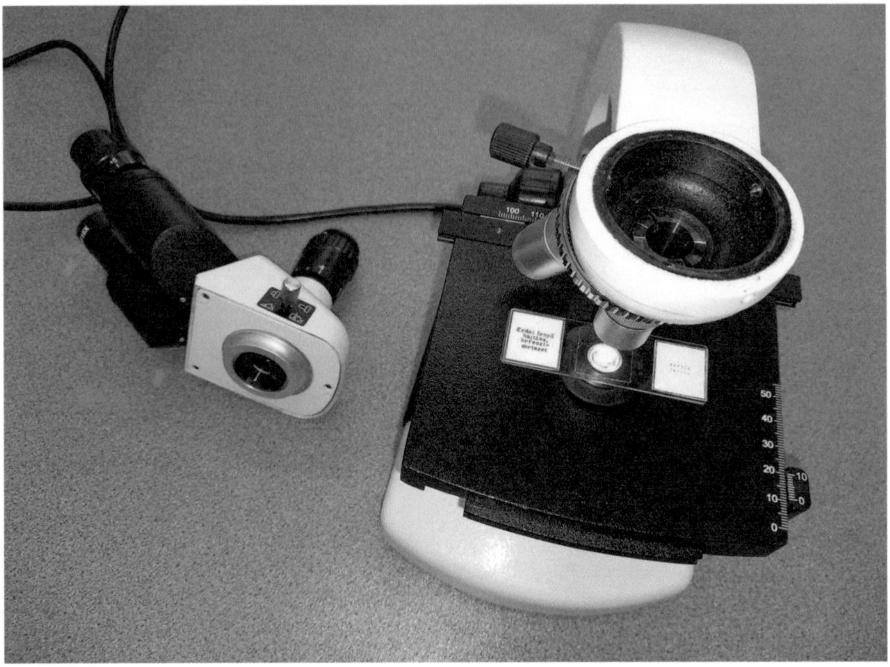

Fig. 1.10 Light microscope with the dismantled head-tube assembly

Unlike stereo microscopes, light microscopes use a single objective lens, so most of the photos taken with them can be made three-dimensional only with post-processing manipulations. This is also true for light microscopes with two eyepieces, because they deliver the same image to the right and left eye.

1.3 Light Microscopy Methods

Oblique Illumination Oblique illumination can be useful when examining translucent objects (e.g. starch granules) with light microscopy (Fig. 1.14).

To do this, after the object has been set up in the light microscope, open the condenser aperture fully and then, using a dark sheet or even a finger under the condenser, make the illumination one-sided. The manipulation results in a 3D-like image of the object under examination appearing in the field of view.

Reactions Under the Cover Slip Often if we want to study a reaction in a covered preparation using a light microscope, in particular for fast processes (e.g. plasmolysis, dissolution of calcium carbonate crystals, staining of starch particles with a Lugol solution, etc.) it is important to introduce the reagent gradually under the cover slip.

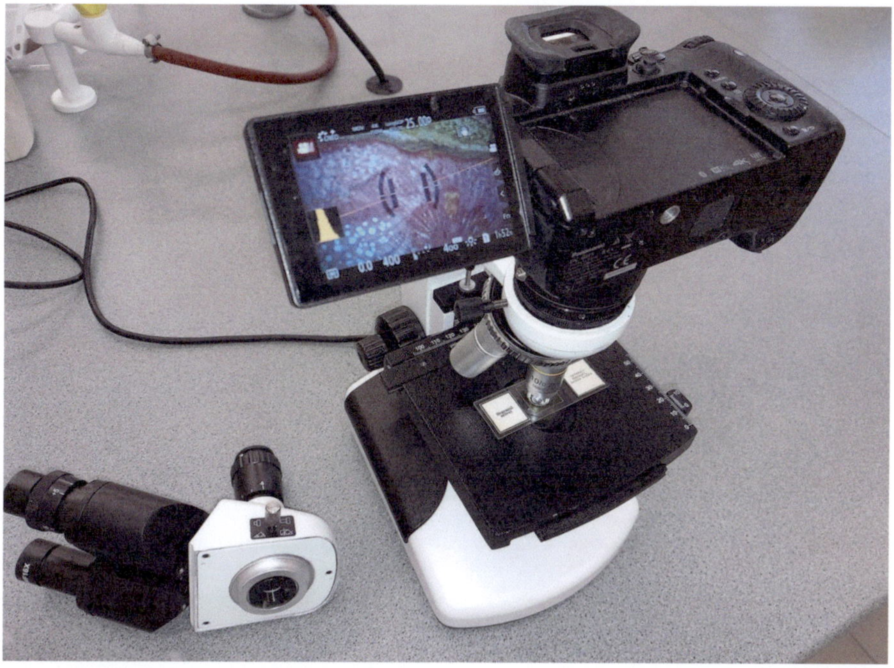

Fig. 1.11 Connecting the camera to the head

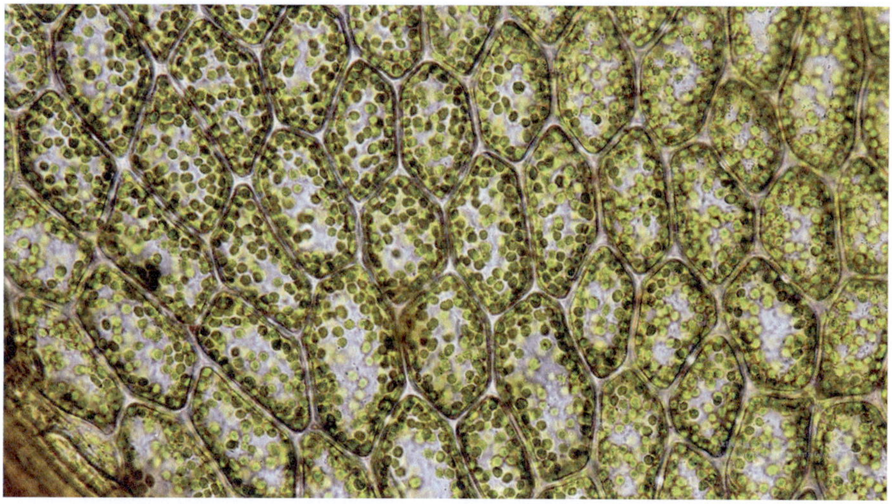

Fig. 1.12 Light microscopy image of a moss leaf with a narrower aperture

In such cases, during the test, drop a reagent onto one edge of the cover slip, then use a strip of filter paper to draw the solution through the preparation from the opposite edge of the cover slip. The reagent then gradually penetrates under the cover

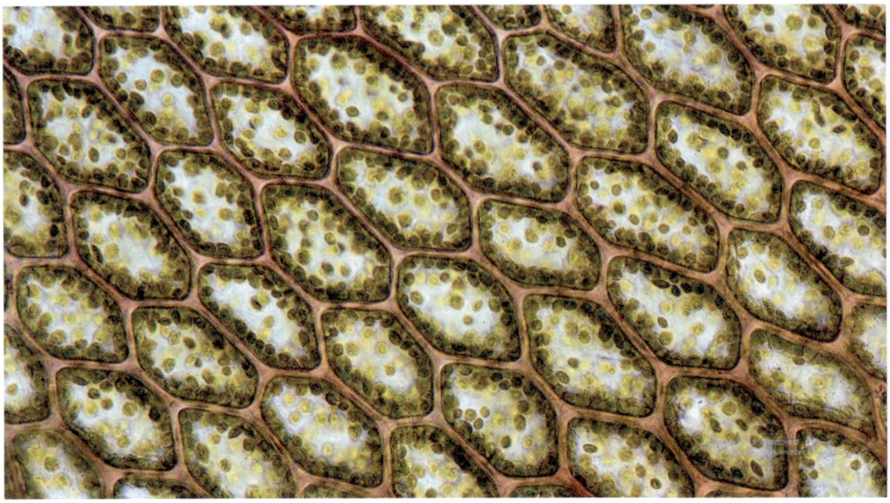

Fig. 1.13 Light microscopy image of a moss leaf with a less narrow aperture

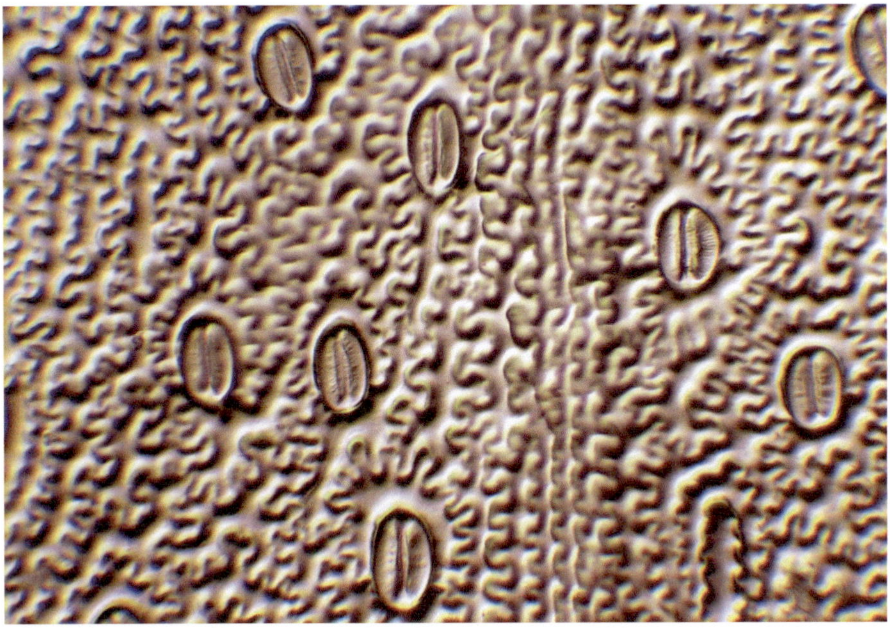

Fig. 1.14 Oblique illumination image of male fern (*Dryopteris filix-mas*) leaf epidermis

slip, so that the change occurs more slowly and gradually. This is particularly important when testing multi-step reactions, such as pH-dependent colour changes of anthocyanins, where the slow infiltration of alkali into the red cabbage tissue causes the appearance of not one but several colours (Fig. 1.15).

Fig. 1.15 Discolouration of a section of the upper coloured part of a red cabbage (*Brassica oleracea* var. *capitata* f. *rubra*) leaf by sodium hydroxide solution

Optical Staining Products examined with a light microscope are usually less colourful and therefore offer less aesthetic pleasure. Complex staining processes are also deterrent because of their complexity. Here we describe a method—optical staining—that is a very simple way to make our microscopic preparations magically coloured (Fig. 1.16).

Cut out a 1–2 cm disc from a coloured, transparent file folder and cut out the centre with a punch! Glue a disc cut from another colour folder in place of the tiny hole (Fig. 1.17)!

Place the resulting Rheinberg filter, named after its discoverer, under the stage in the path of the light (in the filter holder) so that the smaller, central coloured disc is exactly in the middle. By varying the position of the filter and (if available) the size of the aperture, you can adjust the position so that the colour of the object under examination in the microscope is same as the larger coloured disc and the background is same as the smaller one. Once the filter has been adjusted, it is advisable to fix it with adhesive tape. It is advisable to glue several of the smaller discs in the middle on top of each other to get a more colourful background. The most beautiful colour effects can be achieved with yellow-blue, red-green, and red-blue filters.

Lovas's technique differs from Rheinberg's in that the smaller coloured disc is stuck to a colourless film or glass plate, so that only the background is coloured while the object is in its original colour (Fig. 1.18).

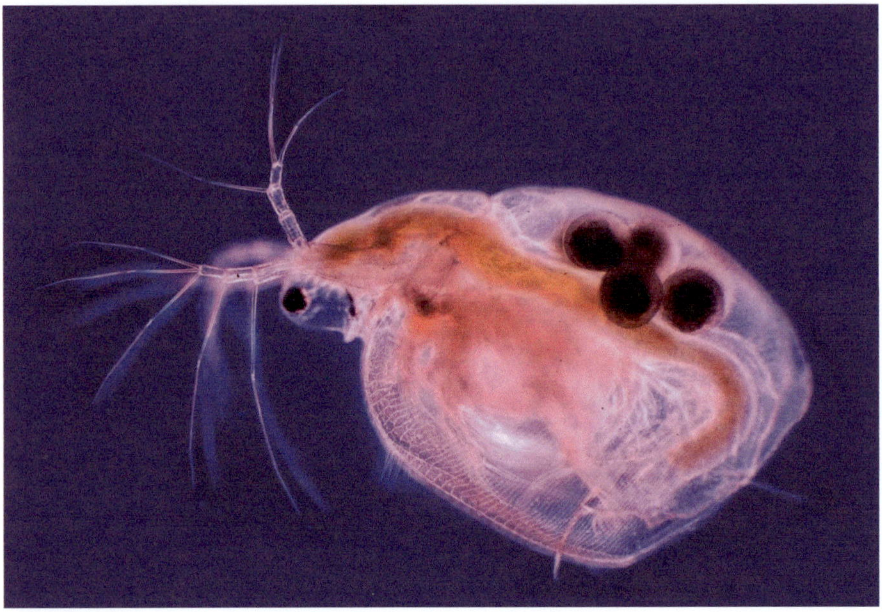

Fig. 1.16 Blue-red Rheinberg filter image of a water flea (*Simocephalus* sp.) (Photo: Béla Lovas)

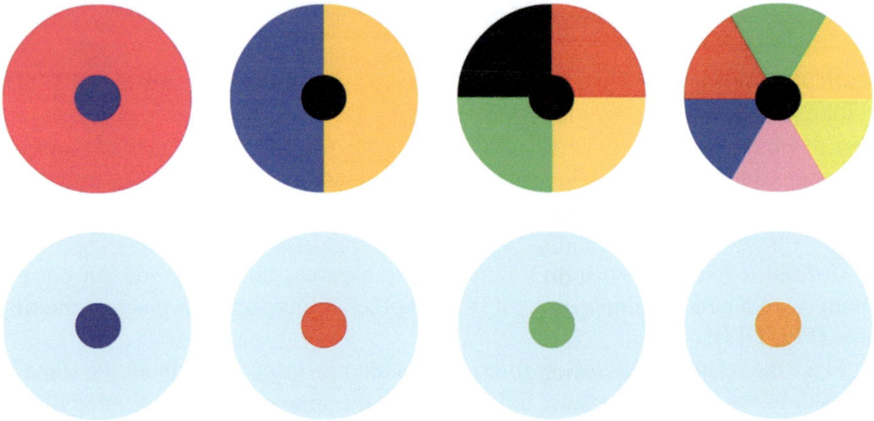

Fig. 1.17 Rheinberg filters (top row) and Lovas filters (bottom row)

Using four- or multi-sector filters to examine transparent crystals, each crystal plane appears in a different colour. The image of the crystal becomes three-dimensional, and the colours change continuously as the filter is rotated (Fig. 1.19).

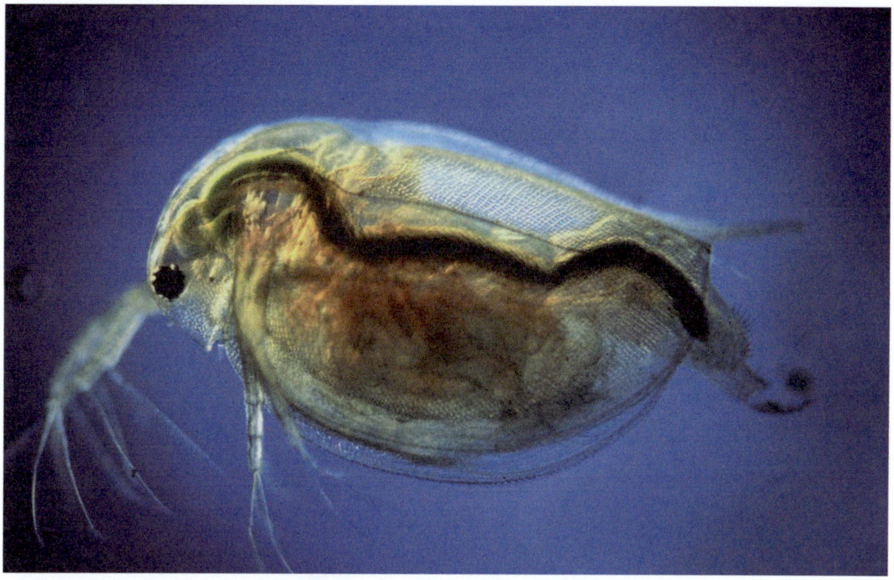

Fig. 1.18 Image of a large water flea (*Daphnia magna*) taken using the Lovas (blue-white) technique

1.4 Stereomicroscope

A **stereomicroscope** (Fig. 1.20) is always binocular, i.e. it has two eyepieces. What makes it special is not that it has two eyepieces, but that it also has two objective lenses. The latter are set at an angle and distance so that they do not give exactly the same image of the object (Fig. 1.21), so that the result of the two slightly different images is a single, but three-dimensional image in the microscope.

The stereomicroscope, unlike the transmission light microscope, is primarily used to examine objects that do not transmit light, but in most cases the instrument also has a bottom illumination option. In the lower illumination, light is projected onto the object by means of a mirror.

Nowadays, their light source provides cold light illumination, which is particularly important when examining living material.

Stereomicroscope lenses are not as good as camera lenses, and only expensive models have an aperture. This leaves something to be desired in the quality and depth of field of the image of the object under examination, especially at higher magnifications (Fig. 1.22). Therefore, for photography and filming, it may be preferable to use macro optics and extension tube instead of a stereomicroscope.

The stereomicroscopes can be trinocular, which means that while the object is viewed through two eyepieces, a camera can be attached to the third. Among stereomicroscopes, those with zoom optics are common, allowing continuous magnification.

Fig. 1.19 Rheinberg filter image of sugar (sucrose) crystals (Photo: Béla Lovas)

1.5 Microscope Photography, Filming, and Projection

Nowadays, digital cameras are also available as accessories for mono-, bi-, and trinocular microscopes, allowing you to capture microscopic subjects.

These can be operated by connecting them to a computer once the corresponding software has been installed. The microscope image and the camera control panel are also displayed on the computer monitor. The installation of the computer software is not always easy and the quality and resolution of the resulting image or film is not very good, which is a barrier to their wider use. It is therefore more practical to connect a digital camera with interchangeable lenses to the computer. This makes it possible to take high-resolution photographs and film. In addition, you can set high sensitivity of the camera (ISO), so that you can take good photos of fast-moving objects such as *Paramecium* using short exposure times (1/60 s) (Fig. 1.23).

We have several options for mounting the interchangeable lens camera on the microscope. For example, you can record through the eyepiece (Fig. 1.24).

Using this method with a stereomicroscope, we can even produce 3D images. In this case, the object under examination is photographed through both eyepieces, and a computer program is used to generate a 3D image (Figs. 1.25 and 1.26) from the resulting pair of images, which can be viewed through red-cyan glasses (Fig. 1.27).

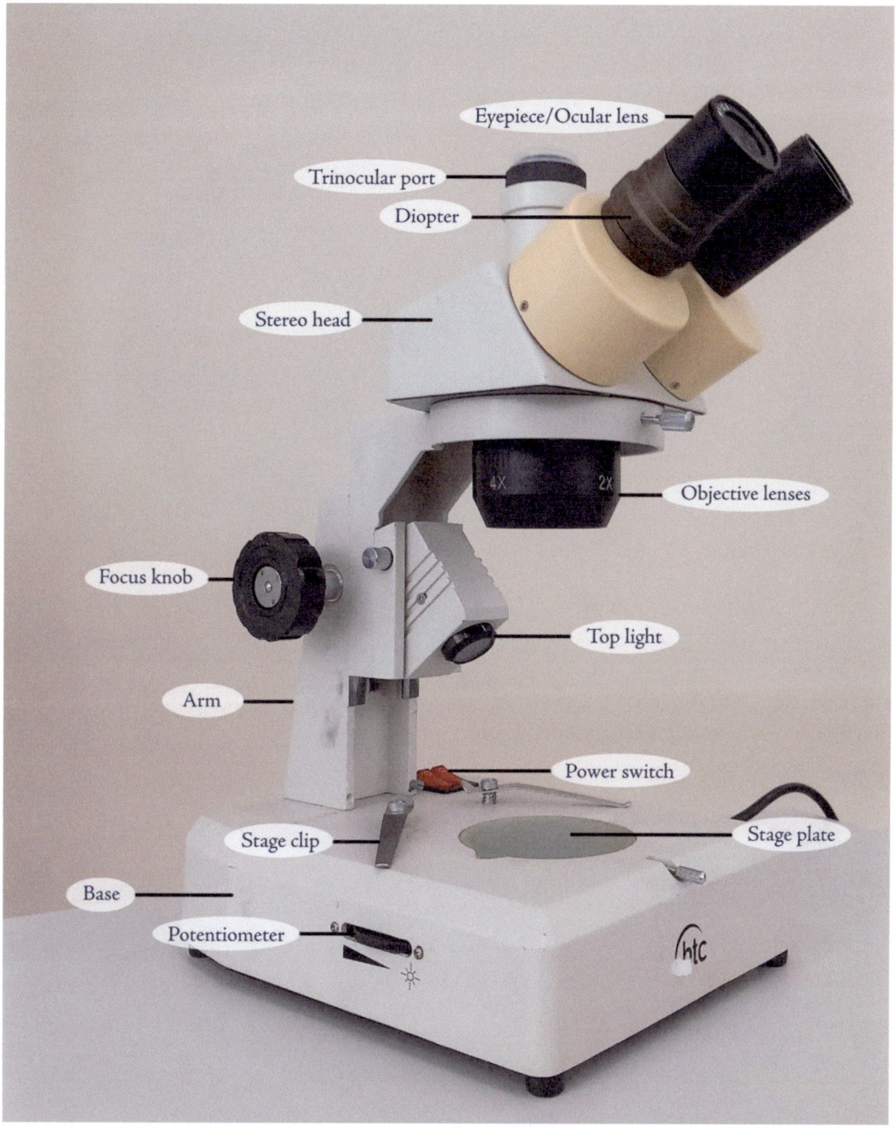

Fig. 1.20 Construction of a stereomicroscope

For monocular and binocular microscopes, the camera can be mounted in place of the head (Figs. 1.10 and 1.11).

At lower magnification, the image may appear circular and not fill the field of view completely.

Fig. 1.21 Bottom view of stereomicroscope objective pairs. The microscope has two magnification levels, so we can see four objective lenses

Fig. 1.22 Stereomicroscopic image of a fumewort (*Corydalis solida*)

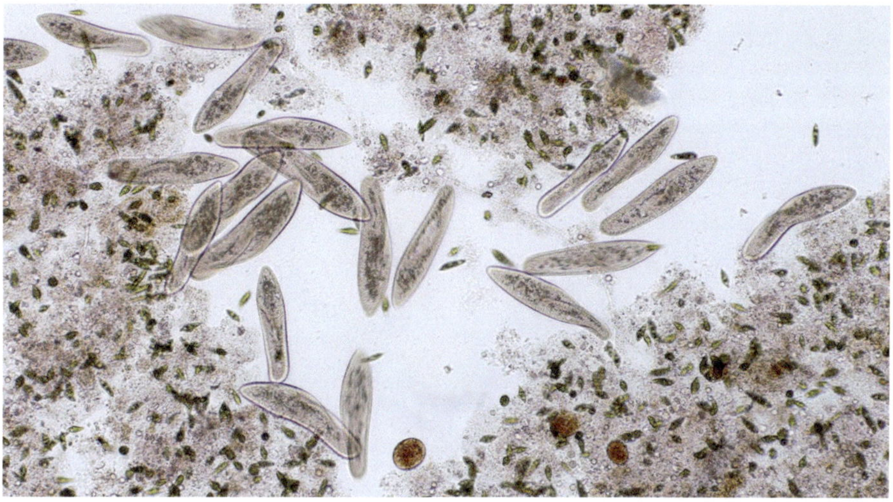

Fig. 1.23 Light microscope image of *Paramecium*

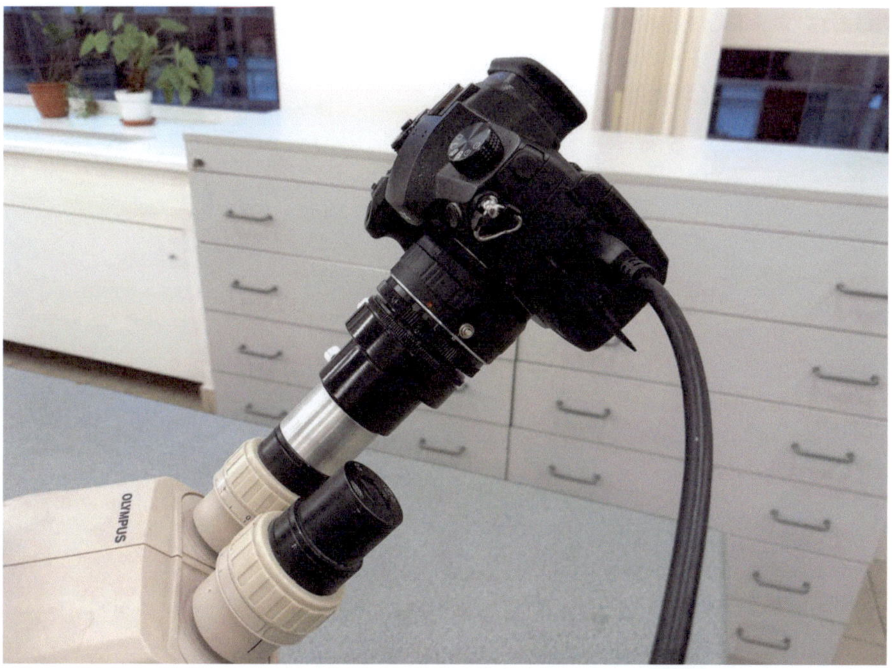

Fig. 1.24 Connecting the camera to one of the eyepieces of the stereomicroscope

Fig. 1.25 Picture of violet (*Viola odorata*) flower taken using the above method

Fig. 1.26 3D image of a violet (*Viola odorata*) flower in red-cyan glasses

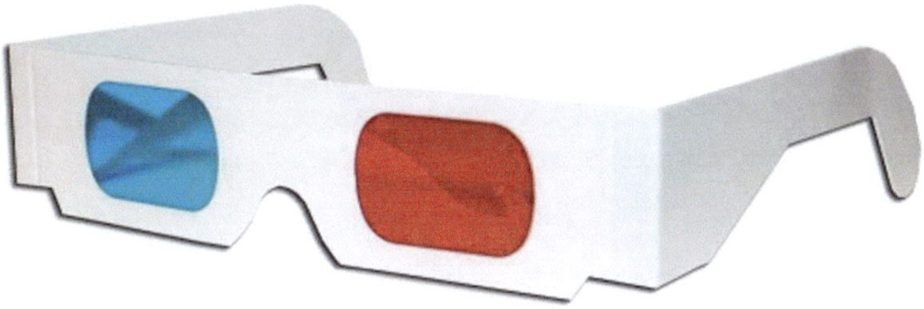

Fig. 1.27 Red-cyan glasses for viewing anaglyph 3D images

This effect can be eliminated by removing the camera away from the microscope by inserting a photo adapter. The photo adapter contains a built-in projection lens, which ensures the creation of a high quality image that completely fills the field of view.

The above phenomenon can also occur in trinocular microscopes, so a photo adapter may be necessary at low magnification (Figs. 1.28 and 1.29).

Adapter rings are used for stable mounting of the interchangeable lens camera to the microscope. One side of them is connected to the bayonet lock of the camera, while the other one provides the connection between the ring and the microscope (Fig. 1.30).

One half of the rings, which are rolled together, can be connected to the camera and the other half to the microscope. When attaching the camera, the two rings are first placed on the camera and then attached to the microscope or to the photo adapter attached to the microscope.

If your camera has an HDMI output, you can connect it to a monitor so you can adjust the sharpness of the image more precisely before shooting (Fig. 1.31).

On the other hand, by connecting the camera to a projector, you can even hold microscopic demonstrations for groups (Fig. 1.32).

1.6 Increasing the Depth of Field of a Light Microscope Photo

When examining preparations under a light microscope, the fact that the depth of field is small, so that not all details of the image are sharp at the same time, is somewhat confusing. This can be easily remedied by turning the fine focus during microscopy, which allows us to scan the different planes of the object in a matter of moments. But what if you want to document your microscopic object in a photo, so that every detail of the image is as sharp as possible? Fortunately, there is an excellent computer solution for this. All you need to do is to take several, even 15, consecutive pictures of the same object, so that each picture has different parts in focus (Fig. 1.33).

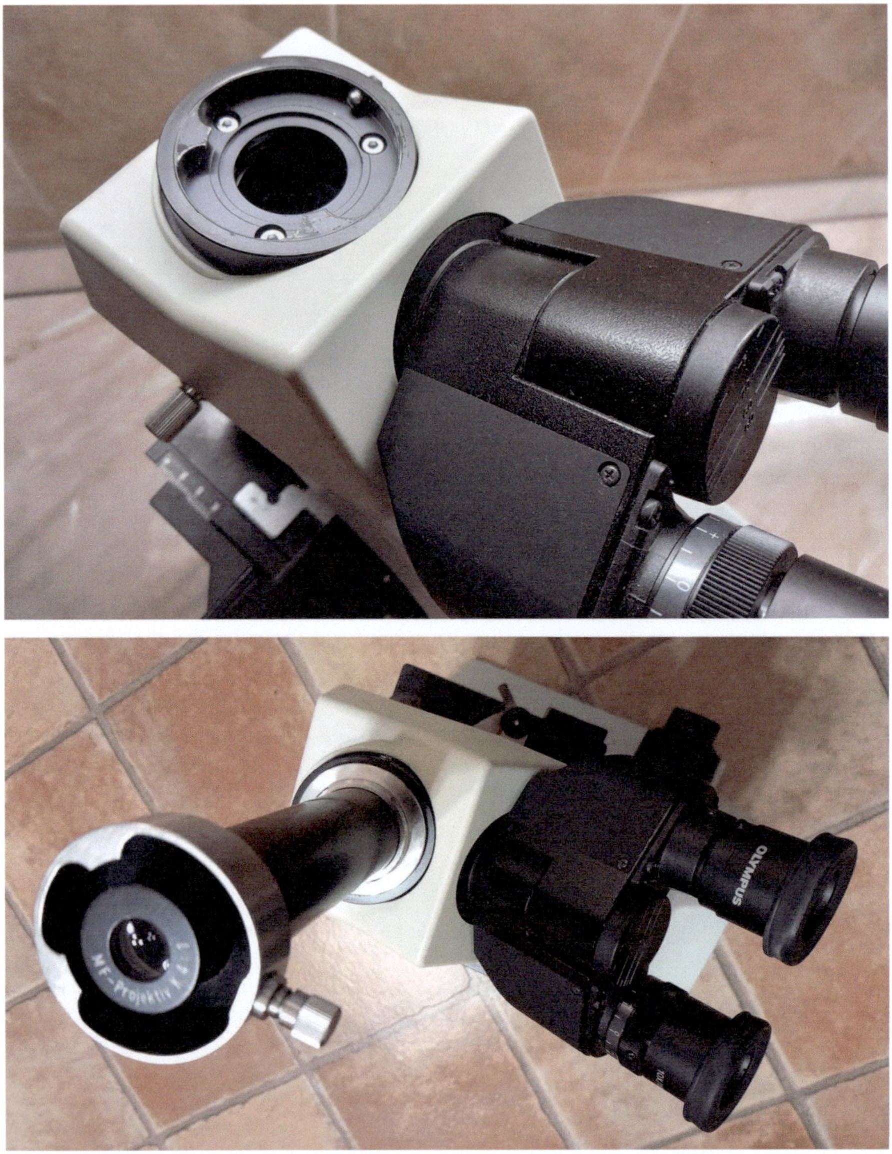

Fig. 1.28 Photomicrographic attachment location on the trinocular light microscope and on the photo adapter. Inside the photo adapter a replaceable projection lens is visible

Then all you need to do is import these photos into a suitable program, and the app will combine the sharp details into a single image. Let us look at an example of this in the case of a cross-section of an elder stem.

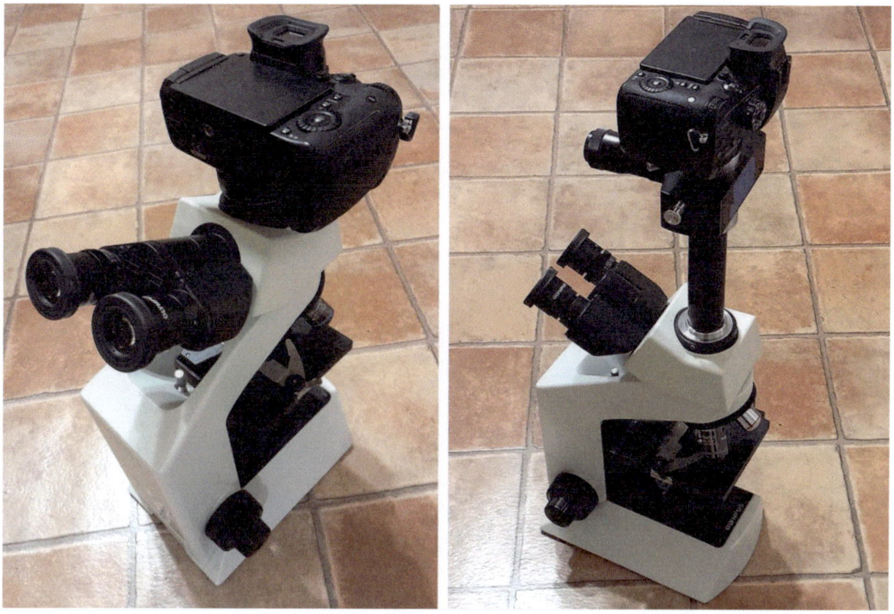

Fig. 1.29 Camera connected directly to a trinocular microscope and via a photo adapter

Fig. 1.30 The front part of the adapter ring is connected to the microscope and the rear one to the camera

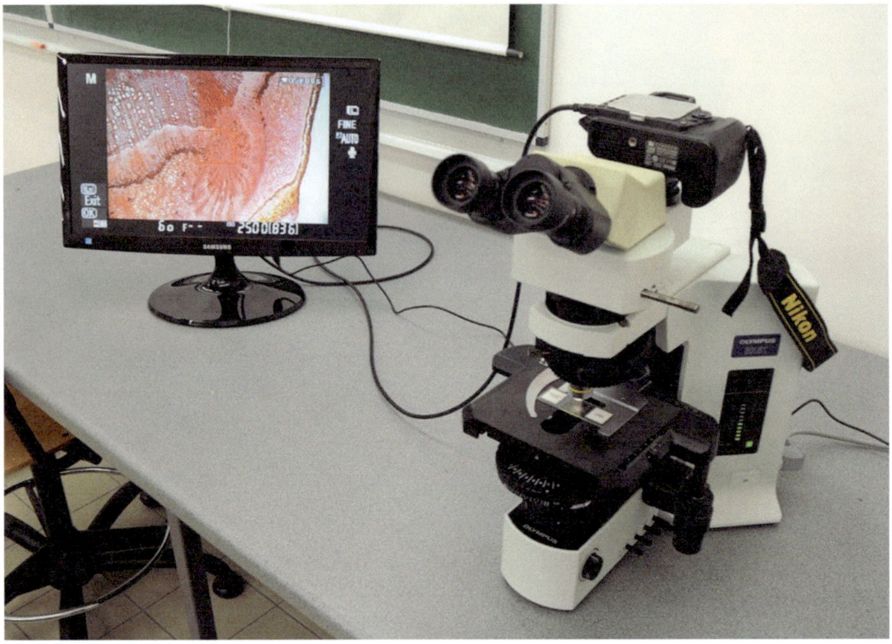

Fig. 1.31 Displaying an image from a camera connected to a microscope on a monitor

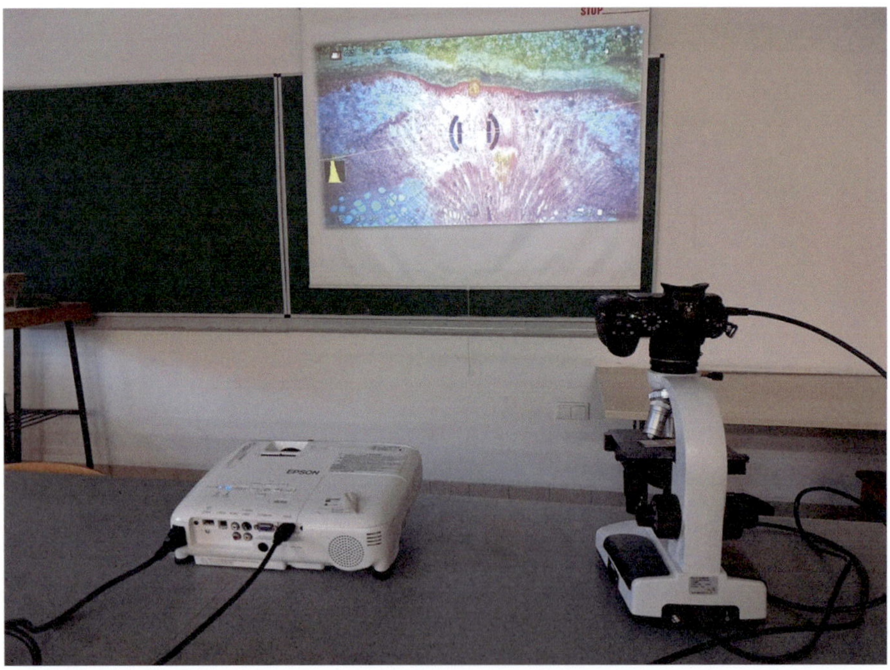

Fig. 1.32 Projected image of a camera mounted in place of a head

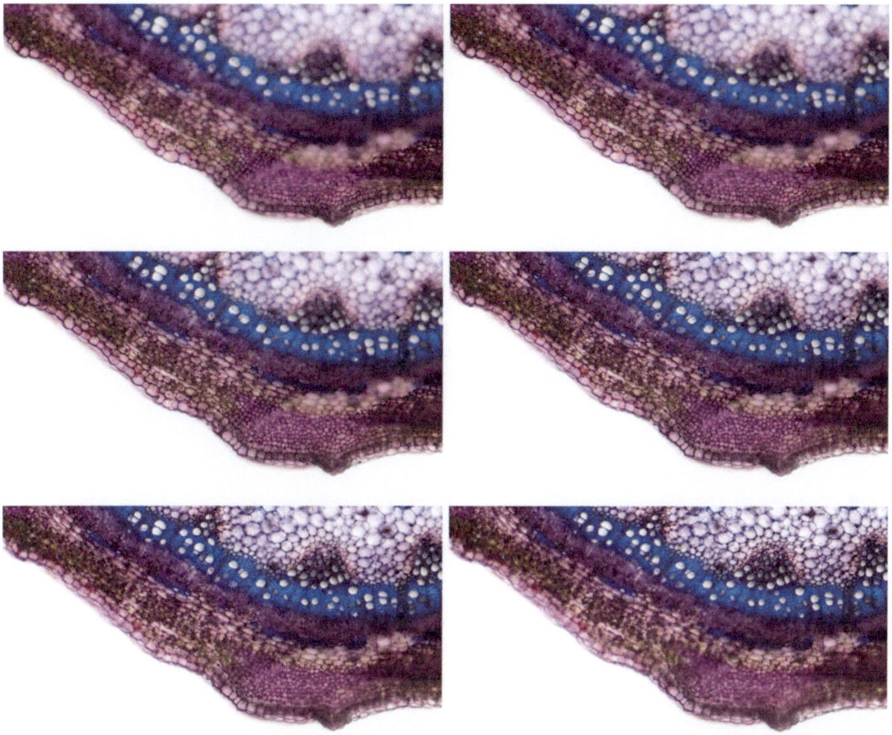

Fig. 1.33 Light microscopy series of photos of elder's (*Sambucus nigra*) cross-section

Take a series of photographs of the preparation (Fig. 1.33) and copy the images into a folder. Next, open the free program CombineZM and open your folder with File/New, select the image files and click "Open". Then use the Macro/Do Stack command to activate the image transformation (Fig. 1.34).

The program performs various calculations and then displays the corrected image, which can be saved using the File/Save Frame/Picture As command (Fig. 1.35).

Photo series can also be merged with other programs.

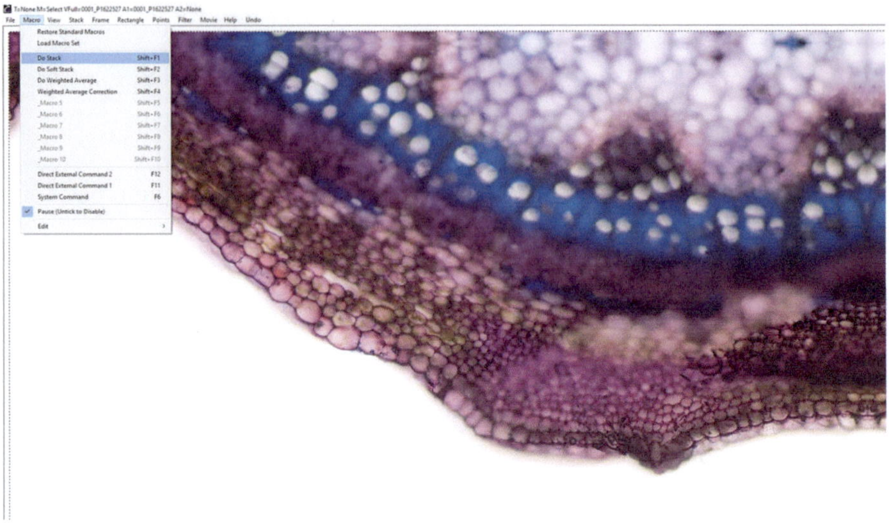

Fig. 1.34 Starting the image conversion

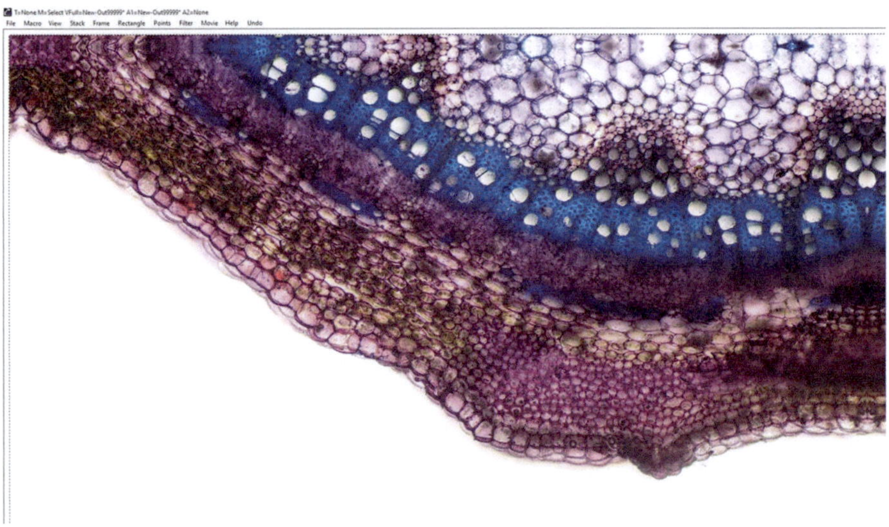

Fig. 1.35 The resulting image in CombineZM

Algae

2

2.1 Blue-Green Bacteria (Cyanobacteria)

Blue-green bacteria (Cyanobacteria) are prokaryotic algae that have no nucleus or chloroplast. They do, however, possess chlorophyll a, which is necessary for photosynthesis and is different from the chlorophyll of photosynthetic bacteria. Unlike bacteria, the photosynthesis of blue-green bacteria also produces oxygen, which is why they are classified as algae.

Blue-green bacteria can be unicellular, form a cell colony surrounded by a mucus envelope, or have a filamentous body organisation (Figs. 2.1 and 2.2).

Blue-green bacterial cells are bordered by a cell wall, which often may have a mucoid sheath. The mucoid sheath coat protects the cells from dehydration and is also involved in movement. When they change position, contractile protein filaments extending from the cytoplasm create the movement of the mucus coat, which carries the blue-green bacteria forward. *Oscillatoria* filaments are often found in water samples from different habitats (Fig. 2.3). It is easy to distinguish them from the bright green filamentous green algae because of their bluish green colour and lack of green chloroplasts. Another important feature is their slow sliding movement, by which they constantly change their position.

The central part of the cellular content is divided into the **nucleoplasm** and the **groundplasm**. The nucleoplasm contains the hereditary material, DNA, while the main part of the groundplasm is made up of the cell membrane intrusion to which photosynthetic pigments are attached. During photosynthesis, blue-green bacteria use light energy to produce organic matter by reducing carbon dioxide, while producing oxygen during water decomposition, similar to plants. Under anaerobic conditions of oxygen deficiency, a significant proportion of them can also use sulphur hydrogen as an electron donor for photosynthesis instead of water. In this case, these organisms excrete elemental sulphur instead of oxygen. In filamentous blue-green bacteria, among the vegetative cells that make up the filament, one can often observe empty, thick-walled cells, called **heterocysts**, which are able to fix atmospheric nitrogen (Fig. 2.4). The nitrogen already fixed is an important nutrient that

G. Kriska, *Teaching Biology*, https://doi.org/10.1007/978-3-031-84085-2_2

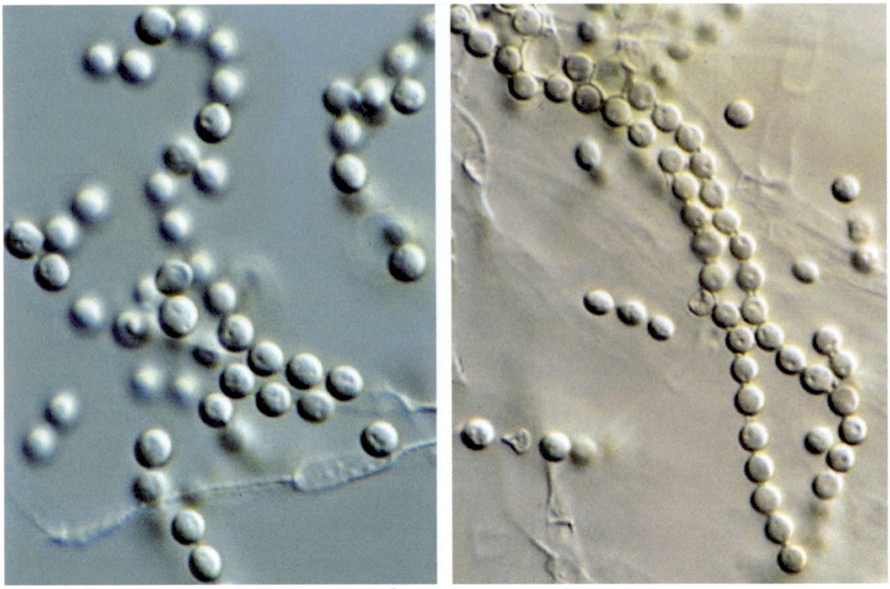

Fig. 2.1 Filamentous blue-green bacteria (*Nostoc*) (Photo: Béla Lovas)

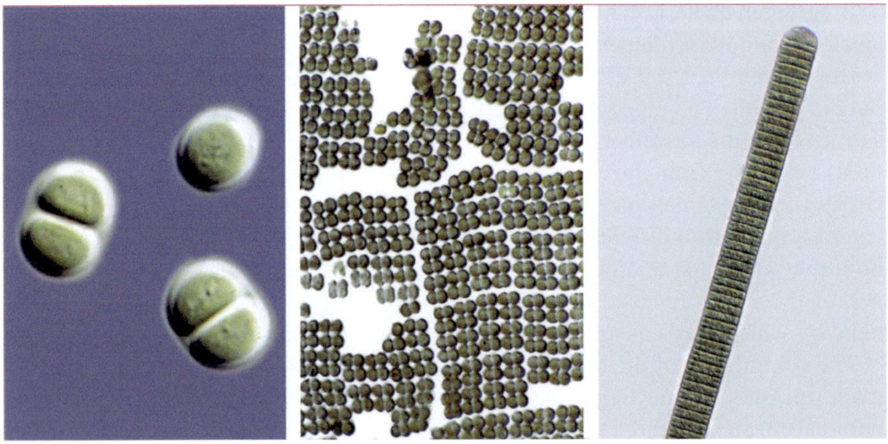

Fig. 2.2 Unicellular (*Chroococcus turgidus*), colonial (*Merismopedia elegans*), and filamentous (*Oscillatoria*) blue-green bacteria (Photo: Béla Lovas)

the cells use in their metabolic processes. The fragmentation of the filaments often occurs at the heterocysts, which is why they are often found at the ends of the blue bacteria filaments.

The asexual reproduction of blue-green algae can occur by cell division or, in some of the filamentous blue-green algae (Hormogonales), by filamentous fragments formed by the disintegration of cell filaments, the **hormogonia** (Fig. 2.5).

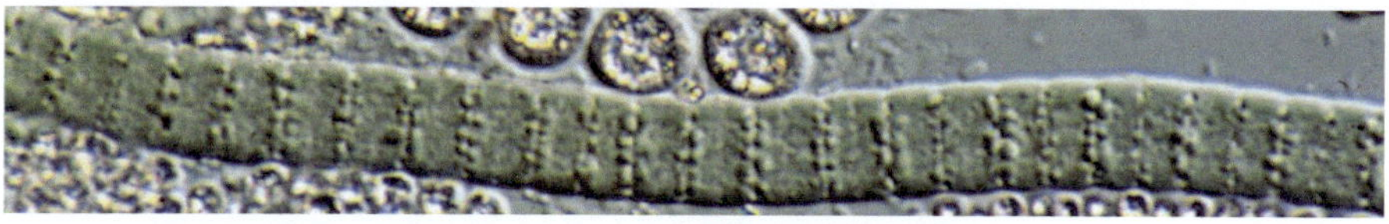

Fig. 2.3 Filamentous blue-green bacteria (*Oscillatoria*) (Photo: Béla Lovas)

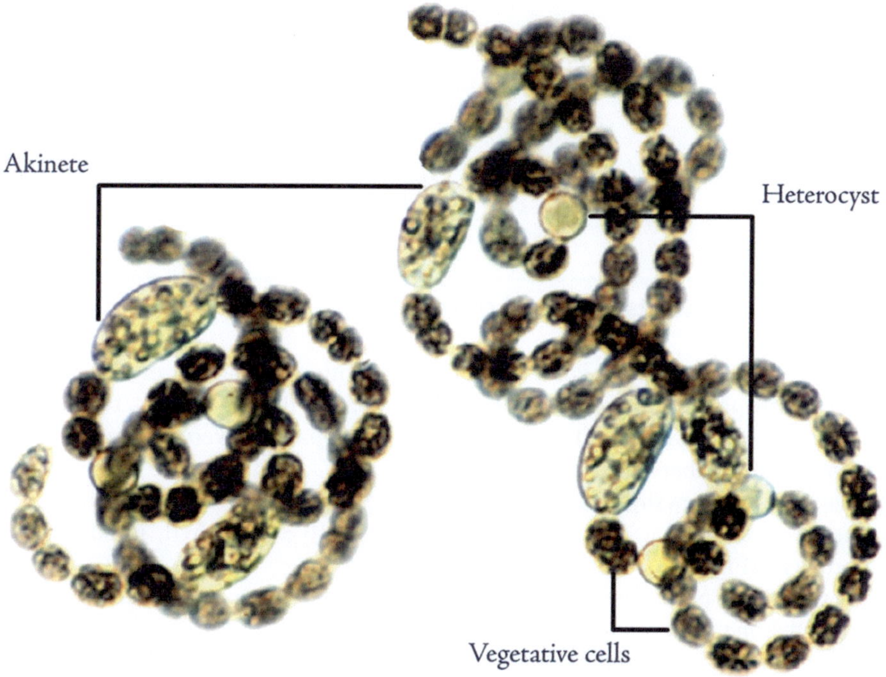

Fig. 2.4 Blue-green filamentous bacterium (*Anabaena circinalis*) (Photo: Béla Lovas)

Another possibility is the formation of a thick-walled spore, the **akinete**. Unlike vegetative cells, the akinete does not contain gas cavities, so it sinks to the bottom and survives adverse periods.

Many species of blue-green algae produce potent toxins (cyanotoxins) that can be dangerous to humans if the algae proliferate. Ingestion of cyanotoxin water can cause nervous system damage and liver damage, and bathing in contaminated water can trigger skin allergies. The mass proliferation of blue-green algae and water blooms usually occurs when nitrogenous plant nutrients in the water are depleted under conditions favourable to plant photosynthesis. In such cases, blue-green algae, which are able to fix nitrogen, have gained an advantage over other algae, which can lead to their mass proliferation.

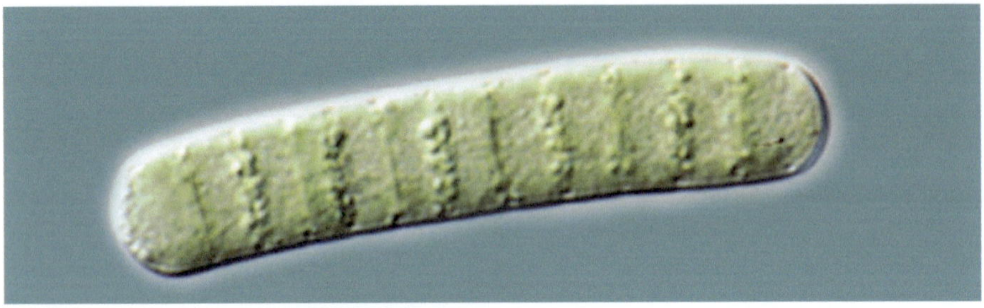

Fig. 2.5 *Oscillatoria* hormogonium (Photo: Béla Lovas)

Film
Filamentous blue-green bacteria (*Oscillatoria*) movement: https://youtu.be/IJcgT6XBYTg

2.2 Heterokonts (Heterokontophyta)

The introduction to eukaryotic algae begins with the classes of the phylum **Heterokontophyta**. The name of the phylum refers to the fact that the two flagella of the primary cells are of different lengths. Their apparently very different species are grouped together because of their shared cellular features. The colour of the **golden algae (Chrysophyceae)** varies from golden yellow to brown, due to the presence of additional pigments in the cells, mainly xanthophyll, in addition to chlorophyll (Fig. 2.7).

A significant proportion of their species swim freely with two flagella of different lengths, the cells often forming colonies. The species of the class golden algae are very diverse in appearance. Most of them are both autotrophic and heterotrophic, photosynthesising during the day and feeding on organic matter, mainly microorganisms and organic molecules, at night. In the latter case, they grab food with their tiny pseudopods or with the help of their flagella. Colourless species can only live heterotrophically. They are highly efficient at absorbing inorganic phosphorus compounds from water and, unlike many other algae, can survive in phosphate-poor waters. The majority of golden algae live in freshwater and their species show a high diversity of shapes. In their asexual reproduction, the cells divide to form zoospores that can swim with their flagella, or immobile autospores, and develop into new algae. A well-known representative of the golden algae is *Dynobrion*, whose colony-forming cells live in a funnel-shaped house (lorica) (Fig. 2.8).

The **diatoms (Bacillariophyceae)** are unicellular or form a loose colony of cells (Figs. 2.6, 2.9, 2.10, 2.11, 2.12, and 2.13).

In addition to the pectin shell of the cell, they have a solid silica skeleton (frustule) consisting of crystalline or amorphous SiO_2. The highly hard frustule, with its ornate pattern, consists of two valves (epitheca and hypotheca), which are

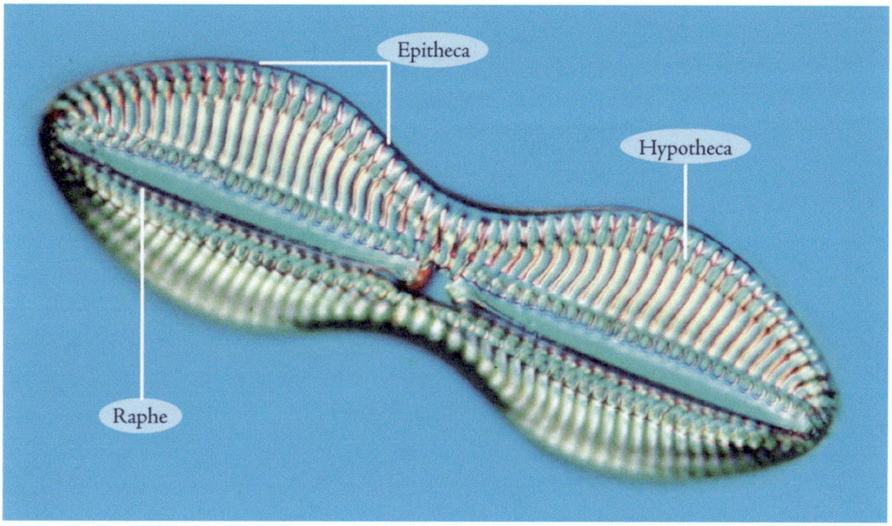

Fig. 2.6 Marine *Diploneis* diatom skeleton (Photo: Béla Lovas)

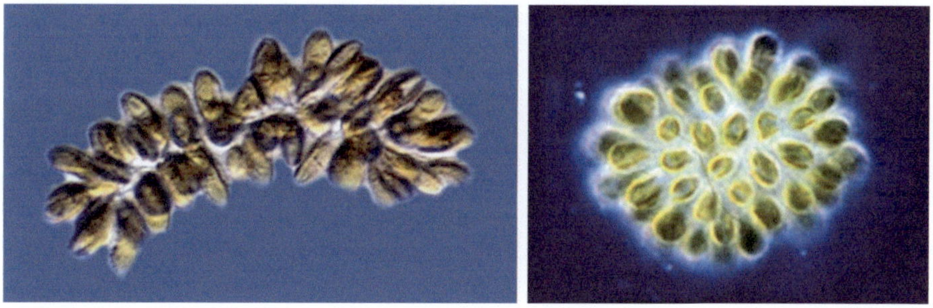

Fig. 2.7 *Synura* golden algal cell colonies (Photo: Béla Lovas)

connected to each other in a Petri dish-like manner. Based on the symmetry of their frustules, we classify the diatoms into two orders.

Species of the order **Centrales** have a radial or triangular symmetry, while species of the order **Pennales** have an axial symmetry or are asymmetrical. In axially symmetrical diatoms, a longitudinal channel (raphe) is often formed on the ventral side of the theca through which the cytoplasm can flow out, forming the characteristic sliding movement of the diatoms. Diatoms are autotrophic, photosynthesising organisms, but they also require an organic substance, vitamin B12, to function. When examining water samples under the light microscope, one almost always encounters elongated, yellowish-brown diatoms of the Pennales order, which move forward with a slow creeping motion.

Diatoms are most abundant on aquatic plants and stones, so if you take a sample from their brownish coating, you will almost certainly find diatoms in it.

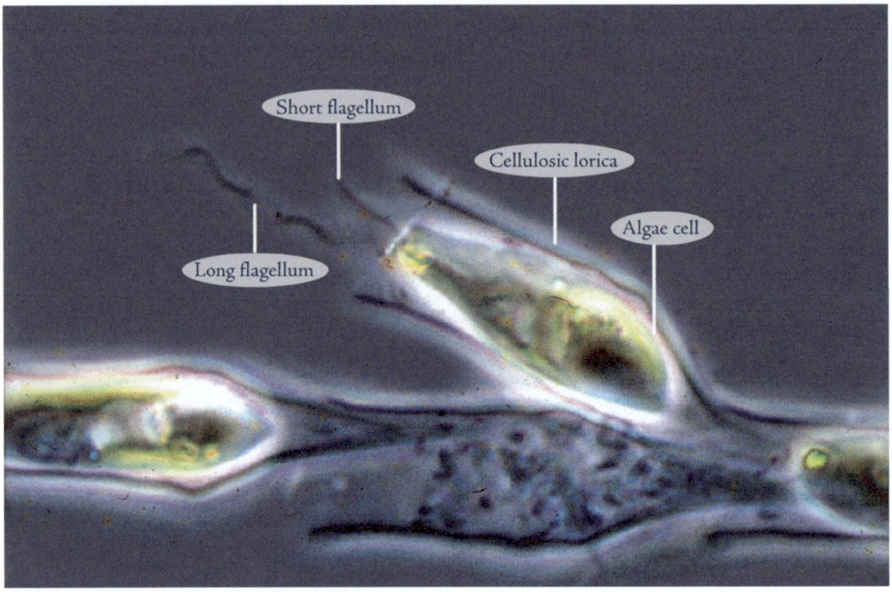

Fig. 2.8 Detail of a *Dynobrion* golden algae "bouquet" (Photo: Béla Lovas)

Some *Fragilaria* diatoms have a peculiar movement. Several of them live together, and their movements are coordinated, sometimes forming a tight accordion-like arrangement, sometimes forming a long thread with their apical parts touching. The transformation between the two forms takes place in seconds.

The frustules of dead marine diatoms can often accumulate in large masses in the sediments, resulting diatomaceous earth of diatoms with a high specific surface area. Diatomaceous earth is used as thermal insulator, in filtration and transport of fluids (dynamite), and as absorbents.

Their colour bodies contain a lot of fucoxanthine, which masks the green colour of the chlorophyll and gives the algae a brownish colour. Their asexual reproduction is a process of dividing, whereby the two progeny cells carry one valve of the frustule and the other valve of the frustule, and then both complete the silica frustule with the smaller half. Successive divisions result in the frustule of one of the progeny cells becoming smaller and smaller. When the size is reduced to 30–40% of the original body size, the alga forms gametes, which fuse to form a zygote. The zygote leaves the frustule and develops into a growing spore (auxospore). The growing spore, after reaching normal cell size, makes a new frustule for itself. Diatoms are able to survive in low-light conditions and can be found in deeper water and under polar ice.

Some planktonic species are able to move vertically in the water column, taking up nutrients in the deeper, while photosynthesising near the surface. Many species are substrate-bound, benthonic creatures that multiply in masses to form a slippery, brownish coating on the surface of rocks and other objects.

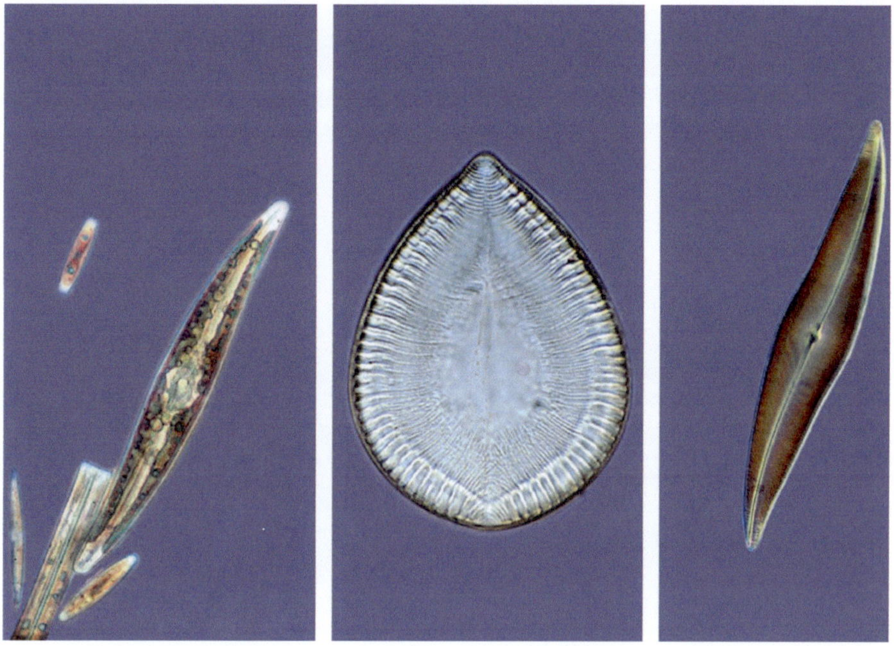

Fig. 2.9 Diatoms of the order Pennales: *Gyrosigma*, *Surirella,* and *Pleurosigma* (Photo: Béla Lovas)

The yellow-green algae (Tribophyceae) are characterised by their richness of form.

Their unicellular species may be two flagellated (**Heterochloridales**), amoeboid (**Rhizochloridales**), or surrounded by a gelatinous envelope (**Heterogloeales**). **Tribonematales** have branched or simple filaments of mononuclear cells. Multinucleate filaments (polyenergid colonies) are formed by species of the order **Botrydiales**. Some species form compact stands on the moist soil surface. Their rounded heads, which rise above the surface, are clearly visible to the naked eye. They are anchored to the substrate by their colourless root-like appendages (Fig. 2.14). Species of the order **Vaucheria** form filamentous colonies of multinucleate cells, often forming continuous grasslands on the water bottom.

Films

Cymbella diatom sliding on the substrate: https://youtu.be/4DfZWzX3Hys

Coordinated movements of interconnected *Fragilaria* diatom individuals: https://youtu.be/6gEkw98efGs

Fig. 2.10 *Cyclotella meneghiniana* and *Campylodiscus noricus* diatoms of the order Centrales (Photo: Béla Lovas)

Fig. 2.11 *Encyonema* and *Gomphonema*, diatoms of the order Pennales. Scanning electron microscope images (Photo: Béla Lovas)

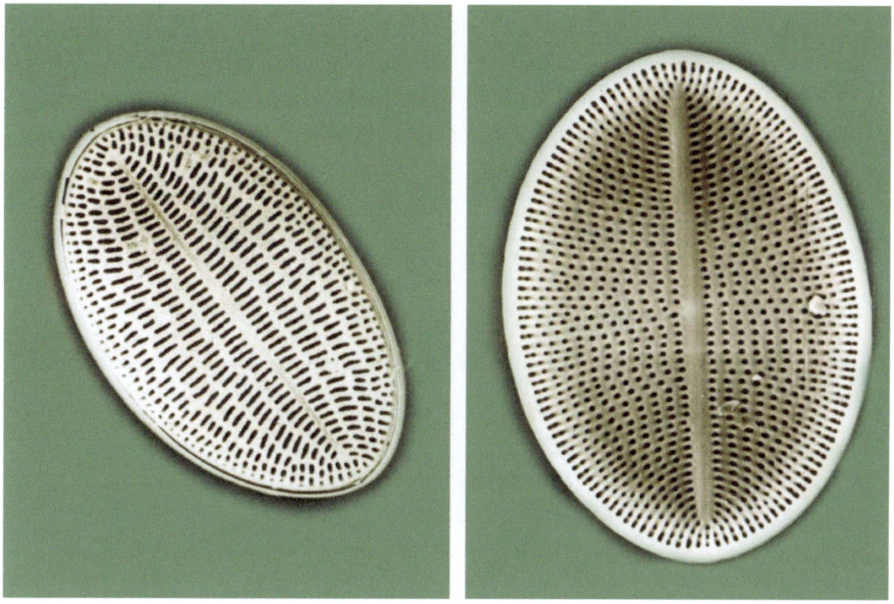

Fig. 2.12 External and internal surfaces of the theca surface of a *Cocconeis* diatom of the Pennales order. Scanning electron microscope images (Photo: Béla Lovas)

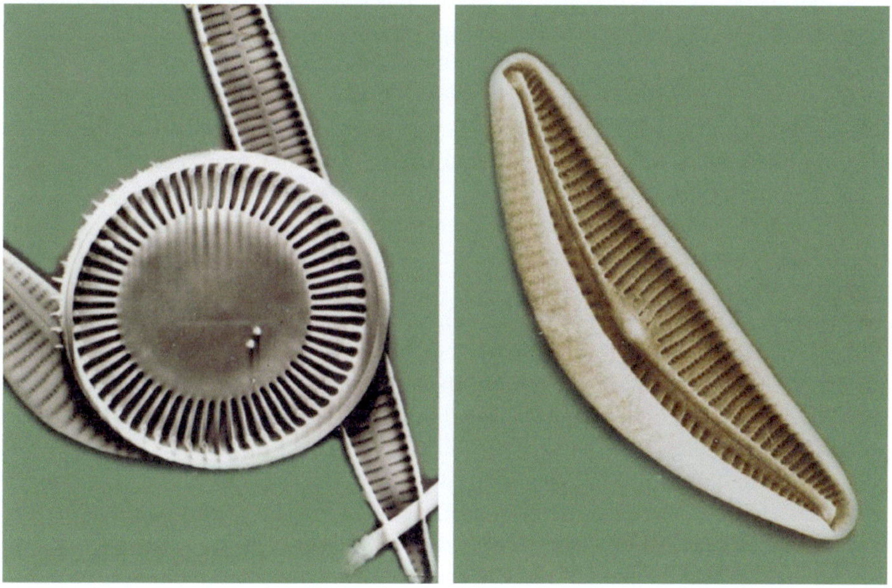

Fig. 2.13 *Cyclotella meneghiniana* diatom with diatom of the order Pennales and a *Cymbella* diatom (Photo: Béla Lovas)

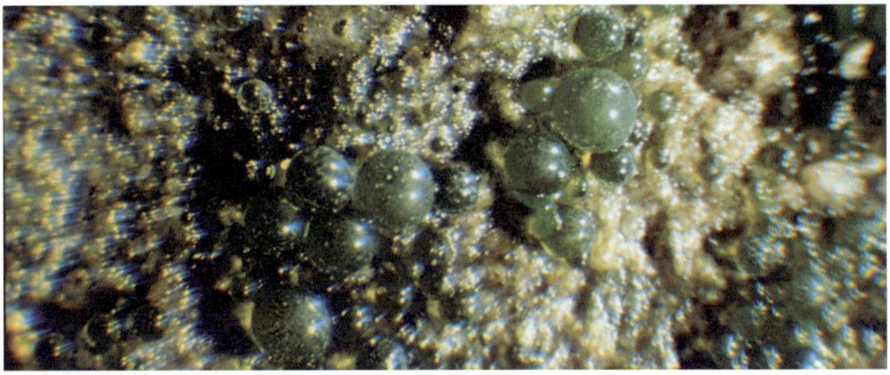

Fig. 2.14 *Botrydium granulatum* on the soil surface (Photo: Béla Lovas)

2.3 Dinoflagellates (Dinophyta) and Euglenids (Euglenophyta)

Among the **Dinophyta** are the fastest swimming algae (Figs. 2.15 and 2.16). Some species are bare, while others protect their cells with a pellicle made of cellulose sheets. They have two flagella, the shorter of which lies in the ridge groove and propels the alga forward, while the longer one, which acts as a "rudder", extends backwards from the longitudinal groove.

Some of the species are photosynthetic, but some of them are also capable of heterotrophic feeding by phagocytosis (mixotrophs). Many species are colourless and are heterotrophic phagocytic. Some of their species may live in symbiosis with corals, sea lilies, flatworms, but also have green algal symbionts can living in their cells. Their asexual reproduction can be achieved by longitudinal or latitudinal dividing, the formation of zoospores and cysts. The latter process is usually triggered by nutrient depletion in the water.

During their sexual reproduction, the male merges with the female to form the zygote. One or two spores develop inside the zygote wall, which develop into a final vegetative cell after their release.

The **euglenids (Euglenophyta)** are unicellular, less often forming a slime colony (Fig. 2.17). Their moving is ensured by one or two flagella. Their carbohydrate-type nutrients are stored in characteristically shaped paramylon bodies. Euglenids do not have a cell wall, but underneath their cell membrane a characteristic spiral skin of protein- and carbohydrate-containing ribbons develops, which plays a crucial role in the cell's high degree of shape variability. The two flagella of the euglenids originate from a reservoir in the anterior part of the cell. One of the two flagella is usually very short and does not protrude from the reservoir. Close to the reservoir is a contractile vacuole and the orange-red eye spot, which is involved in light perception.

They help the euglenids to detect excessively bright light, which drives them away (negative phototaxis), and moderate light conditions, which attract them (positive phototaxis). Euglenids can develop a palmella stage in response to external conditions. The cells drop their flagellum, become rounded, and secrete a jelly sheath around themselves, which protects them from external influences. In this

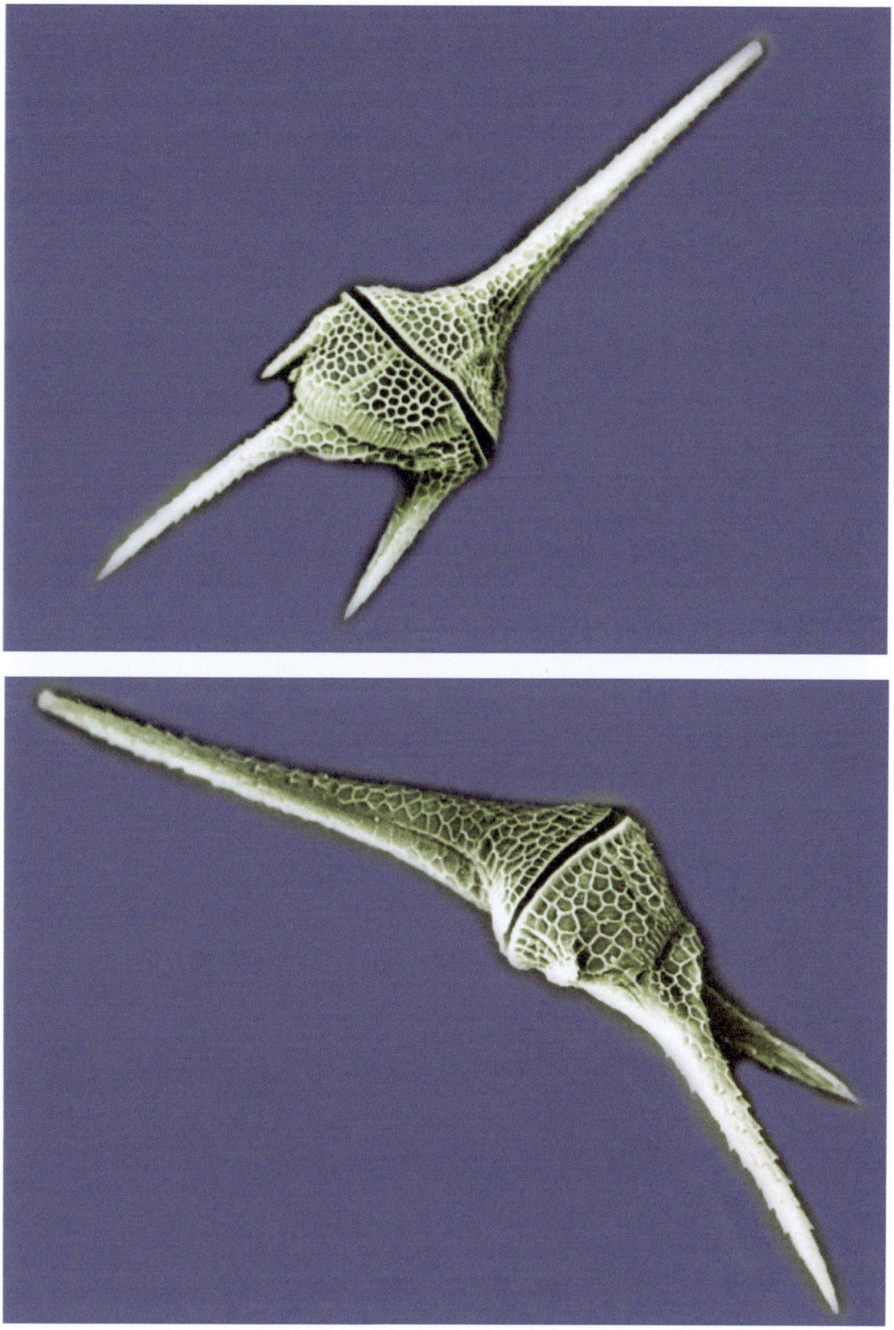

Fig. 2.15 *Ceratium hirundinella* (Photo: Béla Lovas)

Fig. 2.16 *Peridinium* pellicles. Scanning electron micrographs (Photo: Béla Lovas)

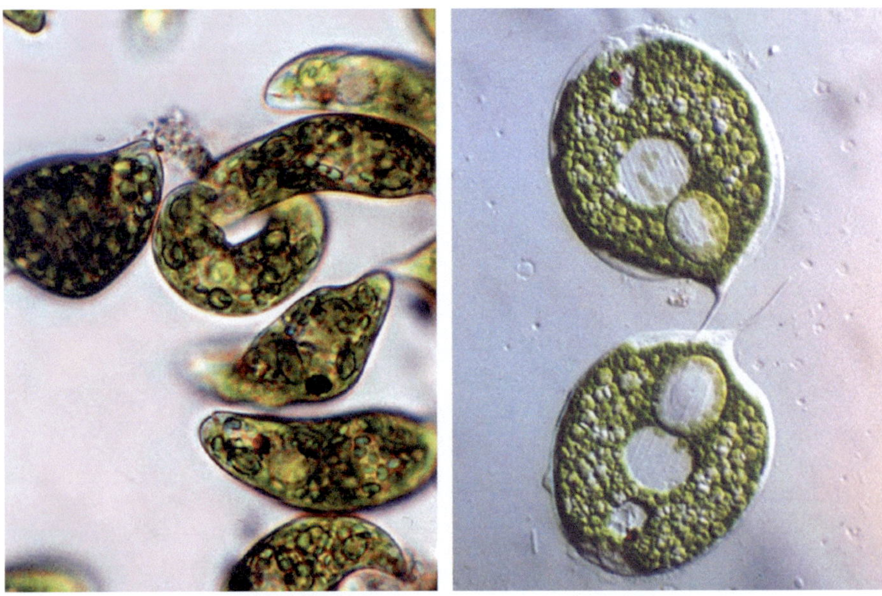

Fig. 2.17 Euglenid with flagellum (Photo: Béla Lovas)

state, the cells remain active, move in one place and can divide. A characteristic feature of euglenids is a special movement. During this process, the cell changes its shape continuously, from the most rounded to the most elongated. Little is known about the role and mechanism of this particular movement. Euglenids tend to occur in large numbers in small waters, where they can cause aquatic blooms when they multiply en masse. They reproduce asexually by longitudinal cell division, and sexual reproduction is not known.

During the light microscopic examination of water samples, we often see individuals of euglenids. The species most commonly encountered are *Phacus* species, whose cells are flattened and terminate in a spine. It often turns around during swimming. In **Peranema** species, the whip is rigid, with only the end rotating like a propeller, which propels the organism forward.

Film

Gymnodinium: https://www.youtube.com/watch?v=MqY5wr0az9A

Euglena: https://youtu.be/n7jr1gDXm1A

2.4 Green Algae (Chlorophyta) and Stoneworts (Charophyta)

The green algae are divided into two large groups, the **green algae (Chlorophyta)** and the **stoneworts (Charophyta)**. Both phyla include algae with a unicellular (Fig. 2.18), cell colony (Figs. 2.19, 2.20, 2.21, and 2.22), and filamentous organisation (Figs. 2.23, 2.24, 2.25, 2.26, 2.27, 2.28, and 2.29). In addition, the Chlorophyta algae include species with a disc-like, while the Charophyta include species with a colonial structure consisting of root and shoot-like parts (Fig. 2.30).

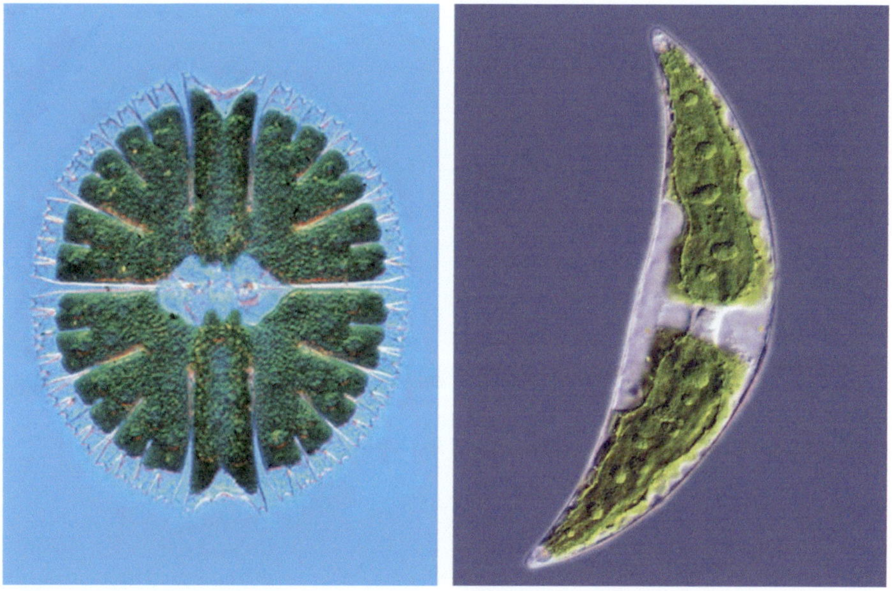

Fig. 2.18 *Micrasterias rotata* and *Closterium* green algae (Photo: Béla Lovas)

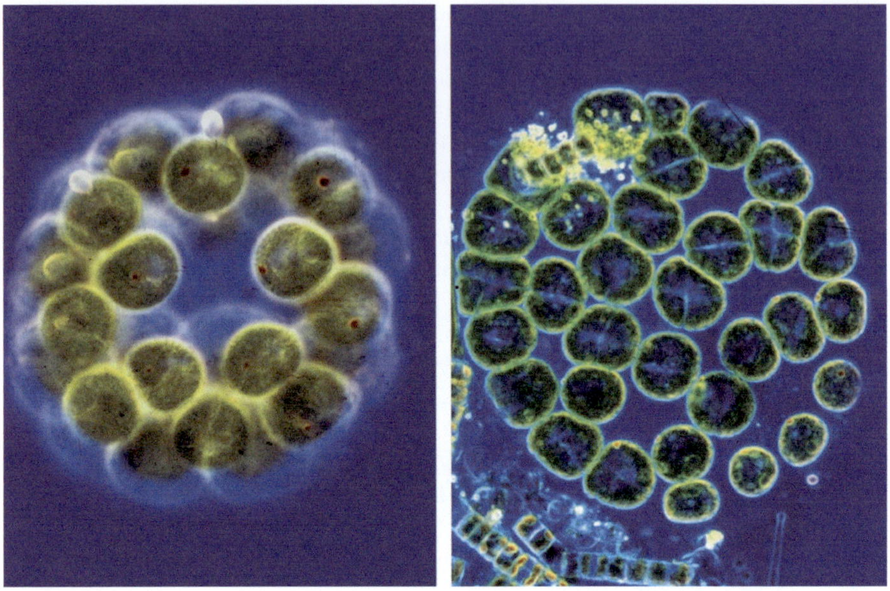

Fig. 2.19 Species of the order Volvocales. Cell colony of *Pandorina morum* surrounded by a jelly sheath, and cells in division in a colony of *Eudorina* (Photo: Béla Lovas)

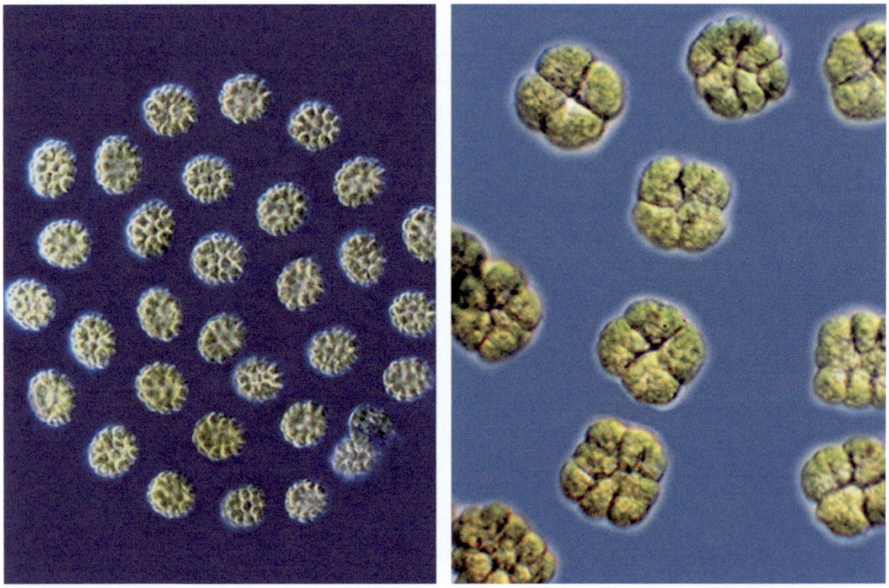

Fig. 2.20 Small *Eudorina* colonies formed as a result of the divide (Photo: Béla Lovas)

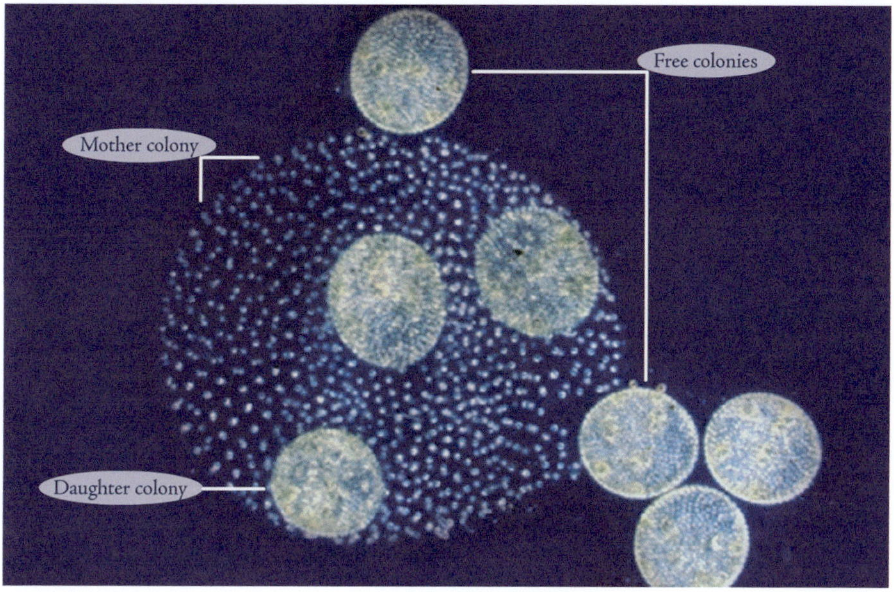

Fig. 2.21 *Volvox aureus* spherical colony (Photo: Béla Lovas)

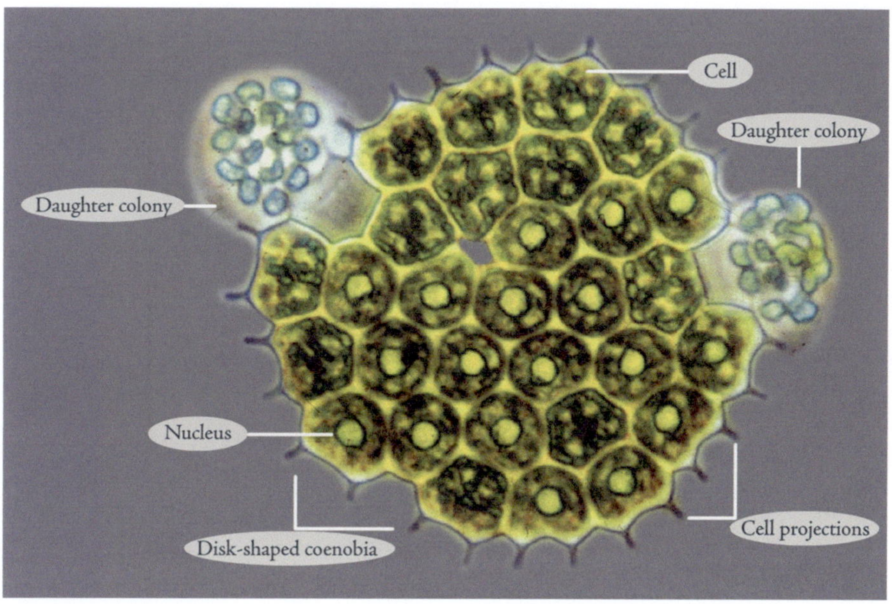

Fig. 2.22 *Pediastrum boryanum* colony (Photo: Béla Lovas)

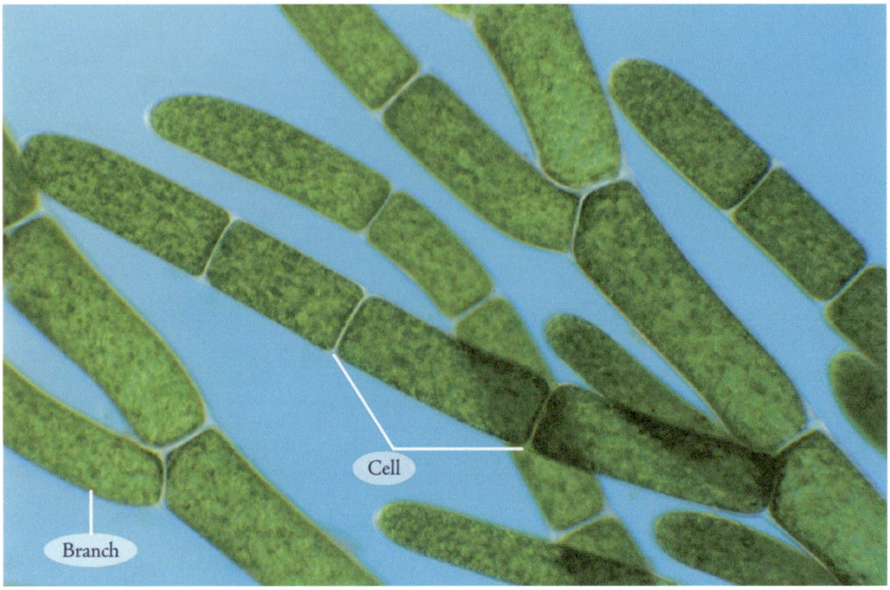

Fig. 2.23 Branched filaments of *Cladophora*

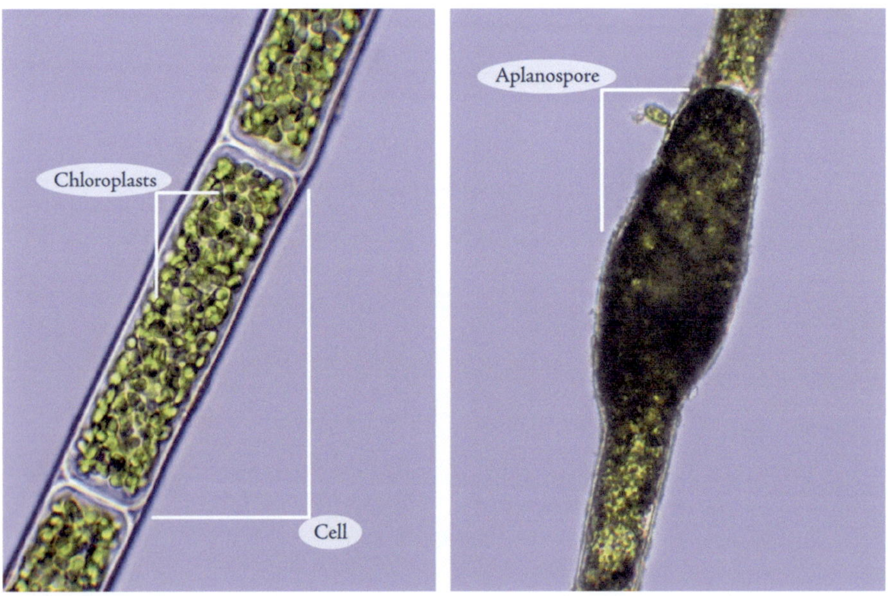

Fig. 2.24 *Cladophora* part and a forming aplanospore

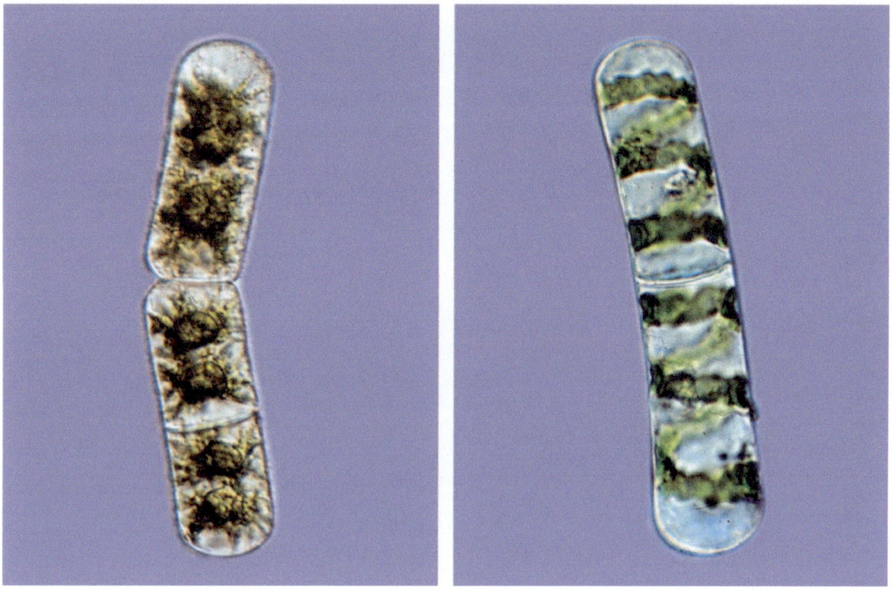

Fig. 2.25 The filamentous green alga *Zygnema* with star-shaped chloroplasts and the filamentous green alga *Spirogyra* with spiral chloroplasts (Photo: Béla Lovas)

Fig. 2.26 *Spirogyra* filament fragments with a spiral chloroplast

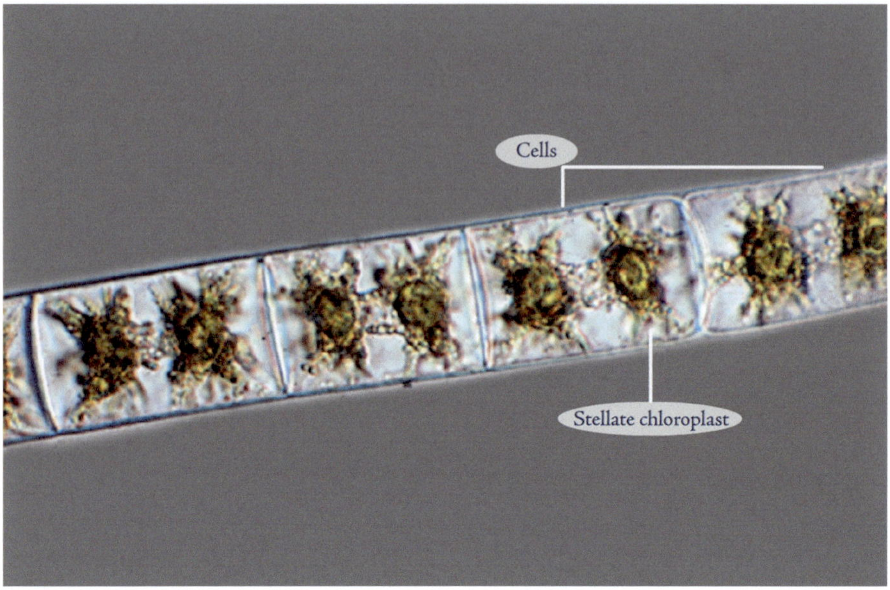

Fig. 2.27 *Zygnema* species with star-shaped chloroplasts (Photo: Béla Lovas)

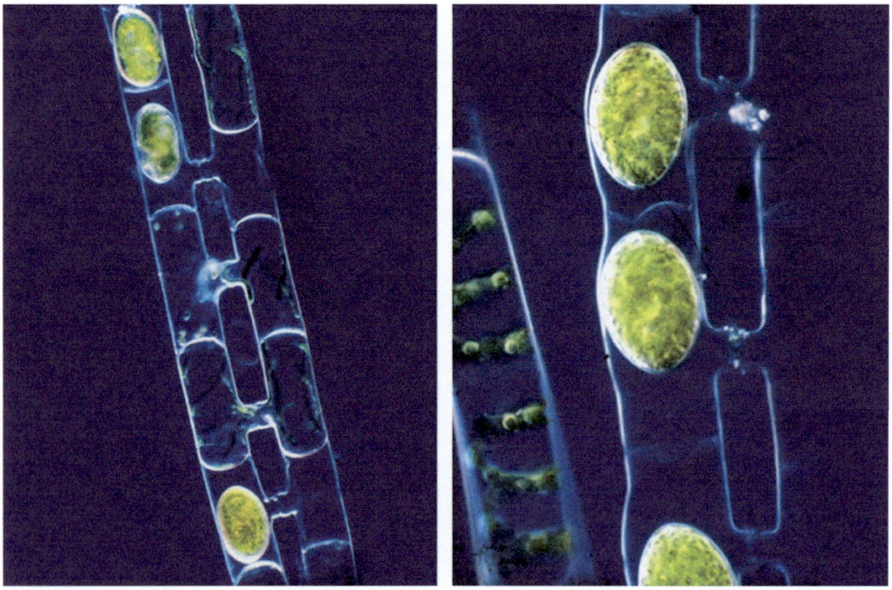

Fig. 2.28 *Spirogyra* zygospores (Photo: Béla Lovas)

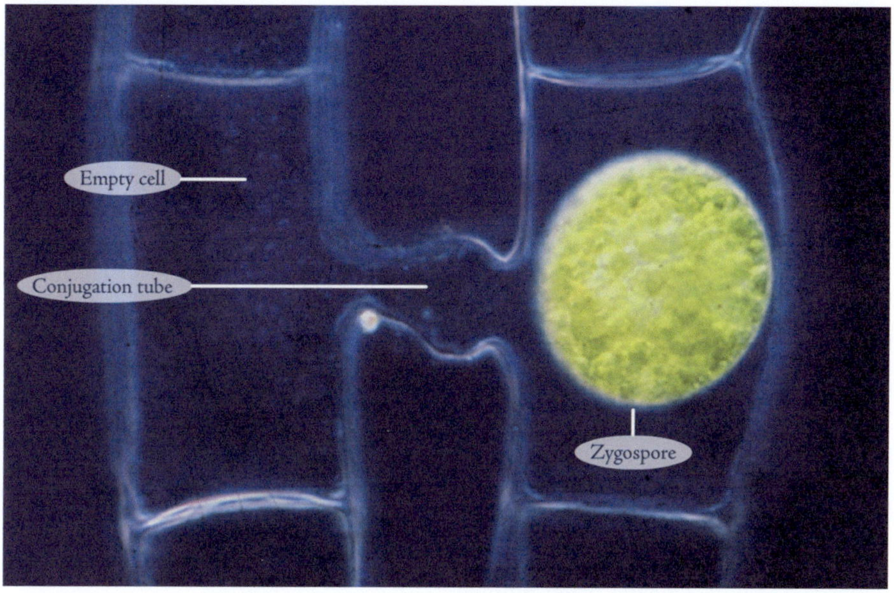

Fig. 2.29 *Spirogyra* zygospore formation after conjugation (Photo: Béla Lovas)

Fig. 2.30 Common stonewort (*Chara vulgaris*). Its branches are borne in whorls, appearing in clusters around the stem

The taxonomic classification of species is based on the flagellum structure of the flagellum forms, cell division patterns, life cycle, and external morphology. The primary pigments of green algae are chlorophyll a and b. They also have carotenoids, which do not mask the grass green colour of chlorophylls. Their cell wall material is cellulose, and their photosynthesis is identical to that of terrestrial plants, suggesting a direct evolutionary relationship. **Pyrenoid** granules are common in their green bodies. This protein-rich part is involved in the storage of the product of photosynthesis, starch. The motile cells swim with two or four flagella and are characterised by migration towards the light, by positive phototaxis. They detect light by means of an eye spot (stigma) located in the green body.

The unicellular green algae reproduce asexually by dividing, the colonial by colony fragmentation, the filamentous ones by filament fragmentation, the shoot-like ones (Charophyta) by detachment of axial and root tubers. The ones forming a spherical cell colony, by the daughter colony forming in the mother colony reproduce asexually (Figs. 2.20 and 2.21).

In addition, they may be characterised by zoospore with flagellum forming in vegetative cells and immobile thick-walled aplanospore, as asexual reproductive cells. Their sexual reproduction can take a variety of forms during their life cycle.

Species of the order **Volvocales** are unicellular with flagellum or form a multicellular colony (coenobia). They are characterised by phototaxis. Their cells with two flagella contain a single large, cup-shaped green body. The cells are located in the outer layer of the spherical coenobia, and the inner part of the sphere is filled with a gelatinous substance. There is a division of labour between the cells and their flagellum-flapping is coordinated.

Most of the cells are vegetative, while some produce two-flagellated gametes.

They are characterised by the formation of daughter colonies which are released to the outside world by the cleavage of the mother colony. Species of the order Volvocales often appear in large numbers in early spring and summer in roadside ditches and pools, causing water blooms.

The vegetative cells of algae of the order **Chlorococcales** are not capable of active movement. Their species are unicellular or form a cell colony. They can live in freshwater, on the soil surface, and in lichens as symbiotic algae.

During the vegetative reproduction of the *Pediastrum* species (Fig. 2.22) that make up the cell colony, zoospores are formed in the mother cells of the colony, which move freely within the cell with their flagella. Later, they attach to each other with their flagella and form a daughter colony.

Then the flagella disappear and the mother cell opens, releasing the daughter colony, which no longer has any cell division but continues to grow in size until it reaches the size typical of the species. The final cell number depends on the number of zoospores that have fused within the mother cell.

The vegetative cells of green algae of the class Ulvophyceae never have flagella. Their body organisation can be unicellular, multicellular, or multinucleate siphonal. The cells have a chloroplast adjacent to a cell wall, which is cup-shaped or ring-shaped. Most of them live in the sea.

The order **Ulotrichales** includes the *Ulothrix* species, which have long, unbranched filaments surrounded by a mucilaginous sheath. A common marine species of the order is the massed sea lettuce of the intertidal zone.

The order **Cladophorales** includes the bottom-dwelling filamentous algae (Figs. 2.23 and 2.24). The multinucleate cells of these algae, which are common in both marine and freshwater environments, form branched filaments.

Their colonies are often anchored to the bottom for their entire lives. In rivers and lowland streams, they can be found in large numbers, serving as substrates for other algae. The filamentous mass is rough to the touch due to the diatoms that colonise it and the secretion of lime ($CaCO_3$).

The order **Zygnematales** includes unicellular, unbranched filamentous and colonial algae. The species of the family Zygnemataceae form non-branching filaments. Most of them live in small stagnant waters and pools. Their cell walls are covered with a fine layer of jelly, which makes them mucous. The genera of the family Zygnemataceae are easily distinguished by their chloroplasts. *Spirogyra* species have ribbon-shaped chloroplasts spiralling along the cell wall (Figs. 2.25 and 2.26), *Zygnema* species have two star-shaped chloroplasts (Figs. 2.25 and 2.27), while *Mougeotia* species have a single twisted chloroplast.

During their reproduction as gametes, the vegetative cells are connected to each other by a direct connection through a conjugation tube (Figs. 2.28 and 2.29).

Through the conjugation tube, the plasma of one cell passes in an amoeboid movement to the other cell, where fusion takes place. The resulting zygote undergoes a meiotic division to form a thick-walled **zygospore**, which, after a period of dormancy, starts to develop and gives rise to the new plant.

The Desmidiaceae species are unicellular or colony-forming algae (Figs. 2.31 and 2.32).

The unicellular form consists of two mirror halves. One is older, because it consists of the mother cell half at division, while the other, the younger half, develops after division. The nucleus is located in the middle of the cell.

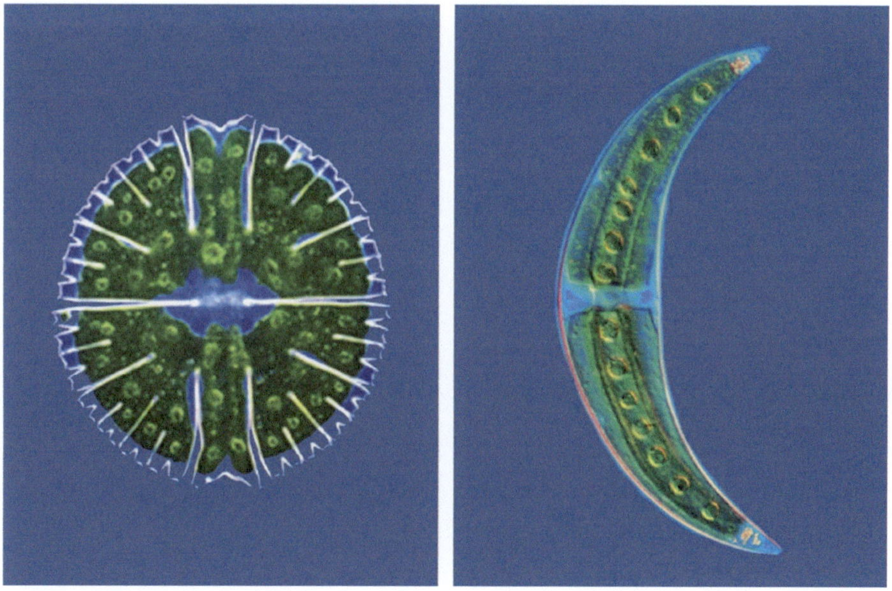

Fig. 2.31 *Micrasterias* and *Closterium* algae (Photo: Béla Lovas)

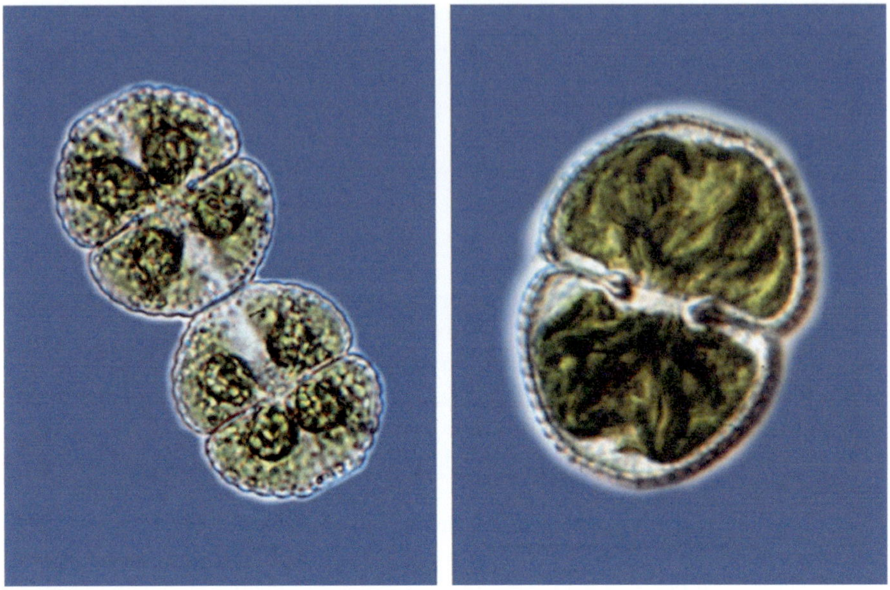

Fig. 2.32 *Cosmarium* at the end of division, and *Cosmarium cucurbita*

Mosses, Lichens, and Ferns

3

3.1 Mosses

Mosses are plants with a simple body structure (Figs. 3.1, 3.2, 3.3, 3.4, 3.5, and 3.6). Their living species today are classified into the liverworts, mosses, and hornworts phyla. They do not have real organs, some of their leaves are often only one cell layer thick. Most of their species are a few centimetres in size. The **moss plant** (gametophyte) is a plate-like colony or a small shoot-like body consisting of a stem and leaves, which is anchored to the substrate by root-like filaments.

Moss plants usually cluster into moss patches consisting of many individuals. The reproductive organs appear at the top or side of the moss plant (Fig. 3.7).

The female reproductive organ contains a single egg cell, from which a sporophyte forms after fertilisation. The flagellated sperms formed in the antheridiums approach the egg cells by swimming. Some leafy mosses form a cup-like structure around the antheridiums, which helps the movement of the male reproductive cells by catching raindrops. Some liverworts develop a stalked reproductive organ holder. The spore capsules, which is a stalkless, single-cell-layered capsule surrounded by a wall, develop from the female reproductive organs. These are the dark capsules located at the bottom of the cap-like structures seen in the above picture (Fig. 3.7).

The sporophyte of mosses consists of the foot, the elongated seta, which may also be missing, and the capsule sitting on it (Figs. 3.8 and 3.9).

The cover that is on the opening of the capsules, which reveals the small teeth standing on the edge of the capsule in mosses after its detachment, is an important identification mark (Fig. 3.10). In some species, the young capsule is covered by the remnant of the wall of the female reproductive organ, the cap (calyptra).

By rubbing the capsule on the slide in a few drops of water with a cover slip, you can examine the escaping spores under a light microscope (Fig. 3.11).

The filamentous or, less commonly, the lamellar protonema is created by the division of spores. Later, on these appear the bud-like formations from which the stem and the leaves located on it develop. Mosses absorb water through their entire body surface; they have not developed an internal water transport system. However,

G. Kriska, *Teaching Biology*, https://doi.org/10.1007/978-3-031-84085-2_3

Fig. 3.1 Star moss (*Tortula ruralis*) and a bicoloured bryum (*Bryum bicolor*)

Fig. 3.2 Structure of moss plants

Fig. 3.3 Neat feather-moss (*Pseudoscleropodium purum*) and haircap moss (*Polytrichum* sp.)

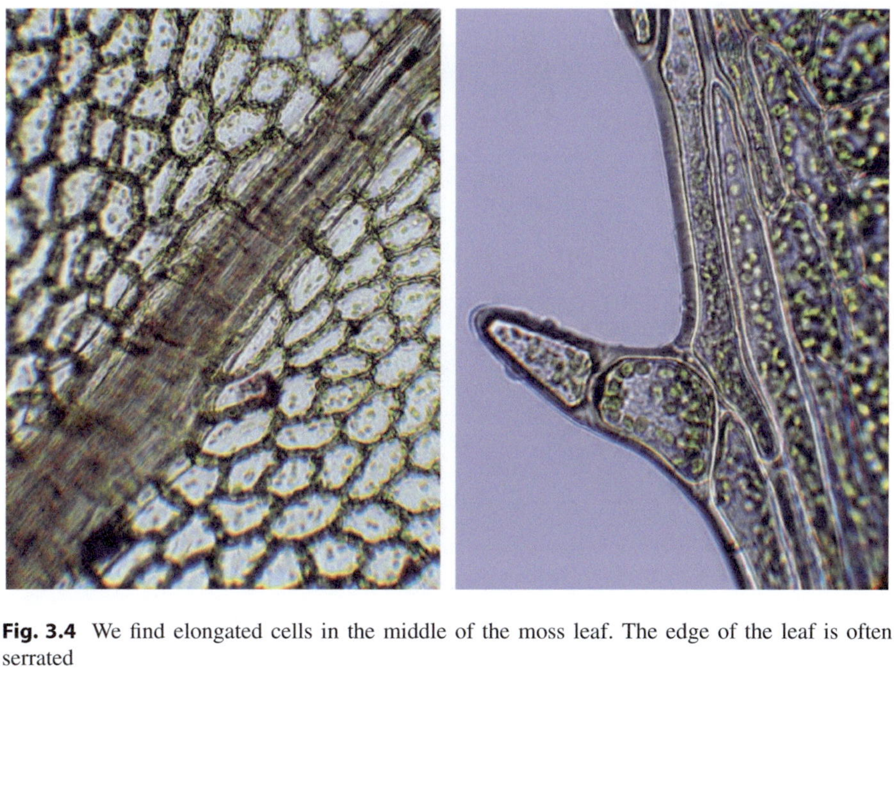

Fig. 3.4 We find elongated cells in the middle of the moss leaf. The edge of the leaf is often serrated

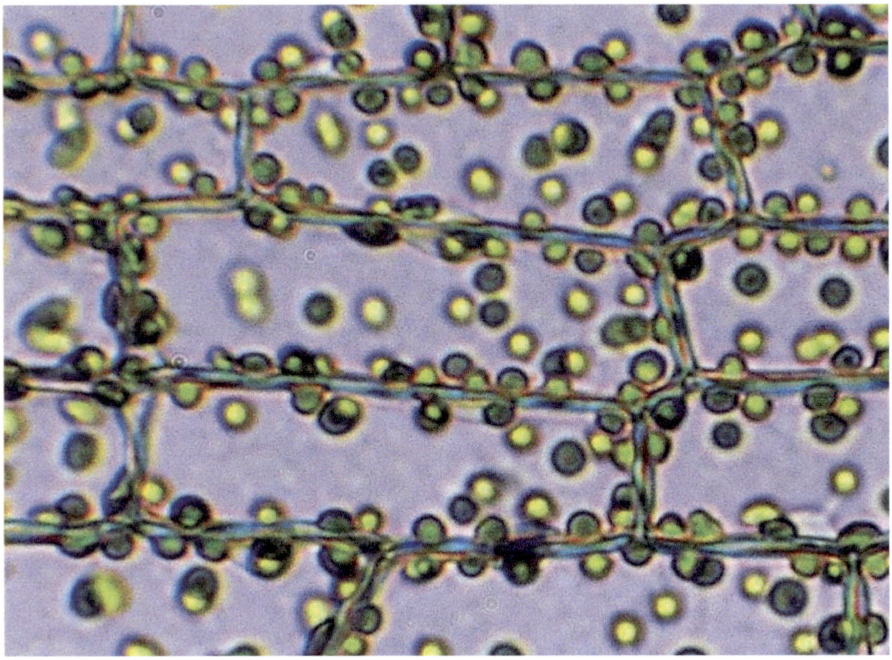

Fig. 3.5 High magnification light microscope image of the cells of a moss leaf

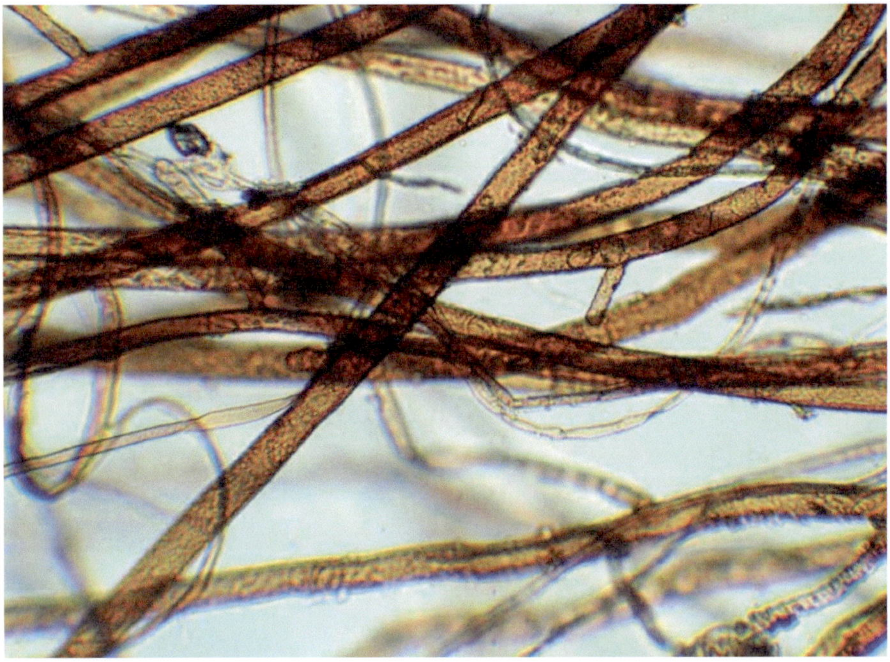

Fig. 3.6 Light microscope image of moss's root-like extensions

Fig. 3.7 Polytrichum moss (*Polytrichum piliferum*) and dotted thyme-moss (*Rhizomnium punctatum*) male reproductive organs

Fig. 3.8 *Brachythecium velutinum* and the great scented liverwort (*Conocephalum conicum*)

Fig. 3.9 Sporophytes of moss

Fig. 3.10 Scanning electron microscope image of a capsule (Photo: Béla Lovas)

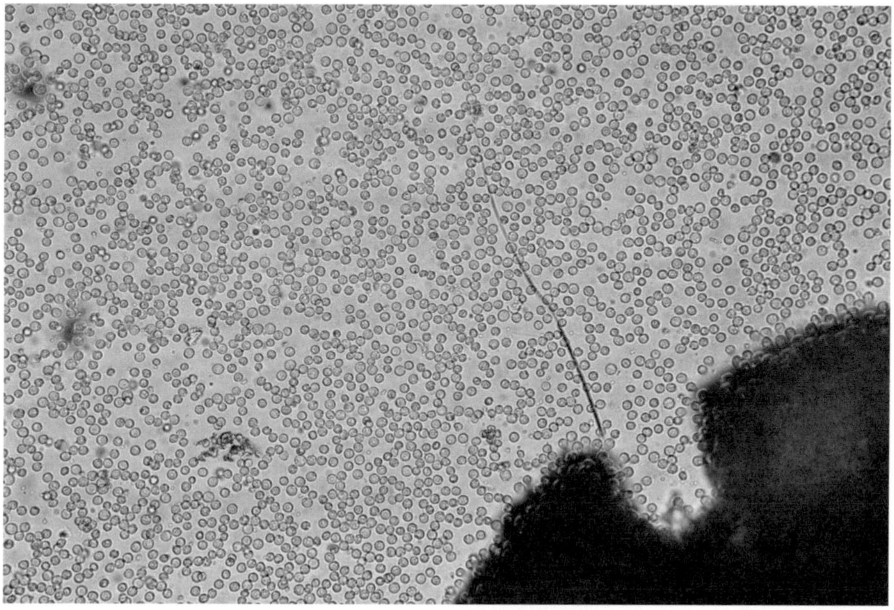

Fig. 3.11 Light microscope image of moss spores released from capsule

the moss patch can bind a large amount of water in the capillary tubes that form between the moss plants, which plays an important role in the water supply of the individuals as well as in the water balance of the soils. When dried, the functions of the moss plant slow down, their cytoplasm turns into a jelly-like state. If it is hit by water again, it turns green and comes to life (Fig. 3.12).

Films

Moss leaves: https://youtu.be/JW7Oqf7MfMo

Cells of moss leaf: https://youtu.be/Bs-s0aUM-Sw

Cells filled with chloroplasts and elongated transport cells in the moss leaf: https://youtu.be/-vnIWn6yNQU

Root-like formations of moss: https://youtu.be/2TlKvq_Jx2U

Developing and mature sporophytes: https://youtu.be/SlgRaXI4y94

Moss spores: https://youtu.be/PVJZjDvP9AE

Star moss water absorption: https://youtu.be/XIZAv0oqqpg

3.2 Lichens

Lichens are colonial plants that arise from the coexistence of green algae or cyanobacteria and filaments of different fungi species. The algae produce organic substances for the fungi through their photosynthesis, while the fungus provides the

Fig. 3.12 Star moss that turned green in moments under the effect of water sprinkling

algae with water-soluble nutrients and carbon dioxide (CO_2), and also protects them from environmental influences.

The internal structure of the colony can be homogeneous or layered. In the former case, the algal cells are evenly distributed among the fungal filaments (Figs. 3.13 and 3.14), while in the latter, the lichen consists of a photosynthesising layer containing algal cells (photobiont layer), a medulla located underneath, consisting of fungal filaments, storing water and nutrients, and a cortex surrounding the colony. When examining Icelandic lichen tea soaked in water with a light microscope, the fungal filaments and the ones between them rounded algal cells can be seen (Figs. 3.13 and 3.14).

Lichens can also cover their water needs from the humidity of the air. Lichens can only reproduce asexually, with thallus fragments or the so-called soredia. The latter are formations containing algal cells and fungal filaments, formed on the surface or inside the thallus. Many of them are sensitive to air pollution, their disappearance indicates increased sulphur dioxide and fluoride content in the air. We call lichens indicator organisms.

Lichens produce special substances, lichen substances. These compounds play an important role in the first step of soil formation, in the weathering of rocks, but their effect on humans can also be significant.

Fig. 3.13 Soaked Icelandic lichen (*Cetraria islandica*) under a light microscope

Fig. 3.14 High magnification light microscope image of Icelandic lichen (*Cetraria islandica*)

The thallus body of the lichen does not show the organs characteristic of shoot plants (root, stem, and leaf), however, the colour and shape of the thallus are characteristic of the given lichen species or group (Figs. 3.15, 3.16, and 3.17). A very common lichen is the common orange lichen (Fig. 3.17), whose rounded, rosette thallus occurs in sunny places on tree bark and rocks. The rounded spore-forming bodies (apothecium) stand densely in the middle of the thalli, their rim is the same colour as the thallus, while their interior is darker, yellowish-brown.

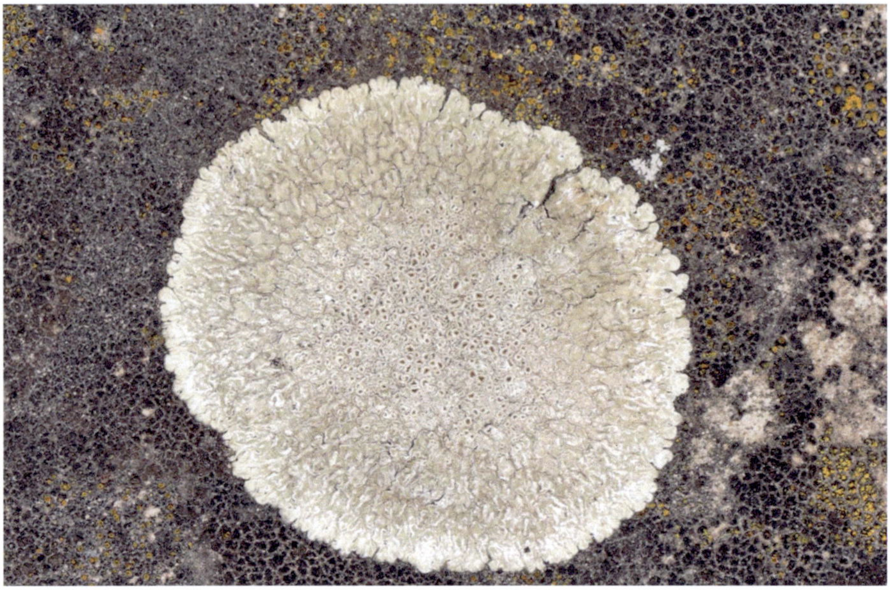

Fig. 3.15 Lichens developed on rock surface

Fig. 3.16 Dog lichen (*Peltigera canina*) living on tree bark

Fig. 3.17 Common orange lichen (*Xanthoria parietina*) on tree bark

3.3 Ferns

The evolution of the ancient ferns could have taken place in two main directions, therefore the group of ferns cannot be considered taxonomically uniform. As a result, the currently accepted classification places their species in the phyla of lycophytes (Lycopodiophyta) and horsetails, whisk ferns, and eusporangiate and leptosporangiate ferns (Monilophyta).

It is uniformly characteristic of them that a prothallus carrying the sexual organs develops from their haploid (half chromosome number) spores. Their fertilisation is tied to water, and their flagellated sperms reach the egg cells by swimming.

Today's **horsetails** are soft-stemmed, low-growing plants. Their shoot is whorled branching, and their leaf arrangement is also whorled (Figs. 3.18, 3.19, 3.20, and 3.21). The ribbed-walled stem segments separated by the stem nodes can be clearly recognised on the stem.

The small leaves originating at the stem nodes have mostly grown together into leaf whorls, which are also called nodal sheaths. A root system is originating from the shoot.

Their spore-producing organs (**strobili**) develop on **sporangiophore**s forming chains, which either appear on the photosynthesising green shoot, or on a separate so-called cone-bearing shoot, which is often brownish due to its lower chlorophyll content.

In the latter case, we call the green shoot that does not contain sporangiophores a sterile shoot (Fig. 3.19).

Fig. 3.18 Sterile (non-reproductive) shoots of the field horsetail (*Equisetum arvense*)

Fig. 3.19 Marsh horsetail (*Equisetum palustre*) and field horsetail

Nutrients accumulate in the plant's underground shoot part, the rhizome, through the photosynthesis of the sterile shoot. The sporangiophore chain (strobilus) consists of hexagonal sporangiophores connected to the central axis by a stalk, which originate six to ten spore-producing organs (sporangia) from their inner side. Before the spores are dispersed, the previously tightly closed sporangiophores turn out and scatter the chlorophyll-containing green spores.

With the rupture of the outer wall of the spores, four arm-like extensions (haptera) capable of hygroscopic movement are formed. These curl up in high humidity, while they extend under dry conditions. The wind can easily catch the arms extended

Fig. 3.20 Segment of marsh horsetail shoot

Fig. 3.21 Strobili of marsh horsetail on shoot tips

Fig. 3.22 Marsh fern (*Thelypteris palustris*)

in dry conditions and allow spores to disperse in clusters, and thus spores can drift further towards a more humid place, which is more favourable for their further development. The spore develops into a fringed, photosynthesising gametophytes, which has several female reproductive organs, so several sporophytes are formed on one gametophytes. Unlike mosses, the sporophyte of horsetails is more developed than the gametophyte.

In **ferns**, leaves differentiated into a large leaf base, leaf stalk and even leaf blade appear (Fig. 3.22), which, when young, protect their apical dividing tissues by curling up (Fig. 3.23).

The sporangium most often appears on the underside of the leaf (Fig. 3.24). In some species, separate photosynthesising green leaves and separate sporangium-bearing leaves develop (e.g. ostrich fern: *Matteuccia struthiopteris*), while in others (e.g. male fern: *Dryopteris filix-mas*), the same leaf performs both tasks. The gametophyte developing from the spore is a green, lamellar formation, which is fixed to the substrate with root-like threads (rhizoids). Its size is much smaller than that of the sporophyte, which develops on the gametophyte for a while, then turns to an independent life after the gametophyte dies. The sporangia are on the underside of the photosynthesising leaf and they form rounded sori (Fig. 3.24).

The perennial male fern is equally common in the undergrowth of deciduous and coniferous forests. Due to its aesthetically pleasing compound leaves, it is often planted in gardens and parks, and its leaves are frequently used by florists in bouquets. The leaves, which can grow up to one metre in length, are attached to the

Fig. 3.23 Young male fern (*Dryopteris filix-mas*)

underground stem, the rhizome, with a large leaf base and a short, yellowish, downy-haired stipe. The stipe extends to the first branching, from where we call the continuation of the stipe and the side branches arising from it the leaf axis. The leaflets (pinnae) are attached to the leaf axis of the side branches. The leaf includes all the leaflets. The edges of the opposite leaflets can be smooth, wavy, or serrated on the same leaf (Fig. 3.25).

The spore-producing formations (**sori**) of the male fern develop on the underside of the leaves in late spring. By examining a leaf fragment, we can study the fern leaflets, on which the sori on the leaf underside appear as small spots. The round sori containing the sporangia are initially light in colour, but darken as they mature (Fig. 3.26).

The sori are covered by a kidney-shaped indusium when young. When it dries out, the indusium shrivels and the sporangia sitting on a stalk underneath appear.

Place a sorus on a slide, rub it with a cover slip, and examine it under a light microscope. On the preparation, the spherical head part of the sporangium, with U-shaped thickened ring and thin-walled parts, can be clearly observed. After the indusium shrivels, the hygroscopic movement of the thick-walled cells tears apart the part made of thin-walled cells, causing the spores to be ejected from the sporangium (Fig. 3.27).

Fig. 3.24 Leaf of the male fern (*Dryopteris filix-mas)* with mature sori on the underside

Fig. 3.25 Structure of the leaf of the male fern (*Dryopteris filix-mas)*

Fig. 3.26 High magnification image of mature sori

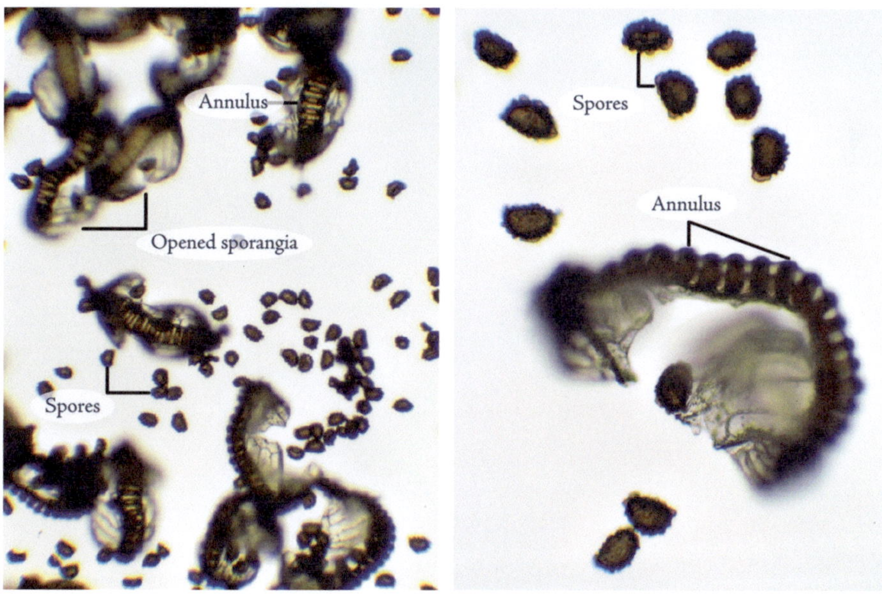

Fig. 3.27 Light microscopic image of the opened sporangia of the male fern

4 Vegetative Organs

The vegetative organs of flowering plants are the root, the stem, and the leaf, which develop from the fertilised egg cell, the zygote. The germ or embryo, consisting of one or more cotyledons, the embryonic stem forming the shoot system, and the radicle forming the root system, develops from the zygote through a series of cell divisions. In ferns, the sexual organs appear on the prothallus that develops from the spore, and after the fertilisation of the egg cell, the fern embryo developing from the zygote begins to develop in the female sexual organ. The embryo develops into a new plant without a dormant state, while the prothallus and the root primordia of the embryo die off and are replaced by roots of shoot origin.

In the case of flowering plants (gymnosperms and angiosperms), the embryo surrounded by nutritive tissue enters a dormant state inside the seed for a while, and only resumes development during germination.

4.1 Germination

4.1.1 Examination of a Plant Germ

Soak dry beans in water for an hour! The mature seed usually shows a scar where the funicle detached, the **hilum**. The hilum through which the **placenta** developing from the inner wall of the ovary nourished the developing ovule, can be observed at the curvature of the bean seed. Once the seed has softened, remove the **seed coat**! Underneath the seed coat, the two large **cotyledon**s appear, between which the embryo from which the plant develops can be found next to the hilum (Fig. 4.1).

The germ is a bipolar formation, at the apex of which is the embryonic stem, from which the shoot (leafy stem) will develop. In the case of beans, the embryonic stem is also surrounded by two small leaves. At the other end of the germ is the **radicle**, from which a root develops. The cotyledons develop on the stem part (mesocotyl) connecting the embryonic stem and the rootlet. In the case of beans, the two large cotyledons contain the nutrients necessary for germination.

G. Kriska, *Teaching Biology*, https://doi.org/10.1007/978-3-031-84085-2_4

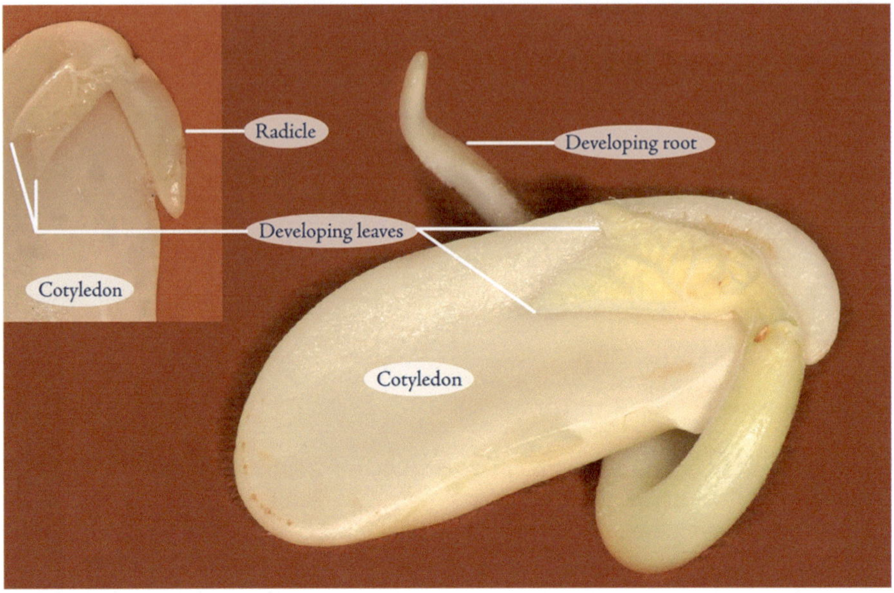

Fig. 4.1 Parts of a bean seed stripped of its seed coat (small picture) and a bean seed that is already germinating

In the formation of a corn grain (caryopsis), the carpel wall also participates alongside the seed coat, therefore this is actually not a seed, but a fruit (Fig. 4.2).

Unlike beans and peas, the stored nutrients of the corn grain are not in the cotyledon, but in the nutritive tissue of the seed (endosperm). The cotyledon is a shield-like formation (scutellum) that separates the embryo from the yellow **endosperm** of the corn.

If any of the conditions for germination (appropriate temperature, presence of water and oxygen, and absence of germination inhibitors) are missing, the seed will not germinate.

4.2 Root

The **root** is a radially symmetrical organ used for anchoring the plant in the soil, transporting inorganic nutrients (water and mineral salts), and storing organic materials. The root appears in most shoot plants, with only a few plant species being rootless. Its tip is covered and protected by the **root cap**. The outer part of this becomes slimy, which aids the growing root's progress through the soil.

The cells of the root cap sense the direction of Earth's gravity through the starch grains inside them, hence they play an important role in the root growing towards the centre of the Earth. The root develops from the apical meristems of the root tip.

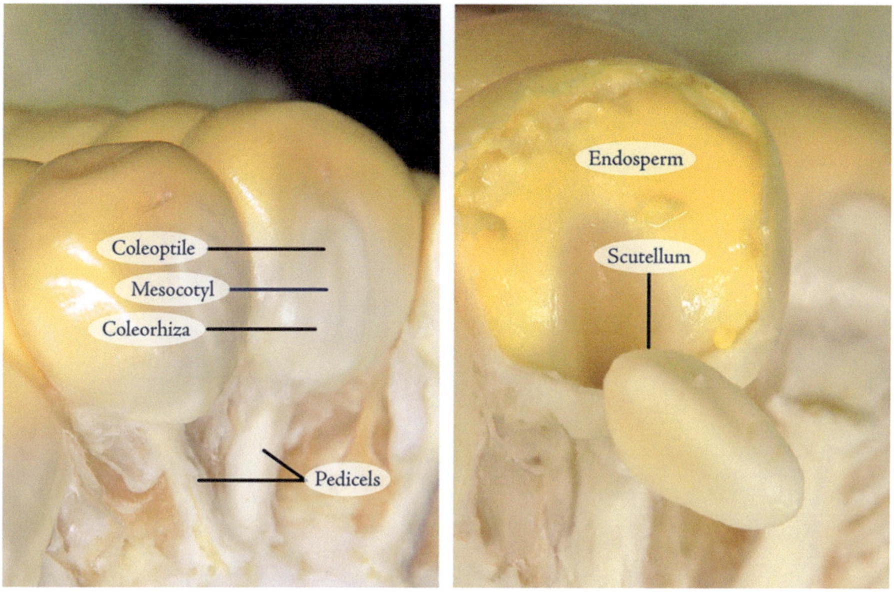

Fig. 4.2 Parts of a corn grain

In the elongation zone of the root, the separation of the tissues of the epidermis, the cortex, and the central cylinder begins, from which the functioning form of the organ develops.

The elongation zone is followed by the maturation zone, which can be recognised by the appearance of **root hair**s (Fig. 4.3).

Root hairs are single-celled formations of the epidermis, which are continuously formed after the elongation zone, while they die off towards the end of the maturation zone. Thus, as the root tip advances, the maturation zone always comes into contact with new parts of the soil, from where it can absorb nutrients. The nutrients taken up by the root hairs, the water and the mineral salts dissolved in the water, reach the vascular tissue through the cortex of the root, from where they are transported to the site of photosynthesis, the shoot. The next part is the branching zone, where the formation of lateral roots begins. In dicots, further from the root tip, the secondary thickening zone follows, which transmits, anchors, and strengthens the plant, and stores nutrients.

In gymnosperms and angiosperms, a main root system develops from the radicle, which consists of the main root and branching lateral roots.

In most monocots, the main root system that develops from the radicle soon dies off and is replaced by an equivalent lateral root system that develops from the shoot (mesocotyl), which generally does not thicken secondarily. Later, as the plant develops from its first nodules, many more adventitious roots develop (Figs. 4.3 and 4.4).

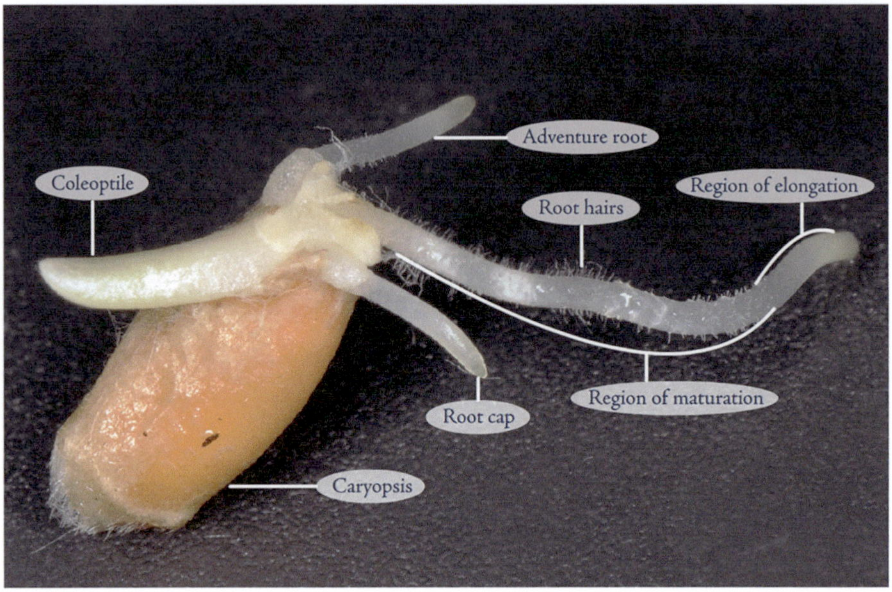

Fig. 4.3 The root parts can be clearly identified on the wheat germ

Fig. 4.4 Root system of a grass clump and an enlarged image of a part of it

In most monocots (grasses), the main root system that develops from the radicle and the mesocotyl of the embryo generally does not thicken secondarily (Figs. 4.3 and 4.4).

Depending on various environmental factors, the shape and function of the roots can change, leading to the appearance of modified roots. The background to minor changes is most often the appearance of the following functions: nutrient storage, plant support, climbing, aeration, defence, and parasitism.

The **aerial roots** are adventitious roots originating from the stem nodes or stem segments, which can be classified into different types based on their function (Figs. 4.5 and 4.6).

True aerial roots are those that do not penetrate the soil, but bind the humidity of the air and the precipitation. Orchids living on trees have such roots. The climbing aerial roots, such as the short roots formed on the stem segments of ivy, only fix the climbing plant's shoot.

The **brace roots** are supporting roots originating from the lower stem nodes, as we have already seen in the case of maize (Fig. 4.6). The nutrient-transporting aerial roots are able to bind precipitation on their surface, and then transport the salts dissolved in the water when they enter the soil.

The lesser celandine is a herbaceous, perennial plant. Its flowers bloom from the end of March until May. By the beginning of June, the above-ground parts of the plant die off and it survives in the soil until the following spring in its small **root tubers**. Their ground tissue contains a large amount of reserve nutrients, starch (Figs. 4.7 and 4.8).

Characteristic root modification of leguminous plants are the **root nodules** (Fig. 4.9). These contain nitrogen-fixing bacteria, which are able to bind the nitrogen content of the air, some of which they provide to the host plant. The activity of the bacteria has a good effect on the development of the leguminous plant and also increases the nitrogen content of the soil, making the soil richer in plant nutrients.

Fig. 4.5 Aerial roots developing on a trumpet vine (*Campsis radicans*) shoot

Fig. 4.6 Appearance of brace roots at the lower stem nodes of maize (*Zea mays*)

Fig. 4.7 The root tubers of the lesser celandine (*Ficaria verna*)

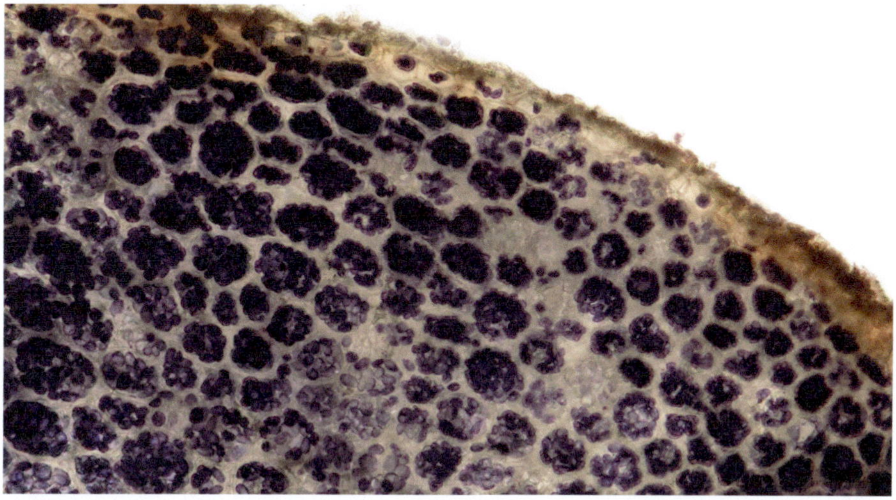

Fig. 4.8 Starch grains stained with Lugol's solution in the cortical parenchyma cells of the root tuber

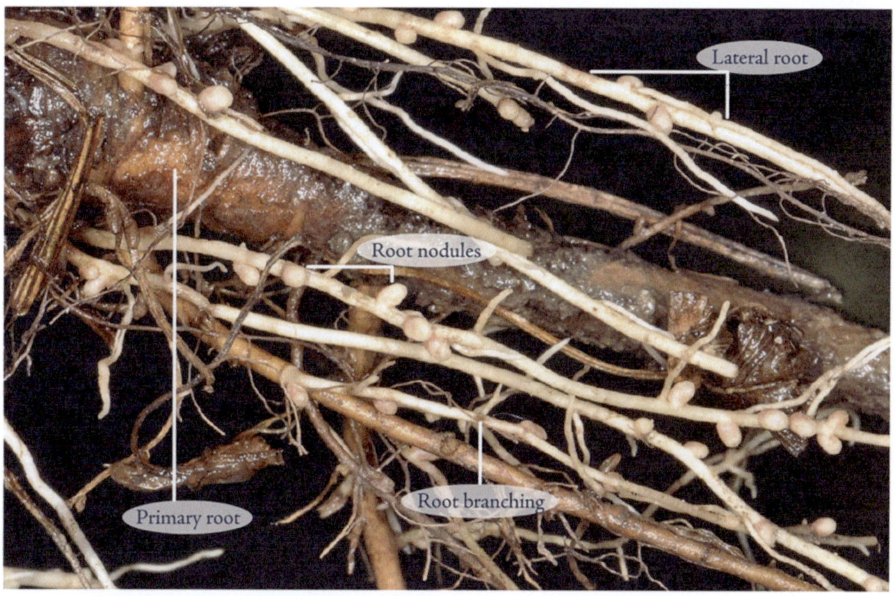

Fig. 4.9 A section of the red clover (*Trifolium pratense*) root with root nodules

4.3 Stem

The shoot axis is the **stem**, and its appendages are the leaves. The **bud** is the embryonic state of the shoot, which consists of the bud axis, the shoot apex located at the top of this, and the leaf- and shoot primordia located laterally. It is usually covered by **bud scale** protecting it against the effects of weather and plant pests (Fig. 4.10).

Some of these, such as the rust mite of grape vine, overwinter under the bud scales and start damaging the plant in spring. The bud scale is a thick, often fluffy or sticky formation.

Instead of the covered buds (covered with bud scales), some plants, such as the ash tree, develop naked buds which are protected by hairy leaves. The herbaceous plants, tropical trees and shrubs do not need serious protection for their buds, so we do not find bud scales on them.

The buds of trees and shrubs usually form in the summer and sprout after surviving the winter.

Initially, only the abaxial (lower) surface of the leaf primordia grows, so they completely cover, hide the shoot apex, which only comes to the surface during leaf growth, at bud burst.

Based on the location of the bud, it can be a **terminal bud**, which is located at the end of the shoot and ensures longitudinal growth, and it can be a **lateral bud**, which is directly under the terminal bud standing on the side replaces the latter in case of its decay.

Often, along with the terminal bud, the lateral buds also sprout, leading to branching of the shoot. The axillary buds are located at the origin of the leaves in their axils. In addition to the above three types of buds (regular buds) (Fig. 4.11), there are also so-called **dormant buds**, which start developing in case of decay of the

Fig. 4.10 The terminal bud of the horse chestnut (*Aesculus hippocastanum*) covered with sticky bud scales and a cross-section of the bud of the common ash (*Fraxinus excelsior*) tree

Fig. 4.11 Buds of the common ash (*Fraxinus excelsior*) tree

Fig. 4.12 The dormant bud of the plane tree (*Platanus* sp.) under the leaf base and the dormant bud of the tree of heaven (*Ailanthus altissima*) above the leaf scar

regular buds. These are hidden under the bark of the stem, or under the base of the leaves (e.g. plane tree) (Fig. 4.12).

The buds developing under the leaf base remain invisible to the external observer until the autumn leaf fall (Fig. 4.12). In many cases, dormant buds are hidden under the bark and only start to develop if the tree needs to develop a new crown due to damage. In cities, you can often see trees that have been so drastically pruned that almost their entire crown has been removed. With these, you can often observe the dormant buds sprouting from under the bark and producing new shoots.

Buds appear not only on above-ground shoots but also on underground ones. Thus, buds can be found on the stem tuber of a potato, which sprout under suitable conditions. This process is often referred to as potato germination, which is a completely incorrect definition, as germination is characteristic of seeds (Fig. 4.13).

The stem's task is to hold various plant organs, the leaves, flowers, and fruits, and ensure their connection with the root.

The plant nutrients absorbed by the root, the salts dissolved in water, reach the leaves, the site of photosynthesis, through the stem, from where the finished organic materials are transported to the other plant organs. The transport of organic materials can also take place in other directions, in the spring, for example, many plants (e.g. snowdrop) draw material and energy from their underground storage organs, the root tuber, bulb, or rhizome, for their sprouting. The transport of materials in plants is ensured by the vascular tissues, which completely network the plant's body. In young stems, the vascular tissues usually form bundles (Fig. 4.14).

The **tree trunk** is a special organ of woody plants. On its cross-section, the outer layer of bark and the inner xylem tissue are clearly separated. The outer part of the bark is given by layers of cork, while the inner part is given by the phloem and the cambium that produces it, located on the inside. Organic materials are transported in the thin, cylinder-like phloem of the tree trunk, so if we remove even a 1 cm ring of a tree's bark, it inevitably causes the plant to die. Underneath the phloem, also in a cylinder-like arrangement, is a thin layer of meristematic tissue (cambium), which develops the phloem outwards and the xylem inwards. Only the thin layer of the

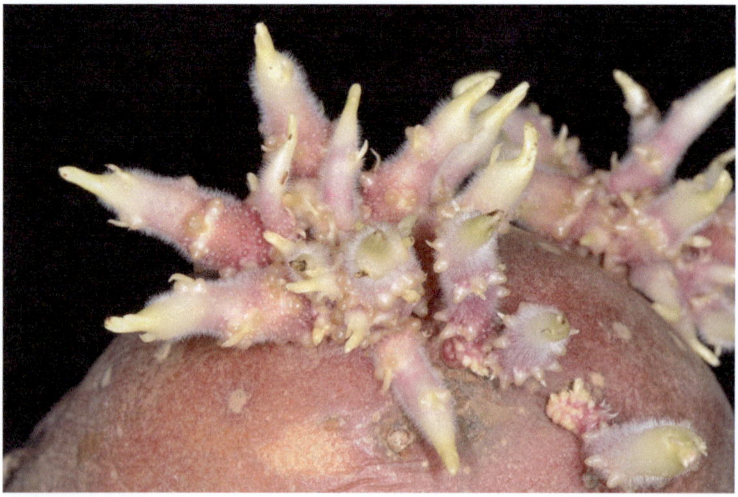

Fig. 4.13 Shoot primordia formed from potato (*Solanum tuberosum*) buds

Fig. 4.14 Detail of the noli me tangere (*Impatiens* sp.) stem under transmitted light

phloem next to the cambium participates in the transport of materials, the layers formed earlier are pushed outwards and become part of the bark by tearing. In contrast, the layers of xylem tissue formed year after year remain. The solution of mineral salts taken up from the soil is transported in the trunk to the shoot.

Not only the parts of the xylem next to the cambium, but also the inner layers participate in the transport of materials, but the central part of the tree trunk is already given by dead xylem tissue. Radial medullary rays run in the xylem tissue, which participate in the nourishment of the transport cells and participate in their material cycle (Figs. 4.15 and 4.16). When trees reach nearly 20 years of age, a process of heartwood formation begins within their trunks.

As a result, the water content of the wood significantly decreases and resin, alkaloids, dyes, and tannins accumulate in the dead cells, making the interior of the tree resistant to fungi and other harmful effects.

The resulting heartwood is darker than the still functioning sapwood with higher water content located next to the cambium. Poplar and willow trees do not form heartwood, so their trunks rot more frequently.

4.4 Leaf

We categorise plant leaves into four types: cotyledons, **bracts**, **scale-**, and **foliage leaves**. The cotyledons can remain in the seed during germination (e.g. corn) and transmit nutrients from the nutritive tissue towards the germ, they can emerge from

Fig. 4.15 Cross-section of a tree trunk

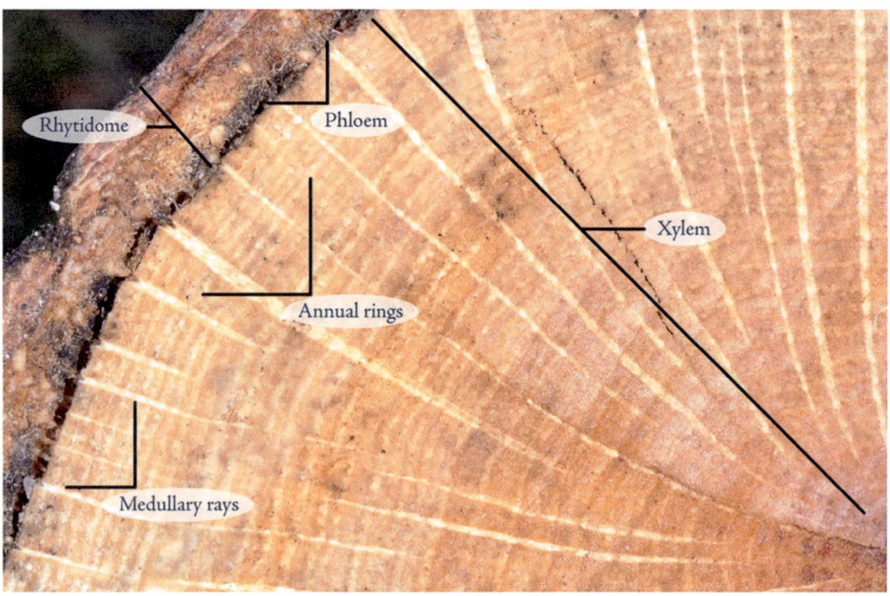

Fig. 4.16 Detail of a tree trunk cross-section

the soil and nourish the developing plant with their stored materials until the first foliage leaves appear (e.g. beans), or they can turn green and photosynthesise (e.g. beech).

In the case of beans, we can also observe the greening of the cotyledons, but their rate of photosynthesis is much lower than that of the radish cotyledons (Fig. 4.17).

Bracts are protective leaves originating above the level of the foliage leaves. Examples include the bracts of hazel, oak, and chestnut, and the bracts at the peduncles (Figs. 4.18 and 4.19).

The flower itself is also considered a shoot, and the floral leaves involved in its formation can be considered as bracts.

Scale leaves originate below the level of the foliage leaves. They include the bud scales protecting the leaf buds, or for example the fleshy leaves of the onion developing underground (Fig. 4.20).

The appearance of onion scale leaves significantly differs from that of foliage leaves, thus, at first glance, anyone may have doubts about their leaf origin. However, there are also microscopic characteristics that clearly indicate that onion leaves are indeed leaves. Such a feature, for example, is that the epidermis of the fleshy leaves contains stomata.

The onion is a short-stemmed underground shoot, from whose apical bud the above-ground part of the plant develops, while from the lower part of the trunk root-origin shoots develop.

The onion is covered on the outside by dry **sheath leaves** (tunic or scaly leaf), within which the overlapping fleshy leaves are located. These surround the terminal bud sitting on the stem.

Fig. 4.17 The first two leaves appear among the large heart-shaped cotyledons of the radish (*Raphanus sativus*)

Fig. 4.18 Bracts on developing acorns of Turkey oak (*Quercus cerris*) and hazel (*Corylus avellana*)

Fig. 4.19 Bracts of sweet chestnut (*Castanea sativa*) and greater burdock (*Arctium lappa*)

Fig. 4.20 Structure of onion (*Allium cepa*)

The foliage leaves are the most important organs of photosynthesis, evaporation, and gas exchange, mostly large-surface, flattened organs (Figs. 4.21, 4.22, and 4.23).

The foliage leaf consists of the broadening **leaf blade**, the **leaf stalk,** and the **leaf base** attached to the stem. The leaf base usually forms a broadened leaf foot (e.g. plane tree), or forms a leaf sheath that encircles the stem (e.g. reed). The appendages of the leaf base are the stipules, which in the case of the rose are fringed, in *Polygonum* form ocrea encircling the stem, and in the false acacia they transform into thorns.

The shape of the leaf blade can be very varied (obcordate, reniform, lanceolate, rhomboid, acicular, etc.). When describing it, we distinguish the leaf apex and lamina base, which can also be diverse.

The edge of the leaf blade is smooth (e.g. lilac, dogwood), or incised. The incision can be serrate (e.g. common hackberry and nettle), biserrate (e.g. hornbeam and elm), dentate (e.g. common barberry and butterbur), and crenate (e.g. ground-ivy) (Figs. 4.24 and 4.25).

In the case of deeper incisions, the leaf blade is segmented, lobed, cleft, divided, or dissected (Figs. 4.26 and 4.27). If the incisions do not reach the midline of the half leaf blade, then the leaf is pinnatifid, as in the case of the oak. The Norway maple has a palmatifid leaf, where the incision reaches the midline of the half leaf blade. The pinnatipartite leaf has lobes with incisions that extend more than halfway to the midrib (e.g. small geranium), and the pinnatisect leaf has lobes with incisions that extend almost to, or up to, the midrib.

Fig. 4.21 Detail of a wild rose (*Rosa canina*) shoot with fruit

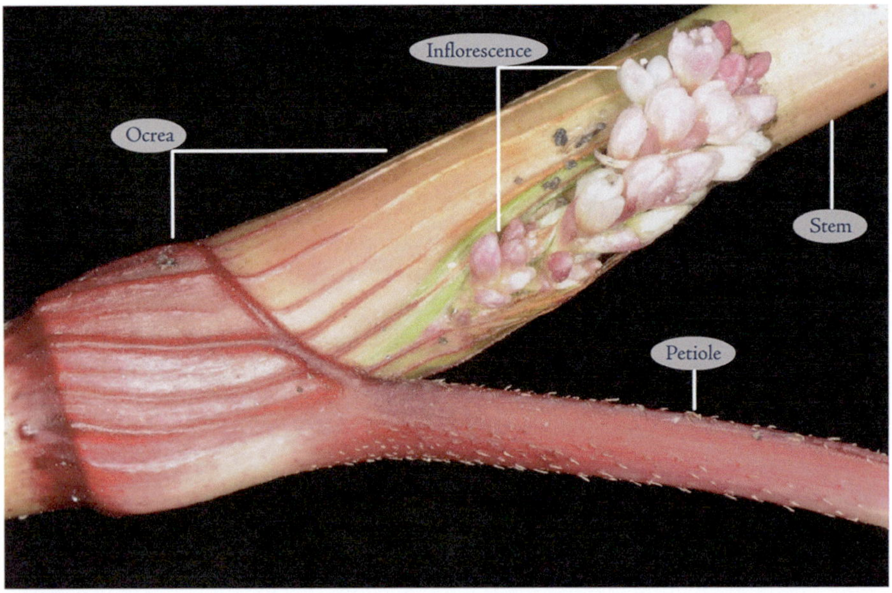

Fig. 4.22 The knotweed (*Polygonum* sp.) flowers emerge from ocrea

Fig. 4.23 Leaf of the dogwood (*Cornus* sp.)

Fig. 4.24 Serrated leaf edge of the common hackberry (*Celtis occidentalis*) and biserrate leaf edge of the hornbeam (*Carpinus betulus*)

Fig. 4.25 Dentate leaf of the common barberry (*Berberis vulgaris*) and crenate edge leaf of the ground-ivy (*Glechoma hederacea*)

Fig. 4.26 Pinnatifid leaf blade of the downy oak (*Quercus pubescens*) and palmatifid leaf blade of the Norway maple (*Acer speudoplatanus*)

The blade of a simple leaf is uniform, while in a **compound leaf**, the leaf blade is divided into separate leaflets.

The axis of the compound leaf is provided by the rachis derived from the main vein, from which the leaflets originate. If the leaflets are arranged in a row on both sides of the rachis, then the leaf is pinnately compound leaf (e.g. false acacia), but if the leaflets branch out from one point of the leaf spine, then we can talk about a

Fig. 4.27 Pinnatipartite leaf of the small geranium (*Geranium pusillum*) and pinnatisect leaf of the meadow buttercup (*Ranunculus acris*)

Fig. 4.28 Odd pinnate leaf of the false acacia (*Robinia pseudoacacia*), and trifoliate compound leaf of wood sorrel (*Oxalis acetosella*)

palmately compound leaf (e.g. *Oxalis* and horse chestnut) (Fig. 4.28). The distinction between the leaflet and a leaf is possible by the fact that we never find a bud in the axil of the leaflets.

Fig. 4.29 Young undivided, older showing initial signs of division, and deeply divided mature leaves of a white mulberry (*Morus alba*)

Fig. 4.30 A section of a shoot with alternate and with whorled leaf arrangement

The mature leaves of a plant can have different appearances, which we call **heterophylly** (Fig. 4.29).

Heterophylly can be caused by the different position of the leaf, for example, the submerged leaves of the white water-crowfoot are feathery, while those rising above the water surface are undivided.

The leaves on the stem are arranged in a regular order, based on which we can talk about alternate, opposite, and whorled leaf arrangements (Fig. 4.30).

In the case of alternate arrangement, only one leaf originates from a stem node (Fig. 4.31). One type of alternate arrangement is the spiral leaf arrangement, where

Fig. 4.31 Spiral leaf arrangement

Fig. 4.32 Decussate leaf arrangements

the leaves are arranged along a spiral on the stem. The angle value, which is enclosed by the successive leaf origin points, is the divergence angle. In the case of opposite leaf arrangement, two leaves originate from a stem node opposite each other (Fig. 4.32). In the case of decussate leaf arrangement, the leaf pairs located underneath each other are rotated 90° relative to each other. In the case of whorled leaf arrangement, 3 or more leaves originate from the stem nodes (Fig. 4.30).

The vascular bundles running in the leaf form the leaf venation. The venation transports the nutrients absorbed from the soil (water and mineral salts) into the leaf, on the other hand, the leaf delivers the organic materials formed during photosynthesis to the rest of the plant.

The final branches of the leaf veins end blindly or merge to form a continuous network.

The leaf venation is not only important for material transport, but also stiffens and strengthens the leaf. The leaf venation can vary among different plants.

Monocot plants have parallel leaf venation. In the latter, several equivalent vascular bundles enter the leaf, between which a network connection is formed (Fig. 4.33).

In dicots, a main vein enters the leaf, from which multiple branching side veins emerge. We call this type of venation reticulated venation.

The leaf is pinnately veined (Fig. 4.34) when the main vein runs through the leaf, while side veins branch off from it.

The leaf is palmately veined (Fig. 4.35), if the main vein continues in radial side veins upon entering the leaf. In the case of the Norway maple, the main vein running in the leaf stalk divides into seven branches upon entering the leaf blade and branches out like a finger towards the leaf tips.

The red pigments accumulating in the leaves are capable of protecting the plants from the harmful effects of too strong sunlight. These dyes also protect the plant from excessive cooling by absorbing sunlight and warming the leaf in extreme cold. This type of protection is primarily needed by the leaves that are still vulnerable, developing from buds at the end of winter. This explains why the new leaves are sometimes not green, but reddish due to the protective pigments accumulated in them. As the leaf develops, the protective substance breaks down, the leaf becomes completely green.

Fig. 4.33 Detail of a lily of the valley (*Convallaria majalis*) leaf at lower and higher magnification

Fig. 4.34 Leaf of one-seed hawthorn (*Crataegus monogyna*)

Fig. 4.35 The numbers on the Norway maple (*Acer platanoides*) leaf indicate the side veins running radially

Fig. 4.36 The young leaves of the hazel (*Corylus avellana*) contain anthocyanin

Fig. 4.37 The green leaf of the hazel (*Corylus avellana*)

The very young leaves of the hazel also contain anthocyanins, which give the leaves a reddish-brown colour (Fig. 4.36). Later, these pigments are absorbed and the leaf turns completely green (Fig. 4.37).

The leaves of the barberry undergo a characteristic colour change from spring to autumn (Fig. 4.38). The leaves are green until the end of flowering, then with the arrival of summer, anthocyanin is produced in the leaves, which turns the foliage purple. The production of anthocyanin becomes particularly intense in leaves

Fig. 4.38 The summer leaves of the barberry (*Berberis vulgaris*), protected from direct sunlight, are green-purple striped and another species of barberry has red leaves in autumn

exposed to strong sunlight, here the leaves turn dark purple. In the shaded areas, we often find leaves that partially retain their original green colour, striped with purple and green. The leaves of the barberry turn red in the autumn, the reasons for which will be discussed later.

Examination of Vegetative Organs

5

5.1 The Dormant Buds of the Common Ash (*Fraxinus excelsior*)

Let us take a close look at a branch of an ash tree! The end of the branch embodies this year's new shoot, which can be recognised by its greenish colour and developed buds located in the leaf axil.

Below this, we can see the part of the branch that developed last year, which no longer bears leaves, but the former connection points of the leaves are clearly recognisable in the form of arched incisions. The trace of the former leaf origins on the stem is called a leaf scar. Dormant buds can be observed in the middle of the leaf scar, which are lateral and axillary buds that have not started to develop, being put into "reserve" (Fig. 5.1). These only sprout if, for some reason, the regular buds die.

5.2 Cherry (*Prunus avium*) Shoots Developed from Dormant Buds

Especially on the trunk of a young cherry tree, we can find dormant buds that have started to develop, from which leafy stems or flowers can also develop (Figs. 5.2 and 5.3). Occasionally, multi-year branches can develop from the newly appeared shoot.

In the above picture, taken in early spring, a 1-year-old branch developed from a dormant bud can be seen, under whose swollen terminal buds the leaf scar remaining at the origin of the previous year's leaves is clearly recognisable.

5.3 Leaf Arrangement and Buds

As we have already seen, axillary buds always develop in the axils of the leaves, so the position of the buds also indicates the position of the leaves after they fall in the autumn.

G. Kriska, *Teaching Biology*, https://doi.org/10.1007/978-3-031-84085-2_5

Fig. 5.1 Dormant buds on the branch of a common ash (*Fraxinus excelsior*). The dormant buds developed from the unbroken axillary buds of the previous year's branch

Fig. 5.2 New shoot development from a dormant bud on a young cherry (*Prunus avium*) tree trunk

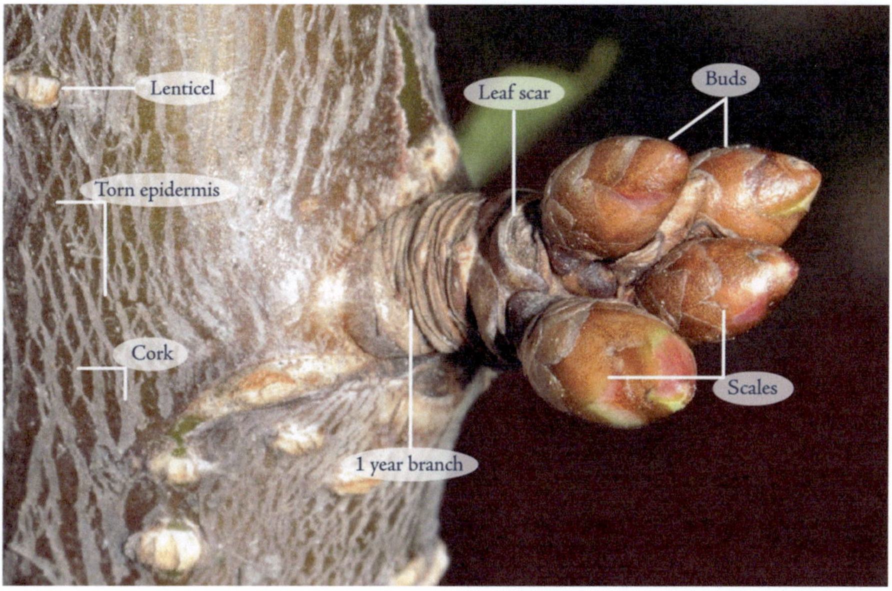

Fig. 5.3 1-year-old side branch developed from a dormant bud of a cherry (*Prunus avium*) tree

Let us examine a common ash branch containing two pairs of axillary buds with a magnifying glass! In the case of the ash, it can be clearly seen in the picture below that the leaves originate opposite each other on the branch, which we call opposite leaf arrangement. We can also observe that the leaf and bud pairs located under each other are also at right angles (90°) to each other, resulting in a decussate leaf arrangement (Fig. 5.4).

5.4 Types of Buds and Bud Burst

New shoots develop from leaf buds, and flowers from flower buds (Fig. 5.5), or an inflorescence consisting of several flowers unfolds. Mixed buds are also very common, from which both new shoots and flowers develop (Fig. 5.6).

With the arrival of spring, buds undergo a spectacular transformation. The initially barely noticeable tiny formations on the branches swell, then with bud burst, young shoots, flowers, and inflorescences emerge and unfold from between the bud scales. The initially completely folded leaves stretch out, the flowers open.

5.5 Bud Burst of the Tree of Heaven (*Ailanthus altissima*)

The tree of heaven is a tree native to China and Korea, which is an invasive, aggressively spreading species also appearing in large numbers in Europe.

Fig. 5.4 Decussate pairs of buds on the branch of a common ash (*Fraxinus excelsior*)

Fig. 5.5 Flowers developed from a cherry (*Prunus avium*) flower bud

Fig. 5.6 The mixed bud of the horse chestnut (*Aesculus hippocastanum*) unfolding in early spring contains several leaf primordia and the inflorescence primordium consisting of many flowers

Some of the buds sitting on the leaf scars of the tree of heaven start to develop in spring, while the others continue to hide on the branches as dormant buds.

The bud starting to develop swells strongly, then after the bud scales stretch out, the first leaves appear. These are initially still crumpled, but during their rapid development, they increasingly take on their mature form (Figs. 5.7 and 5.8).

5.6 Cherry (*Prunus avium*) Branch

Unlike herbaceous plants, in woody plants, the stem undergoes characteristic changes with age that we can observe on the branches. We study these changes on the cherry tree. The very young, smooth-surfaced and still green branches are covered by a single layer of epidermis. It is unable to keep up with the thickening of the branch, so it tears. The plant hormones released from the tearing epidermis cause the cells located deeper to start dividing and create a new, secondary dermal tissue, the periderm, or cork. The new layer reduces the plant's evaporation, protects the plant from parasites and overheating, but also cuts off the stem from the air, so to maintain ventilation, the so-called lenticels develop on the stem. The thin-walled, loosely arranged cells located in these allow the exchange of respiratory gases between the stem and the outside world.

Examine this year's, 1-year-old and 2-year-old cherry branch with a magnifying glass! The current year's branch is light green, and although its epidermis still appears intact, the appearance of lenticels indicates that the formation of the

Fig. 5.7 Buds starting to develop on the branch of the tree of heaven (*Ailanthus altissima*)

Fig. 5.8 Bud burst of the tree of heaven (*Ailanthus altissima*)

Fig. 5.9 Cherry's (*Prunus avium*) current year, 1-year-old (left image) and 2-year-old (right image) branch

periderm has already begun underneath. The 1-year-old, lenticel-covered branch is already brown, and the epidermis is already showing signs of fragmentation in places.

On the 2-year-old branch, the lighter cork emerges from under the fragmented epidermis, and the lenticels are large (Fig. 5.9).

Based on our knowledge about buds, identify the different age branch parts and the buds! In the above picture, this year's new shoot developed from the terminal bud formed last year.

The lateral buds next to the terminal bud did not start to develop, so there is no branching at the junction of this year's and last year's, 1-year-old branch part. In contrast, at the junction of the 1- and 2-year-old branch parts, a double branching can be observed, one of which we have also indicated in the above picture. It is clearly visible that the 1-year-old branch part developed from the terminal bud is much longer than those developed from the lateral buds. On the branch section shown in Fig. 5.10, all the regular buds can be observed.

5.7 The Branches and Trunk of the Tree of Heaven (*Ailanthus altissima*)

On the fresh, still velvety shoots of the tree of heaven, light spots on the epidermis indicate the place of the developing lenticels (Fig. 5.11).

Fig. 5.10 Cherry (*Prunus avium*) branch section

Fig. 5.11 Young tree of heaven (*Ailanthus altissima*) branch

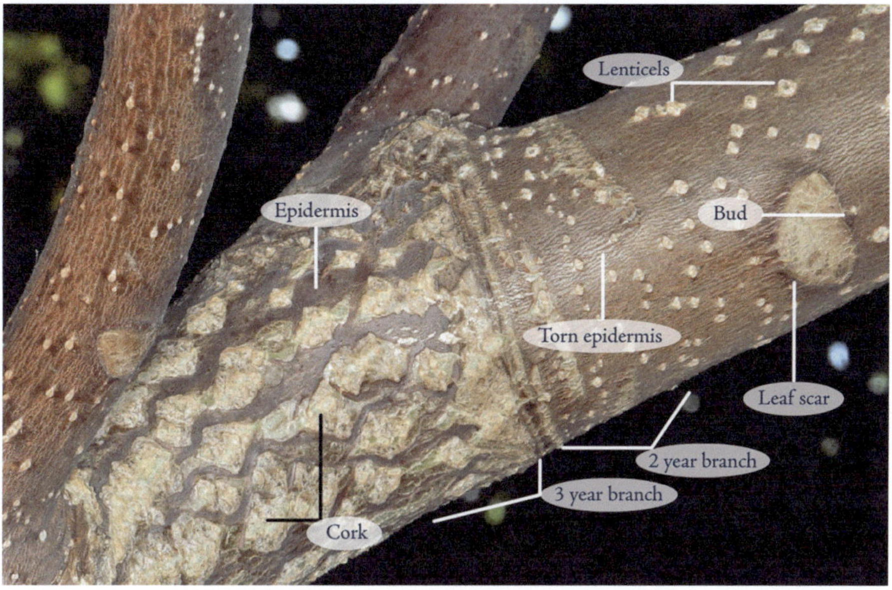

Fig. 5.12 Older branch of heaven tree (*Ailanthus altissima*)

On the older branch, the epidermis is already tearing, revealing the lighter coloured cork layer underneath. With further thickening of the tree trunk, the bark cracks more and more, and the cracks reach the phloem part of the vascular tissue, which is arranged cylindrically under the bark. From then on, the further thickening of the bark is done by the phloem creating the rhytidome (Figs. 5.12 and 5.13).

On the trunk of the old tree, we can see protruding bark parts, the upper layers of which are given by periderm layers, while the lowest layers already form part of the rhytidome. In the cracks of the bark parts, we can see the lighter surface of the rhytidome produced by the phloem.

5.8 Periderm Formations

The spindle (*Euonymus europaeus*) is a common shrub in deciduous forests, but it is also widely planted as an ornamental plant because of its decorative fruit. It got its striped name because there is a characteristic longitudinal striping that can be observed on its branches.

The appearance of stripes is caused by the uneven operation of the cork cambium that creates the cork. In fact, the meristem beneath the stripes produces more periderm, which then rises out of the branch plane in the form of stripes, and even as parallel ridges. On the spindle branch section, the parallel ridges running parallel to each other and the branch, protruding high from the plane of the older branch, can be clearly observed (Fig. 5.14).

Fig. 5.13 Tree of heaven (*Ailanthus altissima*) trunk

Fig. 5.14 Striped spindle (*Euonymus europaeus*) branch with lenticels

Fig. 5.15 Layered cork protrusions on the bark of the common hackberry (*Celtis occidentalis*)

New shoots sprouting from the two buds of the branch, leaves, stipules, and the beginning of the inflorescence can be seen.

We can observe prominent periderm formations on the bark of the common hackberry tree (Fig. 5.15). The protruding bark parts are layered due to the successive layers of cork forming. The rhytidome appears between them.

5.9 Examination of Annual Rings

The formation of lighter and darker concentric circles, the annual rings in the tree trunk can be traced back to the changing of the seasons. In spring, following bud burst, the plant's water demand significantly increases, so at this time the cambium cylinder forms large-cavity xylem elements, which give the lighter coloured layer called earlywood. At the end of summer, after flowering and fruit ripening, the tree's water demand decreases, so narrow-cavity xylem elements are formed, which give the darker coloured latewood. Therefore, one light and one dark ring form in the tree trunk in a year. Moisten the tree trunk section and examine with magnifying glass! Due to the magnification, you can clearly observe the medullary rays running radially in the tree trunk, which continue in the bark. The dark coloured bark can be seen between their light spots. The layers of periderm are clearly visible in the outer part of the bark. The light rings of earlywood and the dark rings of latewood are clearly separated (Fig. 5.16).

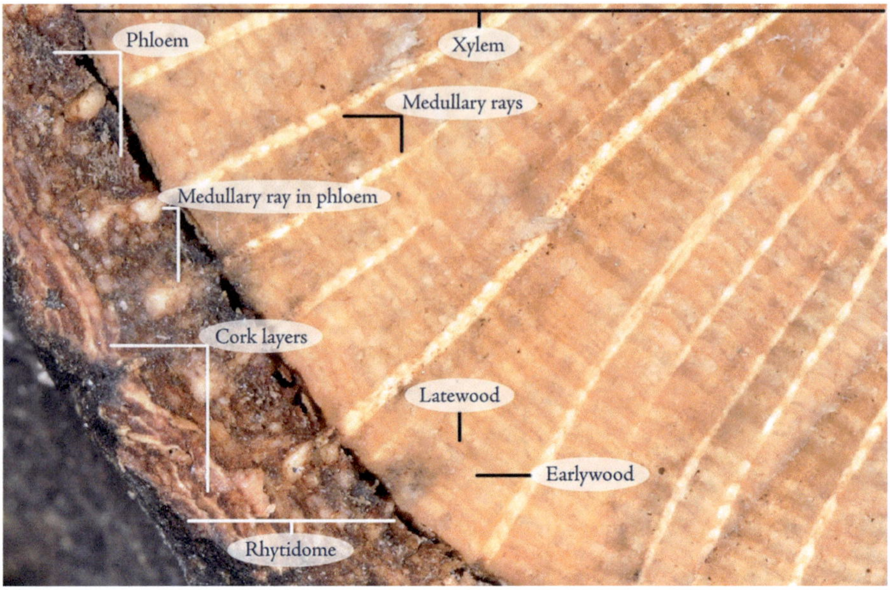

Fig. 5.16 Magnified detail of a tree trunk cross-section

We can also estimate the age of the tree with the help of annual rings, but let us not forget that as we move up the tree trunk, we find fewer and fewer tree rings and the branch found at the top of the shoot axis is only 1 year old. Occasionally, two annual rings can form in the tree trunk in a year. If, for example, severe frost damage or caterpillar feeding destroys the entire crown, new foliage develops from the dormant buds, and in this case two annual rings form in a year. The formation of a new annual ring can also be caused by a prolonged autumn resulting in a second flowering, which again leads to the production of large-cavity xylem elements and the appearance of earlywood.

5.10 Linden (*Tilia* sp.) Branch

When discussing the formation of bark, it was mentioned that due to the thickening of the trunk of older trees the bark deeply cracks, although there are some exceptions. The bark of very old linden trees also remains smooth, which is because the outwardly pressed and torn dead bark parts are filled by radially arranged medullary rays. We can study these on the 1–1.5 cm thick cross-section of the linden tree branches (Fig. 5.17).

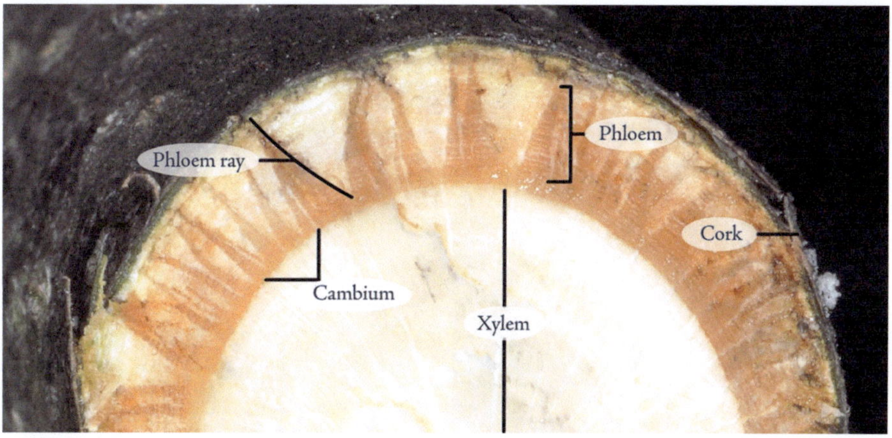

Fig. 5.17 Cross-section of a linden (*Tilia* sp.) tree

5.11 Detection of Starch Stored in Potato (*Solanum tuberosum*) Tubers

The potato tuber plays an important role in human nutrition. This underground modification of the shoot stores a large amount of starch.

Let us make a thin section from the potato tuber and examine it under a magnifying glass, with transmitted light! In the nutrient-storing cells of the potato tuber, the starch grains form smaller groups, which can be recognised in the potato tissue even without staining. The detection of these can be even more successful if we drop Lugol's solution onto the cut surface of the potato tuber. The iodine present in the solution interacts with the helical components of the starch, forming a blue molecular complex (Fig. 5.18).

After dropping, wash off the brown iodine solution from the surface of the potato, thus the groups of starch grains located in the cells, which have turned blue, become clearly visible (Fig. 5.19).

As a result of the staining, we can also see that most of the starch is found inside the potato tuber, so peeling the potato does not mean a significant loss of nutrients for humans. The starch stored in the cells of the potato tuber is found in amyloplasts, which can turn into chloroplasts under the influence of light. If the potato tuber is exposed to light, the aforementioned transformation begins after a few days, as a result of which the cell layers under the potato skin turn green. The green potato tuber can cause poisoning similar to the green shoot, so it should not be consumed! Such a potato should be peeled so thickly that it does not contain green parts.

Fig. 5.18 Cut surface of a potato and its section stained with Lugol's solution

Fig. 5.19 Groups of starch grains stained blue with iodine solution can be observed in the cells on the section of the potato

Fig. 5.20 Spring aspect of a beech forest and older beech (*Fagus sylvatica*) seedling

5.12 Examination of Beech (*Fagus sylvatica*) Seedlings

In early spring, we can often find developing seedlings in beech forests, which are created by the germination of beech seeds (Fig. 5.20).

In younger plants, photosynthesis is still only performed by the large green cotyledons, while in older ones, the first two leaves also photosynthesise. Later, the leaves also are able to support the plant, therefore the unnecessary cotyledons dry up and fall off. Let us observe the different stages of seedling development and compare the cotyledons with the foliage leaves!

5.13 The Vein Pattern of the Walnut (*Juglans regia*) Leaf

The walnut has odd-pinnate compound leaves, therefore in its case it is not the leaf, but the vein pattern of the leaflet that we examine. From both sides of the main vein of the walnut leaflet, side branches originate, between which the bridging veins are densely lined up (Fig. 5.21).

5.14 The Vein Pattern of the White Mulberry (*Morus alba*) Leaf

The mulberry tree is a tree originating from Central Asia, which is widespread throughout Europe. The young leaves of the white mulberry were formerly collected in large quantities because they were used to feed the silkworms kept for silk.

Fig. 5.21 Detail of walnut (*Juglans regia*) leaflet

Let us examine the mulberry leaf under a Triple Bug Viewer with transmitted light! The mulberry leaf has a palmate vein pattern, the main vein entering the leaf divides into five radial side veins. The leaf veins, branching multiple times and interconnecting, completely network the leaf blade (Figs. 5.22 and 5.23).

During the study of the leaf vein pattern, we can observe that several leaf veins merge directly under the leaf surface at the base of the leaf edge lace. In humid, wet weather, when plants cannot release water by evaporation, leaf droplet secretion takes place here. At this time, water is separated from the vascular bundles, which drips off the lacy leaf edge. The plant needs droplet secretion because in the absence of evaporation, this is the only way it can maintain the transport of nutrients absorbed through the root towards the leaves.

Fig. 5.22 The leaf of the white mulberry (*Morus alba*)

Fig. 5.23 A higher magnification image of the white mulberry (*Morus alba*) leaf

6 Flower, Inflorescence, and Fruit

6.1 Flower

The **flower** is the limited growth, reproductive shoot of angiosperms and gymnosperms. Restricted growth means that unlike the leafy stem, the size of the flower does not change with the size of the plant, the flower of the large tree is not larger than that of the smaller one. Flowers stand alone or form groups. The flowers developing from the same bud form the **inflorescence**. This attracts pollinating insects much more than a solitary flower, which explains the significant territorial expansion of inflorescences. They can be unisexual, in this case the individual flowers forming the inflorescence are either staminate or pistillate. The flower first evolved in gymnosperms, which include pines. In female flowers of gymnosperm, the carpels did not fuse into a closed **ovary**, their ovules develop freely, not in a closed ovary. The fruit only develops from a closed ovary, so gymnosperms do not have fruit. The ovulate **cone** is nothing more than the female inflorescence of the pine (Fig. 6.1)!

The male flowers of pines also form cone (pollen cone), which produce pollen in large quantities (Fig. 6.2).

The "ripe" pine cone is not a fruit, but a dried female inflorescence. In this winged seeds are formed in the cone, so the spread cone scales reveal the wings (Figs. 6.3 and 6.4).

By gently tapping the pine cone placed on its top, the winged seeds fall out, on which the seed part and the associated membranous wing can be clearly seen (Fig. 6.5).

The flower axis of the angiosperms consists of the **pedicel** and its upper, widened part, the **receptacle**. The pedicel usually originates from a supporting leaf base, which is called bract. This structure can be well observed in the case of the white campion (Fig. 6.46).

The flower leaves originate from the receptacle, which can be protective- and sexual leaves. The protective leaves (floral envelope) protect the sexual region, and often with their striking colours attract the insects participating in pollination. The

G. Kriska, *Teaching Biology*, https://doi.org/10.1007/978-3-031-84085-2_6

Fig. 6.1 Black pine (*Pinus nigra*) cone

Fig. 6.2 Black pine (*Pinus nigra*) ovulate and pollen cones and the magnified image of the pollen cone

Fig. 6.3 Ripe cone of scots pine (*Pinus sylvestris*)

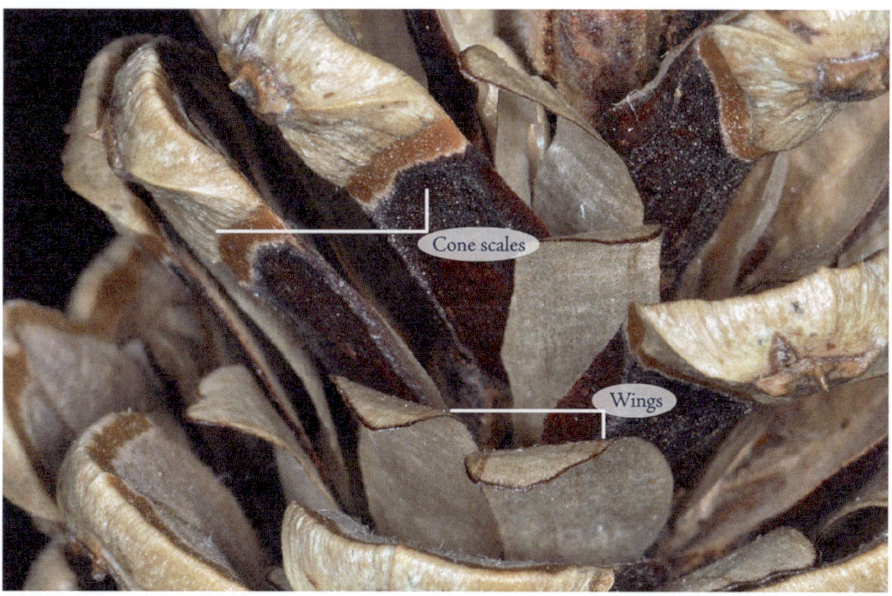

Fig. 6.4 Detail of a ovulate cone collected at the end of summer

Fig. 6.5 Winged seeds of the pine

Fig. 6.6 Outer and inner tepals of the tulip (*Tulipa gesneriana*)

protective region can be double: the mostly green **sepal** leaves forming the **calyx** and the colourful **petal** leaves forming the **corolla** or uniform **tepal** leaves forming the **perigone** (e.g. tulip, Fig. 6.6).

The calyx protects the flower, but also photosynthesises and transpires. The sepals can fuse. In this case, the number of calyx lobes connected to the calyx tube can determine how many sepals the calyx has grown together from.

The sepals in some plants develop in a special way. In the case of the trumpet creeper, the calyx is not green, but orange like the corolla. These colourful sepals can contribute to attracting insects participating in pollination (Fig. 6.7).

The trumpet creeper is a climbing shrub belonging to the Bignoniaceae family. This North American native plant is often planted as an ornamental plant as well. Its flowers bloom from June to September. The plant is poisonous. The floral envelope is made up of a corolla tube and orange sepals. The inner stripes of the corolla tube lead the pollinating insect into the flower. The ovary is superior, its style is elongated.

A common Ranunculaceae in arable fields and dry weed communities is the *Nigella*. The flower of this annual plant, similar to the meadow buttercup, has radial symmetry, but its structure is quite different. Interestingly, its enlarged and coloured sepals have taken over the role of attracting insects from the petals, while the original petals have transformed into characteristic green, yellow, and blue-striped **nectaries** that produce nectar (Fig. 6.8).

In some plants (e.g. common poppy, Fig. 6.9), the sepals fall off the flower early, while in other cases they become larger during fruit ripening. An example of this is the bladder cherry, whose large inflated calyx hides the red berry fruit (Fig. 6.10).

The sepals are arranged in a circle, but in some plants, such as the sand spurrey and the mallows, an “outer calyx” or “false calyx” can develop from the bracts (Fig. 6.11).

Fig. 6.7 Flowers of the trumpet creeper (*Campsis radicans*)

Fig. 6.8 Top view of a wild fennel flower (*Nigella arvensis*)

Fig. 6.9 Common poppy (*Papaver rhoeas*)

The flower of the oriental knight's spur, which belongs to the Fabaceae family, has five blue sepals and two blue petals. The upper sepal bears a nectar-producing spur (Fig. 6.12).

Fig. 6.10 Bladder cherry (*Physalis alkekengi*)

Fig. 6.11 The false calyx (epicalyx) of the mallow (*Malva* sp.)

Like the sepals, the petals can also fuse to form a corolla lobes and corolla tube (e.g. common comfrey, Fig. 6.13). The entrance of the corolla tube is called the corolla throat.

A characteristic formation of the corolla is the **corona**. This is a growth formed on the inner side of the corolla, which in the case of the borage (Boraginaceae) plants resembles small cushions (Fig. 6.14). The monocot daffodil has a somewhat similar structure surrounded by tepals, the osmophore.

Fig. 6.12 Oriental knight's spur (*Consolida orientalis*) flower in side and top view

Fig. 6.13 Flowers of the common comfrey (*Symphytum officinale*) with purple corolla

Fig. 6.14 Flowers of the hound's tongue (*Cynoglossum officinale*) and the common bugloss (*Anchusa officinalis*)

The fusion of the tepals forms a **perianth** tube and perianth lobes (Solomon's seal and grape hyacinth, Fig. 6.15).

The floral envelope leaves in insect-pollinated plants can play a role in attracting insects and occasionally trapping them (e.g., hawthorn, Fig. 6.16).

The colour of the corolla and the calyx can be attributed to several factors. The white colour can be caused by the total reflection of light in air-filled intercellular spaces, or by the accumulation of starch under the epidermis. The yellow colour is caused by flavonoids and coloured plastids, while the blue, purple, and red colours are mostly the result of anthocyanins accumulating in the cell sap of epidermal cells, which have acid-base indicator properties.

Anthocyanins, for example, cause the red colour of geraniums and common poppies, the blue colour of the common chicory (*Cichorium intybus*), and the purplish flower colour of the common bugloss, the violet and the purple gromwell (Figs. 6.17 and 6.18).

The androecium of flowers is made up of **stamen**s, which consist of the **filament** and the **anther**s. The filament is not always filamentous, it can also be flattened (e.g., Bouché's star of Bethlehem, Fig. 6.19).

The anthers consist of two anther halves, which are connected to the filament by the **connective**. The connective can sometimes elongate and become similar to the filament (e.g. meadow sage, Fig. 6.20).

Parts of the stamens can fuse together. This can be seen in legumes, and in the tube-staminate flowers of the mallow family (Fig. 6.21).

Fig. 6.15 Flowers of the Solomon's seal (*Polygonatum* sp.) and the grape hyacinth (*Muscari neglectum*)

Fig. 6.16 Birthwort (*Aristolochia clematitis*) flowers

Fig. 6.17 The red colour of the common poppy (*Papaver rhoeas*) is caused by the accumulation of peonin, an anthocyanin, in the petal cells

Fig. 6.18 Different aged/coloured flowers of the purple gromwell (*Aegonychon purpurea-coeruleum*) and a hoverfly feeding on nectar

Fig. 6.19 Flower of the Bouché's star of Bethlehem (*Ornithogalum boucheanum*)

Fig. 6.20 Flower of the meadow sage (*Salvia pratensis*)

Fig. 6.21 Flower of the mallow (*Malva* sp.) with ants feasting inside

In the flowers of the composite plants, the anthers fuse together to form an **anther cylinder** (Fig. 6.22).

Some plants' stamens play an active role in pollination due to their special formation or movement (e.g. meadow sage and common barberry). If the stamens sitting in the fold of the inner petals of the barberry are touched with a blade of grass, the stamens suddenly lean towards the pistil (Fig. 6.23). This movement of the stamens is triggered by insects visiting the flower under natural conditions, thus they are more likely to transfer the pollen to other barberry flowers.

The entirety of the **pistil** leaves is called the **gynoecium**. In angiosperms, a hollow organ, the pistil, develops from the pistil leaves, which consists of the **ovary** hiding the **ovules**, the **style,** and the **stigma**. The pistil can also develop from a single pistil leaf (e.g. lesser celandine, Fig. 6.24), but it usually develops from several carpels.

During the formation of the pistil, the fusion of the pistil leaves can vary in degree, therefore, several styles and stigmas can belong to one ovary, but the opposite can also be true, several ovaries can have a single style and stigma.

If the pistil is located at the top of the flower axis, and the stamens and the flower cover leaves originate below it, then the ovary is superior (e.g. white campion, Fig. 6.25).

If the ovary is partially embedded in the receptacle, then the ovary is half-inferior (e.g. cherry, Fig. 6.26). In this case, the flower cover leaves and the stamens originate in the middle of the ovary.

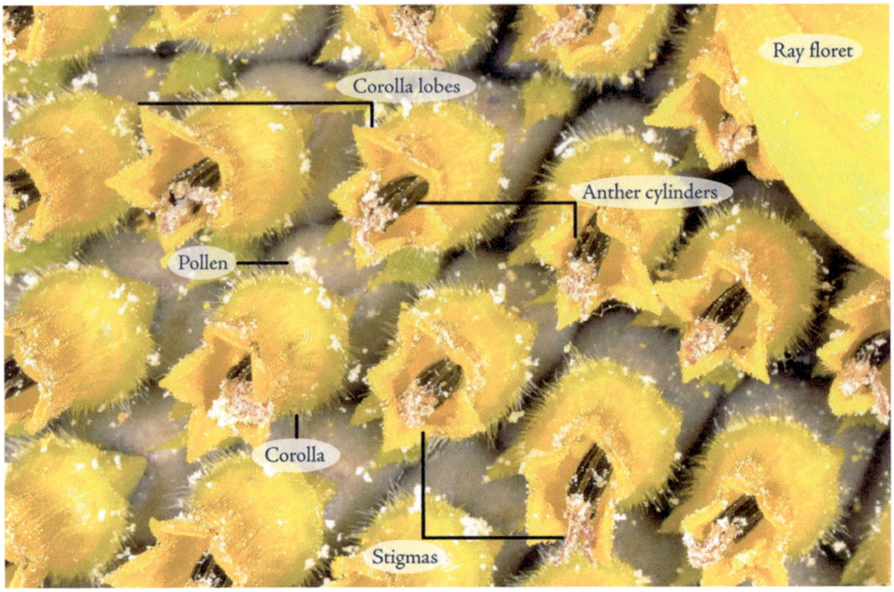

Fig. 6.22 High magnification image of the disc florets of the sunflower (*Helianthus annuus*)

Fig. 6.23 Flowers of the common barberry (*Berberis vulgaris*)

Fig. 6.24 The gynoecium of the lesser celandine (*Ficaria verna*) consists of pistils formed from a pistil leaf

Fig. 6.25 The female flower of the white campion (*Silene latifolia*) flower, both intact and opened

Fig. 6.26 Cherry (*Prunus avium*) flower and the fruit developing from the half-inferior ovary

If the receptacle completely surrounds the ovary, then the pistil is inferior (e.g. pear, Fig. 6.27). In the latter case, during ripening, the fleshy wall of the receptacle participates in the formation of a pseudofruit.

Not only the receptacle, but other parts of the flower (e.g. sepals and petals) can also participate in the formation of the pseudofruit (e.g. bladder cherry, Fig. 6.10).

The uppermost part of the pistil is the stigma, which plays a role in the reception of pollen.

Its surface is usually sticky and covered with surface-increasing protrusions (papillate). Its shape can vary, especially in wind-pollinated species it is large (e.g. walnut, Fig. 6.28).

The first part of reproduction is pollination, when the pollen reaches the stigma. The adhesion of the pollen is made possible by the microscopic spikes and other protrusions that form on its surface, as well as the sticky surface of the stigma. As the first step of fertilisation, the cell in the pollen divides into two. One cell drives a pollen tube (Fig. 6.29), while the other cell forms two sperms as a result of another cell division. The pollen tube is growing into the ovary. Here, in the embryo sac of ovule fertilisation happens.

If the flower is pollinated by its own pollen, we speak of self-pollination. In this case, the pollen and the ovule located in the ovary both reach maturity at the same time, and pollination often occurs before the flower opens (e.g. violet). In other cases, such as in composite flowers, the neighbouring flowers in the inflorescence pollinate each other (neighbour pollination). During cross-pollination, the gametes of different plant individuals meet, resulting in offspring with new genotypes. This

Fig. 6.27 Pear (*Pyrus communis*) fruits starting to develop after flowering

Fig. 6.28 The female flowers of the walnut (*Juglans regia*)

Fig. 6.29 White lily (*Lilium candidum*) flower and its pollen tube

is the most advantageous form of reproduction for adaptation to the environment and the survival of the species. Various factors can facilitate cross-pollination. In some species, the stamen and pistil mature at different times (e.g. simplestem bur-reed, Fig. 6.30).

In flowers where the stamen and pistil mature simultaneously, the distinctive formation of the stamen and pistil often prevents self-pollination.

One form of this is **heterostyly**, where the length of the stamens and pistils can vary in the flowers of different individuals of the species.

For example, the primrose family has two types of flowers in terms of the length of the stamen and pistil (heterodistyly). In one, the stamens are long and the pistil is short, while in the other, the situation is reversed, with short stamens and a long pistil.

The shorter stamens produce smaller pollen grains, which bind to the shorter pistils covered with smaller papillae, while the larger pollen of the longer stamens binds to the longer pistils with larger papillae.

Thus, pollination will be successful if the pollen formed in the longer stamens goes to the longer pistil, while the pollen formed in the shorter stamens goes to the shorter pistil.

The flowers of the purple loosestrife (Fig. 6.31) can have three different lengths of stamens and pistils (heterotristyly), so they can have short and medium stamen, long pistil, medium and long stamen, short pistil, and short and long stamen, and medium pistil flowers. Perfect exclusion of self-pollination occurs when plant individuals develop either only staminate (male) or only pistillate (female) flowers (e.g. white campion, Figs. 6.25 and 6.43).

Fig. 6.30 Mature pistillate and staminate inflorescences of the simplestem bur-reed (*Sparganium erectum*)

Fig. 6.31 Purple loosestrife (*Lythrum salicaria*) flower with short and medium stamens and with long pistil

In bisexual flowers, there are both stamens and pistils, while in unisexual flowers, the stamens and pistils develop in separate flowers (e.g. simplestem bur-reed, white campion). If the staminate (male) and pistillate (female) flowers are found on the same plant, then the plant is monoecious (e.g. maize), or if they are different, then they are dioecious (e.g. white campion) plants. Several flowers can develop from a single bud, which form an inflorescence. The presence of many flowers in one place attracts the attention of nectar-collecting insects, which increases the chances of pollination. Inflorescences can be unisexual (e.g. simplestem bur-reed) or bisexual (e.g. composites).

The floral leaves can be arranged spirally or cyclically in the flower. Two types of arrangement can appear in one flower, in the buttercup family the reproductive leaves are spiral, and the protective leaves are cyclic.

The symmetry of the flower is determined by the number, shape, and arrangement of the floral leaves around the flower axis. If it has more than one axis of symmetry, which divide it into mirror-symmetrical halves, then it has radial symmetry (actinomorphic), if it has one axis of symmetry, then the flower is monosymmetrical (zygomorphic). Originally radially symmetrical flowers can become bilateral by the disappearance of some floral leaves (e.g. crucifers, Fig. 6.32).

The flowers of the grass family are most often bisexual, containing both female and male reproductive leaves. The ovary can be observed in the central part of the grass flower, which has a long protruding feathery stigma (Fig. 6.33). The large surface stigma plays an important role in grasses, as it captures the pollen spreading on the wind. The number of stamens is 2 × 3 (6), 3, 2, or 1.

Fig. 6.32 The flower of the garlic mustard (*Alliaria petiolata*)

Fig. 6.33 High magnification image of a grass inflorescence

6.2 Inflorescence

The flowers stand alone or form groups. The flowers developing from the same bud form the inflorescence (Fig. 6.50).

They can be unisexual, in this case the individual flowers forming the inflorescence are either staminate or pistillate. This solution also contributes to reducing the chances of self-pollination for the same reasons we have already seen in unisexual flowers. On a simplestem bur-reed, the pistillate inflorescences always open before the staminate ones, so the plant's staminate flowers are unable to pollinate the pistillate flowers, this can only be done by the pollen of another plant. Thus, with this solution, the simplestem bur-reed excludes the possibility of self-pollination (Fig. 6.30).

We distinguish two main types of inflorescences, the **racemose** (indefinite) and **cymose** (definite) **inflorescence**s. The main axis of racemose inflorescences is capable of growing for a long time, so the flowers forming the lateral axes do not outgrow the main axis, the individual flowers open towards the apex, or from the outside inwards (e.g. composites). In cymose inflorescences, however, where the main axis usually ends in a flower, the flowers located on the side outgrow the main axis. Racemose inflorescences: The raceme inflorescence consists of flowers with shortening pedicels towards the apex of the main axis (e.g. false acacia, Fig. 6.34).

The spike inflorescence has sessile flowers (e.g. common agrimony, Fig. 6.35).

The spadix is actually a spike whose inflorescence axis has thickened (e.g. flamingo flower, Fig. 6.34). The spadix inflorescence of the flamingo flower is surrounded by a bract.

Fig. 6.34 Inflorescence spadix of flamingo flower (*Anthurium* sp.) and raceme of false acacia (*Robinia pseudoacacia*)

Fig. 6.35 Inflorescence of common agrimony (*Agrimonia eupatoria*)

Fig. 6.34 Inflorescence spadix of flamingo flower (*Anthurium* sp.) and raceme of false acacia (*Robinia pseudoacacia*)

Fig. 6.35 Inflorescence of common agrimony (*Agrimonia eupatoria*)

Fig. 6.33 High magnification image of a grass inflorescence

6.2 Inflorescence

The flowers stand alone or form groups. The flowers developing from the same bud form the inflorescence (Fig. 6.50).

They can be unisexual, in this case the individual flowers forming the inflorescence are either staminate or pistillate. This solution also contributes to reducing the chances of self-pollination for the same reasons we have already seen in unisexual flowers. On a simplestem bur-reed, the pistillate inflorescences always open before the staminate ones, so the plant's staminate flowers are unable to pollinate the pistillate flowers, this can only be done by the pollen of another plant. Thus, with this solution, the simplestem bur-reed excludes the possibility of self-pollination (Fig. 6.30).

We distinguish two main types of inflorescences, the **racemose** (indefinite) and **cymose** (definite) **inflorescence**s. The main axis of racemose inflorescences is capable of growing for a long time, so the flowers forming the lateral axes do not outgrow the main axis, the individual flowers open towards the apex, or from the outside inwards (e.g. composites). In cymose inflorescences, however, where the main axis usually ends in a flower, the flowers located on the side outgrow the main axis. Racemose inflorescences: The raceme inflorescence consists of flowers with shortening pedicels towards the apex of the main axis (e.g. false acacia, Fig. 6.34).

The spike inflorescence has sessile flowers (e.g. common agrimony, Fig. 6.35).

The spadix is actually a spike whose inflorescence axis has thickened (e.g. flamingo flower, Fig. 6.34). The spadix inflorescence of the flamingo flower is surrounded by a bract.

Fig. 6.36 Inflorescences of willows (*Salix* sp.) and spica of ribwort plantain (*Plantago lanceolata*)

Fig. 6.37 The compound inflorescences of ivy (*Hedera helix*)

The catkin inflorescence usually has short-pedicelled male flowers (Fig. 6.36).

In the corymb inflorescence, the pedicels starting from different levels of the main axis shorten as they go upwards, resulting in the flowers being arranged in one plane. The umbel inflorescence has nearly equal length flower pedicels starting from almost one point.

Similar to the umbel inflorescence is the head inflorescence, in which the flowers appear forming a spherical surface (Fig. 6.37).

Fig. 6.38 The pistillate and staminate inflorescences (catkins) of alder (*Alnus glutinosa*)

The staminate flowers of the alder form a double catkin. The flowers have short stalks (Fig. 6.38).

The capitulum axis is flattened, thickened, with the flowers opening from the outside in. Its flowers are ligulate (ray floret) and tubular flowers (disc floret), which can occur separately and together in the capitulum.

The sterile ray florets originate on the edge of the sunflower's capitulum (Fig. 6.39). These are incomplete flowers, as they do not contain sexual organs. Their only task is to attract insects participating in pollination to the inflorescence with their bright yellow corolla. The disc florets are located inward from the ray florets (Fig. 6.22), from the ovary of which an achene pseudofruit forms.

In the capitulum of the thistle, there are only ray florets, but unlike the sunflower, they contain sexual organs, stamens and pistils (Fig. 6.40). The lower part of the broadened inflorescence axis of the capitulum is covered by involucral bracts (Fig. 6.41), which at first glance resemble sepals, but do not let this deceive anyone, as the similarity is only apparent.

Simple cyme inflorescences can be classified into three major groups: single cyme, double cyme, and multiple cyme. In the case of a single cyme, only one lateral branch originates from the main axis, which ends in a flower similar to the main axis (e.g. St. John's wort, Fig. 6.42). This can have another lateral branch, which overgrows the first lateral branch.

In the case of a double cyme, two lateral branches originate under the apical flower (e.g. white campion and bladder campion, Fig. 6.43).

Fig. 6.39 The opened capitulum of the sunflower (*Helianthus annuus*)

Fig. 6.40 Top view of the common chicory's (*Cichorium intybus*) capitulum

Fig. 6.41 Bottom view of the common chicory's (*Cichorium intybus*) capitulum

Fig. 6.42 The compound cyme (monochasium) inflorescence of the St. John's wort (*Hypericum perforatum*). The bright yellow flower is visible below the already bloomed ones

Fig. 6.43 The simple dichasia inflorescence of the white campion (*Silene latifolia*) and bladder campion (*Silene vulgaris*) is a double cyme

In the case of a multiple cyme, several lateral branches of equal length originate in a whorl under the apical flower (e.g. cypress spurge, Fig. 6.44). In the case of a simple inflorescence, the lateral axes (pedicels) always end in a flower (these were discussed above).

In the case of a compound inflorescence, there is a multi-order branching system developing on the secondary, tertiary, or higher order axes. Compound inflorescences consist of multiple, often different types of branching.

For example, the compound umbel consists of identical, racemose branching types, in which secondary umbel branches develop from the base of the bract leaves located at the ends of the primary umbel branches, on which flowers sit. This type of compound inflorescence is common among the umbelliferous (Fig. 6.45).

The butterbur's capitula, consisting only of tubular flowers, form loose groups, the name of the compound inflorescence is capitulate raceme (Fig. 6.46).

Two types of racemose branching can be recognised in the yarrow's compound inflorescence. The small capitula sit on the ends of the corymb inflorescence frame, hence the name of this compound inflorescence is capitulate corymb (Fig. 6.47).

On the edge of the yarrow's capitula, white ray florets can be seen, while inside are radially symmetrical disc florets. The yellow anther cylinder can be clearly observed inside the white corolla of the disc florets (Fig. 6.48).

The compound inflorescence can also consist of two types of cyme inflorescence, as in the case of the elder, where a compound cyme umbel forms.

In the picture depicting the elder's compound inflorescence, numbers indicate the buds sitting on the first, second, third, fourth, and fifth order side branches relative to each other Different types of cymose and racemose branching types can also form compound inflorescences (Figs. 6.49 and 6.50).

Fig. 6.44 The multiple cyme inflorescence of the cypress spurge (*Euphorbia cyparissias*)

Fig. 6.45 The compound inflorescence of the umbelliferous plants is the compound umbel

Fig. 6.46 The capitula of the butterbur (*Petasites hybridus*) consisting of tubular flowers form a capitulate raceme

Fig. 6.47 The yarrow's (*Achillea millefolium*) capitulate compound corymb inflorescence

Fig. 6.48 Top view of the yarrow's (*Achillea millefolium*) capitula

Fig. 6.49 The elder's (*Sambucus nigra*) compound cyme umbel. The numbers on the buds indicate which order side branches they originate from

Fig. 6.50 The inflorescence of the great celandine (*Chelidonium majus*) and the red deadnettle (*Lamium purpureum*)

6.3 Fruit

The beginning of a new life in animals, humans, and flowering plants is marked by the meeting and union of an egg cell and sperm cell, which we call fertilisation. From the two gamete, a unicellular zygote will be, from which the embryo is created through a series of cell divisions, and then with its further development, the individual capable of independent life. The fusion of the egg cell and the sperm cell in flowering plants (gymnosperms and angiosperms) takes place inside the ovule. The ovules of angiosperms develop inside the ovary of the flower. The ovules are connected to the nourishing placenta inside the ovary by a funiculus (Fig. 6.51).

After this, the only question is how the sperm cell fertilising the egg cell. The sperms develop in the pollen, after pollination.

During pollination, the pollen is carried to the stigma of the pistil by the wind or insects, then grows a microscopic tube through the style into the ovary. The sperm cell formed in the tube finally reaches the inside of the ovule, where it fuses with the egg cell to form the zygote. The zygote divides to form the embryo, which, unlike animals and humans, does not develop into a mature individual in flowering plants, but remains in a primitive state of development. After the flower blooms, the **fruit** develops from the ovary of the pistil, in which the ovules have already developed into mature, germinable **seed**s.

Fig. 6.51 Shepherd's purse (*Capsella bursa-pastoris*) ovules inside the developing fruit

The seed is therefore a reproductive structure that contains a plant embryo inside, which sprouts and forms the mature plant when the seed germinates under favourable conditions.

The true fruit develops from the ovary of the pistil, if other parts of the flower (e.g. perianth and receptacle) also participate in the formation of the fruit, we speak of a **pseudofruit** (e.g. strawberry and apple). During fruit ripening, the fruit wall developing from the wall of the ovary becomes low in water content, dry, or juicy and fleshy, based on which we talk about dry and fleshy fruits. Some dry fruits, once fully ripe, open and scatter their seeds, while others remain closed even after ripening.

The opening of dry fruits is caused by the tension that develops between the parenchymal and sclerenchymal layers of the fruit wall during drying.

Dry fruits that open (dehiscent fruits) include the follicle, the pod (legume), the silique, the silicle, and the capsule (Figs. 6.52, 6.53, 6.54, 6.55, 6.56, 6.57, and 6.58).

The fruit of follicle-bearing plants develops from a single carpel and there are many carpels in the flower (Ranunculaceae), so a cluster of follicles develops in one flower. The fruit opens along the suture of the carpel fusion, moving from top to bottom (Fig. 6.52).

The pod fruit also develops from a single carpel, but splits into two equal parts: along the dorsal main vein, where the main vascular bundle runs, and at the ventral suture, where the carpel fused (Fig. 6.53).

In the case of the Japanese pagoda tree (*Sophora japonica*), the ripe pod does not open, but disintegrates.

Fig. 6.52 Common peoni (*Paeonia officinalis*) follicle cluster. A single seed develops in the opened follicles

Fig. 6.53 Structure of a legume

Fig. 6.54 Parts of the silicule fruit of the honesty (*Lunaria annua*)

Fig. 6.55 Jimsonweed (*Datura stramonium*) and developing capsule fruit

Fig. 6.56 The castor bean's (*Ricinus communis*) capsule fruit opens along the fusion of the carpels and the main veins, while the morning glory's opens (*Ipomoea* sp.) with a cap

Fig. 6.57 Flower of the great celandine (*Chelidonium majus*), its capsule fruits and seeds with food-rich appendages (elaiosomes)

Fig. 6.58 The orange seeds covered with seed coat protrude from the opened capsule fruits of the spindle (*Euonymus europaeus*)

During pollination, not all ovules usually fertilise, the remnants of which can be observed in the fruit. In the inner two layers of the fruit wall, the cellulose fibres running in the cell walls are perpendicular to each other, which results in the twisting of the valves and the opening of the fruit when drying.

The silique develops from two carpels, and opens with two valves from bottom to top. Inside, the placenta forms a partition wall (septum), on both sides of which the seeds sit (Fig. 6.54).

If the length and width of the fruit are roughly the same, then we call the fruit a silicle. The silique and silicle fruits are characteristic of cruciferous (Brassicaceae) plants.

The capsule fruit develops from two or more carpels, is one or multi-chambered.

The capsule fruit can split at the fusions of the carpels (e.g. jimsonweed) (Fig. 6.55), or at their central veins (e.g. onion) but with a cap (morning glory) (Fig. 6.56), can open with teeth (white campion) and holes (poppy).

The great celandine has a silique-like capsule fruit, which has no partition. An interesting feature of the plant is that its tiny poppy-seed-sized seeds carry umbilical appendages, white elaiosomes, which ants like to consume, so they carry the seeds into their underground passages (Fig. 6.57). The seeds can also have a seed coat, as in the case of the spindle (Fig. 6.58).

The dry, closed fruit (indehiscent fruit) is the acorn, which develops from two or more carpels (Fig. 6.59). The fruit wall is hard, woody, and not fused with the seed coat. The fruit is often surrounded by a cupule formed from bracts, which can be

Fig. 6.59 Fruiting acorns of pubescent and downy oak (*Quercus pubescens*)

leaf-like (hazelnut), membranous (hornbeam), hard cap-like (oak), or spiky (chestnut, beech).

The characteristic pseudofruit of composite flowers is the achene formed from two carpels (Figs. 6.60 and 6.61). At the apex of each fruit, a hairy dispersal device, a pappus, forms as a remnant of the calyx. With its help, the mature fruits are blown far by the wind. The sunflower also has an achene (sunflower seed), which should not be confused with the plant's seed. Inside the hard fruit wall, we find the seed, which is covered by a membrane formed from the seed coat and the thin nutrient tissue. Inside this, there are two oil-containing cotyledons and the embryo. Hooks (e.g. common cocklebur), or teeth (e.g. *Bidens tripartita*), which get caught in the fur of animals, helping to spread the fruits, can often develop on the surface of the achenes.

The characteristic fruit of grasses is the caryopsis, which usually develops from one carpel (Fig. 6.62). A characteristic of the fruit is that the carpel wall and the seed coat are fused, which contributes to the fact that superficial observers identify the wheat grain as a seed instead of a fruit.

The samara is a winged nut that develops from two carpels (Fig. 6.63). The achene is a one-chambered fruit that develops from one carpel (Fig. 6.64). Most often, the individual achenes form a fruit group. The thin fruit wall did not fuse with the seed coat. The style of the pistil can turn into a flying device (old man's beard).

We categorise separately those dry fruits that fall apart during their ripening. An example of this is the achenes characteristic of plants with umbellate inflorescences (Apiaceae), which splits into two parts (Fig. 6.65).

Fig. 6.60 Achenes of a composite flowers

Fig. 6.61 The achenes of common cocklebur (*Xanthium strumarium*) and three-lobe beggartick (*Bidens tripartita*)

Fig. 6.62 The caryopses of barley (*Hordeum vulgare*)

Fig. 6.63 Samaras of tree of heaven (*Ailanthus altissima*) and elm (*Ulmus minor*)

Fig. 6.64 Hooked achene fruits of wood avens (*Geum urbanum*) and hairy achene fruits of old man's beard (*Clematis vitalba*)

Fig. 6.65 Double achenes of wild parsnip (*Pastinaca sativa*) and double samara of box elder (*Acer negundo*) fruits

Among the labiates (Lamiaceae), the nutlet fruit is common, which often develops under the protection of the drying sepals (Fig. 6.66). The fruit splits into four mericarps when ripe.

Fleshy fruits include the berry, the pepo, the drupe, and the hesperidium (Figs. 6.67, 6.68, 6.69, 6.70, 6.71, 6.72, and 6.73). The berry (Figs. 6.67 and 6.68) is a multi-seeded fruit formed from two or more carpels, from a superior ovary. The outer layer of the fruit wall is skin-like, the middle and inner layers are fleshy.

The gourd has a pseudofruit as the pumpkin, watermelon, and cucumber (Fig. 6.69). The ovary develops from three to five carpels, and the ovary is inferior. The outer layer of the pseudofruit wall is skin-like, in the case of the bottle gourd it becomes woody. The inner pseudofruit walls are fleshy, this is where the oily seeds are.

The outer wall of the drupe fruit is skin-like, the middle layer is fleshy, rich in nutrients, and often colourful. The inner fruit wall is a hard formation made of cells with thickened walls, which includes the seed covered with a thin seed coat (e.g. cherry, plum, and peach) (Figs. 6.70 and 6.71).

The outer fruit wall of the hesperidium is yellow or orange, containing spherical oil holders. The middle fruit wall is white and spongy, the inner one is the skin-like part covering the segments, from which juicy hairs penetrate and fill the inside of the fruit (Figs. 6.72 and 6.73).

The aggregate fruits consist of several dry or fleshy fruits. We have already encountered the follicle cluster in the case of the meadow buttercup. The strawberry forms achenes, while the blackberry and raspberry form a group of drupes (Fig. 6.74).

Fig. 6.66 Developing nutlets of red deadnettle (*Lamium purpureum*) and mature nutlets of woodland sage (*Salvia nemorosa*)

Fig. 6.67 Grape (*Vitis vinifera*) clusters

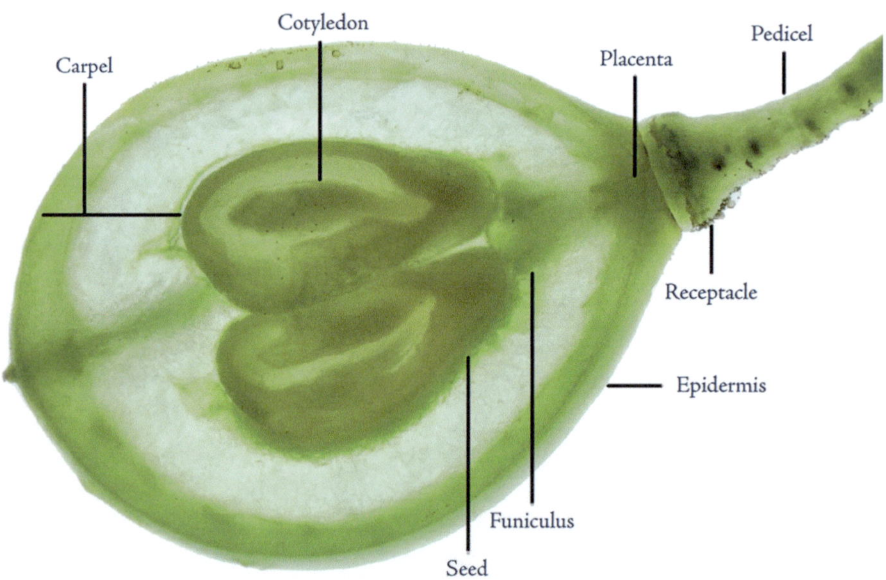

Fig. 6.68 Cross-section of a grape (*Vitis vinifera*)

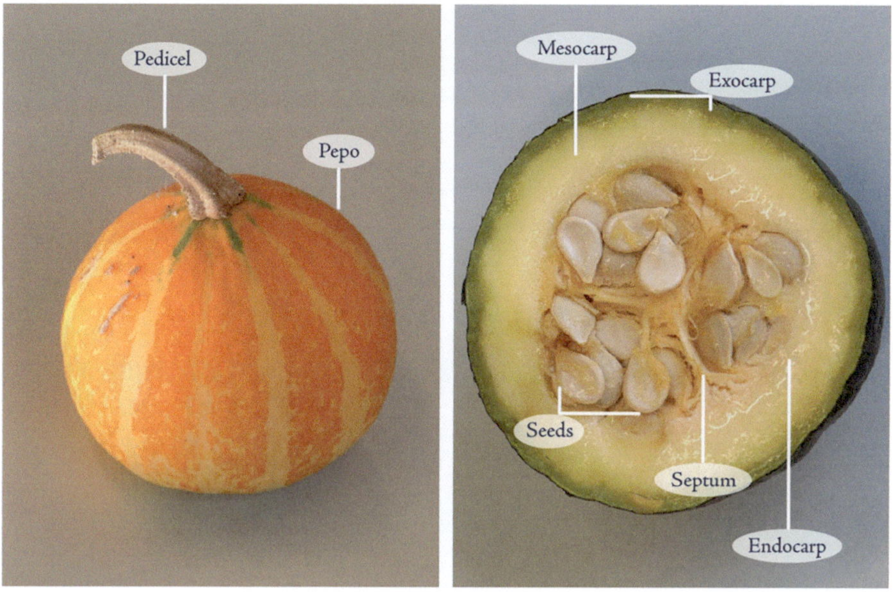

Fig. 6.69 Parts of the pseudofruit of an ornamental gourd

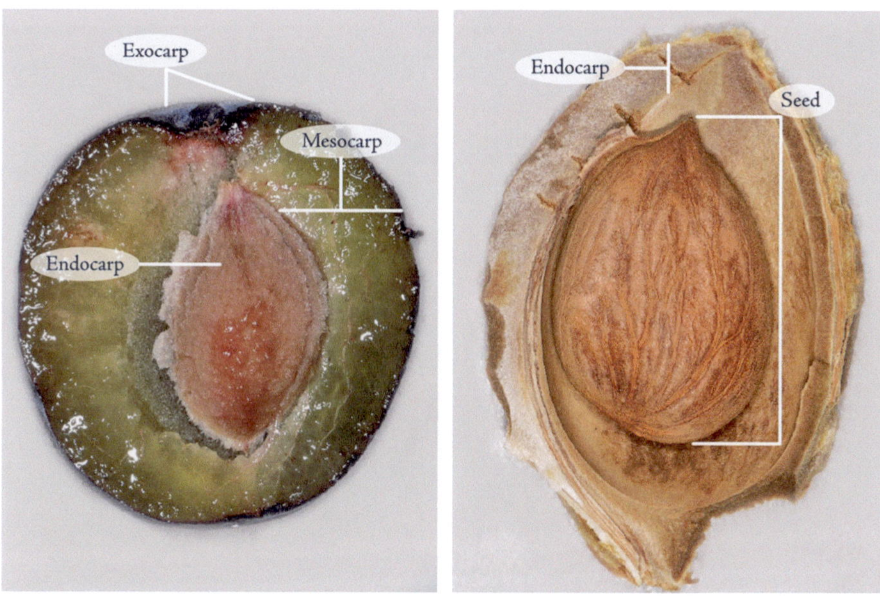

Fig. 6.70 Inside of a plum (*Prunus domestica*) fruit and its seed

Fig. 6.71 When the walnut (*Juglans regia*) fruit ripens, the outer and middle layers of the fruit wall open and the seed surrounded by the hard inner fruit wall falls out

Fig. 6.72 Lemon (*Citrus lemon*) and mandarin (*Citrus reticulata*) tree with hesperidium fruits

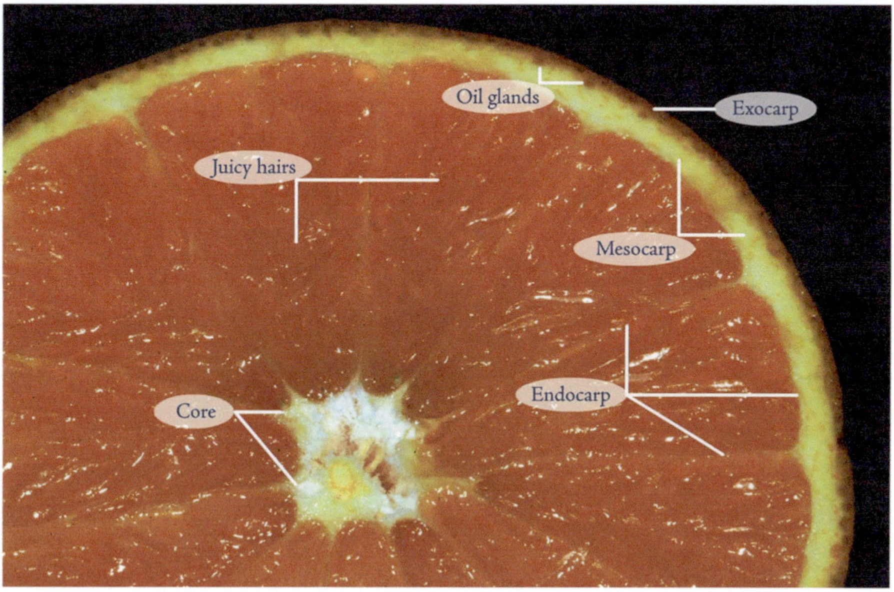

Fig. 6.73 Parts of a cross-cut orange (*Citrus sinensis*) fruit

Fig. 6.74 Aggregate drupe fruits of raspberry (*Rubus idaeus*) and blackberry (*Rubus* sp.)

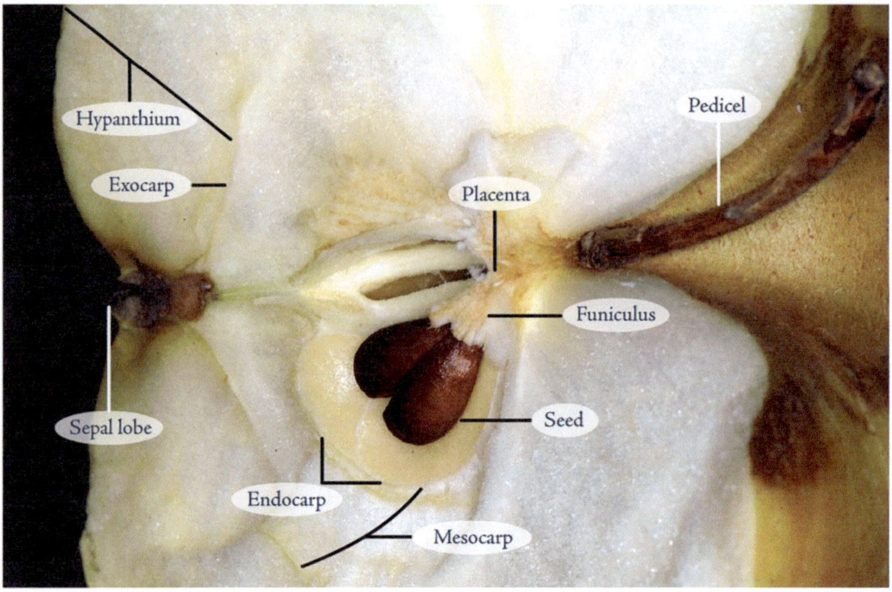

Fig. 6.75 Parts of a split pome (*Malus domestica*)

Inside the pome (apple), five membranous follicles form a follicle fruit group, which is surrounded by the fleshy part developing from the core and the base of the calyx (Fig. 6.75).

The medlar and the pomegranate's false fruit are considered special pomes (Figs. 6.76 and 6.77). The wall of the medlar's follicles hardens, becoming similar to the drupe. The wall of the pomegranate's false fruit hardens, while the fleshy seed coat provides the edible part of the fruit.

In certain plant species, the entire inflorescence becomes a fruit. An example of this is the mulberry tree (Fig. 6.78). The fruit of the mulberry tree is a special formation. Based on its similarity to the blackberry and raspberry, at first glance, it seems that the fruit of the mulberry tree is also a fruit group, but in fact, it is a multiple fruit (infructescences). The mulberry tree has very small achenes (similar to strawberries), whose fleshy husk has developed from the flower's petal leaves. Each little sphere of the mulberry fruit is a false fruit, consisting of the inner achene and the surrounding fleshy part originating from the petal.

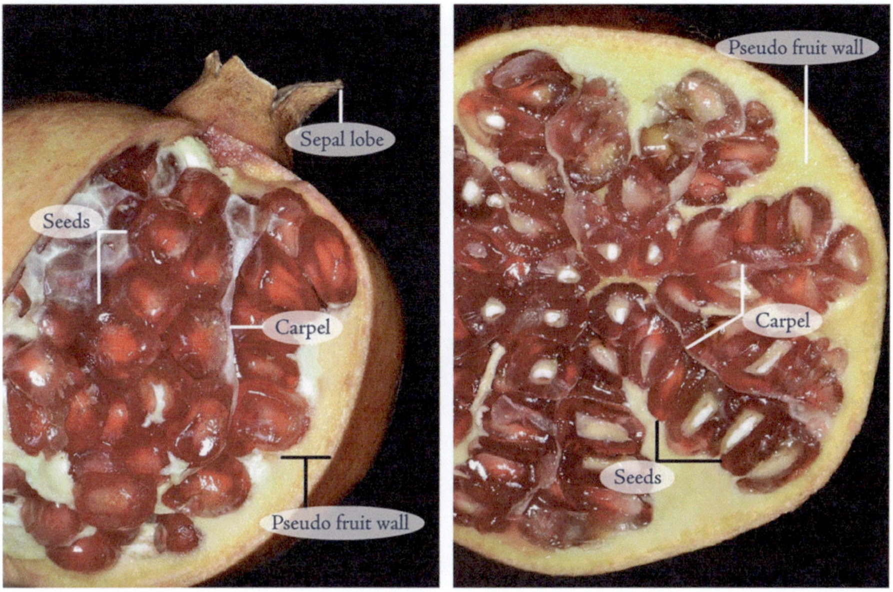

Fig. 6.76 Pomegranate (*Punica granatum*) fruit

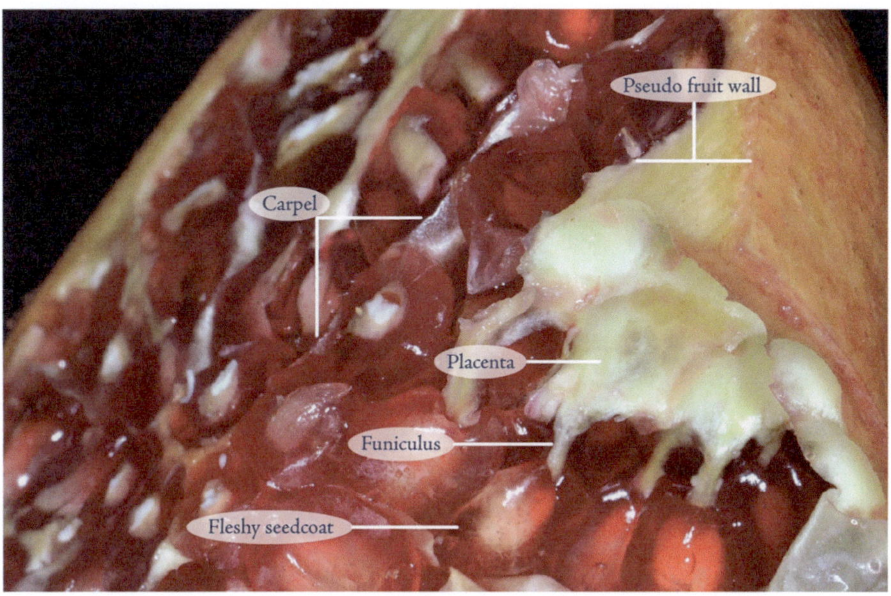

Fig. 6.77 Pomegranate (*Punica granatum*) fruit

Fig. 6.78 The false fruit clusters of the mulberry (*Morus alba*)

Examination of Flowers and Inflorescences

7

7.1 Buttercup (Ranunculaceae) Flowers

The meadow buttercup (*Ranunculus acris*) is a common plant of wet meadows. Its flower has five sepals and five petals in alternating positions. The stamens and the separate carpels are arranged spirally (Fig. 7.1). Achene fruits develop from the carpels' ovaries and form cluster of achenes (Fig. 7.2).

The oriental knight's spur (*Consolida orientalis*), classified among the buttercups, is a common appearance in arable fields, on the edges of crops, and in dry weed communities (Fig. 7.3). Its flower has developed in a special way. The oriental knight's spur flower shows zygomorphic symmetry, and the usually green sepals cannot be recognised on it. The reason for this is that its five sepals are attached to the two petal leaves similarly blue in colour. The upper sepal of the flower forms a nectar-producing spur (Fig. 7.4).

7.2 Birthwort (*Aristolochia clematitis*, Aristolochiaceae)

The birthwort (*Aristolochia clematitis*) grows nearly half a metre, or even a metre tall as a perennial herbaceous plant (Fig. 7.5).

Its flower traps tiny flies for pollination. This common plant in floodplains, shrubs, and edge vegetation blooms from May to September. In the flowers of the birthwort, the floral leaves have grown together into a tepal tube. If we open this tepal tube, we can observe downward-facing woolly hairs covering its inner wall, which hold the insects that have entered the flower until they have completed pollination, or the flower has wilted. The gynoecium is located at the bottom of the flower with the attached stamens (Fig. 7.6).

G. Kriska, *Teaching Biology*, https://doi.org/10.1007/978-3-031-84085-2_7

Fig. 7.1 Freshly opened flower of the meadow buttercup (*Ranunculus acris*)

Fig. 7.2 In the opening flower of the meadow buttercup (*Ranunculus acris*), the achene cluster forms in place of the gynoecium

Fig. 7.3 Flowers of the oriental knight's spur (*Consolida orientalis*)

Fig. 7.4 Parts of the oriental knight's spur (*Consolida orientalis*) flower

Fig. 7.5 Birthwort (*Aristolochia clematitis*) population on the roadside

Fig. 7.6 Structure of a birthwort (*Aristolochia clematitis*) flower

7.3 Rose (Rosaceae) Flowers

The subfamily of pome fruits includes trees and shrubs. Let us examine the pear flower! The pear (*Pyrus communis*) has five sepals and five petals. Red anthers and separate filaments are generally characteristic of the pear. The very young stamens bend into the flower and only straighten out later (Figs. 7.7 and 7.8). The pear's ovary is sunk into the receptacle, so the pear is a pseudofruit, in which not only the ovary but also the fleshy receptacle participates in its formation.

The European dewberry (*Rubus caesius*) is a common shrub of dry grasslands and floodplains. Let us examine its flower! The bramble's perianth is made up of five sepals and five alternating petals. Further in, we can see the many stamens and the gynoecium consisting of separate carpels. Small drupes develop from the ovaries of the pistils, which sit on the conical receptacle (Fig. 7.9).

The tormentil (*Potentilla arenaria*) blooms from March to May and often in the autumn as well on sandy soil meadows, grasslands, and roadsides (Fig. 7.10). The *P. arenaria* has five sepals and five petals, and the number of stamens is a multiple of 5. The gynoecium clearly consists of several carpels. A special feature of the flower is the five bracteoles located outside the sepals, which are called an epicalyx. These are in an opposite position to the petals (Fig. 7.11).

Let us examine the cherry (*Prunus avium*) blossom! The ovary of the cherry fruit is half submerged in the cavity, from the edge of which the flower covering leaves (sepals and petals) and the stamens originates. Therefore, the ovary remains hidden during the study of the flower (Figs. 7.12 and 7.13).

Fig. 7.7 Pear (*Pyrus communis*) blossoms

Fig. 7.8 The pear's (*Pyrus communis*) freshly opened flower

Fig. 7.9 Flowers of the European dewberry (*Rubus caesius*)

Fig. 7.10 Flower of the tormentil (*Potentilla arenaria*)

Fig. 7.11 The flower of the tormentil (*Potentilla arenaria*) from above and below

Fig. 7.12 Cherry (*Prunus avium*) blossoms

Fig. 7.13 Cherry (*Prunus avium*) flower

7.4 Legume (Fabaceae) Flowers

In the legume flower, the upper part of the corolla forms a large protruding banner, under which the two lateral wings and the keel they enclose are located. The keel consists of two petal leaves, which are connected by their epidermis or just their cuticle. The pistil and around it ten stamens are in the keel, of which nine have grown together the stamen filament, while one stamen is separate (Figs. 7.14 and 7.15).

The false acacia (*Robinia pseudoacacia*), originating from North America, is a drought-resistant, fast-growing tree. Its white flowers, blooming in clusters, open in May and June (Fig. 7.16). Let us examine the legume flower of the acacia! In the structure of the corolla, the large protruding banner at the top of the flower, under it the two lateral wings and the keel located between them participate. Inside the keel, we find the pistil and around it the ten stamens, of which nine have grown together the stamen filament, while one stamen is separate.

The crown vetch (*Securigera varia*) is common in shrubs, semi-dry grasslands, and roadside weed communities. The plant's legume flowers form a spherical cluster inflorescence (Fig. 7.17). The banner part of the corolla is lilac-pink, the wings and the keel are mostly white.

The red clover (*Trifolium pratense*) is one of the most valuable fodder plants next to alfalfa, so it is widely cultivated. Its tiny flowers are almost exclusively pollinated by bumblebees (Fig. 7.18).

Fig. 7.14 Flower of the spiny restharrow (*Ononis spinosa*)

Fig. 7.15 Flower of the large yellow vetch (*Vicia grandiflora*)

Fig. 7.16 The legume flowers of the false acacia (*Robinia pseudoacacia*)

Fig. 7.17 The spherical cluster flower of the crownvetch (*Securigera varia*)

Fig. 7.18 The legume flowers of the red clover (*Trifolium pratense*) form a head inflorescence

7.5 Purple Loosestrife (*Lythrum salicaria*, Lythraceae)

The purple loosestrife (*Lythrum salicaria*) is a common perennial plant of wet meadows, reed edges, and marshes (Fig. 7.19). Its flowers open in groups at the base of the spikelets.

The sepals have fused into a ribbed surface cup, which carries longer and shorter lobes. The pointed 6–12 mm petals are purple. The stamens stand in two circles of six. The ovary of the pistil is in upper position.

The flowers of the purple loosestrife can have three different lengths of stamens and pistils (heterotristyly), so they can have short and medium stamen, long pistil, medium and long stamen, short pistil, and short and long stamen, and medium pistil flowers (Figs. 7.20, 7.21, and 7.22).

Successful pollination is only possible between reproductive leaves (stamens and pistils) of the same length. This fact prevents self-pollination of plants, as only flowers with different lengths of stamens and pistils can be found on a plant. In the habitat of the purple loosestrife, for example, on the bank of a stream, we can find many flowering plants on which, with a little luck, we can find all three types of flowers.

Fig. 7.19 Purple loosestrife (*Lythrum salicaria*) on the riverbank

Fig. 7.20 The flower of the purple loosestrife (*Lythrum salicaria*) with a long pistil

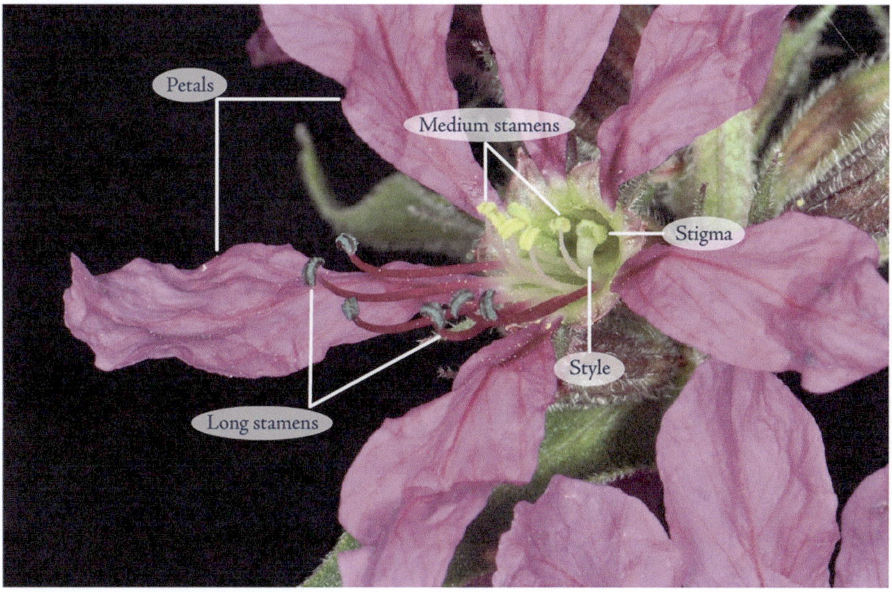

Fig. 7.21 The flower of the purple loosestrife (*Lythrum salicaria*) with a short pistil

Fig. 7.22 The flower of the purple loosestrife (*Lythrum salicaria*) with a medium pistil

7.6 Norway Maple (*Acer platanoides*, Sapindaceae)

The soapberry family includes trees and shrubs. Maple species have spread in the temperate climates of the northern hemisphere. Their flowers are radially symmetrical, their floral envelope is not conspicuous, greenish in colour, sometimes missing. Their stamens stand in two circles, in fours or fives. Winged samaras form from their two-chambered ovary.

Among the maple trees, the Norway maple (*Acer platanoides*) is most often planted in tree lines. The tree's flowers bloom from March to May, before or at the same time as leafing. The top view of the flower clearly shows the five blunt sepals, which alternate with the five oval petals. The stamens stand in fours in two circles. A nectar ring is attached to the ovary part of the pistil (Fig. 7.23). In older flowers, the beginning of the winged samaras forming from the ovary can already be clearly seen (Fig. 7.24).

7.7 Umbellifers (Apiaceae)

The umbellifers include plants of similar appearance, which occur in almost every meadow and weed community. These herbaceous plants have a compound umbel inflorescence, which consists of several umbels. Most of the tiny flowers are radially symmetrical, but those on the edge of the umbel can also be monosymmetrical. In

Fig. 7.23 Flower of the Norway maple (*Acer platanoides*)

Fig. 7.24 In the older flower of the Norway maple (*Acer platanoides*), the developing samaras can already be observed

the latter, the petals facing the edge of the umbel are larger. A nectar ring is attached to the pistils. From the bottom view of the compound umbel inflorescence, we can observe that small umbel inflorescences sit on the ends of the umbel branches originating from one point of the inflorescence axis. So, the framework of the compound inflorescence is given by a large umbel, which has small umbels at the ends of its branches (Figs. 7.25, 7.26, and 7.27).

7.8 Field Scabious (*Knautia arvensis*, Caprifoliaceae)

The field scabious (*Knautia arvensis*) blooms from May to October in semi-dry grasslands, along roadsides and in meadows. Its flowers form a head inflorescence, arranged on the widened part of the receptacle. The corolla of the flowers is composed of four violet petals. The flowers located on the edge of the inflorescence are larger, and their sizeable outer corolla lobes extend to the side (Fig. 7.28).

7.9 Linden (*Tilia* sp., Malvaceae)

The linden (*Tilia* sp.) species are mainly widespread in the tropics, in Europe they appear in mixed forests of mountains and hills. A bract grows on their inflorescence axis, which acts as a flying device for the fruit branch, and thus plays a role in the spread of seeds. Inside the large sepals, the stamens can be seen, while the pistil is visible in the innermost part (Figs. 7.29 and 7.30).

Fig. 7.25 Side view of a compound umbel inflorescence of sickleweed (*Falcaria vulgaris*)

Fig. 7.26 Bottom view of a compound umbel inflorescence of European wild carrot (*Daucus carota*)

Fig. 7.27 Umbel inflorescences with bloomed flowers of wild parsnip (*Pastinaca sativa*)

Fig. 7.28 Top and bottom view of the flower of the field scabious (*Knautia arvensis*)

Fig. 7.29 Dried flowers of the linden (*Tilia cordata*)

Fig. 7.30 A magnified image of the dried linden (*Tilia cordata*) flower

7.10 Borage (Boraginaceae) Flowers

The hound's tongue (*Cynoglossum officinale*) is a frequent plant of dry grasslands and weed communities. The biennial plant's violet flowers bloom from May to August. A characteristic feature of its corolla is the corona. This is an outgrowth formed on the inner side of the corolla, which in the case of the borage ones resembles small cushions (Fig. 7.31).

The perennial common comfrey (*Symphytum officinale*) blooms from May to July in wet habitats, moist meadows, and along stream banks. The corolla of its flowers can be lilac, pink, and white. The plant is hepatotoxic when consumed internally, but externally it is used to treat hard-to-heal wounds and tendonitis. Observe the calyx with long lobes, and the fused petals. The five corolla lobes indicate that the corolla tube grew from five petal leaves. Looking inside the flower, and we can observe the stamens grown to the inside of the corolla tube (Figs. 7.32 and 7.33).

The common bugloss (*Anchusa officinalis*) is a blue-flowered plant of dry meadows, trampled grasslands, fallows, and weed communities (Fig. 7.34). In May and June, we can often encounter the flowering plants with branching and upright stems along the roads in cities.

On the inner side of the bugloss, a light-coloured, cushion-like growth can be observed, which we call a corona. The colour of the flowers is caused by anthocyanins, which give the very young flowers a purplish colour, while the older ones are given a bluish colour.

Fig. 7.31 Flowers of the hound's tongue (*Cynoglossum officinale*)

Fig. 7.32 White corolla flowers of the common comfrey (*Symphytum officinale*)

Fig. 7.33 White and purple flowered shoot detail of the common comfrey (*Symphytum officinale*)

Fig. 7.34 Common bugloss (*Anchusa officinalis*)

Fig. 7.35 True forget-me-not (*Myosotis scorpioides*) flowers

The true forget-me-not (*Myosotis scorpioides*), which blooms from May to July, is common in wet habitats and gardens. The petals of the flower buds are almost white. The petals have fused, around the opening of the short corolla tube yellow coronas can be seen (Figs. 7.35 and 7.36).

7.11 Labiate (Lamiaceae) Flowers

The stem of the labiate plants is always quadrangular, and their leaf arrangement is decussate. In their bilaterally symmetrical flowers, the sepals and the petals have fused. The calyx of the golden (*Lamium galeobdolon*) and red deadnettle (*Lamium purpureum*) formed from five sepals has five lobes. The upper lip of the flower fused from two, while the lower from three petals, so these are two- and three-lobed. The flowers have two shorter and two longer stamens. From their superior ovary, four nutlets are formed. Their four stamens and the pistil are covered by the upper lip (Figs. 7.37 and 7.38).

The upright bugle (*Ajuga genevensis*), which blooms in May and June, is common in pastures, mountain meadows, and dry oak forests. Let us examine the upright bugle flower! The lower lip of the corolla with lateral and medial lobes is well developed, while the upper lip is small, rudimentary. From the centre of the upper part of the flower, the filament, the two longer and the two shorter stamens extend (Fig. 7.39).

Fig. 7.36 True forget-me-not (*Myosotis scorpioides*) flowers

Fig. 7.37 Golden deadnettle (*Lamium galeobdolon*) flowers

Fig. 7.38 Red deadnettle (*Lamium purpureum*) flowers

Fig. 7.39 The flower parts of the upright bugle (*Ajuga genevensis*)

Fig. 7.40 Ground-ivy (*Glechoma hederacea*) flower

The ground-ivy (*Glechoma hederacea*) blooms from March to the end of summer in forests, meadows, and roadside weed communities. It often settles in gardens, where its shoots spread quickly. Up to six flowers can open in the leaf axils, so when it multiplies, it covers the ground surface with a flower carpet. Let us examine the ground-ivy flower! The lower lip of the two-lipped corolla is larger, with well-developed lateral lobes and a medial lobe can be observed. The pistil extends from the upper part of the flower (Fig. 7.40).

The meadow sage (*Salvia pratensis*), which blooms from May to July, is a common labiate plant of dry meadows and pastures. The filament extends long from the upper, sickle-shaped lip of the corolla. The two stamens protrude from the older, already pollinated flowers, while in the young ones these remain hidden in the upper lip (Fig. 7.41). The stamens in the flowers usually only appear when a nectar-seeking insect crawls into the flower (Fig. 7.42).

7.12 Figwort (Scrophulariaceae) Flowers

The orange mullein (*Verbascum phlomoides*) is a biennial plant of dry meadows and roadside verges. In its first year, the plant develops a large, 20–30 cm leaf rosette, from which the 1–2 m flowering axis, which becomes woody over time, emerges in the second year. The light yellow petalled flowers sit in groups on the spike inflorescence. Let us examine the flower of the orange mullein!

Fig. 7.41 Parts of the older flowers of the meadow sage (*Salvia pratensis*)

Fig. 7.42 The stamens emerge due to the effect of meadow sage (*Salvia pratensis*)

The flower's sepals and petals have grown together into a sepal and corolla tube. The corolla has a short tube and is divided into five lobes, of which the two upper ones are smaller than the lower three. Of the stamens, the three upper ones are short and hairy, while the lower two are long and bare. In the middle of the flower, the long style can be seen (Figs. 7.43 and 7.44).

Fig. 7.43 The flower of the orange mullein (*Verbascum phlomoides*) in top view

Fig. 7.44 The flower of the orange mullein (*Verbascum phlomoides*) in side view

Fig. 7.45 The flowers of the common toadflax (*Linaria vulgaris*)

The common toadflax (*Linaria vulgaris*) is common in dry weed communities and on arable land. Its monosymmetrical flowers open in a dense cluster on the 10–20 cm plants, from July to September. The calyx is fused from five sepals, and the lobes are triangular. The two-lipped corolla is pale yellow, at its opening, a the corolla throat is orange (Figs. 7.45 and 7.46).

In April and May, the germander speedwell (*Veronica chamaedrys*) blooms on layers at the edge of forests. Among the speedwells, several species are common in weed communities, so we can often encounter them in urban conditions. Their flowers contain two stamens, which makes them easy to recognise. Let us examine the 6–7 mm flower of the germander speedwell! The floral envelope consists of four green, narrow sepals and four medium blue and dark blue veined rounded petals. The two stamens and the pistil extend from the flower. The ovary of the pistil is superior, so the ovary can be clearly recognised in the middle of the flower (Fig. 7.47).

7.13 Great Celandine (*Chelidonium majus*, Papaveraceae)

The great celandine is a flowering plant from April to October. It is a perennial weed that is common in forest edges, fallow lands, and gardens. When damaged, it releases a yellow milky sap that turns red upon contact with air. The flower has four golden yellow petals surrounding numerous stamens, with the elongated, green pistil visible in the middle (Fig. 7.48).

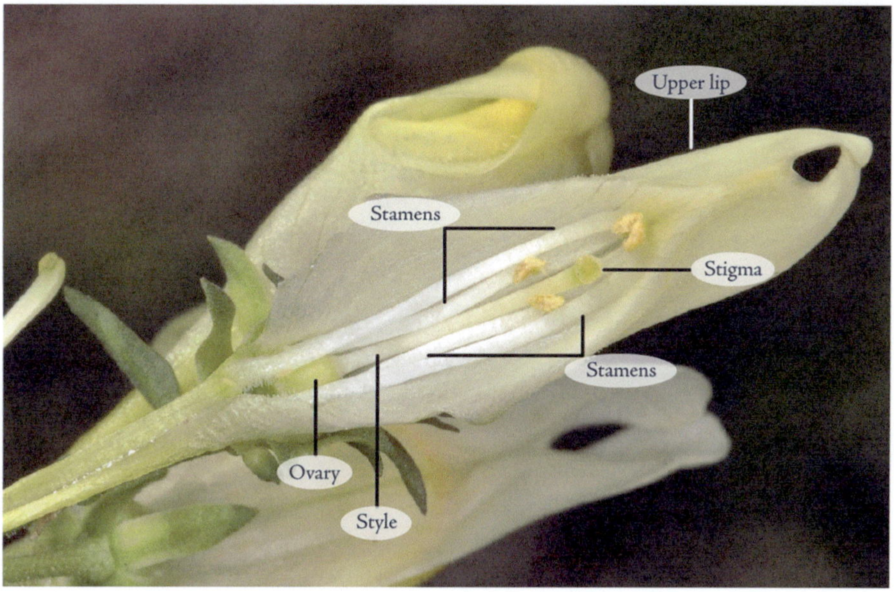

Fig. 7.46 The interior of the common toadflax (*Linaria vulgaris*)

Fig. 7.47 Flower of the germander speedwell (*Veronica chamaedrys*)

Fig. 7.48 The flower of the great celandine (*Chelidonium majus*)

On the flowering plant, we can often find the developing capsule fruits as well. Collect some of these more withered (mature) specimens, then examine the seeds taken from them under a magnifying glass! The tiny shiny seeds are attached to food appendages (elaiosomes), which are a favourite delicacy of ants (Fig. 7.49). Therefore, the ants transport the seeds into their underground passages, contributing to the spread of the plant.

7.14 Charlock Mustard (*Sinapis arvensis*, Brassicaceae)

The cruciferous plants are common members of the meadow and weed associations, where their yellow and white petalled representatives live. At first glance, their flower appears to have radial symmetry, but examining the position of the stamens, we can determine that it has bilateral symmetry. In the outermost circle of the flower, four sepals are crossed, in relation to which the petals, which are also crossed, are alternately positioned. In the outer stamen circle, only two shorter stamens can be seen (hence the flower has bilateral symmetry), while the inner stamen circle consists of four longer stamens. The pistil is located in the innermost part (Figs. 7.50 and 7.51).

Fig. 7.49 The oil-bodied seeds of the great celandine (*Chelidonium majus*) on the plant's leaf

Fig. 7.50 The flowers of the charlock mustard (*Sinapis arvensis*)

Fig. 7.51 The flower of the charlock mustard (*Sinapis arvensis*) from above

7.15 Composite (Asteraceae) Flowers

The giant goldenrod (*Solidago gigantea*) is an invasive weed spreading on forest edges, wet meadows, and floodplains, and has small, yellow-coloured inflorescences (Fig. 7.52). In these, we can find both the sterile ray florets and the central disc florets (Fig. 7.53).

The common sunflower (*Helianthus annuus*) has a flattened, thickened axis of the inflorescence, on which the flowers open from the outside in (Fig. 7.54).

The sterile ray florets originate on the edge of the inflorescence, which do not contain reproductive leaves. Their task is to attract insects to the inflorescence with their bright yellow corolla. The disc florets are located inward from the ray florets, from which achene pseudofruit is formed. Under stronger magnification, the corolla tube and corolla lobes of the disc florets can be clearly recognised. These surround the dark-coloured anther cylinder protruding from the flower, which has formed from the fusion of the anthers (Fig. 7.55).

The garden cosmos (*Cosmos bipinnatus*) is a globally widespread invasive species. It is often planted in gardens for its beautiful colourful inflorescences. This annual plant, originating from Central America, is frost-sensitive, so its seeds are scattered in the spring. Its capitulum, appearing from mid-summer to late autumn, has sterile ray florets on the edge, while disc florets are in the centre (Fig. 7.56).

Some species of goat's beard (*Tragopogon*) populate fallow lands and weed communities in large numbers, so we are likely to encounter their yellow capitulum in the spring and early summer. Their inflorescence consists of ray florets, which

Fig. 7.52 Giant goldenrod (*Solidago gigantea*)

Fig. 7.53 Giant goldenrod (*Solidago gigantea*) inflorescences

Fig. 7.54 The unfolding inflorescence of the sunflower (*Helianthus annuus*)

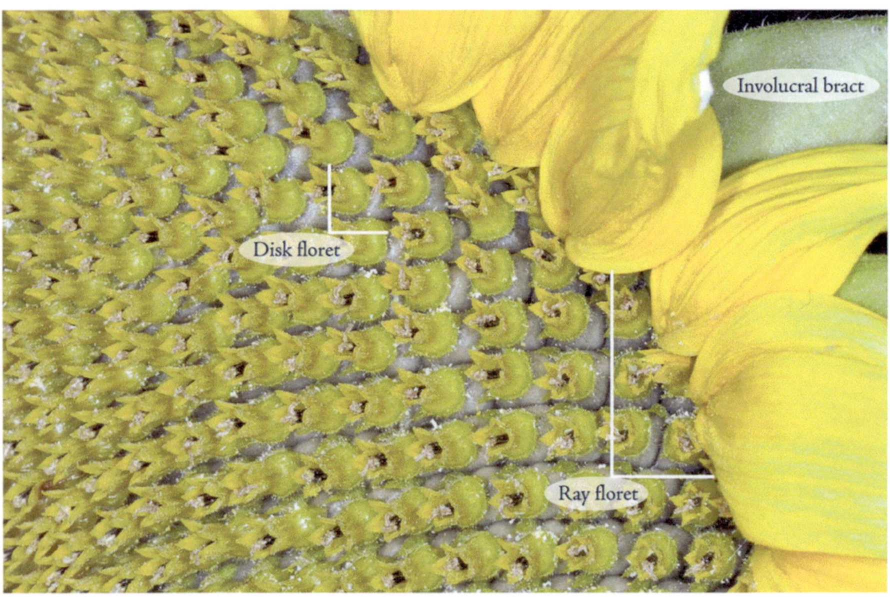

Fig. 7.55 The disc and ray florets of the sunflower (*Helianthus annuus*)

Fig. 7.56 High magnification image of garden cosmos (*Cosmos bipinnatus*) inflorescence and disc florets

contain sexual organs, stamens and pistils (Figs. 7.57 and 7.58). In a Triple Bug Viewer, we can observe the curled stigmas protruding from the anther cylinders of individual flowers. In their flying cypsela the achene fruit's hairs are provided by the dried calyx leaves. The small umbrella-like pseudofruits form a tuft attached to the inflorescence axis (Fig. 7.57).

The well-known dandelion (*Taraxacum officinale*) is also a plant with a capitulum. Its yellow inflorescence consists only of ray florets (Figs. 7.59, 7.60, and 7.61), and its flying cypsela fruits form a tuft.

The cornflower (*Centaurea cyanus*) is common on the edges of wheat fields. The plant's sterile ray florets are light blue and bell-shaped, while a few disc florets located inside the inflorescence are lilac-coloured. In the latter, we can observe the protruding anther cylinder (Figs. 7.62 and 7.63).

7.16 Carnation (Caryophyllaceae) Flowers

The most common carnation of weed communities and roadside is the white campion (*Silene latifolia*). The perennial plant's snow-white petalled flowers fully open at dusk to attract pollinators. The plant is dioecious, on its individuals, either only staminate (male), or only pistillate (female) flowers can be found. Therefore, self-pollination is impossible for them. The simple inflorescence is of a double cyme type. In this case, two lateral branches originate under the terminal flower, whose

Fig. 7.57 The cypselas and yellow capitulum of the oriental goat's beard (*Tragopogon orientalis*)

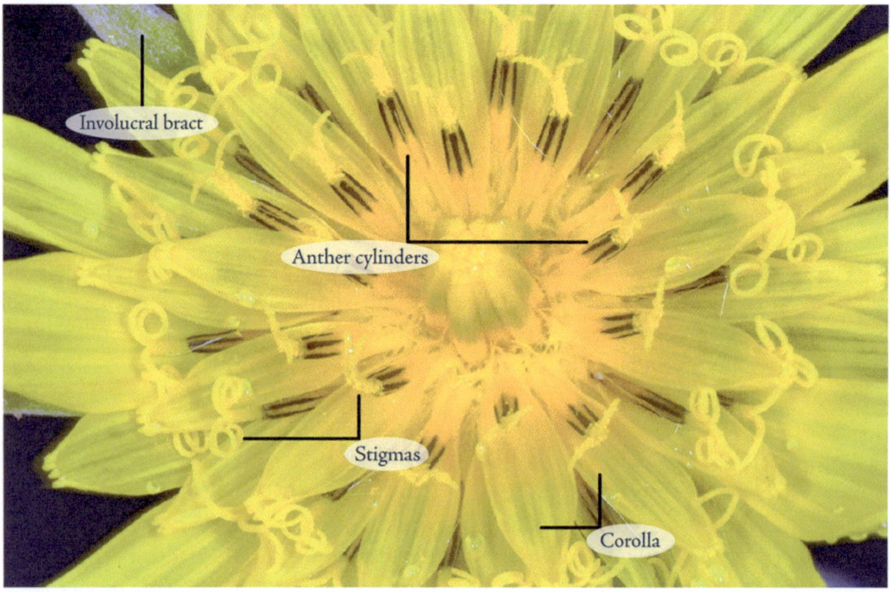

Fig. 7.58 Capitulum of the oriental goat's beard beard (*Tragopogon orientalis*)

Fig. 7.59 Capitulum of the dandelion (*Taraxacum officinale*)

Fig. 7.60 Bottom view of the dandelion (*Taraxacum officinale*) inflorescence

Fig. 7.61 Side view of the dandelion (*Taraxacum officinale*) inflorescence

Fig. 7.62 Capitulum of the cornflower (*Centaurea cyanus*)

Fig. 7.63 High magnification image of the cornflower (*Centaurea cyanus*) capitulum

flower stalks outgrow the main axis (Fig. 7.64). The sepals of the flower are fused, and only the five sepal tips are free (Fig. 7.64).

The lower, fused parts of the petals form a corolla tube, while the upper parts form corolla lobes. With the opening of the corolla tube, the stamens appear inside the male flower (Fig. 7.65).

The lower part of the pistillate flower is widened and inside it a large ovary can be found (Fig. 7.66). From the centre of the pistillate flower, five twisted stigmas extend. The stigma filaments are directly connected to the ovary, from which the capsule fruit containing the tiny seeds develops (Fig. 7.67).

The chickweed (*Stellaria media*) is a common weed of nutrient-rich soils. This carnation blooms from early spring to late autumn, so we can easily obtain its flowering specimens for our studies. Let us study the structure of the flower! The flower of the chickweed is radially symmetrical, and its structure is characterized by the number five. Inside the five sepals, in alternating positions, there are five white petals which have a pinched centre, so at first it seems, as if there were ten petals in the flower. Inside the stamens are standing in two circles in fives, which surround the pistil fused from the five carpels. At the top of the pistil, five stigmas can be seen and this indicates the number of carpels (Figs. 7.68 and 7.69).

Fig. 7.64 Double cyme inflorescence of the white campion (*Silene latifolia*) with staminate flower

Fig. 7.65 The stamens appear in the opened male flower of white campion (*Silene latifolia*)

Fig. 7.66 The opened pistillate flower of the white campion (*Silene latifolia*)

Fig. 7.67 The opened pistillate flower of the campion (*Silene latifolia*) with opened ovary

Fig. 7.68 The tiny flowers of the chickweed (*Stellaria media*)

Fig. 7.69 The flower of the chickweed (*Stellaria media*)

7.17 Walnut (*Juglans regia*, Juglandaceae)

The walnut family includes large deciduous trees. They are generally characterised by pinnately compound leaves and wind pollination. Their leaves are mostly essential oil-containing, and the wind-pollinated flowers are unisexual and simplified. The staminate flowers form large, drooping catkin inflorescences (Fig. 7.70). The pistillate (female) inflorescences have large fringed stigmas that capture the pollen carried by the wind (Fig. 7.71).

7.18 European Water-Plantain (*Alisma plantago-aquatica*, Alismataceae)

The European water-plantain (*Alisma plantago-aquatica*), flowering from July to October, is a perennial marsh plant, whose tuberous rhizome is mostly in water. The number three is characteristic of its monocot flower, the number of its sepals and petals is 3–3, and the number of its stamens is 6. Their pistil consists of numerous separate carpels, which contain a well-visible stigma filament (Fig. 7.72).

Fig. 7.70 Detail of the staminate catkins of the walnut (*Juglans regia*)

Fig. 7.71 High magnification image of the pistillate flowers of the walnut (*Juglans regia*)

Fig. 7.72 Flower of the European water-plantain (*Alisma plantago-aquatica*)

7.19 Garden Star-of-Bethlehem (*Ornithogalum umbellatum*, Asparagaceae)

The garden star-of-Bethlehem (*Ornithogalum umbellatum*), a perennial bulbous plant that produces its snow-white flowers in April and May (Fig. 7.73). It appears in many places, most commonly in dry forests, grasslands, and weed communities can we encounter its individuals.

Let us study the flower structure characteristic of monocot plants! The star-of-Bethlehem's flower cover consists of six tepals, which are arranged in two circles, alternating with each other. The position of the six stamens is similar. The strongly widened stamen filaments surround the pistil (Fig. 7.74).

7.20 Tulip (*Tulipa gesneriana*, Liliaceae)

The flower structure of the tulip (*Tulipa gesneriana*), classified in the lily family, well demonstrates the structure characteristic of monocots. Its symmetry is radial. It has six tepals, which are arranged in pairs in two circles (Fig. 7.75).

Within these, we find the 3–3 stamens standing in two circles and the innermost superior ovary, which has grown together from three carpels. This is indicated by the three-branched stigma and the ovary divided into three compartments, from which a multi-seeded capsule develops (Figs. 7.76 and 7.77).

Fig. 7.73 Garden star-of-Bethlehem (*Ornithogalum umbellatum*)

Fig. 7.74 Flower of the garden star-of-Bethlehem (*Ornithogalum umbellatum*)

Fig. 7.75 Bottom view of a tulip (*Tulipa gesneriana*) flower

Fig. 7.76 Inner part of a tulip (*Tulipa gesneriana*) flower

Fig. 7.77 Magnified view of the pistil

Fig. 7.78 Light microscopic image of the tulip's round pollen in a hanging drop during pollen tube germination

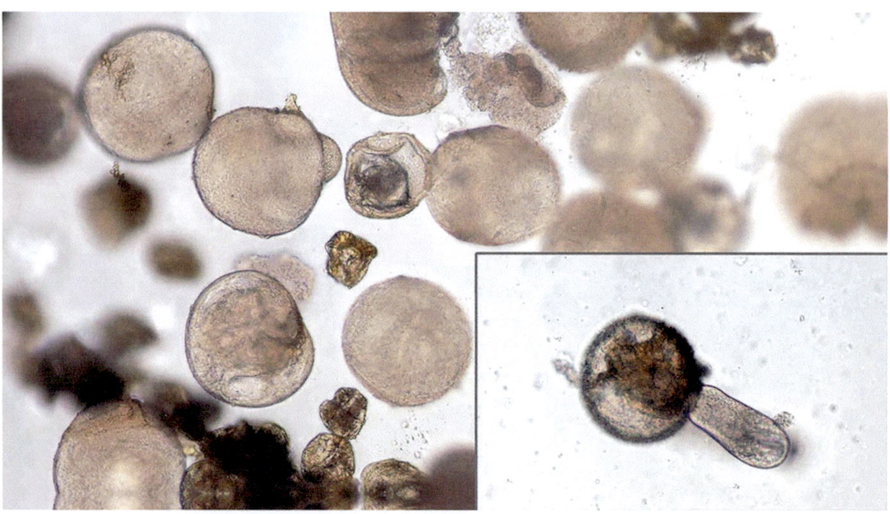

Fig. 7.79 The pollen tube growth begins at one of the tulip's pollen grains

With the round pollen grains scraped off the tulip's stamens, we can even perform a pollen tube germination test. This can be implemented in a moist chamber created on the slide (Figs. 7.78 and 7.79).

By making a cross-section from the tulip's ovary, we can also study the position and structure of the ovules with a light microscope (Figs. 7.80 and 7.81).

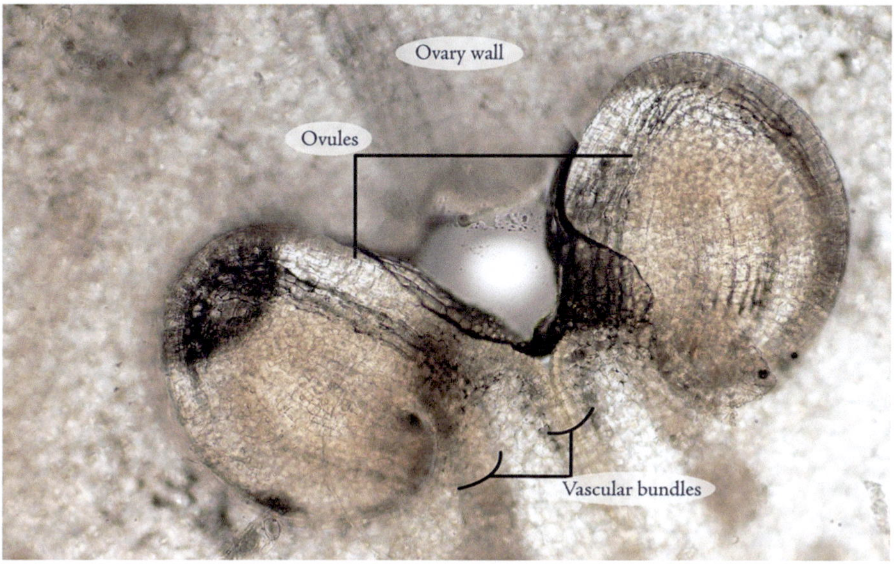

Fig. 7.80 Position of the ovules in the ovary

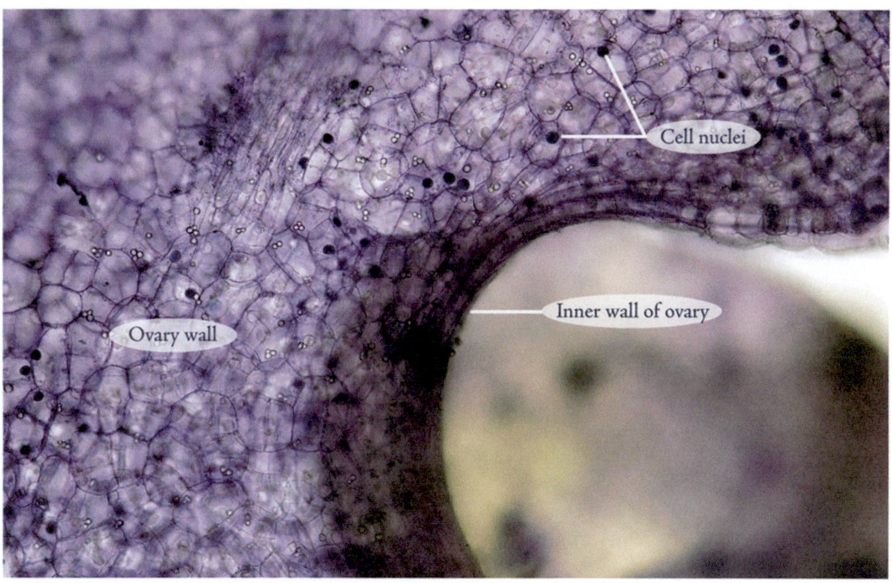

Fig. 7.81 Light microscopic image of a section of the ovary stained with toluidine blue

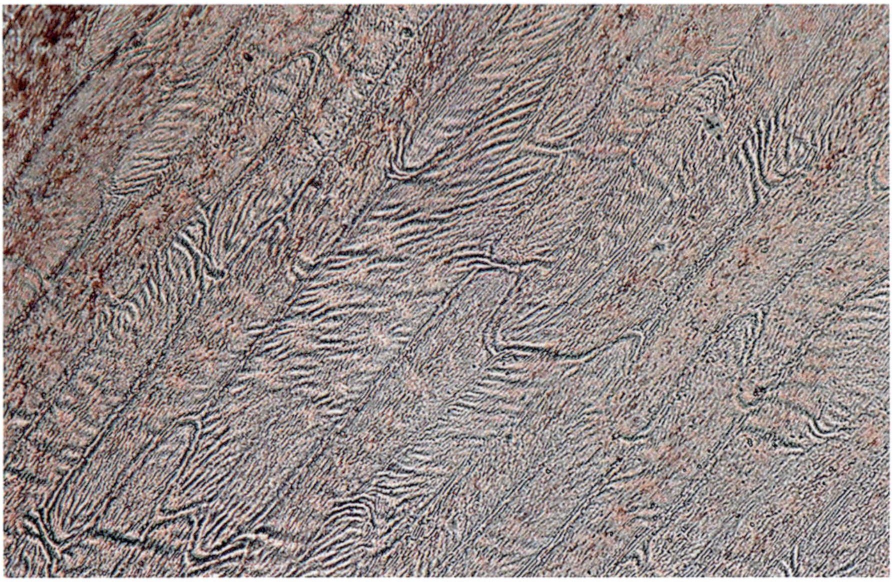

Fig. 7.82 Wavy cuticular pattern on the epidermis of the tepal

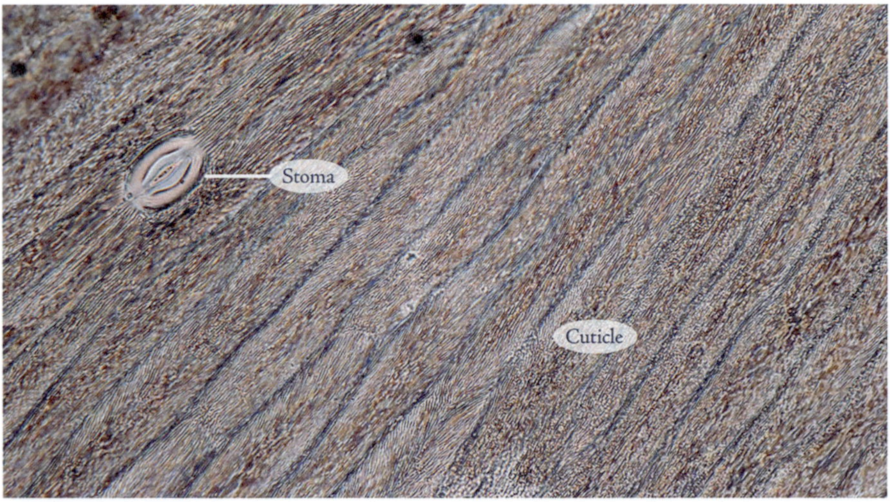

Fig. 7.83 We only sporadically find stomata on the epidermis of the tepal

On the tepal, we can clearly observe the characteristic cuticular patterns of the epidermal cells with a light microscope (Fig. 7.82) and here and there we can also see a few stomata. There are far fewer of these here, than on the leaf surface (Fig. 7.83).

Fruit Examination

8

8.1 The Tomato's (*Solanum lycopersicum*) Berry Fruit

Let us make a thin longitudinal section of the tomato berry and examine it under transmitted light! The tomato is covered externally by a membranous epidermis, beneath which the fleshy fruit wall can be seen. This, especially its outer part, contains a lot of red pigment (lycopene). In the middle, the placenta can be seen, to which the seeds are connected by a funiculus. The slimy seed coat of the tomato seeds forms the high water content jelly layer located under the fruit wall. This often completely fills the space between the inner surface of the fruit wall and the placenta. Make a horizontal section of the tomato as well and examine this under a magnifying glass! On the horizontal section, it can be clearly seen that the partitions of the fruit wall divide the inner space of the berry into three compartments (the number of partitions can vary in different varieties of tomatoes). The placenta's rays extend into the three fruit compartments, to which the slimy-coated seeds are connected (Figs. 8.1, 8.2, 8.3, and 8.4).

8.2 The Inflated Berry Fruit of the Pepper (*Capsicum annuum*)

The pepper species are plants living in South and Central America, the first specimens of which arrived in Europe after Columbus's second voyage, in 1494. Due to its colourful fruit, it was initially kept as an ornamental plant, then when Napoleon declared a continental blockade in 1806, it began to be used as a substitute for Indian pepper.

Like the tomato, the pepper has a berry fruit, in which there is an air-filled cavity between the seeds sitting on the placenta and the inner surface of the fruit wall (Figs. 8.5, 8.6, and 8.7).

The fruit of the pepper sits on the widened receptacle, to which the pedicel is connected at the bottom, while the fused calyx is connected at the top. The pepper's

G. Kriska, *Teaching Biology*, https://doi.org/10.1007/978-3-031-84085-2_8

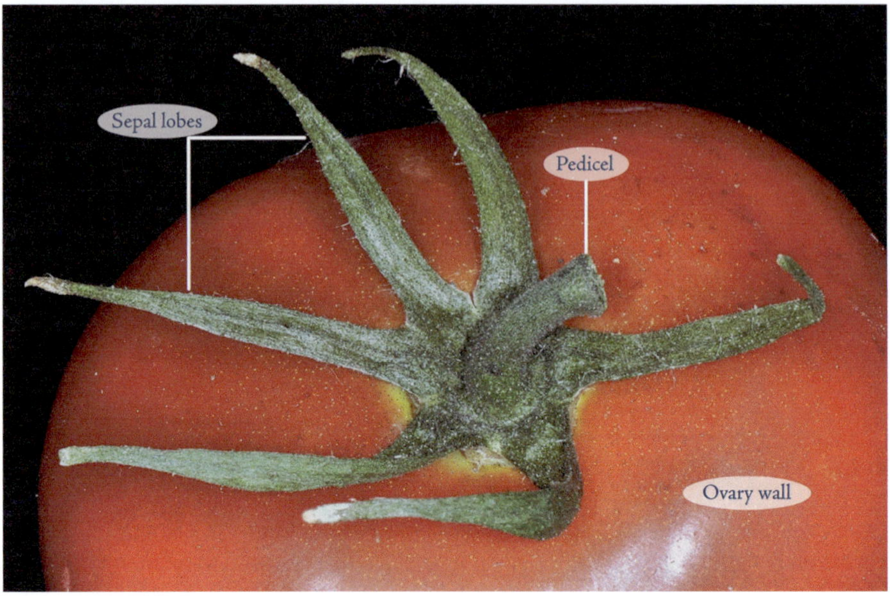

Fig. 8.1 Bottom view of the tomato's (*Solanum lycopersicum*) berry fruit

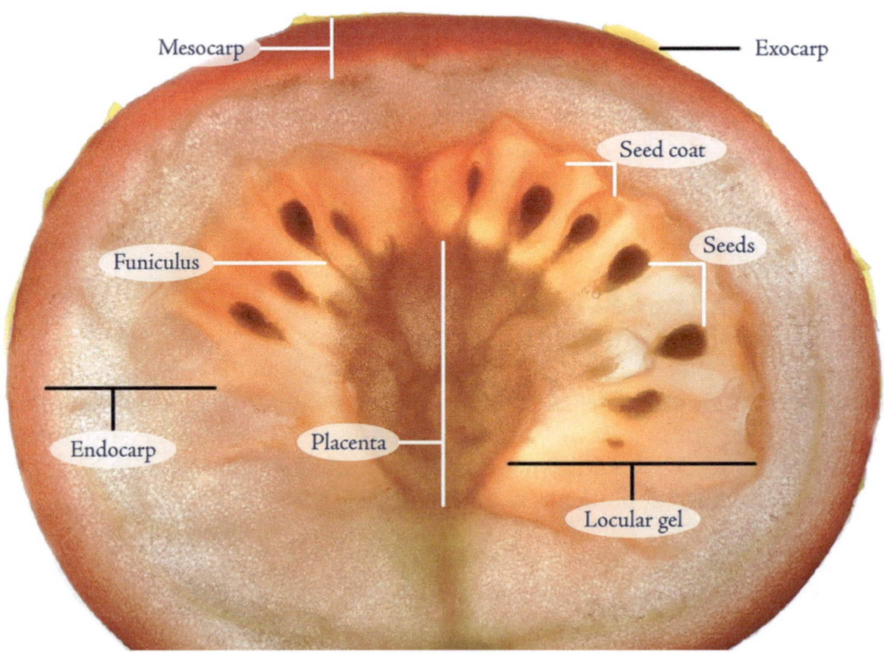

Fig. 8.2 Longitudinal section of the tomato (*Solanum lycopersicum*)

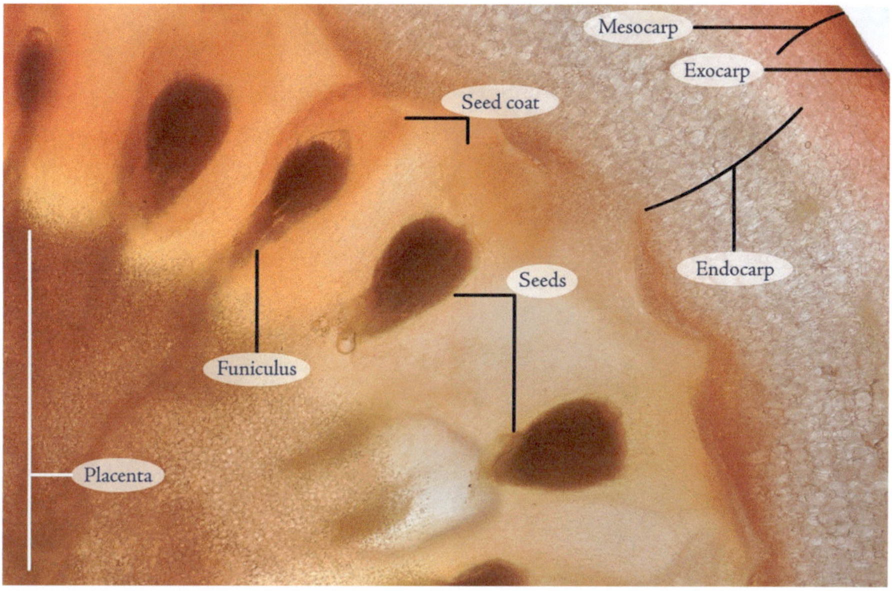

Fig. 8.3 High magnification image of the inside of a tomato fruit (*Solanum lycopersicum*)

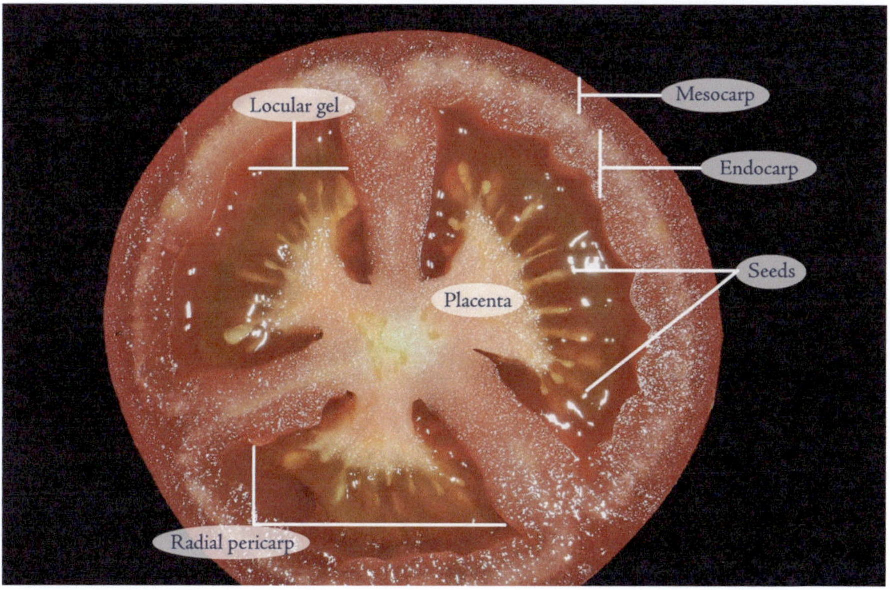

Fig. 8.4 Top view image of the horizontal section of a tomato fruit (*Solanum lycopersicum*)

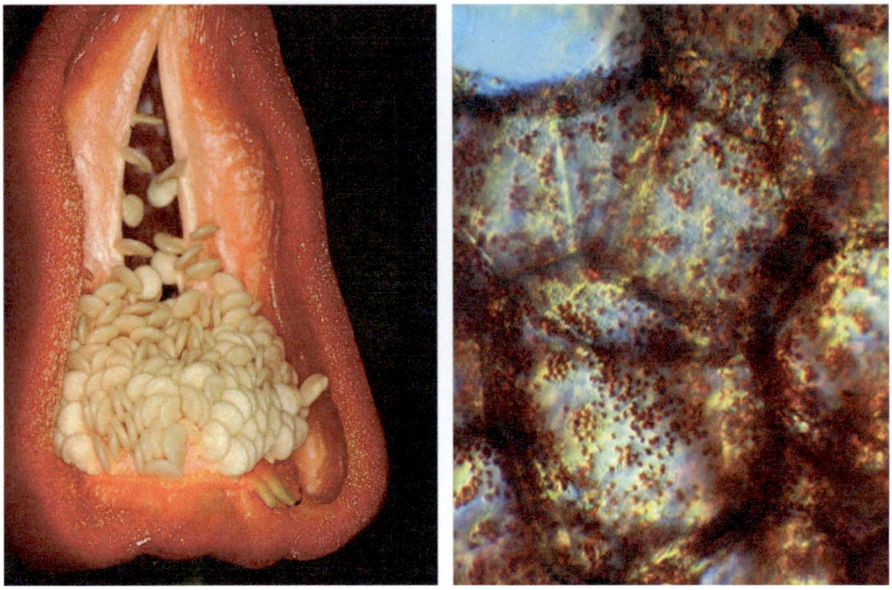

Fig. 8.5 Red pepper (*Capsicum annuum*) and carotene granules in its fruit wall

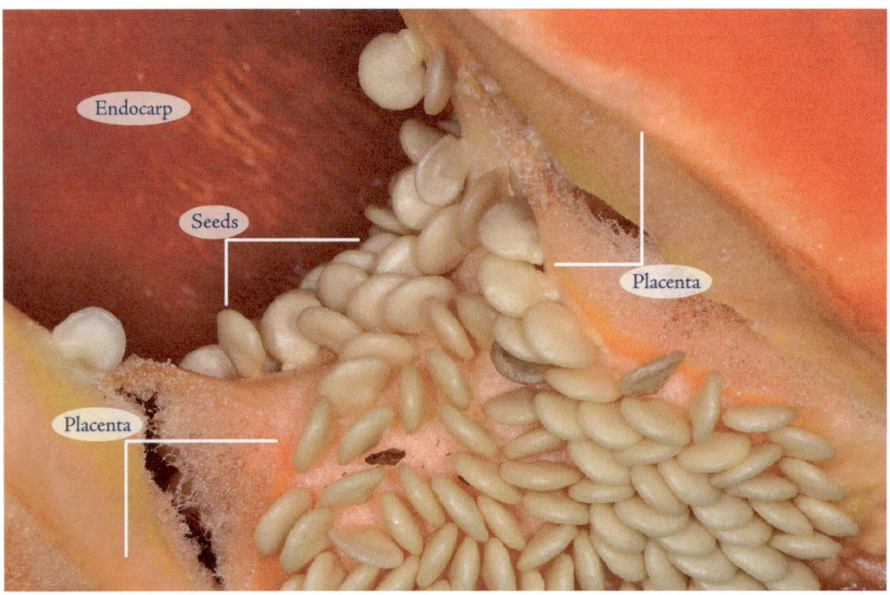

Fig. 8.6 Inside of a pritamin pepper (*Capsicum annuum*)

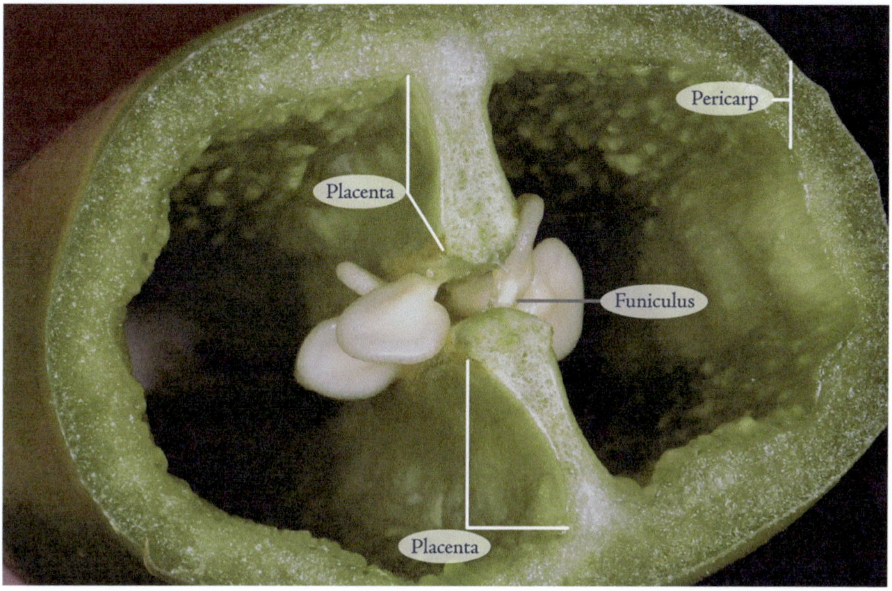

Fig. 8.7 Cross-sectional image of a green pepper (*Capsicum annuum*)

berry fruit is covered externally by a membranous epidermis, similar to the tomato. Beneath it, the juicy fruit wall can be seen. Let us make a section from the middle part of the pepper and examine it under a magnifying glass!

On the section of pritamin pepper (Fig. 8.6), the spongy placenta located on the upper part of the pepper's tip can be clearly seen with a large number of seeds on it. In the background, the red surface of the inner fruit wall appears. In some peppers, remnants of the fruit wall may appear in the inner space of the fruit, near the seed bed. The spiciness of hot pepper is caused by a substance called capsaicin in the placenta. The placenta mostly forms part of the pepper stalk, which we usually do not consume, but the placenta also grows onto the inner wall of the fruit wall, creating the veins of the pepper, which makes the pepper spicy. We can often observe the seeds on the veins, which supports the fact that these are part of the placenta.

8.3 The Drupe Fruit of the Cherry (*Prunus avium*)

Many of our popular fruit trees (e.g. cherry, sour cherry, peach, apricot, plum, and walnut) have drupe fruit. Its outer fruit wall is thin and leathery, the middle is fleshy and juicy, while the inner is bone-hard (Fig. 8.8). If we crack the inner fruit wall, we find the seed inside covered with a thin seed coat (Fig. 8.9). The leathery outer fruit wall, the fleshy middle fruit wall, and the seed covered with a hard inner fruit wall can be identified on the halved cherry.

Fig. 8.8 Parts of the drupe fruit of the cherry (*Prunus avium*)

Fig. 8.9 Cherry (*Prunus avium*) with seed

8.4 The Fruit of the Kiwi (*Actinidia chinensis*)

The kiwi is a native plant in China, originally called Chinese gooseberry. It was introduced to New Zealand in the early twentieth century, where it was first called a small melon, but since the United States imposed import duties on melons, to avoid the duty, the producers named the fruit after New Zealand's national bird. Make a cross-section of the kiwi and examine it under a magnifying glass, with direct and transmitted light! The structure of the fruit wall and the tiny seeds connected to the centrally located placenta by the umbilical cords can be well observed on the cross-sections of the kiwi fruit (Figs. 8.10 and 8.11).

8.5 The Cucumber's (*Cucumis sativus*) Gourd Pseudofruit

The pumpkin, melon, and cucumber have gourd pseudofruits. Their ovary is inferior, completely sunk into the receptacle, so in addition to the wall of the ovary, the receptacle also participates in the formation of the fleshy fruit wall of the pseudofruit. Cut a thin cross-section from the cucumber and examine it under a magnifying glass, with transmitted light (Fig. 8.12)!

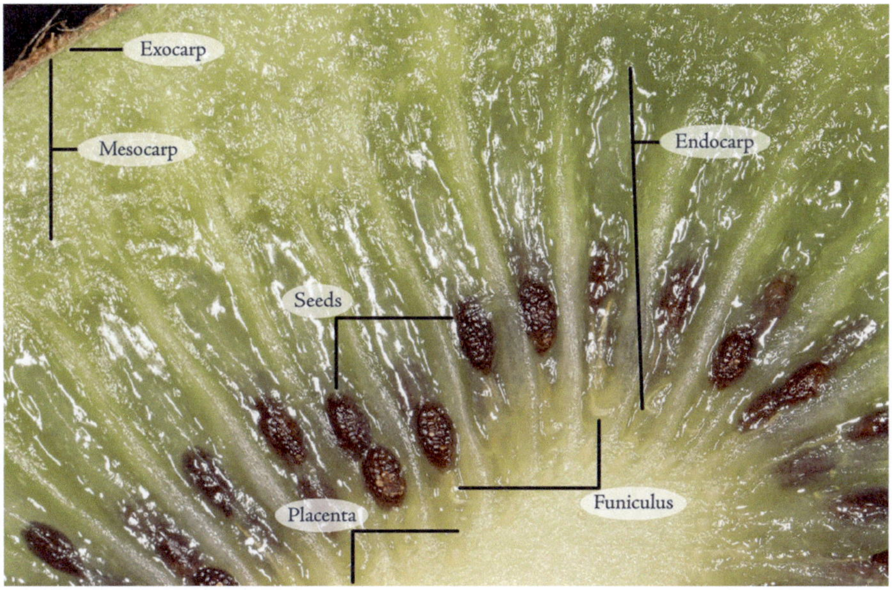

Fig. 8.10 The cross-section of the kiwi (*Actinidia chinensis*)

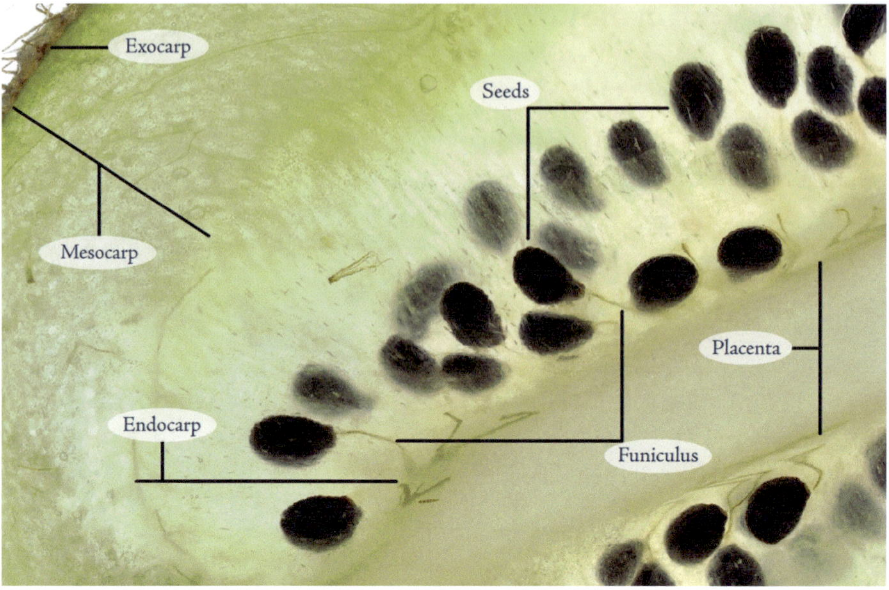

Fig. 8.11 The cross-section of the kiwi (*Actinidia chinensis*) under transmitted light

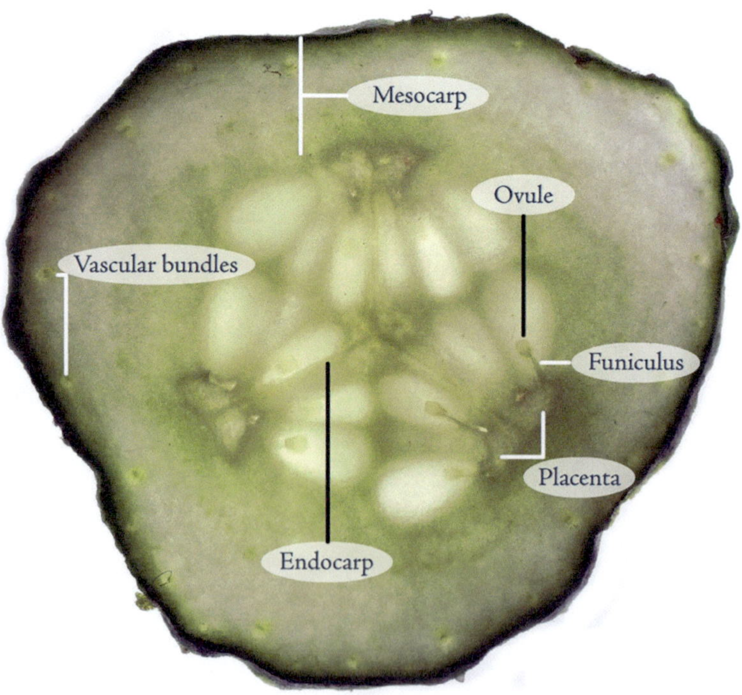

Fig. 8.12 Cross-section of a snake cucumber (*Cucumis sativus*) under a magnifying glass, with transmitted light

8.6 The Strawberry's Pseudofruit, the Strawberry (*Fragaria* × *ananassa*) Fruit

It may seem unbelievable at first hearing, but the strawberry has a dry fruit. The strawberry flower has many separate ovaries, from which single-seeded, dry achenes are formed. These, sitting on the surface of the fleshy receptacle, form pseudofruit, the strawberry fruit. The achenes on the surface of the receptacle can be observed with a magnifying glass in the cross-section of the strawberry, with the remains of the stigma on their top (Figs. 8.13, 8.14, and 8.15).

8.7 Wood Avens (*Geum urbanum*) Achene Fruit Group

Its achene fruits have a hook-ended spine, which often gets stuck in the fur of forest mammals, and thus the plant's fruits travel far. Removing some of the achenes reveals the receptacle to which the fruits are attached (Figs. 8.16 and 8.17).

8.8 Goatsbeard (*Tragopogon* sp.) Achene

Various structures that help spread the fruit can be attached to the achene fruit. Most commonly, as in the case of the goatsbeard, an umbrella-like flying device, the pappus, is formed from the drying sepals at the top of the fruit. In the fruit group formed in place of the inflorescences, we can see the umbrellas of the pappus touching each other from above (Fig. 8.18).

Fig. 8.13 Strawberry (*Fragaria* × *ananassa*) plant with pseudofruits

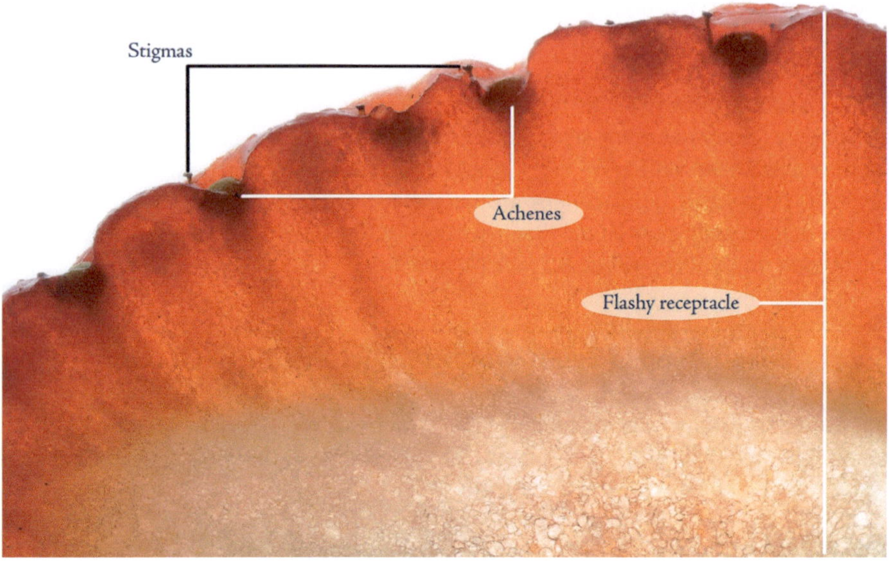

Fig. 8.14 Longitudinal section of strawberry (*Fragaria* × *ananassa*) in the magnifying glass, with transmitted light illumination

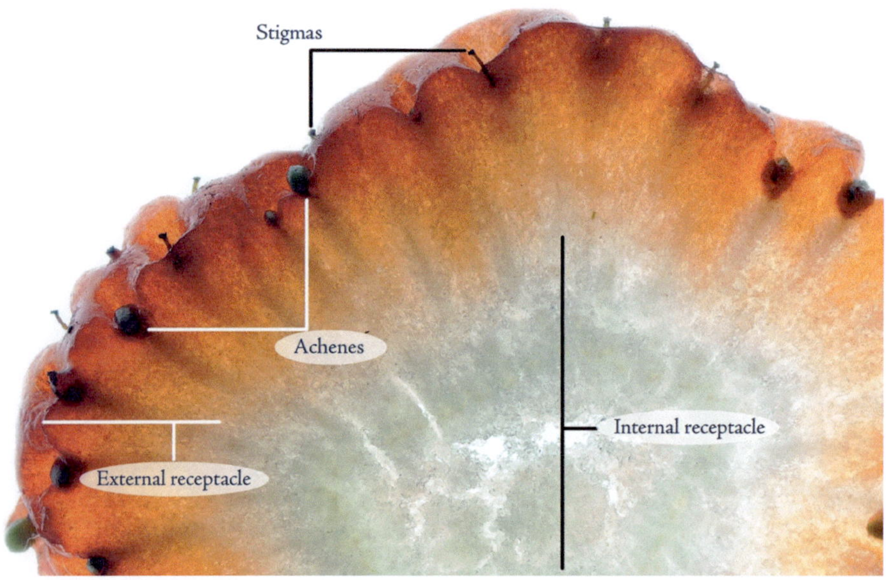

Fig. 8.15 Strawberry (*Fragaria* × *ananassa*) section in the magnifying glass, with transmitted light illumination

Fig. 8.16 The achene fruit groups of the wood avens (*Geum urbanum*)

Fig. 8.17 The fruit group after removing some of the achenes

Fig. 8.18 Top view of the goatsbeard (*Tragopogon* sp.) achene group

Fig. 8.19 Parts of the goatsbeard (*Tragopogon* sp.) achene group

Their thin, radial threads are densely hairy. If we remove some fruits, the fruits attached to the receptacle become visible (Fig. 8.19).

8.9 Maize (*Zea mays*) Caryopsis

The annual maize has a spike inflorescence, whose inflorescence axis has become a thick, elongated, and fleshy structure. The sessile flowers are arranged in dense rows and are covered by an inflorescence bract.

From the ovaries of the sessile flowers, a stigma of various lengths grows out, which continues to grow until pollen lands on it. The entire length of the stigma is suitable for catching pollen. After the flower is pollinated, one or two days, the stigma dries out. If fertilisation does not occur, the corncob will be gap-toothed (Fig. 8.20).

Break the still tender corncob in half and examine the kernels! Under their thin, transparent pericarp, the single cotyledon of the seed, which separates the rest of the embryo from the yellow-coloured nutritive tissue (endosperm) that forms the base of the corn kernel, is outlined in white. The cotyledon remains inside the seed during germination and its main task is to transmit nutrients to the developing plant. The shield-like cotyledon surrounds the mesocotyl, which has a budlet protected by the coleoptile at one pole and a rootlet protected by the coleorhiza at the other pole (Fig. 8.21).

8.10 Capsular Fruits: Wild Poppy (*Papaver rhoeas*), White Campion (*Silene longifolia*), and Goldenrain Tree (*Koelreuteria paniculata*)

Among the dry dehiscent fruits, the legume and the capsule are the most common. The poppy's capsules develop many tiny seeds, which are scattered into the outside world through holes opening under the flat stigma (Figs. 8.22 and 8.23).

The white campion's capsules open with teeth at their apex (Fig. 8.24).

Fig. 8.20 The remains of the stigmas can be observed on the maize (*Zea mays*) caryopses

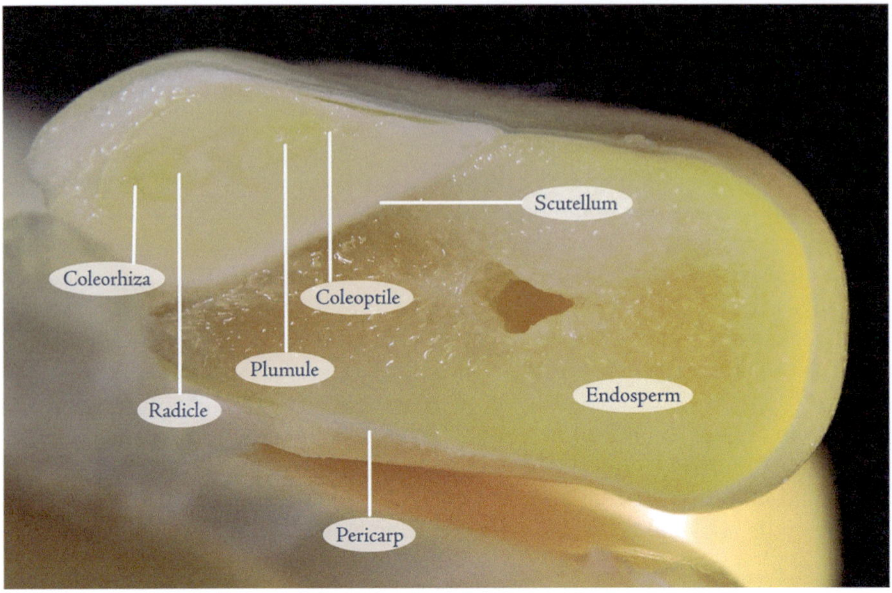

Fig. 8.21 Longitudinal section of the maize (*Zea mays*) grain

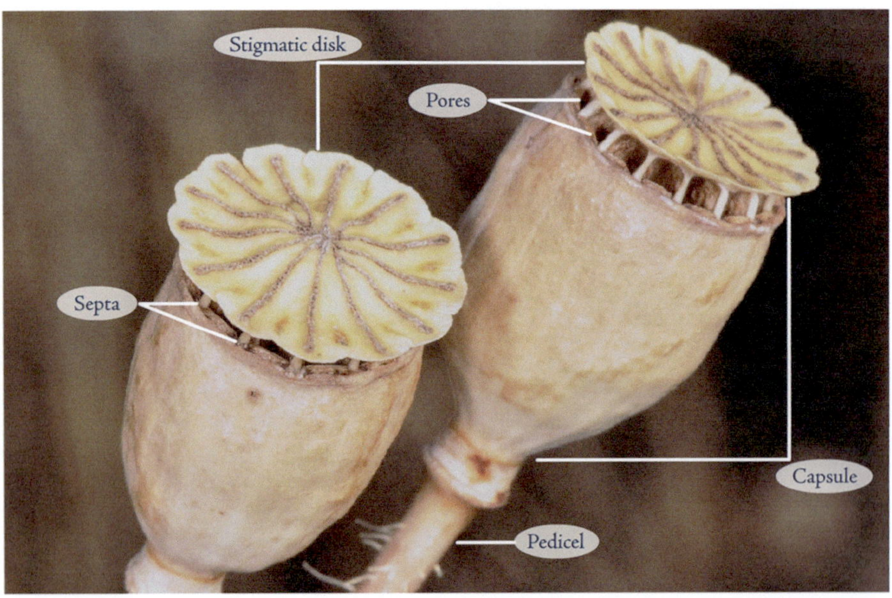

Fig. 8.22 The wild poppy's (*Papaver rhoeas*) capsules

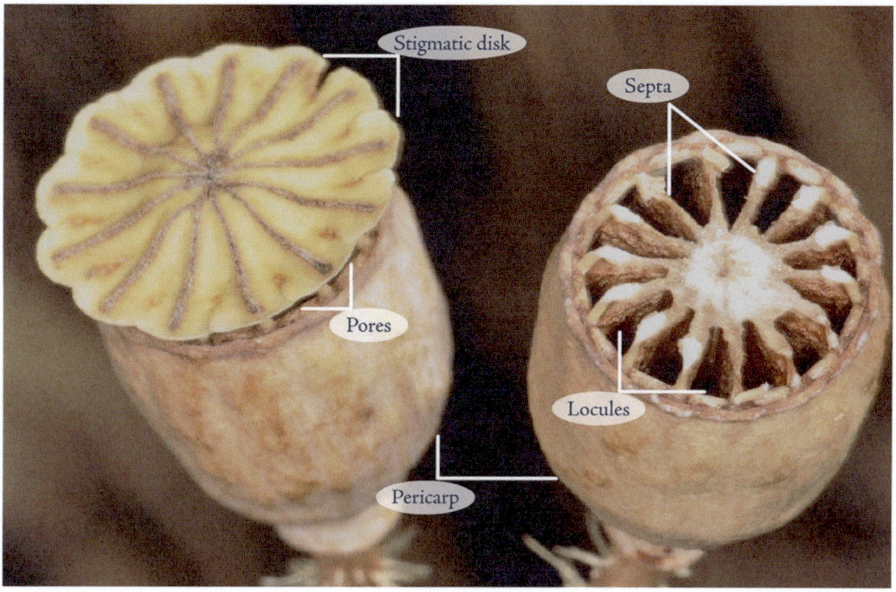

Fig. 8.23 The intact and halved capsules of the poppy (*Papaver rhoeas*)

Fig. 8.24 The opened capsules of the white campion (*Silene longifolia*)

Fig. 8.25 Branch of the goldenrain tree (*Koelreuteria paniculata*) with immature capsules

The goldenrain tree originating from East Asia was introduced to Europe by the Turks in the sixteenth century. The tree planted in parks and along roads has a 3–6 cm, three-edged fruit. The immature capsule is green (Figs. 8.25, 8.26, and 8.27), which turns orange by the end of summer, then dark brown. The bladder-like fruits remain on the tree all winter after the autumn leaf fall. Inside the mature capsule, there are 5–8 mm black seeds. In the longitudinal section of the fruit, we can observe the seeds connected to the central placenta, and the partitions. The internal partitions of the fruit only partially separate the compartments of the capsule. By cutting the fruit horizontally, we can get an accurate picture of the internal structure of the capsule.

8.11 Legumes: False Acacia (*Robinia pseudoacacia*) and Honey Locust (*Gleditsia triacanthos*)

The legume is a dry, dehiscent fruit. The elongated fruit opens lengthwise on its dorsal side. Let us examine the dry legume of the false acacia under a magnifying glass! The legume is connected to the dried floral axis by a pedicel, to which other fruits are attached (Figs. 8.28 and 8.29). At the base of the fruit, we can observe the dried calyx, while at the apex the remains of the style can be seen. Let us open the legume and examine the seeds inside. Inside the fruit, the placenta is located along the dorsal side, to which the seeds are attached by the funiculus. Unfertilised ovules can be recognised as tiny dots.

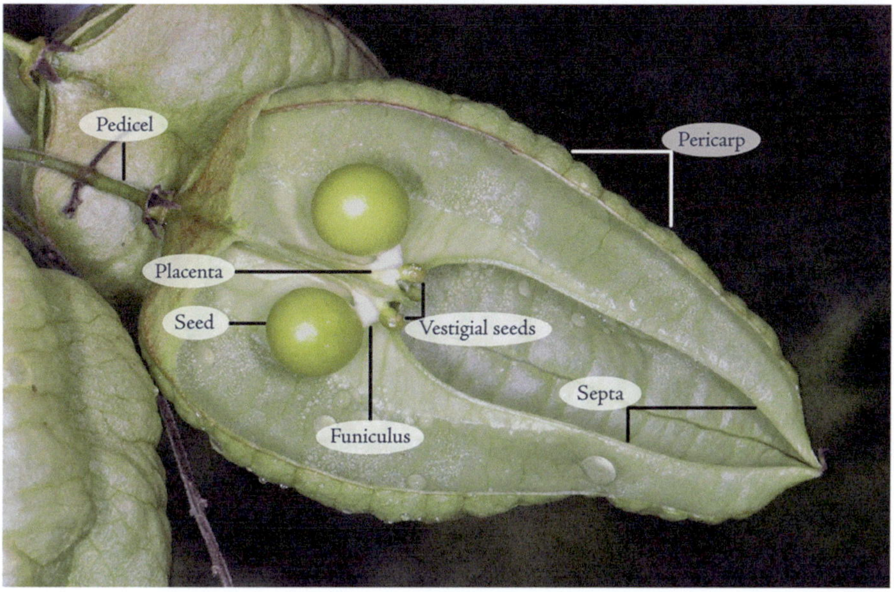

Fig. 8.26 Longitudinal section of the goldenrain tree's (*Koelreuteria paniculata*) capsule in the bug viewer

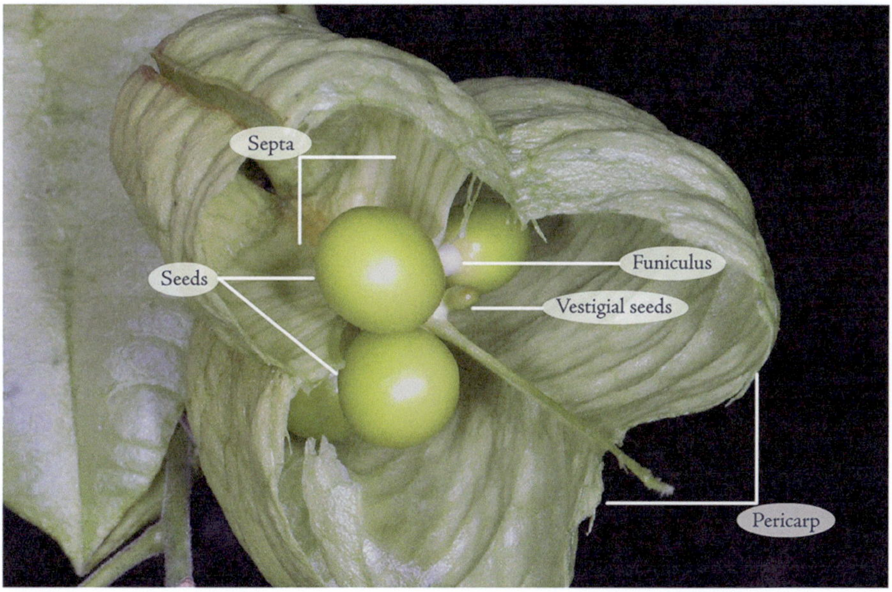

Fig. 8.27 Horizontal section of the goldenrain tree's (*Koelreuteria paniculata*) capsule

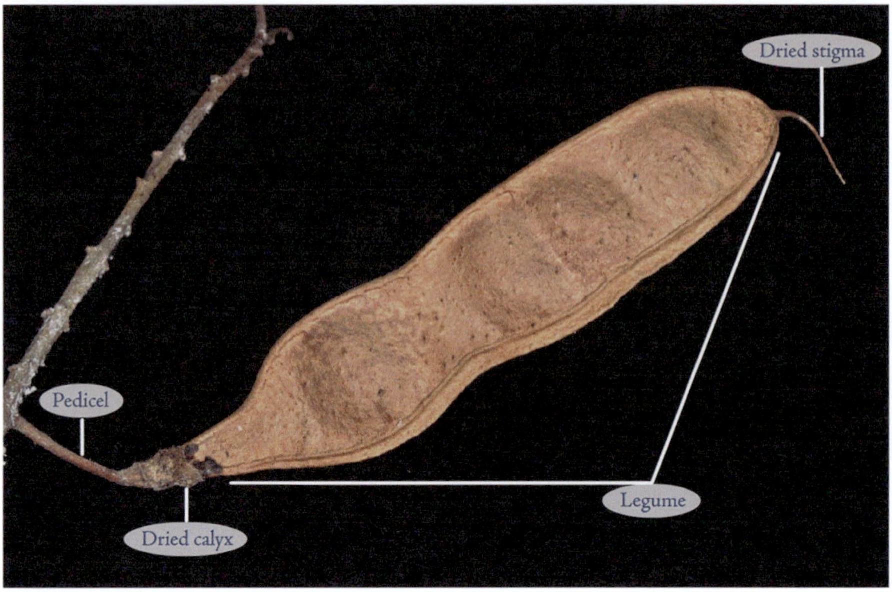

Fig. 8.28 The dried legume of the false acacia (*Robinia pseudoacacia*)

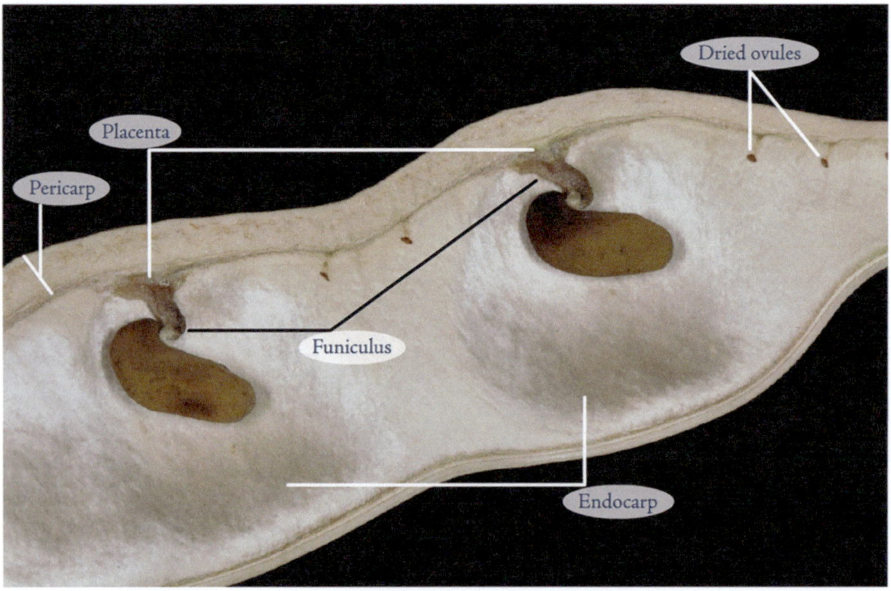

Fig. 8.29 The inside of the acacia's (*Robinia pseudoacacia*) legume

Fig. 8.30 Detail of the honey locust's (*Gleditsia triacanthos*) legumes

Fig. 8.31 Detail of the inside of the still immature honey locust fruit in the bug viewer

The twisted legumes of the honey locust can be up to 20 cm long. Let us examine a section of the fruit under a magnifying glass! In the not yet fully ripe legume, the green seeds are curled up with the funiculus attached to the placenta (Figs. 8.30 and 8.31).

9 Plant Tissues

9.1 Histological Examinations

Below, we review some of the microscopic examination methods. We have already dealt with the descriptions of the generally used experimental tools and methods. The spatial position, size, and structural relationships of a tissue or its components are most often studied on sections made in certain directions. First of all, there is a cross-section made in a plane perpendicular to the longitudinal axis of the organ to be examined (Fig. 9.1), as well as a radial (Fig. 9.2) passing through the longitudinal axis of the organ, and a tangential longitudinal section parallel to the longitudinal axis.

Some tissues or tissue systems can be examined without sectioning, such as the skin tissue system formed on the surface can be easily peeled off, detached from the young stem, leaf, etc., (peeling).

We usually examine our histological preparations with a normal light microscope, in transmitted light, but we can also use top illumination if necessary.

We always start the examination of our preparations with a small magnification (approx. 50–60×) to get an overview of the observable tissue types; then, according to the purpose of our examination, we examine the tissue with a larger (250–500×) magnification in detail.

Sectioning with a razor blade: For light microscopic examination, we usually need to prepare very thin, translucent sections, which can be done most easily with a razor blade.

During sectioning, we first need to make a smooth cut surface on the test object, then we need to cut off thin layers with further cuts below the surface. Thin sections can be made most easily from not too hard objects with a razor blade. These include most soft stems, fleshy fruits, some storage organs (apple, potato tuber, carrot, etc.), or the thick leaves of plants from the Mediterranean region (e.g. *Nerium oleander*, *Ficus*), which are suitable for making sections in different directions. The swollen, hard leaves of succulent plants seem to be a good subject for sectioning at first glance, but when cut, the cells with mucous content disintegrate, the leaf becomes

G. Kriska, *Teaching Biology*, https://doi.org/10.1007/978-3-031-84085-2_9

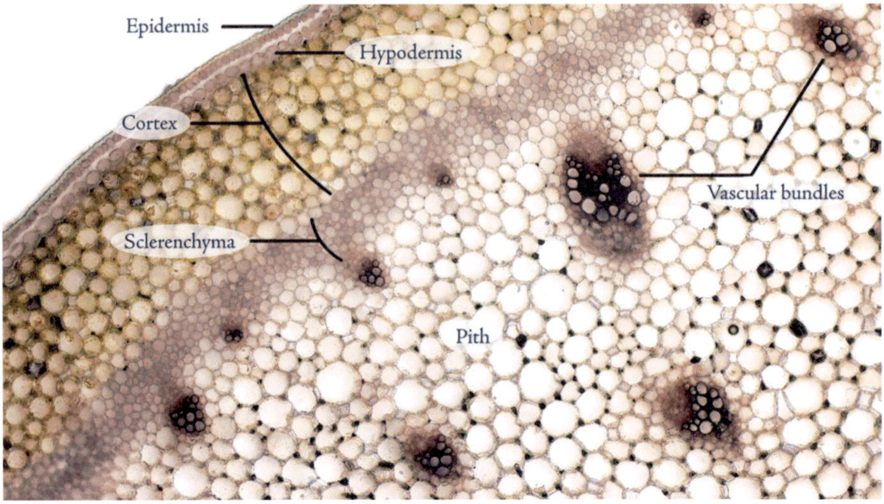

Fig. 9.1 Cross-section of a tulip (*Tulipa gesneriana*) stem

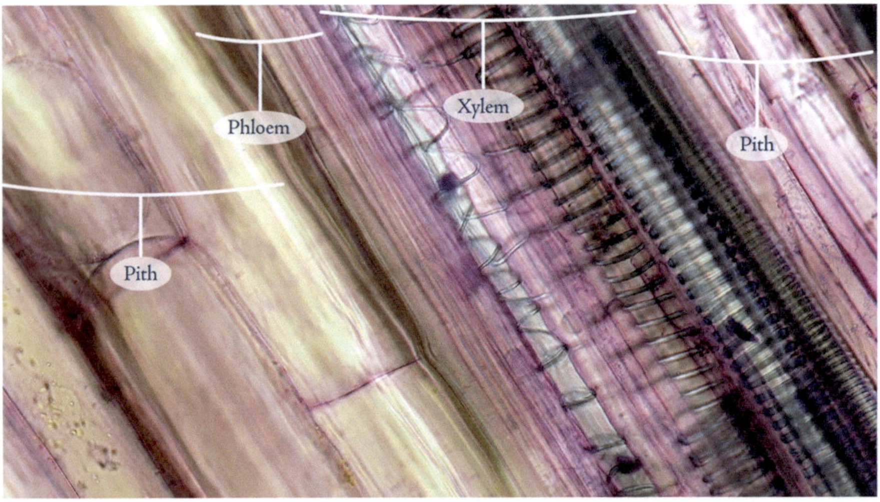

Fig. 9.2 Radial longitudinal section of a tulip (*Tulipa gesneriana*) stem

flaccid, and it is difficult to cut further. It is not easy to make sections from thin and flexible leaf blades with a razor blade, but in certain cases, we can achieve good results indirectly. For example, cross-sections can be made indirectly from reed leaves by making sections from the young shoot above the stem node. Then the stem besides, thin sections are also made from the leaves surrounding it, which can be well examined with a light microscope. A tree branch is usually too hard to make a full section from it, while sectioning rarely results in larger, equal thickness, yet sufficiently thin, thus translucent sections. In such cases, the use of top-illumination technique can be useful, which does not require a thin section to be made from the branch, even a 1–2 mm thick specimen can be used to study the tissue structure of the branch. First, cut the branch in half with a razor blade, then make another section 1 mm from the cut surface. After the more difficult bisecting of the branch, cutting off the individual layers is an easy task. As a result of our work, we get about 1 mm thick smooth surface parts, which can also be examined with a light microscope under top lighting.

Place the slice in toluidine blue solution for 10–15 s, then place it on a slide, and examine it without a cover with a light microscope under top lighting. The top lighting can be achieved with a light source containing multiple power LEDs. Such a light source can provide the sufficiently strong and cold light needed for the examination. Concentrate the light of the lamp on the section, then start the light microscopic examination (Fig. 9.3).

Staining the sections with toluidine blue solution: The sections composed of various tissue system elements (e.g. stem, root, and leaf cross-sections) are often quite complex in structure, so it is not always easy to identify their components. In such cases, staining the specimen with toluidine blue solution can be a great help, which stains the different tissues in different colours (Fig. 9.4).

The different types of tissues can be particularly well studied when the sections are stained with specific cell wall dyes. The cell walls are often chemically different for each tissue type, and accordingly they are stained in different colours, which greatly facilitates the recognition of the more important tissues. For example, in a short time in aqueous solution of toluidine blue, the cellulose wall turns purple or

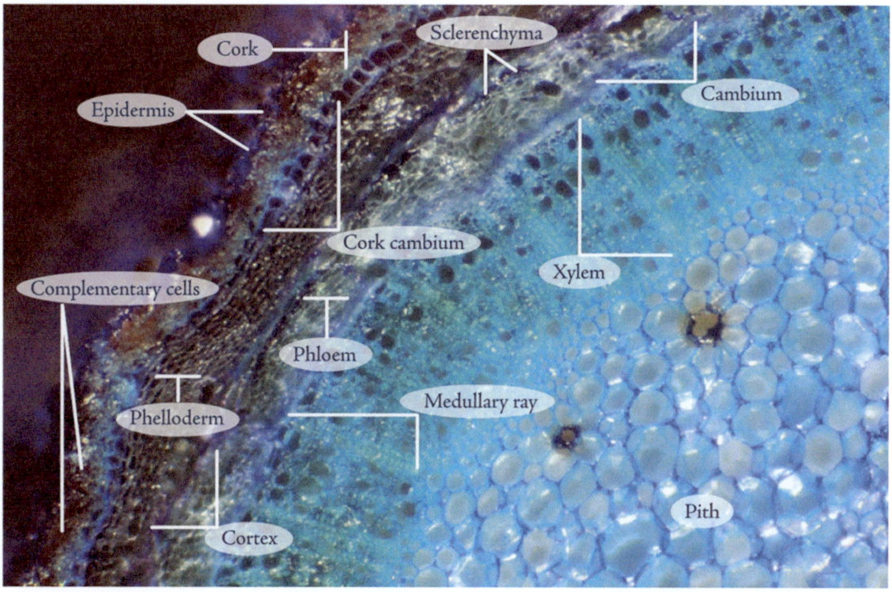

Fig. 9.3 Top-lit image of a toluidine blue stained elderberry (*Sambucus nigra*) branch cross-section

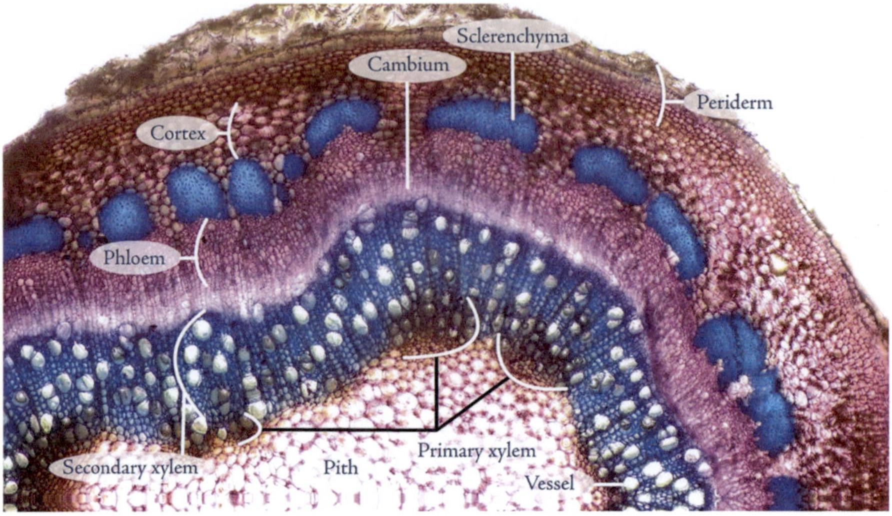

Fig. 9.4 Cross-section of silver poplar (*Populus alba*) branch stained with toluidine blue

purplish-blue, the woody wall turns bluish-green, and the pithy wall turns greenish-brown. However, the colour effect varies somewhat depending on the plant species.

To prepare the dye solution, dissolve enough toluidine blue powder in a few cm^3 of water to get a dark purple coloured solution. The resulting dye solution can be used for staining the sections in the future. Place the section on a slide in a drop of water, then drop 1–2 drops of toluidine blue solution on it. Using a dissecting needle, gently press the section into the solution so that the dye completely permeates its tissues.

After 30–50 s, use an eyedropper to remove the dye solution from the specimen, then use another clean eyedropper to rinse it with clean water. It is important not to over-dye the specimen, otherwise it will turn blackish-purple.

Preparation of epidermis peel: We usually obtain epidermis samples from leaves, the characteristics of which determine the method we should use when preparing the peel. In the case of fleshy, thicker leaves (e.g. succulents, *Sansevieria*, onion leaf), it is advisable to carve a square with a dissecting needle on the leaf surface, then by lifting the corner of the square, we can pull off the piece of skin tissue with tweezers.

For thin leaves, the skin tissue sample can be obtained by scraping or tearing.

When scraping, we need to scrape off one of the leaf's skin tissues and the leaf's mesophyll with a scalpel or razor blade, as a result of which the desired epidermis remains. This method is well applicable, for example, in the case of reed.

In the tearing method, we need to tear the leaf diagonally with a decisive movement, as a result of which a few millimetres wide epidermis edge often remains on the edge of the leaf pieces, which we can cut off with scissors and examine under a microscope (Fig. 9.5).

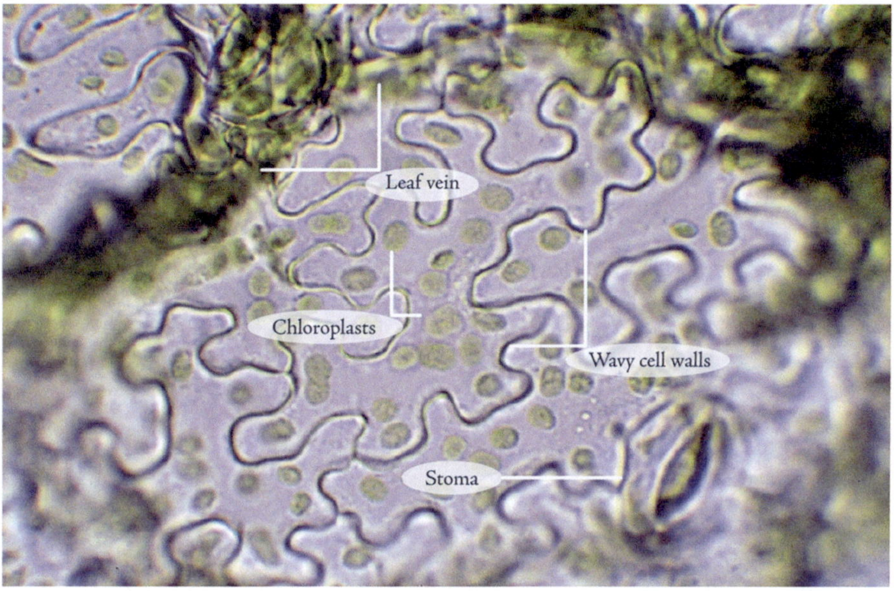

Fig. 9.5 Light microscopic image of epidermis made by tearing from the leaf of small balsam (*Impatiens parviflora*)

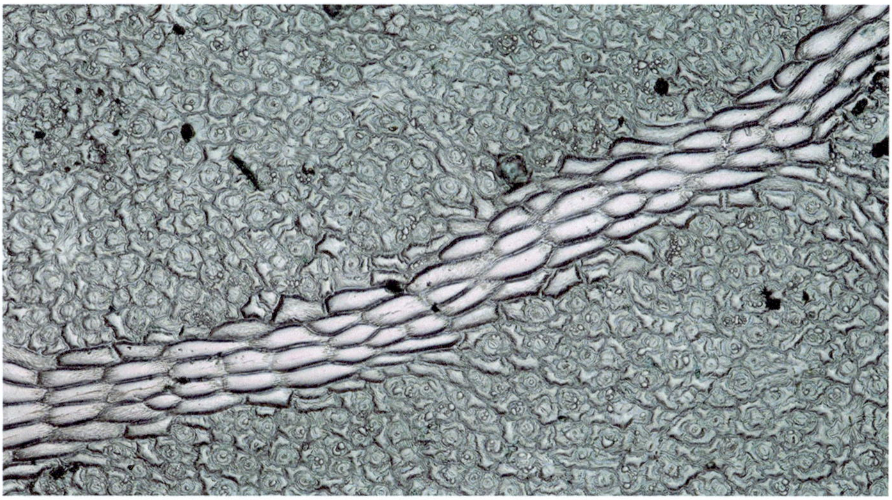

Fig. 9.6 Imprint of a leaf vein of wood violet (*Viola odorata*)

Preparation of imprints: The preparation of imprints is excellent for the light microscopic examination of epidermis and other surfaces (Fig. 9.6).

In this process, we need to apply a thin layer of colourless nail polish, for example, on a leaf surface, then after 1–2 min, when the polish has dried, we can carefully lift it with a needle and pull off the coating with tweezers. The colourless specimen obtained in this way should be examined under a light microscope with oblique illumination, smoothing it onto a slide. The surface of the imprint is stickier, so it is advisable to press it onto the slide. Imprints can only be made from smooth objects, because for example, a hairier leaf sticks to the polish in such a way that it is impossible to remove after drying.

Preparation of scrapings: For the examination of certain plant products, such as plant trichomes, the preparation of scrapings is the suitable method. In this process, we scrape the hairs covering the plant parts onto a slide with a scalpel. The sample can be examined dry, after covering with a cover slip.

9.2 Plant Tissues

The bodies of ferns, gymnosperms, and angiosperms are built up of tissues. Tissues consist of cells of common origin, developing and functioning in harmony. The cells of **simple tissue**s are of one shape and function, while **complex tissue**s contain cells of various shapes and functions. Tissues performing the same tasks are classified into tissue systems. The growth of plants is ensured by meristems. Initially, these completely fill the embryo (primary meristems), then, by creating primary tissues, they build the plant. Mature or differentiated tissues that have developed from the meristems can regain their ability to divide and thus, for example, become secondary meristems. The formation of secondary tissues can contribute, among other things, to the secondary thickening of the stem and root, the appearance of periderm, and the wound-closing callus that forms at the site of injuries. The meristems are usually found in significant quantities in growing, developing young body parts (e.g. root and shoot apex), and in smaller quantities in thickening organs (e.g. in the root, stem and trunk). Their cells mostly fit closely together, are thin-walled, filled with cytoplasm, have large nuclei, and remain in a dividing state with smaller or larger interruptions. Their main task is to increase the plant with new cells and tissues.

Some of the cells and cell groups derived from meristems eventually stop dividing and their function gradually changes to perform some special task, they become permanent. The largest part of the body of a mature multicellular plant is built up by simple tissues consisting of cells of the same shape and function and complex tissues consisting of cells of various shapes and functions. The supporting tissues include sclerenchyma/collenchyma, a simple tissue consisting of cells of the same shape and function (Fig. 9.7).

Xylem is a complex tissue containing parenchyma and sclerenchyma in the form of fibres, in addition to transport elements (Fig. 9.8).

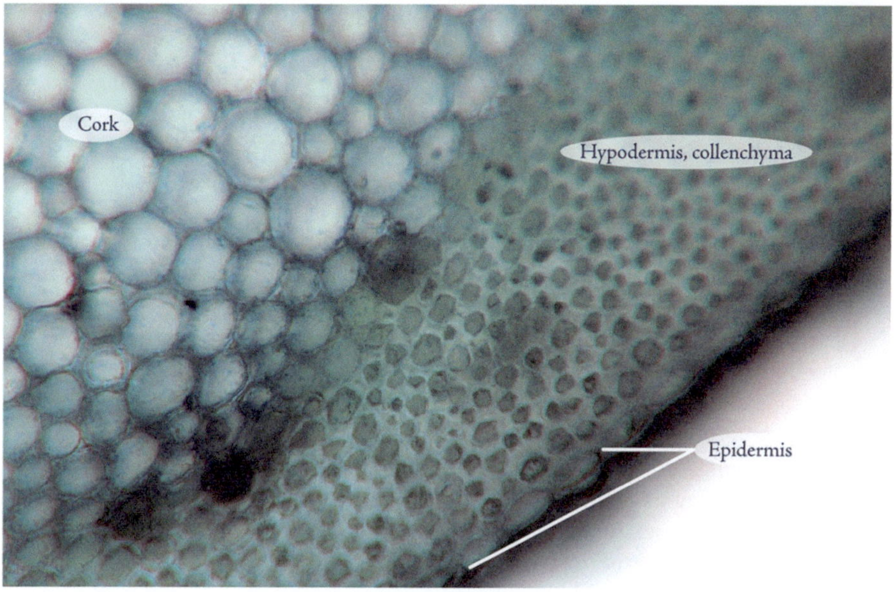

Fig. 9.7 Collenchyma is a simple tissue consisting of cells of the same shape and function

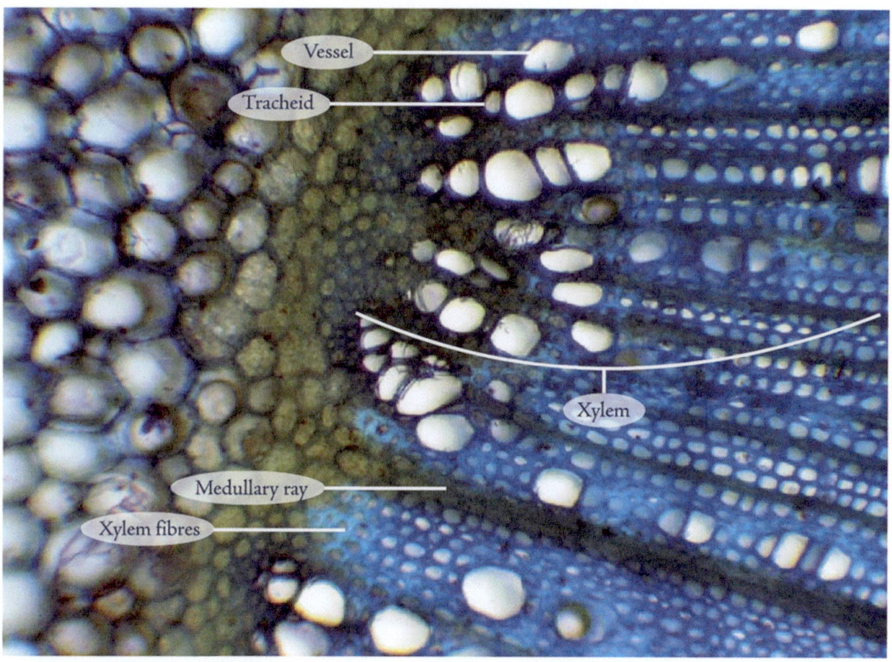

Fig. 9.8 Xylem is a complex tissue

9.2.1 Meristems

The continuously dividing group of cells formed from the zygote is called embryonic meristem, from which the meristems remaining in the shoot and root apexes are formed (Fig. 9.9).

From these, primary meristems (histogens) are formed, from which the primary plant body is formed.

In addition to the primary meristems, there are secondary meristems too, which form secondary tissues and the secondary plant body constructed from these (Fig. 9.10).

Depending on their location, there can be three types of meristems. The apical meristems, which are found in the shoot and root tips, serve the longitudinal growth of the shoot and root. The intercalary meristems mainly ensure the longitudinal growth of stems and flower stalks and are derived from apical meristems.

The thickening of the stem and root is made possible by the lateral meristems located in a cylinder-like mantle, which can be primary meristems derived from apical meristematic tissues, or secondary meristems. There are three types of lateral meristematic tissue: the pericycle, the cambium, and the cork cambium. We will review their operation through one example each.

9.2.1.1 Pericycle

The **pericycle** is the outermost, one or two cell layer cylinder-like layer of the root's vascular cylinder (stele), which participates in lateral root formation (Fig. 9.11).

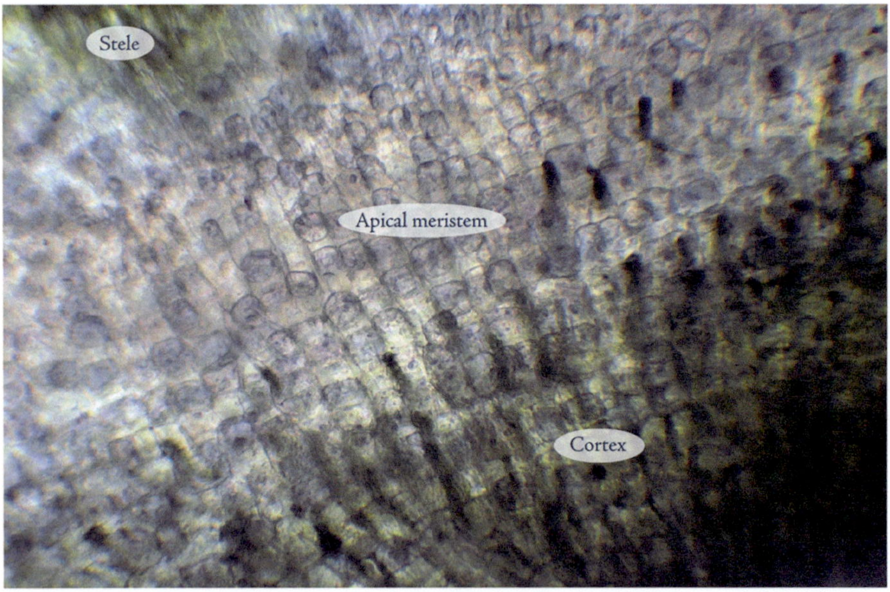

Fig. 9.9 Maize (*Zea mays*) root apex

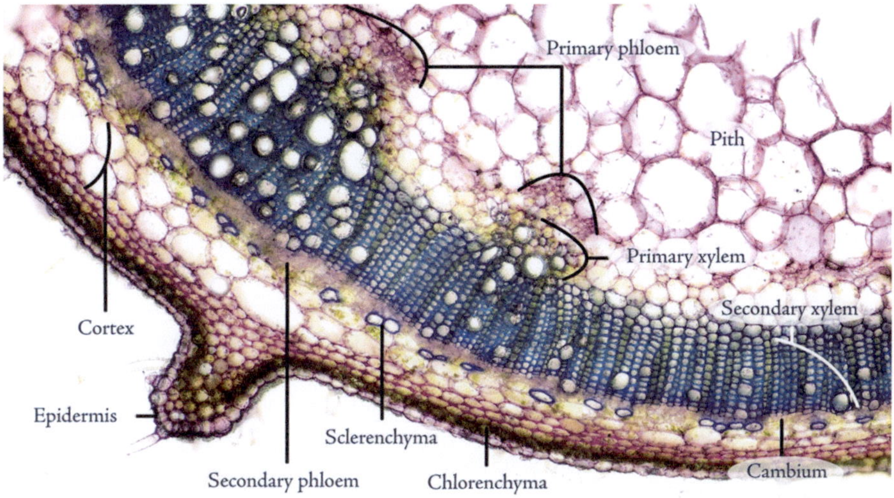

Fig. 9.10 Formation of secondary xylem and phloem in the stem of black nightshade (*Solanum nigrum*)

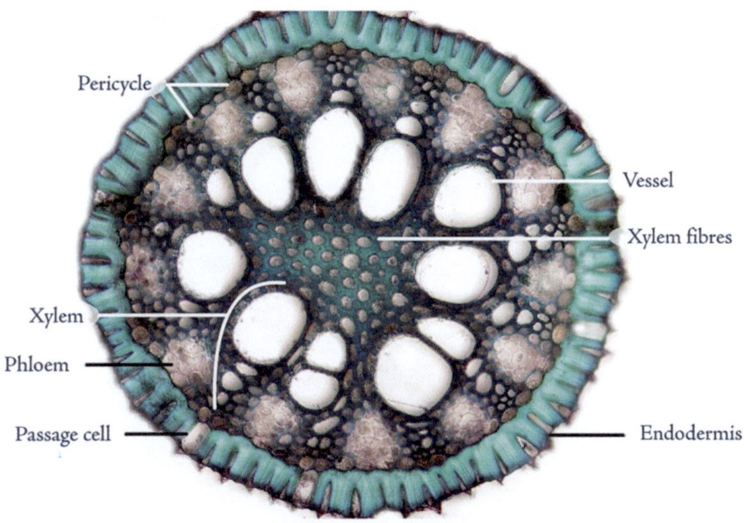

Fig. 9.11 High magnification image of the stele of the *Iris* root

In addition, it can participate in the secondary thickening of the root and the formation of the root's periderm together with secondary meristems formed from ground tissue cells. It is directly connected to the root tip's ancestral meristematic tissue, so the pericycle is a primary meristem.

9.2.1.2 Cambium

The **cambium** thickens the root and stem with secondary vascular tissues (Fig. 9.4). It can be both primary and secondary meristems. The primary cambium originates from the meristem of the shoot tip that produces vascular tissues, the procambium. The procambium is one type of the tissue-forming meristems (histogens). The formation of the primary cambium is made possible by the fact that the meristem does not fully differentiate into vascular tissue, but remains capable of division and forms a continuous cylinder between the primary xylem and phloem (Fig. 9.12), or becomes fascicular cambium when vascular bundles form (Fig. 9.13). In monocots, the meristem of the shoot tip that produces vascular tissues fully differentiates into vascular tissue, so their vascular bundles are cambium-free closed vascular bundles (Fig. 9.1).

The secondary cambium is formed by the dedifferentiation of parenchyma cells between the vascular bundles, so it is called interfascicular cambium. This interfascicular cambium in stems closes with vascular cambium and creates secondary xylem and phloem, as well as parenchyma cells (Fig. 9.10). In the above manner, secondary cambium can also form in the roots of dicots, which, together with the pericycle, forms the heterogeneous cambium cylinder, which is heterogeneous because both primary and secondary cambium participate in its formation (Fig. 9.14).

In perennial woody monocots, such as the dragon tree (*Dracaena draco*), secondary cambium can also form through the dedifferentiation of parenchyma cells, which creates separate vascular bundles and parenchyma inwardly (Fig. 9.15).

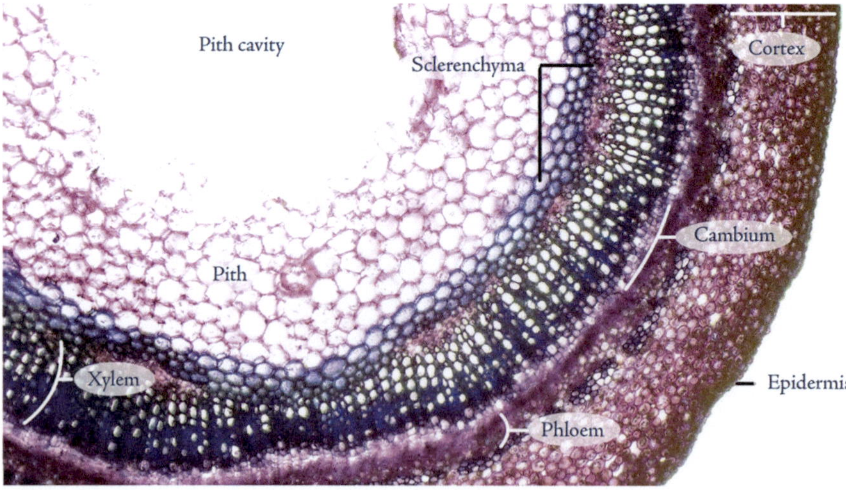

Fig. 9.12 Cross-section of *Forsythia*

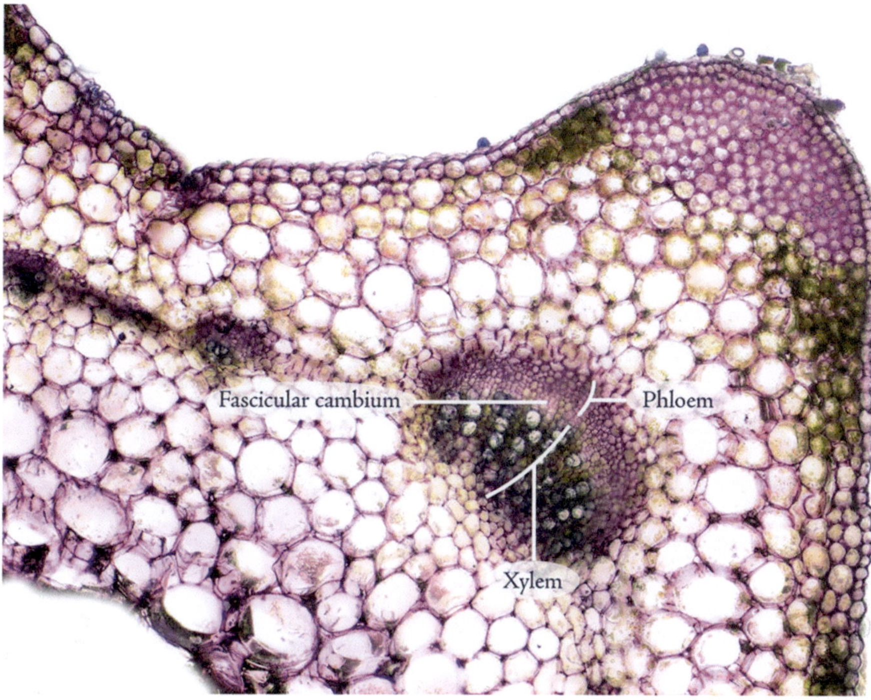

Fig. 9.13 Vascular bundle in the stem of golden deadnettle (*Lamium galeobdolon*)

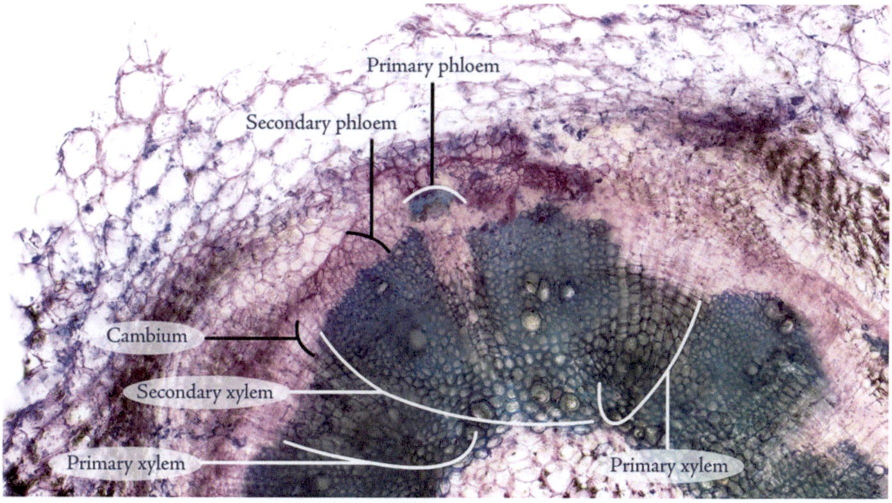

Fig. 9.14 Cross-section of a bean (*Phaseolus vulgaris*) root

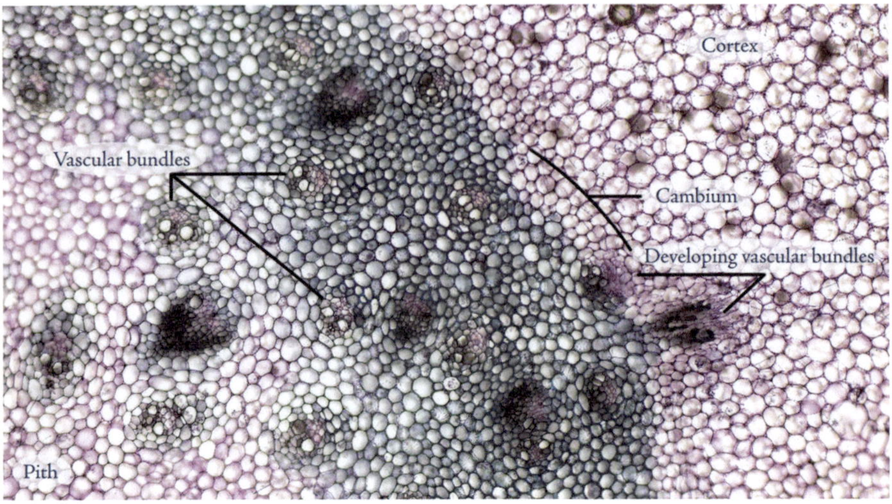

Fig. 9.15 Cross-section of a dragon tree (*Dracaena draco*) stem

9.2.1.3 Cork Cambium

The **cork cambium** (phellogen) is also a secondary meristem, which is formed by the dedifferentiation of cortex, phloem, or epidermis cells. It creates periderm and tertiary dermal tissue or rhytidome.

In the secondary thickening of a dicot stem, cork cambium most often forms in the cortex under the epidermis, which, as a cylinder, creates phelloderm inwardly and cork (phellem) outwardly.

The three layers together form the periderm (e.g. elderberry).

In many cases, the cork cambium only creates cork, in which case the periderm is two-layered (Fig. 9.16).

Rarely, cork cambium can also form through the dedifferentiation of epidermis cells (Fig. 9.17), which forms three- and two-layered periderm (e.g. *Nerium oleander* and *Solanum dulcamara*).

The formation of periderm is well indicated by the lenticels appearing on young branches with still intact epidermis. Under these, the cork cambium produces clusters of living, thin-walled cells instead of suberised cells, which, breaking through the epidermis, create the lenticels ensuring the air permeability of the branch covered with periderm (Fig. 9.18).

During the several-year-long thickening of the stem, cortex cells dedifferentiate into new cork cambium to produce new periderm layers. Thus, year-by-year, the cork cambium is initiated in deeper and deeper layers and it gives rise to regular secondary dermal tissue till there is any cell layer left in the cortex. Afterwards, only phloem parenchyma cells can dedifferentiate into meristem. From this very moment, phloem elements are also present in the produced dermal tissue, which is now called tertiary dermal tissue or rhytidome. Rhytidome is the characteristic surface tissue of perennial woody plants (Fig. 9.19).

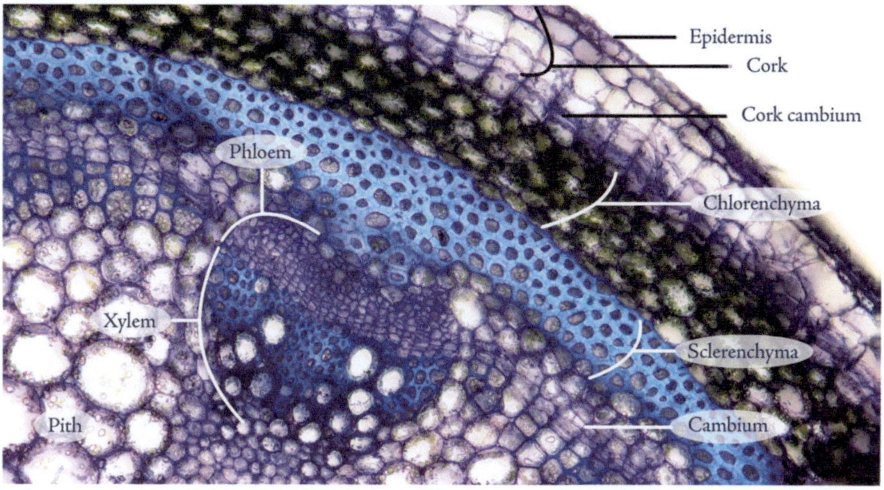

Fig. 9.16 Cross-section of geranium (*Pelargonium zonale*) stem with a two-layer periderm

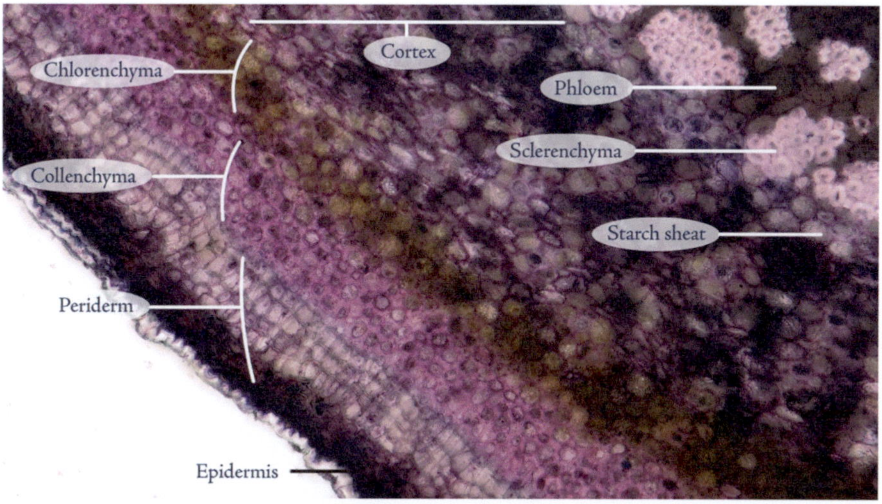

Fig. 9.17 Cross-section of oleander (*Nerium oleander*) stem with a two-layer periderm

The wound cambium is also a secondary dividing tissue, which forms at plant injuries through the dedifferentiation of parenchyma cells. A wound callus made of parenchyma cells forms on the injury, on the edge of which cork cambium can also form, which delimits the injury with a wound cork. The grafting used in horticulture is based on the formation of wound callus and secondary vascular tissue.

Fig. 9.18 Lenticels on a young elderberry (*Sambucus nigra*) branch

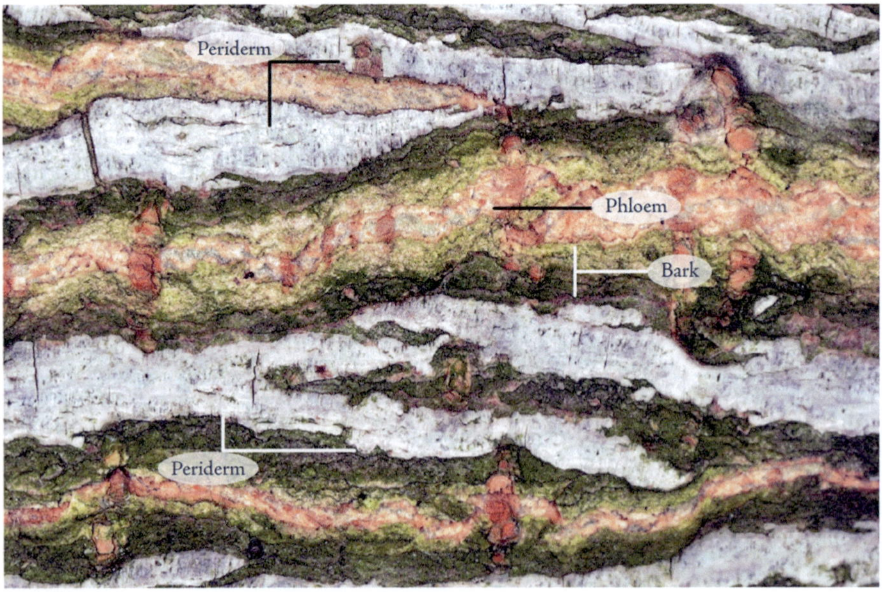

Fig. 9.19 On the outer surface of the cracked bark of the birch, the white, leafy periderm can still be observed, and the phloem appears deep in the cracks

9.2.2 Dermal Tissue System

The dermal tissue system covers and protects the surface of the plant body against external influences and can also participate in gas exchange, evaporation, and absorption. It covers the young parts of the shoot, the leaves, and the stem, usually one cell layer thick, and is called the **epidermis**. The epidermis usually cannot keep pace with the thickening of the shoot axis, so secondary dermal tissue, periderm, forms in place of the torn epidermis.

The growth and division of dermal cells can sometimes keep up with the thickening of the stem (e.g. mistletoe), so we find the same epidermis on the surface of thickened shoots that are several years old. The dermal tissue of the young root is the **rhizodermis**, whose cells develop into long, tubular, thin-walled extensions, root hairs. These penetrate between soil particles and carry out the uptake of water and mineral substances. The rhizodermis is short-lived, only covering the young, tip-near part of the root, the maturation zone. The surface of the older parts of the root is covered by the **exodermis**, which forms from the outer cell rows of the primary cortex becoming corky, replacing the decayed rhizodermis. On several-year-old, thickening roots, a periderm forms (Fig. 9.20).

However, unlike the stem, this periderm does not originate from the cork cambium formed in the cortex, but from certain parts of the pericycle and dedifferentiated parenchyma cells, forming a dividing tissue cylinder sheath.

In summary, we can conclude that, similar to dividing tissues, the dermal tissue system also partly consists of primary tissues and partly of secondary tissues that

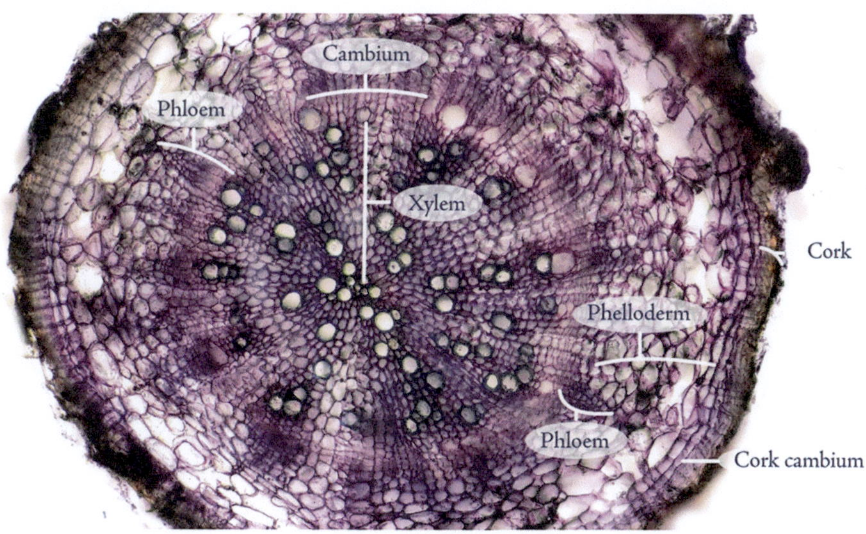

Fig. 9.20 Thickened secondary root of common purslane (*Portulaca oleracea*) covered with a three-layered periderm

form later. Thus, we can talk about primary dermal tissue (epidermis and rhizodermis) and secondary dermal tissue (periderm).

9.2.2.1 Epidermis and Rhizodermis

The epidermis is usually a single-cell layer, but can occasionally be multi-layered (e.g. *Ficus elastica*). Its cells usually fit tightly together. Most of these are made up of epidermal cells, which in monocots and dicots stem are elongated in one direction when viewed from above (Fig. 9.21), while in dicots leaves they are irregularly shaped and have a wavy outline (Fig. 9.22).

When examining the epidermis in cross-section, the cells are of varying shapes, their outer wall is usually thick, covered with **cuticle** or wax layer, while the rest are usually thinner.

The cell nuclei protrude the middle part of the elongated epidermal cells, the **stomata** are located between the narrowing ends of the elongated epidermal cells, arranged in rows.

The epidermis cells, with few exceptions (ferns, aquatic plants, and *Impatiens parviflora*), do not contain chloroplasts, but leucoplasts sometimes occur in them (e.g. *Sansevieria*) (Figs. 9.23 and 9.24).

The epidermis cells are alive, their cytoplasm is usually located in a thin layer along the cell wall, and the large part of the cell is occupied by the vacuole filled with cell sap (Fig. 9.25).

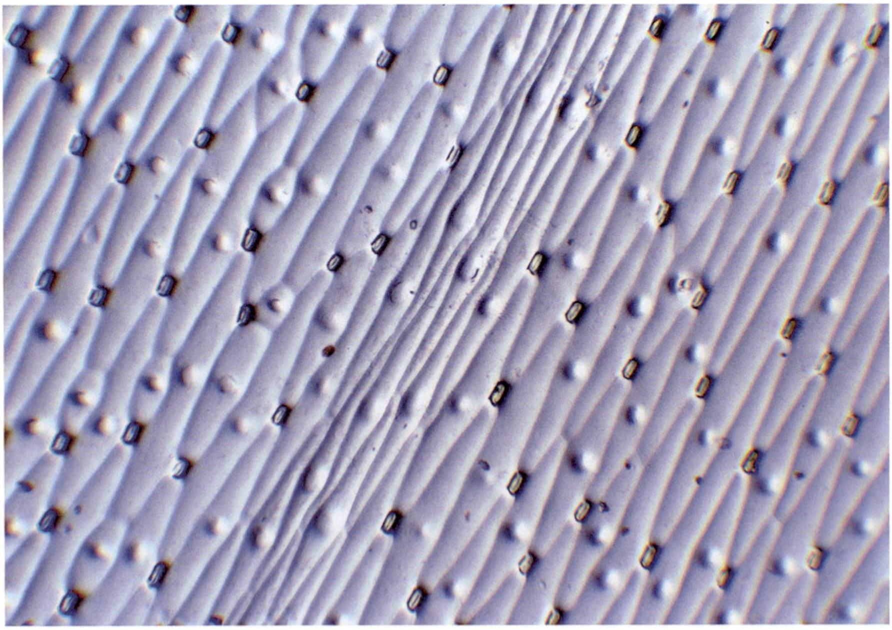

Fig. 9.21 Oblique illumination light microscopic image of the leaf imprint of the *Iris* sp.

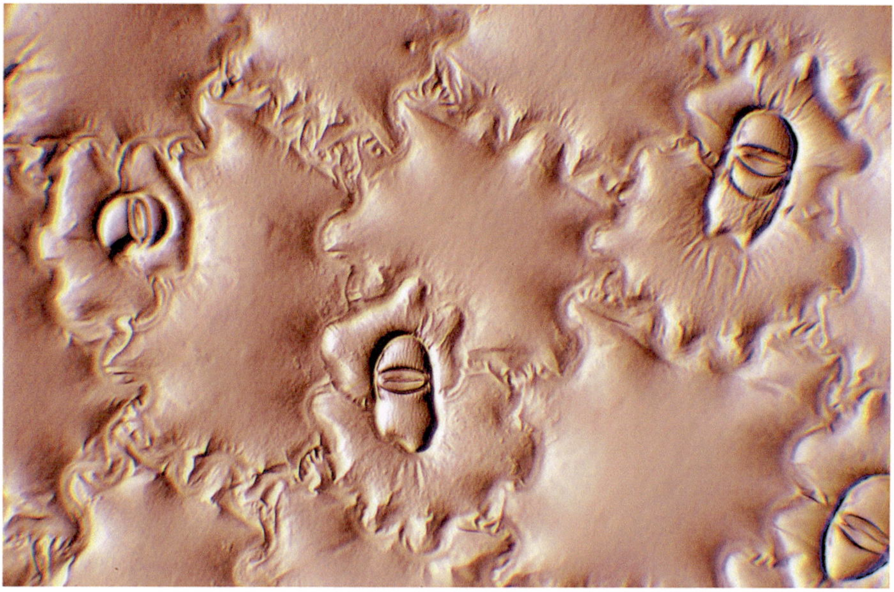

Fig. 9.22 Epidermis imprint of rose moss (*Portulaca grandiflora*) with paracytic stomata

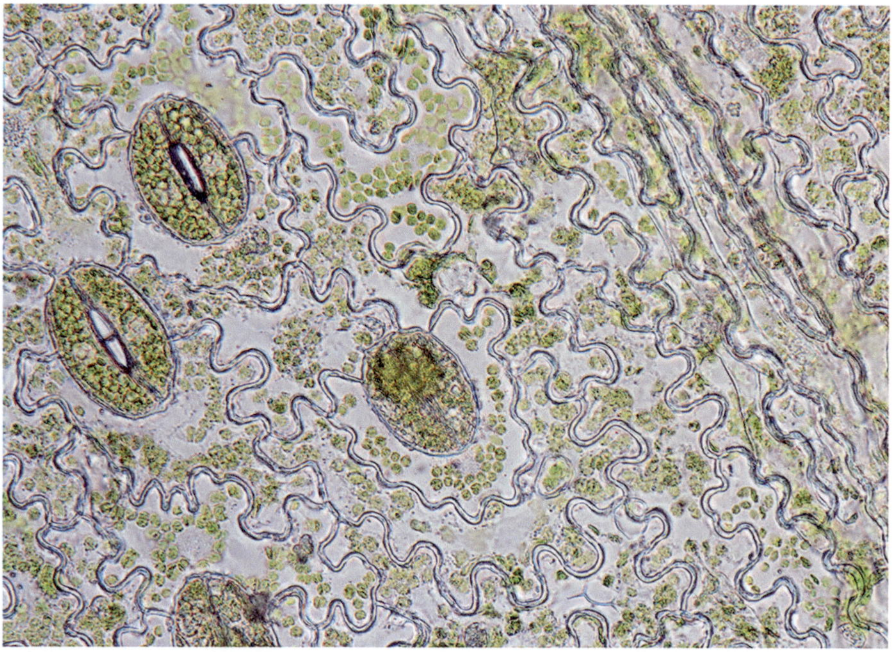

Fig. 9.23 The epidermal cells of male fern (*Dryopteris filix-mas*) contain chloroplasts

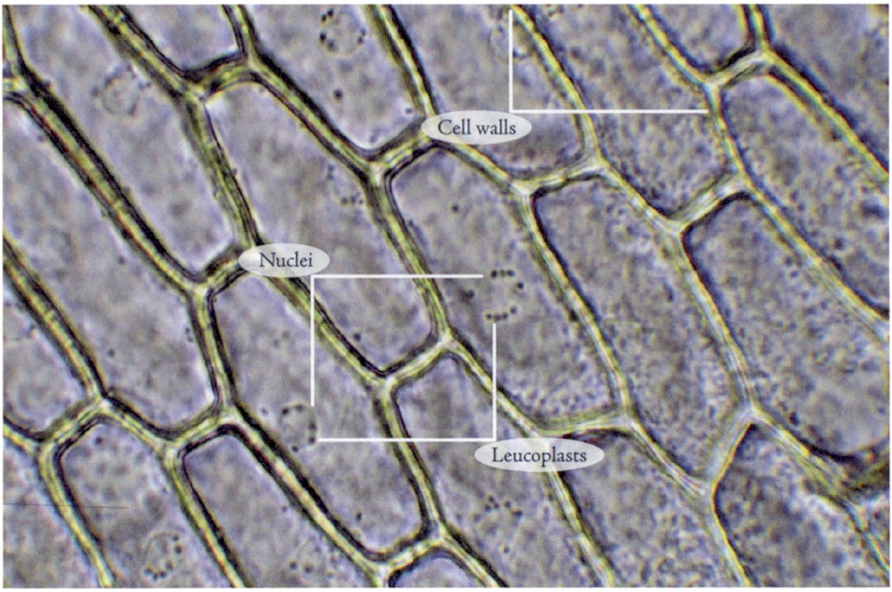

Fig. 9.24 *Sansevieria* epidermis with leucoplasts around the nucleus.

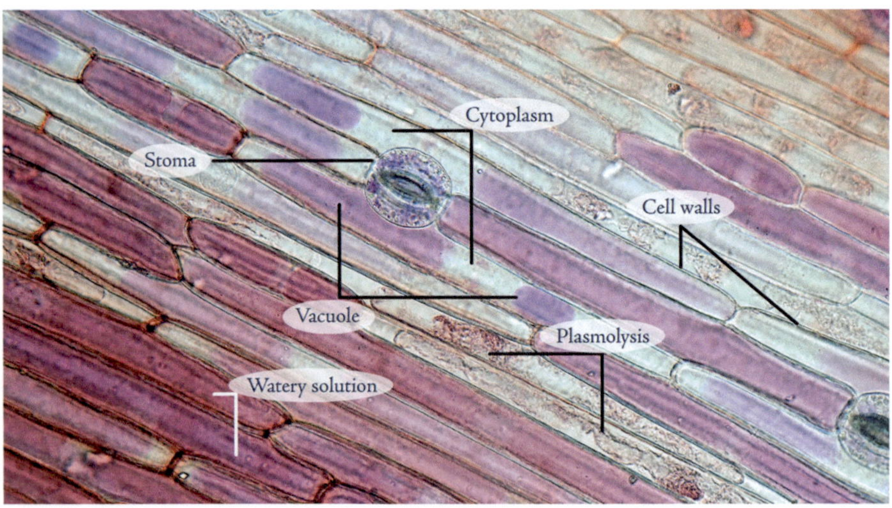

Fig. 9.25 Daffodil (*Narcissus* sp.) epidermis scrape stained with neutral red. The vacuole filled with cell sap dyed purple and the light cytoplasm are clearly separated

The characteristic formations of the epidermis are the stomata located between its cells, which perform and regulate the plant's gas exchange and transpiration (Fig. 9.26).

Fig. 9.26 The stomata can be clearly observed on the epidermis imprint of a lily (*Lilium* sp.)

The plant needs water, minerals, and carbon dioxide for photosynthesis. It takes up water and the mineral salts dissolved in it from the soil with its roots, while carbon dioxide comes from the air to the photosynthesising cells through the stomata.

In addition to the uptake of carbon dioxide, the stomata are important not only for the release of oxygen produced during photosynthesis, but also for the plant's ability to transpirate water through them. This cools the leaf, preventing it from overheating, and transpiration ensures continuous water replenishment from the root.

The stomatal pore is bordered by two bean-shaped (dicotyledons) (Figs. 9.22 and 9.26), or two dumbbell-shaped (grasses) (Fig. 9.27) **guard cell**s, which lead to the cavity inside the leaf, the air chamber.

From this air chamber, the air passages that network the inside of the leaf can start, which introduce carbon dioxide into the inside of the leaf and expel water vapour into the environment during transpiration. Often around the guard cells subsidiary cells of different shape from the other epidermal cells can be seen. The number and arrangement of **subsidiary cell**s is a hereditary characteristic, therefore it is an important property in the classification of plants. The stomatal pores, the guard cells, and the subsidiary cells form the functional and structural unit of the stomata.

In the following, we will get acquainted with the most common types of stomata. In some plants, the cells surrounding the guard cells do not differ from the other epidermal cells, so subsidiary cells are missing (anomocytic stomata). We can encounter this in the case of *Iris* (Figure 9.21) and *Pelargonium* genus. In those plants where the guard cells are surrounded by two subsidiary cells, there can be

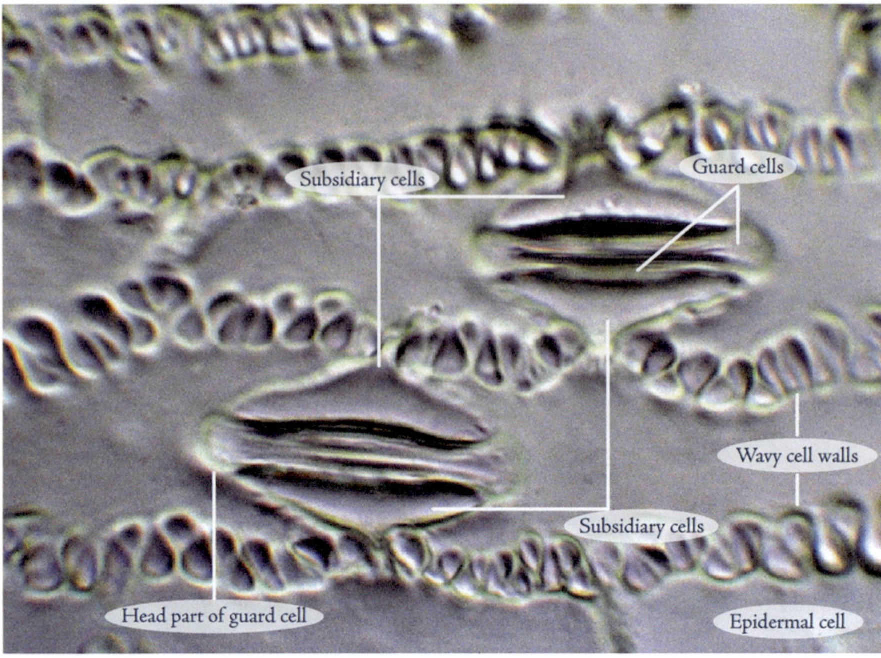

Fig. 9.27 Maize (*Zea mays*) epidermis extract with dumbbell-shaped guard cells

two types of arrangements. In the case of paracytic stomata, the two subsidiary cells are parallel to the stomatal pore (e.g. *Epipremnum* sp., corn, *Portulaca grandiflora*) (Figs. 9.22, 9.27, and 9.28), while in the case of diacytic stomata, the subsidiary cells are perpendicular to the **stomatal pore**.

Corn, like other grasses, has two subsidiary cells that are parallel to the stomatal pore. The tightly closing, wavy-walled epidermal cells, the tiny silica cells, which are almost completely filled with silicon dioxide (SiO_2), and the stomata can be clearly seen on the imprint of the corn epidermis.

In the latter, the two dumbbell-shaped guard cells, which are arranged parallel to each other, are surrounded by two triangular subsidiary cells.

When water is absorbed, the head part of the guard cells swells, so the stem parts of the guard cells move away from each other, and stomatal pore opens (Fig. 9.27).

Examining an imprint made from the tiny fleshy leaf of the *Portulaca grandiflora*, we can see a smaller and a larger subsidiary cell next to the guard cells, which are parallel to the stomatal pore. The stomata of the succulent plant are recessed (Fig. 9.22).

In the epidermis of the crucifers, the guard cells are surrounded by three subsidiary cells (**anisocytic stomata**). In addition to the crucifers, the above arrangement also occurs in other plant families. Very good test subjects are, for example, the *Begonia* and the *Sedum maximum*, from whose leaf we can easily make a peel. In the epidermis of the *Begonia* leaf, usually 4–5 stomata (guard cells and their subsidiary cells) form a group, which are surrounded by epidermal cells. Around a guard cell, the three subsidiary cells are marked with numbers (Figs. 9.29 and 9.30).

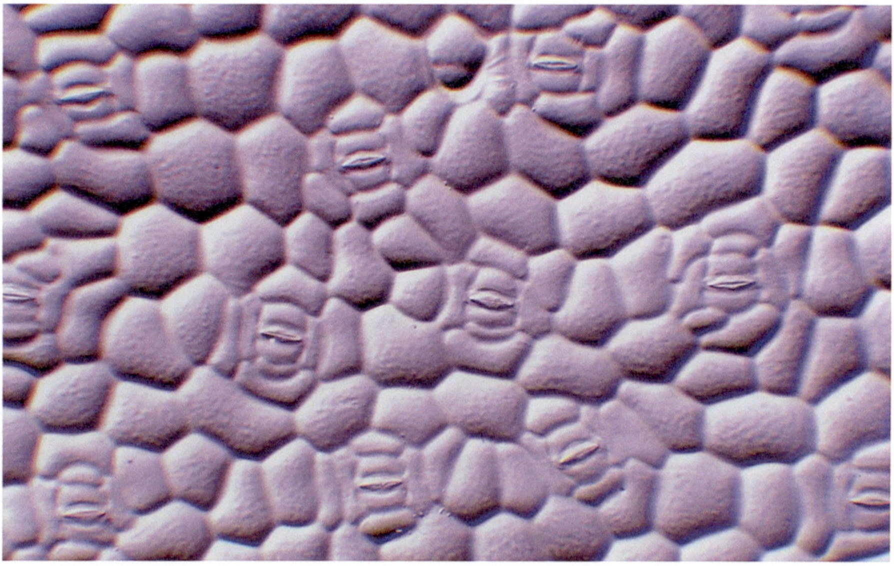

Fig. 9.28 Golden pothos (*Epipremnum*) epidermis imprint with paracytic stomata

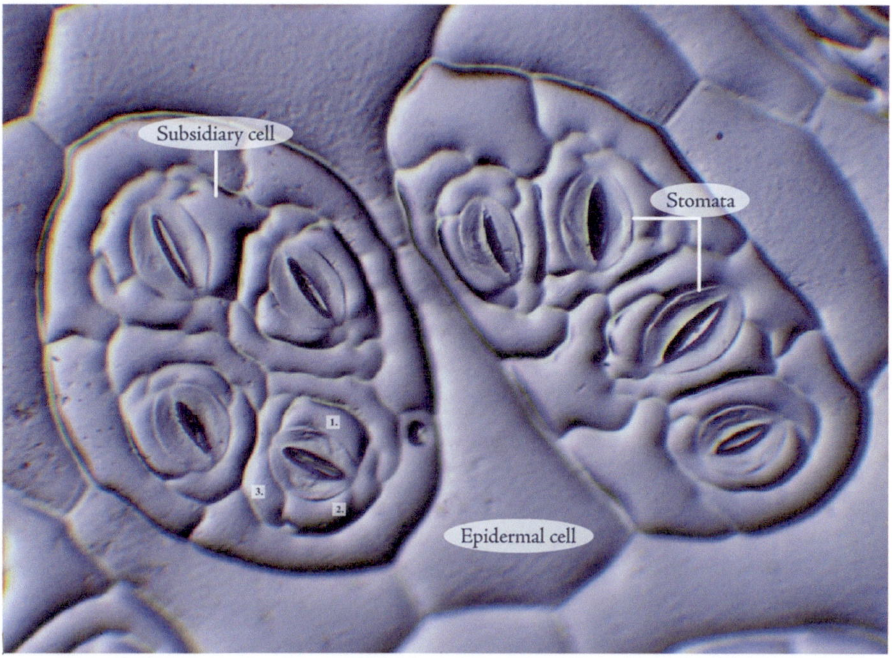

Fig. 9.29 *Begonia* epidermis imprint with anisocytic stomata

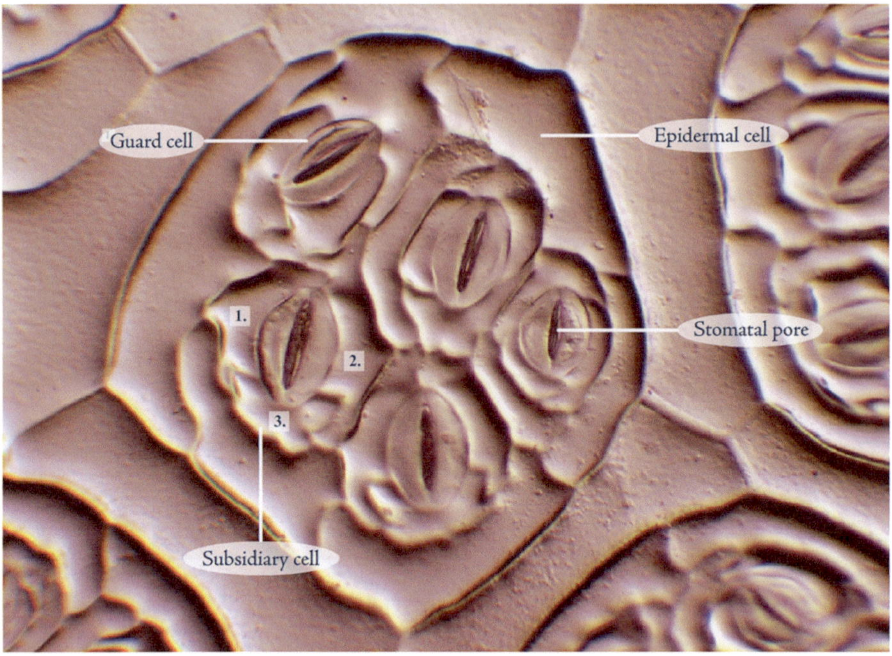

Fig. 9.30 *Begonia* epidermis imprint with anisocytic stomata

In the **tetracytic stomata**, four subsidiary cells are located around the guard cells. This type characterises the epidermis of the leaf of *Tradescantia*, which is popular as a houseplant (Fig. 9.31).

The **actinocytic stomata**, characterised by many elongated subsidiary cells, is typical for the Apocynaceae family. The operation of the stomata is influenced by several factors. One of these is the adequate water supply of the plant, as a result of which the guard cells absorb water, expand and the stomatal pore between them opens, through which the plant can release water by transpiration. In periods of water shortage, the guard cells lose water, the stomatal pore closes, and the amount of water released by transpiration decreases. The two states of the stomata can also be artificially created with solutions of different concentrations.

In the epidermis of grasses, smaller guard cells can be observed between the wavy wall epidermal cells (Fig. 9.32). Their cell walls, which contain silicon dioxide, strengthen the plant's leaf.

In the epidermis, we can sometimes observe cell groups or cell pairs performing the water secretion, these are the hydathodes, which are often very similar to the stomata. They also consist of two guard cells, but the space between them is always open.

The appendages of the epidermis are the papillae, the plant hairs, and the colleters. Papillae are the surface enlarging protrusions of the epidermal cells on the leaf and the petals (Fig. 9.33).

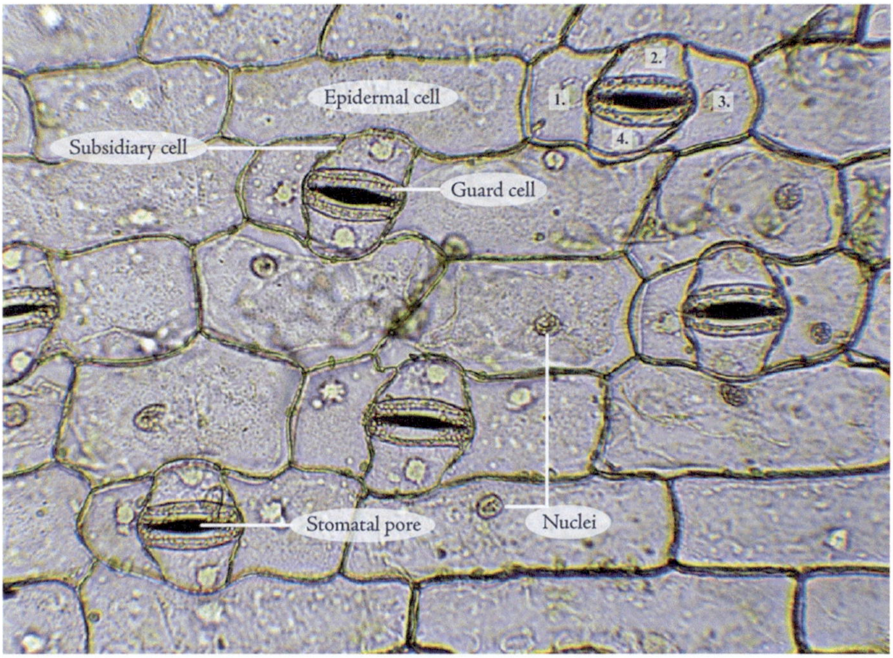

Fig. 9.31 *Tradescantia* epidermis with anisocytic stomata

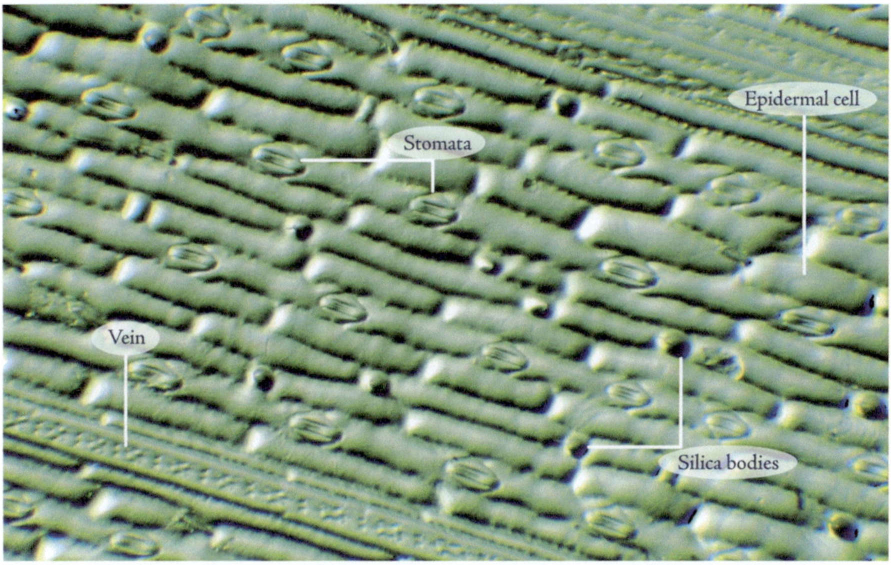

Fig. 9.32 The guard cells can be clearly observed on the epidermis of a grass

Fig. 9.33 The papillate tepal of the daffodil (*Narcissus*)

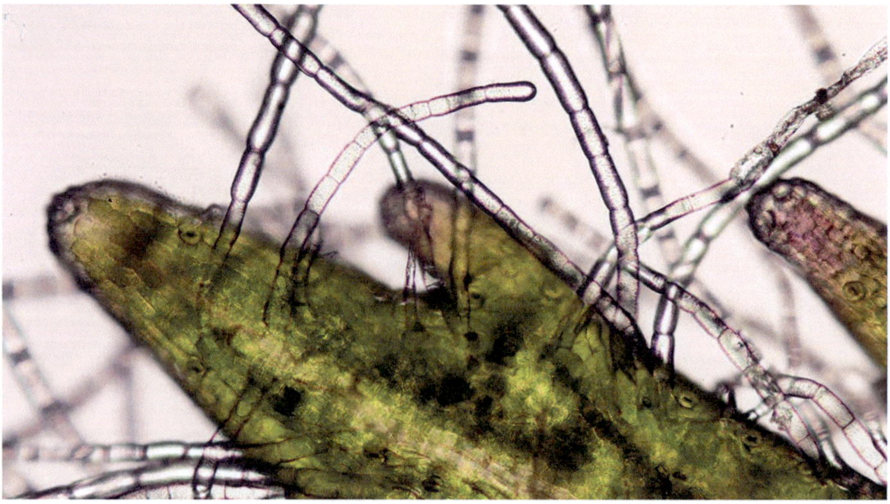

Fig. 9.34 Multicellular hairs on the bud of the spring ragwort (*Senecio vernalis*)

Especially in plants living in dry areas, the leaf underside, the lower epidermis is densely covered with plant hairs, which protect the leaf from overheating and excessive water loss.

The **hair**s (**trichomes**) provide mechanical protection for the plant, reduce transpiratory water loss, and protect the leaf from excessive heating (Figs. 9.34 and 9.35). The **glandular hair**s secrete and store various substances (Fig. 9.36).

Star-shaped cover hairs are the **stellate** and **scale hair**s, which, when dead, also give the plant a silvery appearance (e.g. *Elaeagnus angustifolia*) (Fig. 9.37). The stellate hairs consist of one or more base cells and the radial cells connected to them in a radial direction, lying in the plane of the epidermis. In the case of scale hairs, the large number of radial cells interconnect to form a shield-like structure (Fig. 9.38).

The **bristle hair** serves to defend against herbivores (e.g. *Papaver rhoeas*, Boraginaceae), the bifid hair for anchoring (*Humulus*), and the spreading hair for seed dispersal (e.g. *Gossypium*). A special epidermal formation is the unicellular

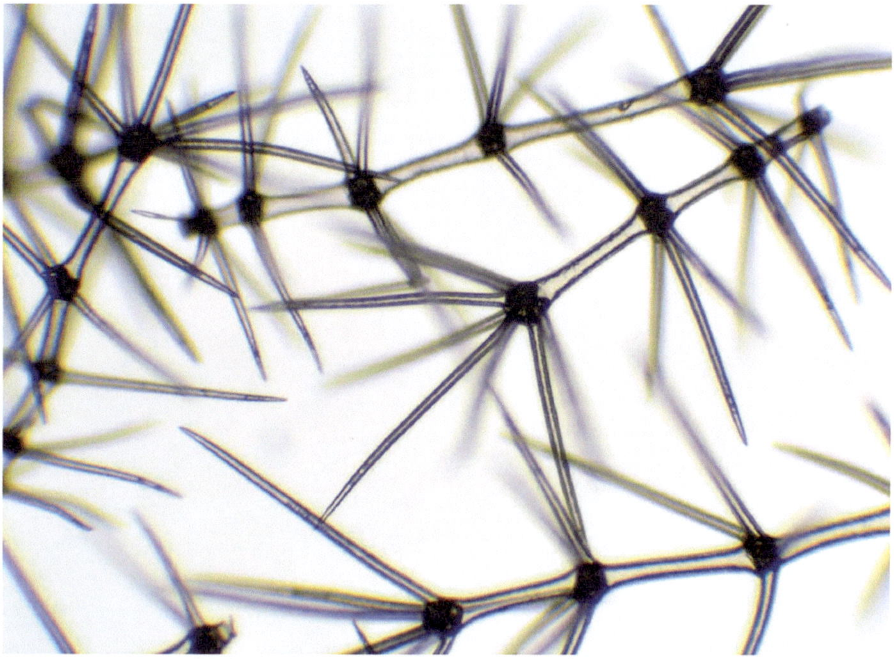

Fig. 9.35 Hairs of orange mullein (*Verbascum phlomoides*) leaf

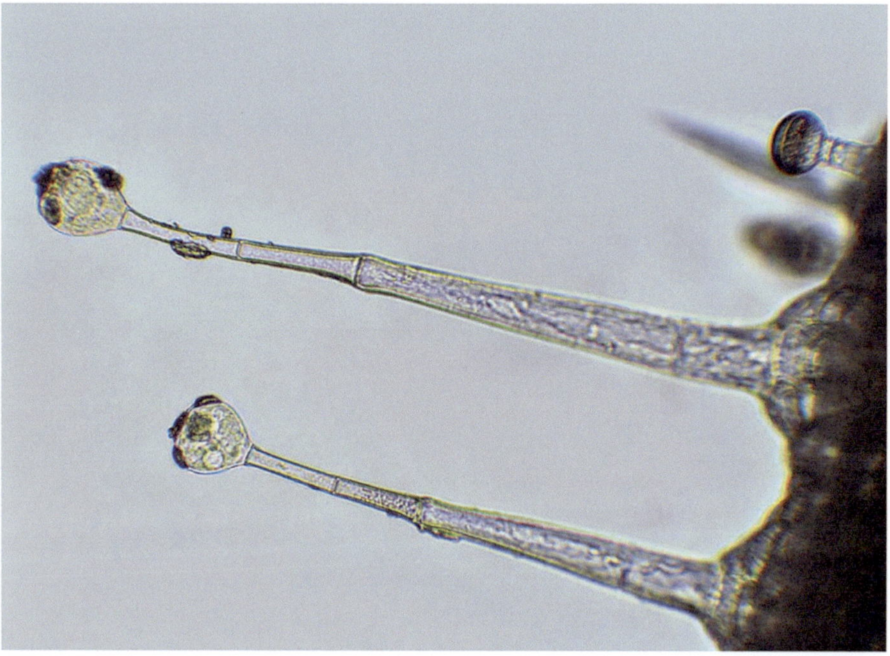

Fig. 9.36 Short and long-stalked glandular hairs of geranium (*Pelargonium*)

Fig. 9.37 Silver berry (*Elaeagnus angustifolia*) branch made silvery by scale trichomes

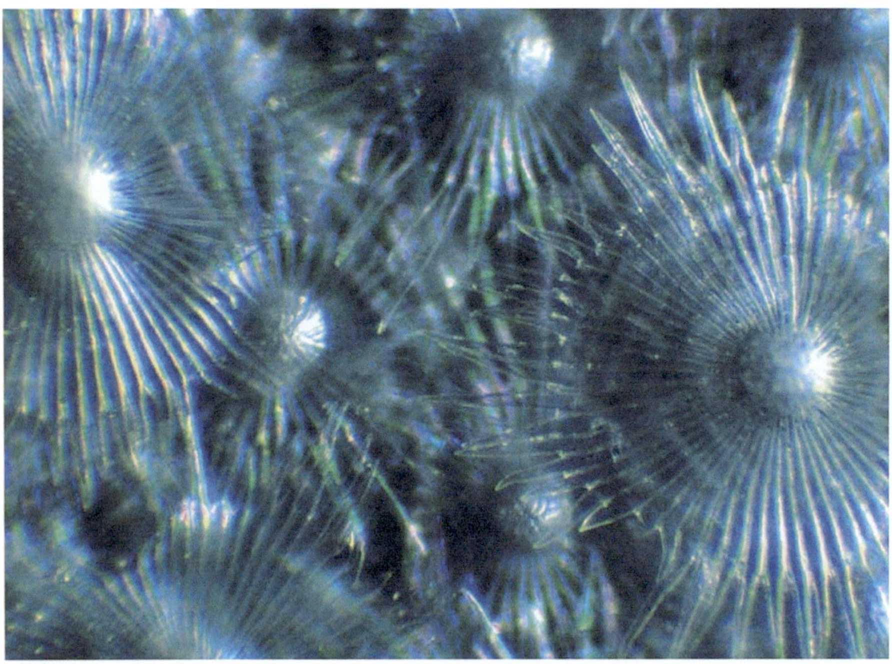

Fig. 9.38 Scale hairs covering the leaf surface of silver berry (*Elaeagnus angustifolia*) in multiple layers

Fig. 9.39 Stinging hair of common nettle (*Urtica dioica*)

stinging hair (e.g. *Urtica dioica*) (Fig. 9.39). Its widened base part dips into an epidermal tissue cushion, from which it extends long.

The cell wall of its narrowing end is glass-like fragile due to the accumulated silicon dioxide inside it, therefore it breaks even at the slightest touch. The broken, sharp stinging hair easily penetrates the human epidermis, while injecting its cell sap containing formic acid and histamine, which cause skin irritation, into the skin.

Emergences are those formations of the plant body in which the epidermis and the tissues underneath participate in their formation (e.g. rose thorn) (Fig. 9.40).

The root hairs are single-cell formations of the rhizodermis, which are continuously formed after the elongation zone, while they die towards the end of the maturation zone (Fig. 9.41). Thus, as the root tip advances, the maturation zone always comes into contact with new soil parts from where it can take up nutrients. The nutrients taken up by the root hairs, the water and the mineral salts dissolved in the water reach the vascular tissues through the root's cortex, from where they get to the site of photosynthesis, to the shoot.

9.2.2.2 The Secondary Dermal Tissue (Periderm)

On the thickening stem and root, the primary dermal tissue system is sooner or later replaced by a secondary dermal tissue the **periderm**. The periderm also occurs on many herbaceous, dicot older shoots and roots, but we rarely encounter secondary dermal tissue on monocots.

With the secondary thickening of the stem and the root, after a while the dermal layer tears, while plant hormones are released from its cells. These hormones trigger the formation of a new, secondary dividing tissue, the cork cambium, from the row of parenchyma cells under the tearing epidermis.

The **cork cambium** produces a protective tissue consisting of thick-walled cells outward (**cork**), and parenchymatous cells inward (**phelloderm**).

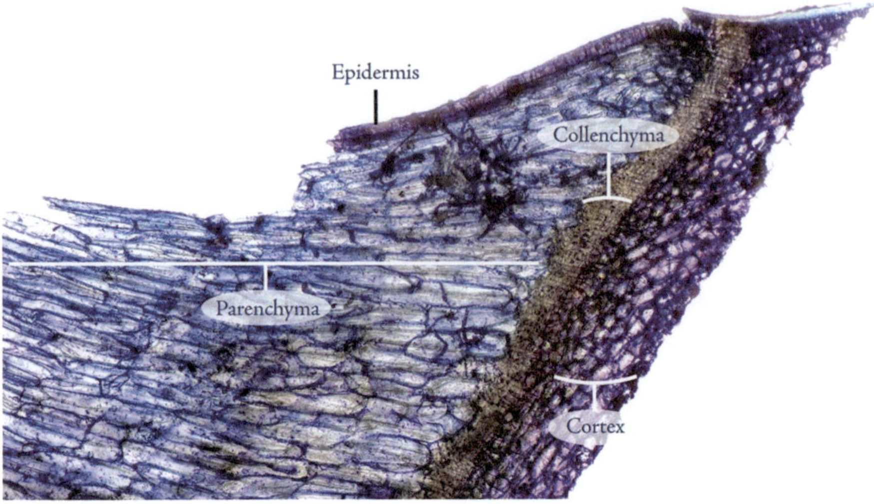

Fig. 9.40 Tissue structure of a rose thorn

Fig. 9.41 European frog-bit (*Hydrocharis morsus-ranae*) root hairs

The cork, the cork cambium, and the phelloderm are collectively referred to as secondary epidermal tissue or periderm (Figs. 9.20 and 9.42).

The phelloderm does not form in every case, in such cases the periderm only consists of the cork and the cork cambium (Fig. 9.16). The formation of the

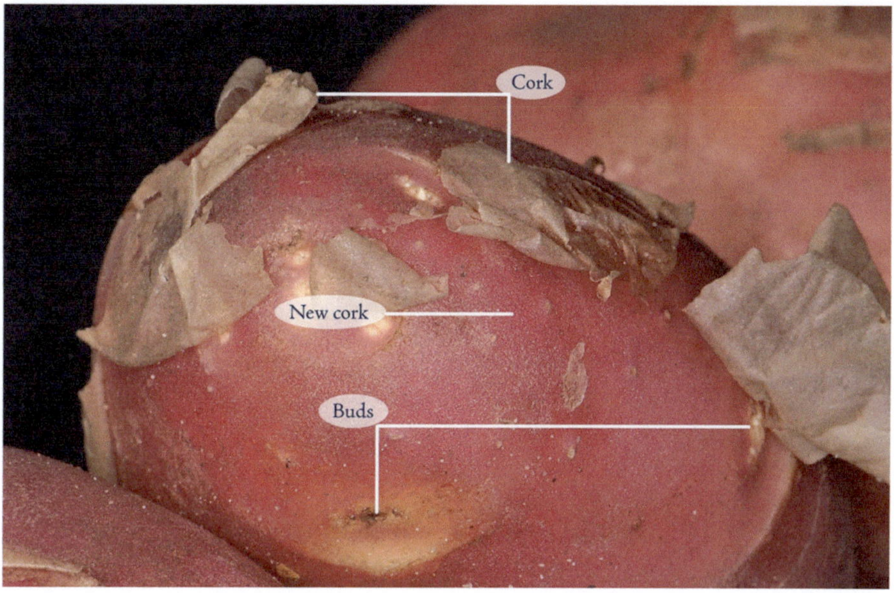

Fig. 9.42 The periderm can forms on underground stem modifications (e.g. potato tuber)

periderm in many plants begins in the epidermis itself, thus the primary dermal tissue transforms into cork cambium (Fig. 9.17).

The periderm reduces the plant's transpiration, protects the plant from parasites and overheating, but also isolates the stem from the air, therefore, to maintain ventilation, the so-called **lenticel**s develop on the stem. Lenticels mostly form under the stomata in such a way that the parenchyma cells bordering the air space lose their chloroplasts and, by dividing, form **complementary cell**s. These cells break through the skin tissue at the stomata and reach the surface. Thus, the lenticels visible to the naked eye form on the branches. Later, the formation of lenticels continues with the tearing of the epidermis and the formation of the periderm, but now the complementary cells are formed by division from the formed cork cambium. The periderm forms a barrier to air exchange, but air can enter the stem's interior through the lenticels. In the autumn, the cork cambium produces tightly closing cell layers instead of loosely placed complementary cells, which close the lenticels for the winter. The protective layer is broken through by the newly formed complementary cells in the spring (Fig. 9.18). The secondary skin tissue rarely remains functional for years (e.g. beech), usually a new periderm forms every year. In some tree species, such as cherry and birch, where the cork cambium closes into a ring, the surrounding cork forms a closed cylinder, and a so-called ringed bark forms.

In other cases, such as with the plane tree, the newly formed cork layers rely on the older ones, which peel off the tree trunk in a scale-like manner. The separation of the periderm layers is generally made possible by a separating layer consisting of parenchyma cells.

9.2.2.3 Tertiary Dermal Tissue

The annually regenerated phloem ring pushes out the previous year's part of the phloem, which no longer performs material transport, and its cells are compressed, clogged, hardened, and die. The outwardly pressed dead phloem part cracks due to the peripheral expansion, and the surrounding parenchyma cells are pushed into the resulting cracks.

In the case of the linden, the cracks in the phloem part are filled by the radially located parenchyma cells, forming **medullary rays**, which are clearly visible to the naked eye on the cross-section of the branches (Fig. 9.43).

These prevent the cracking of the bark and the dead cork part during the thickening of the stem, thus the outer surface of the tree trunk remains smooth. In other trees, the periderm breaks due to the tensile forces of the thickening stem, and the newer cork cambium develops new periderm. After a while, with the depletion of the cortex, no new cork cambium can form in the cortex, so at this time, meristem is formed from the parenchyma cells of the phloem, which produces the tertiary dermal tissue, the **rhytidome**. This is the bark of the tree in the everyday sense, which also contains dead phloem in addition to the cork. In botanical terms, the bark also includes the living phloem. The appearance of secondary and tertiary dermal tissue can often be observed on the same plant. The thicker branches of the old birch tree have ringed bark, covered with periderm, while the trunk is covered with cracked rhytidome. The rhytidome may contain tannins, crystals, resin, and various biologically active substances (Fig. 9.19).

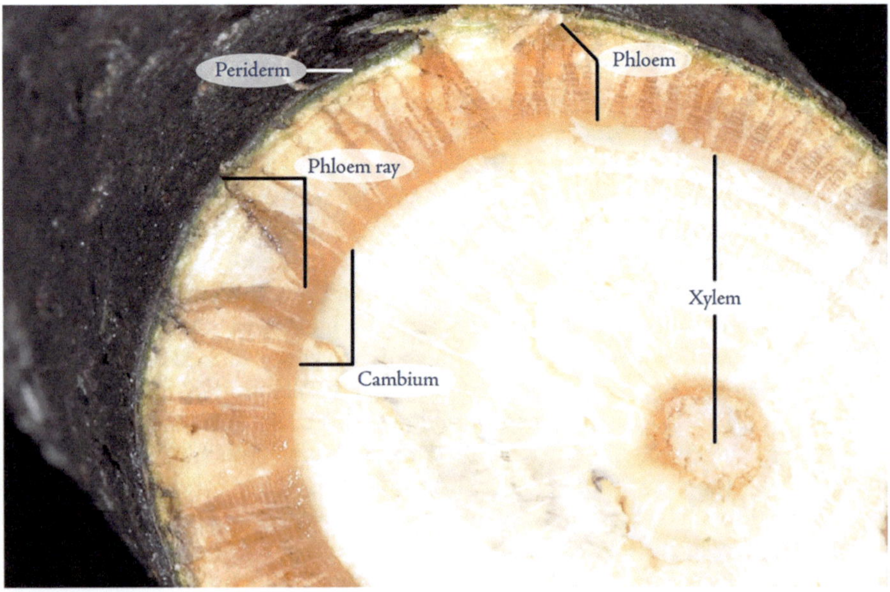

Fig. 9.43 Cross-section surface of a linden (*Tilia* sp.) tree branch

9.2.3 Ground Tissue System

9.2.3.1 Parenchyma

Parenchyma is with rich intercellular spaces and its cells have generally equal extension in different spatial directions (isodiametric) (Fig. 9.44).

They easily regain their ability to divide and transform into secondary meristem, or transform into other tissues with cell wall thickening. We categorise them into different groups according to their function: photosynthesising, storing, water-holding, and ventilating.

The assimilatory parenchyma or **chlorenchyma** is located under the epidermis of the leaf and stem. Its cells filled with chloroplasts give the green colour to the plant (Figs. 9.45 and 9.46). In stems and the leaves of grasses, its cells are isodiametric, while in the leaves of dicots, they form palisade or spongy parenchyma.

The **storage parenchyma** is found in underground organs, seeds, and fruit walls. Its cells store nutrients mostly in leucoplasts (Fig. 9.47). The water storage parenchyma characterises succulent plants living in dry areas. The cell cavity in its cells is filled with mucus, which can bind a large amount of water (e.g. *Portulaca grandiflora*, *Sedum* sp.).

The **aerenchyma** rich in intercellular spaces appears in the submerged organs of aquatic plants. Its task is to deliver oxygen to organs isolated from the air (Fig. 9.48).

Special parenchyma cells form the innermost cell row of the primary cortex of stems, the **starch sheath**, which can be considered as storage parenchyma (Fig. 9.49).

The single-layered **endodermis**, composed of cells with peculiarly thickened walls, is the innermost cell layer of the primary cortex of young roots, underground shoots, and aquatic plants (Fig. 9.44).

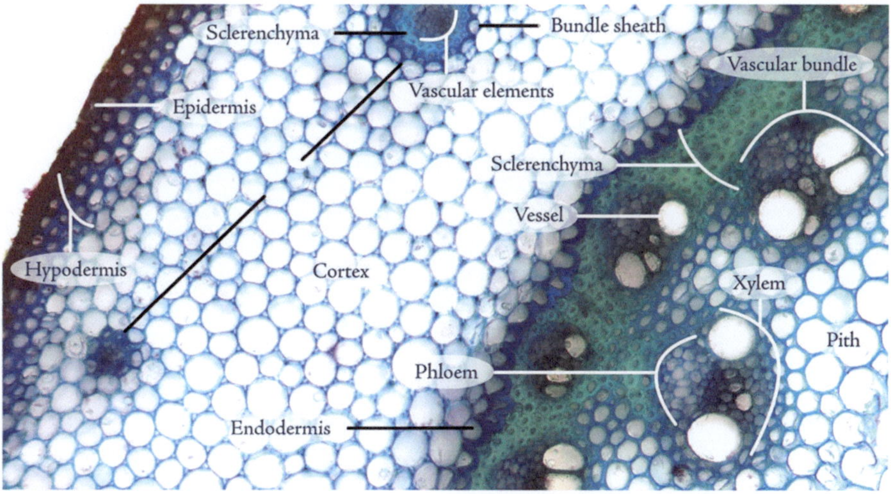

Fig. 9.44 The cortex and pith of the reed (*Phragmites australis*) rhizome is formed by parenchyma

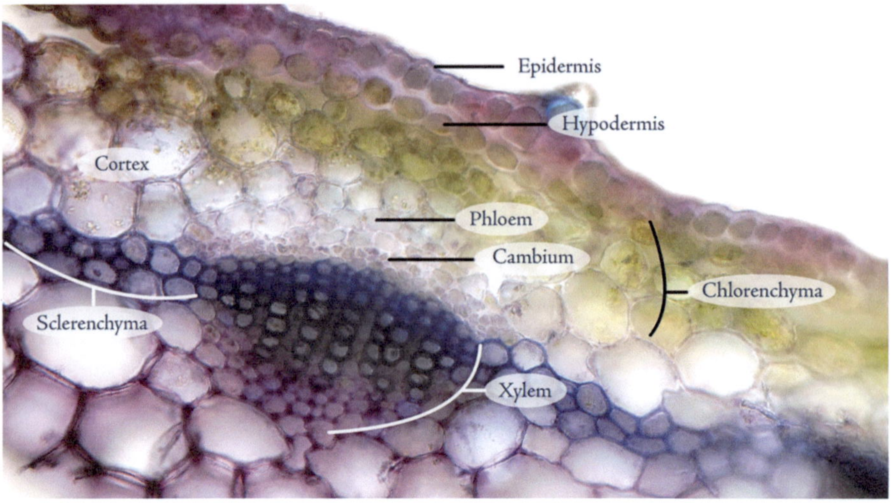

Fig. 9.45 Isodiametric parenchyma cells (chlorenchyma) under the epidermis of red deadnettle (*Lamium purpureum*)

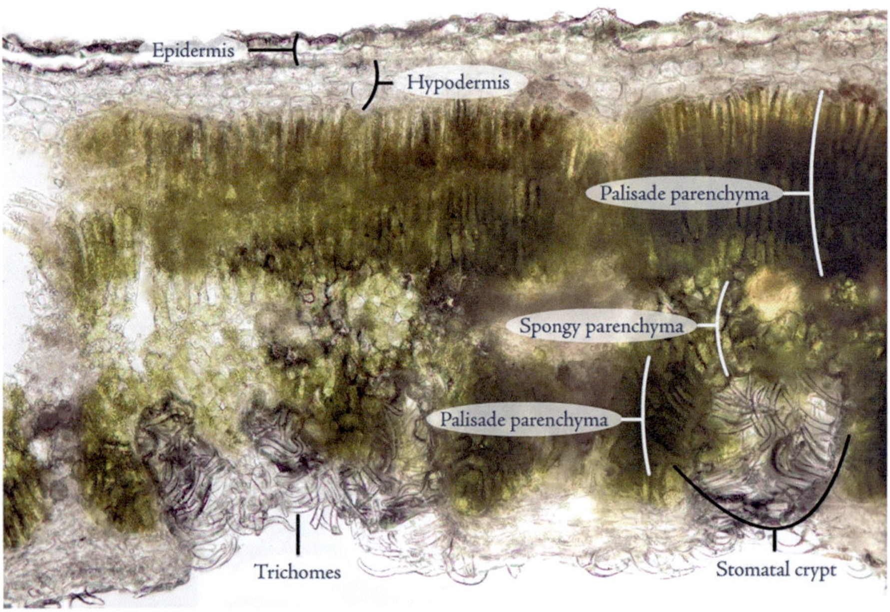

Fig. 9.46 Palisade and spongy parenchyma in a cross-section of oleander (*Nerium oleander*) leaf

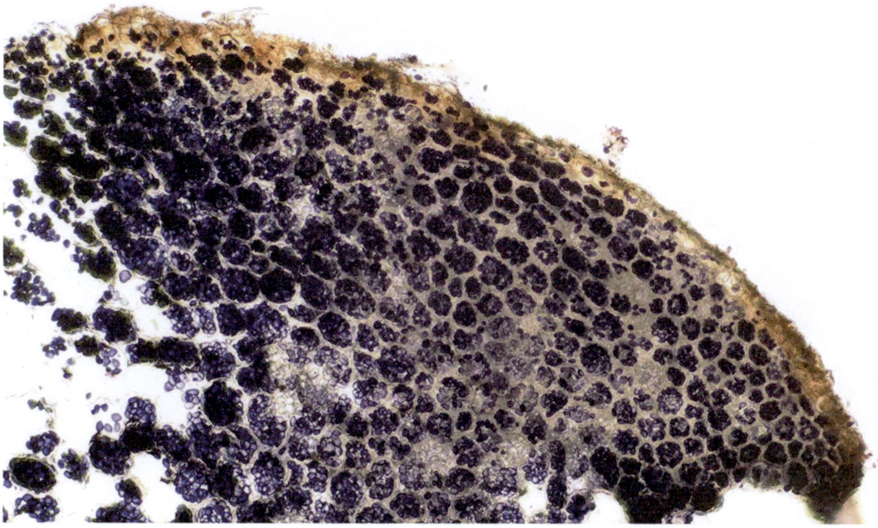

Fig. 9.47 The thick cortex part of lesser celandine (*Ficaria verna*) root tuber is made up of storage parenchyma cells packed with starch grains. The starch grains were stained blue with iodine solution

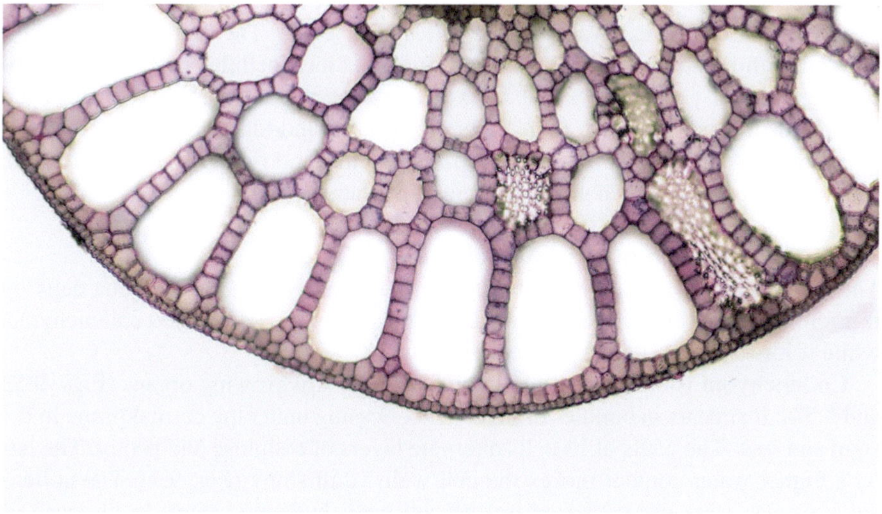

Fig. 9.48 Aerenchyma in the cortex of crisp-leaved pondweed (*Potamogeton crispus*)

9.2.3.2 Secretory Parenchyma

The excretory function of plants can be intracellular, when the metabolic products concentrate in the cell wall or in the cell cavity (e.g. crystal) (Fig. 9.50).

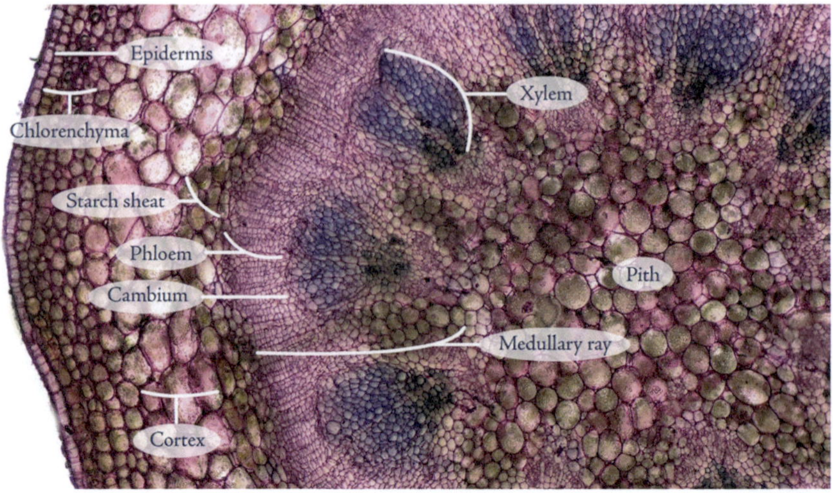

Fig. 9.49 Cross-section of jimsonweed (*Datura stramonium*) stem

The other option is when groups of cells forming **secretory parenchyma** deliver their products to the intercellular spaces, which can appear on the surface of the plant (hydathodes, salt glands, nectaries, and osmophores) (Fig. 9.51) and inside (secreting cells, secretory cavities, and laticifers). The latter include **laticifer**s, as well as **resin duct**s and lysigenous cavities (Fig. 9.52). The plant's body is networked with branching unicellular **latex tube**s or multicellular latex vessels containing white or yellow latex. The latex has many components. Some of these are contained in dissolved form (e.g. sugars, enzymes, and alkaloids). Other components are present as tiny droplets (e.g. oils, proteins, and rubber), or in solid form, such as microscopic starch grains present in the latex.

9.2.3.3 Supporting or Mechanical Ground Tissues

The supporting tissues are basic tissues formed from young parenchyma cells by thickening of the cell wall. If the thickening occurs locally, it is called collenchyma, while if it is uniform, it is called sclerenchyma.

Collenchyma is the supporting tissue of young, still growing organs (Figs. 9.53 and 9.54). It appears in bundles or in a cylinder casing under the dermal tissue in the stem and root. The walls of its cells alternate layers of cellulose and pectin. The latter's higher water content makes the cell walls' dull shiny (Fig. 9.7). The collenchyma types were named based on the cell wall thickening type. In the case of angular collenchyma, the longitudinal cell wall thickenings appear in cross-section in the form of angular thickenings. In the case of lamellar collenchyma, the walls parallel to the surface are thickened, while the radials are thinner.

In dicots, a lamellar collenchyma often forms under the single-layered epidermis, which forms a uniform strengthening cylinder called the hypodermis (Fig. 9.45).

Fig. 9.50 Columnar calcium oxalate crystals in the dry scale-leaf of onion (*Allium cepa*)

Fig. 9.51 The petals of black cumin (*Nigella sativa*) was transformed into secretory structures (nectaries)

Fig. 9.52 Yellow exudation (latex) flows from the freshly cut stem of great celandine (*Chelidonium majus*)

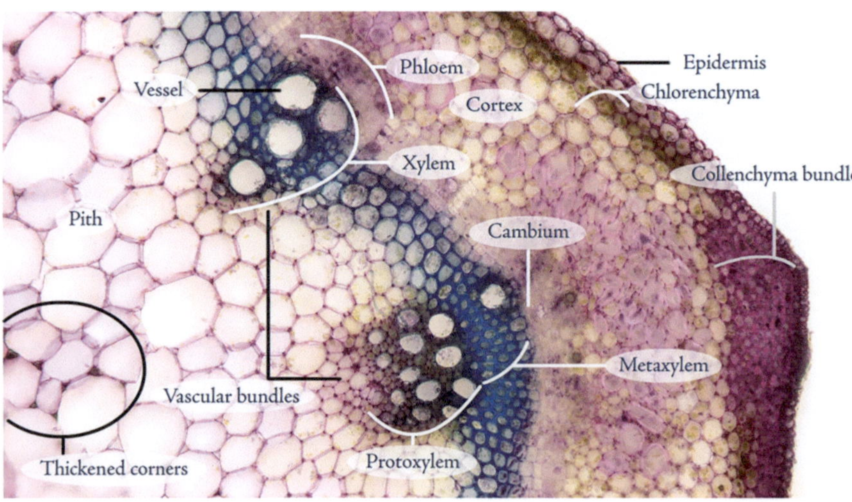

Fig. 9.53 Cross-section of common nettle (*Urtica dioica*) stem

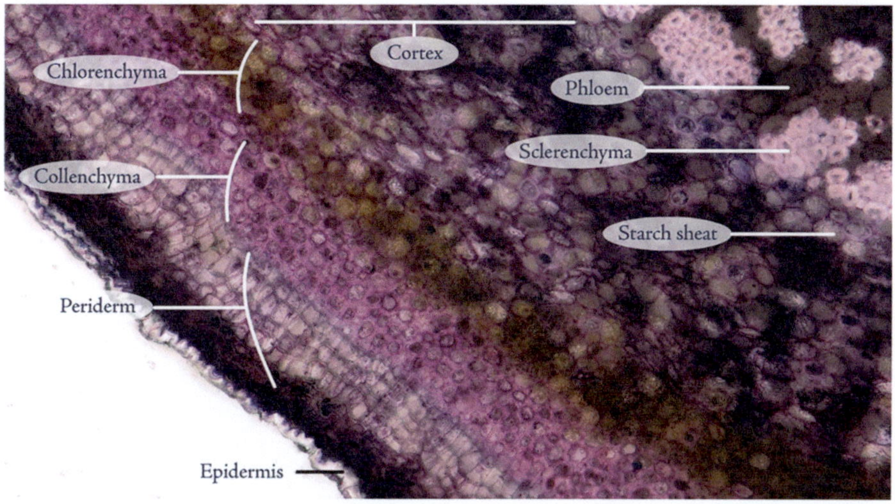

Fig. 9.54 Cross-section of oleander (*Nerium oleander*) stem with annular collenchyma

In monocots, the hypodermis under the skin tissue can be provided by sclerenchyma tissue (Fig. 9.44).

In the case of lacunar collenchyma, the part of the cell walls facing the intercellular spaces is thickened. The wall of the cells of the annular collenchyma is thickened circularly, like the **sclerenchyma**, while at the same time the cell walls perpendicular to the longitudinal axis remain and their cells are alive (Fig. 9.54). The wall of the cells of the sclerenchyma is thickened evenly, incapable of elongation in the mature state, and consists of dead cells. Its two cell types are the elongated fibre (Figs. 9.44 and 9.45) and the generally isodiametric sclereids, e.g. **stone cells**. In the cells of organs in various places, channels and pits can form in the cell wall (Fig. 9.55).

9.2.4 Vascular Tissue System

The transportation of water in plants is ensured by vascular tissues, which completely network the plant body. Vascular tissues are largely composed of elongated cells, tubes formed by the fusion of cells, and strengthening fibres.

The **xylem** of the vascular tissue system transports water and salts from the root towards the leaves, while the phloem transports organic nutrients towards the living cells. Both xylem and **phloem** are complex tissues containing parenchyma and strengthening tissue, sclerenchyma, in addition to transport elements (Fig. 9.8). The elongated transport elements of the xylem consist of lignin-containing thick-walled cell frameworks in their mature state. However, the transport elements of the phloem are provided by living cells, whose walls are less thickened and rarely contain lignin.

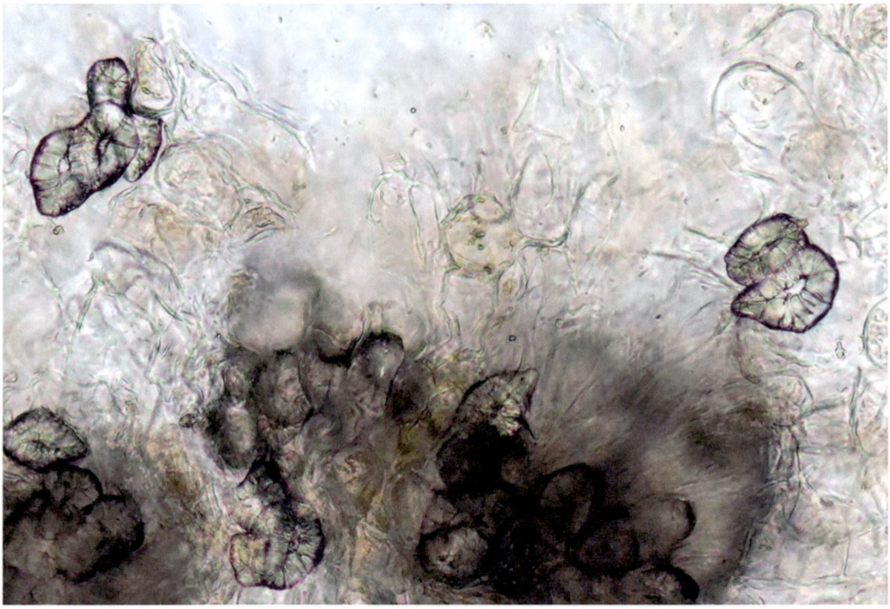

Fig. 9.55 Stone cells from the mesocarp of a pear (*Pyrus communis*)

The primary xylem is formed from the apical meristem. The first part of the xylem to form is the **protoxylem**, which is located centrally in the stems, internally, and peripherally in the roots, externally. Due to the growth of the organ, it soon tears and compresses, thus only participating in material transport for a short time. Its place is taken over by the **metaxylem** (Fig. 9.56).

In organs capable of secondary thickening, secondary xylem is formed from cambium (Figs. 9.4, 9.10, 9.14, 9.53, and 9.57).

The cellular elements of the xylem are the **water-transporting cell**s (**tracheid**s), the **water-transporting tube**s (**vessel** and **trachea**), the xylem parenchyma cells, and the xylem fibres.

The tracheids are angular in cross-section, cytoplasm-free, obliquely ending elements. They have characteristic cell wall thickenings (annular, spiral, reticulate, scalariform, and pitted). In pteridophytes and gymnosperms, with a few exceptions, they are the only water-transporting elements (Fig. 9.58).

The vessel is formed from several cells elongated in the direction of transport (trachea segments) in such a way that the separating transverse walls are partially or almost completely perforated. The trachea in its mature state does not contain cytoplasm, and its cell wall characteristically thickens (Fig. 9.59).

In the trachea, due to the absorption of the transverse wall, the migration of substances takes place much faster than in the water-transporting cells.

Beside or between the tracheids and vessels, we find the living, cytoplasm-containing, thinner-walled parenchyma cells of the xylem.

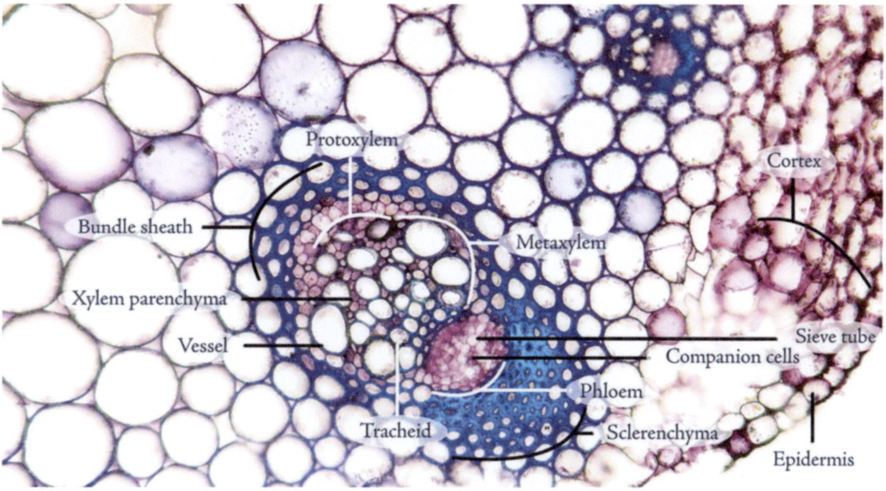

Fig. 9.56 Protoxylem and metaxylem in the vascular bundle of meadow buttercup (*Ranunculus acris*)

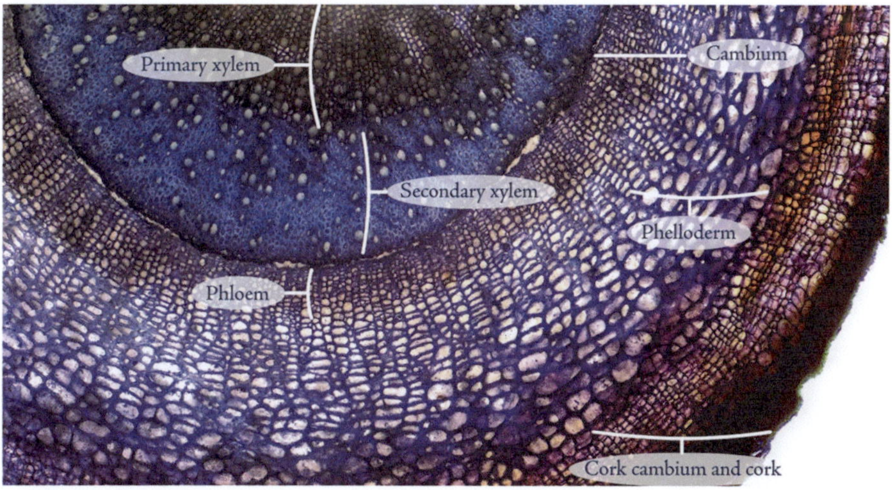

Fig. 9.57 Primary and secondary xylem in the secondary thickening root of the garlic mustard (*Alliaria petiolata*)

Each xylem parenchyma element consists of a few elongated cells arranged in a longitudinal row, the terminal cells often taper. This can mainly be observed in longitudinal section. The xylem parenchyma transports and partly stores the finished nutrients.

Solidifying fibres are often found in the xylem. These are long, pointed cells with thicker cell walls. Once fully formed, they do not contain cytoplasm, thus they are dead. They are typically absent in pteridophytes and gymnosperms.

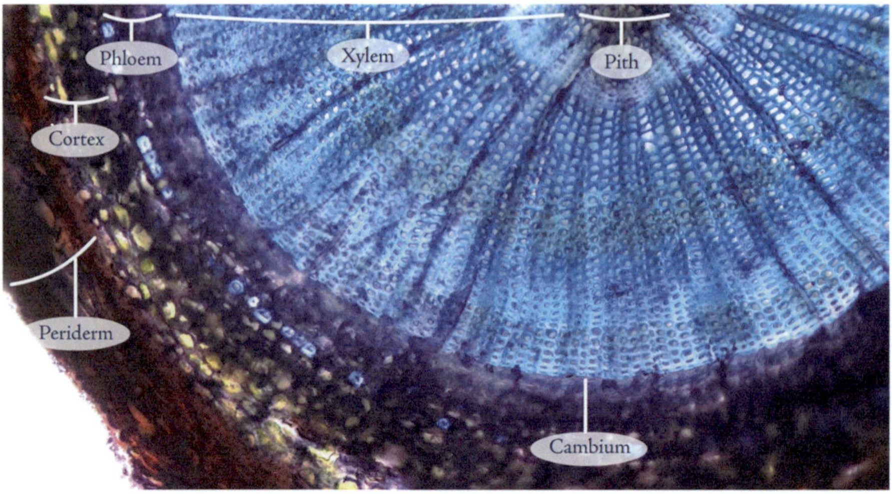

Fig. 9.58 Cross-section of *Thuja* branch

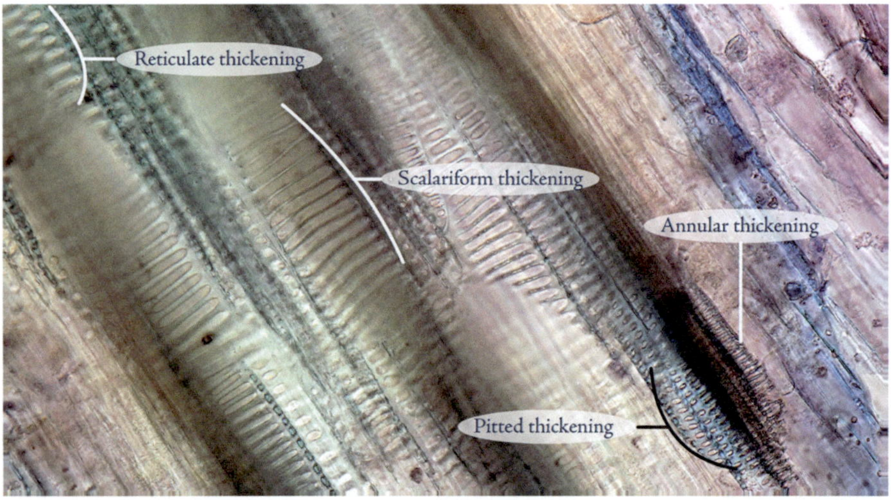

Fig. 9.59 Cell wall thickens in lily (*Lilium*) root vessels

In the **phloem** part of the transport tissue system, we often find the phloem elements performing transport, storage, and support, according to the triple function. The transport elements of phloem differ significantly from the transport elements of the xylem in both function and structure.

First, we mention the **sieve cell** as a more primitive form. It consists of a single cell. It is located next to or above its fellow cells. The sieve cells are elongated, sometimes narrowing at their ends, have thin cellulose walls, always contain cytoplasm, and their side walls are characteristically networked or pitted in places. The ready-made nutrients migrate from one sieve cell to another through cytoplasm threads penetrating the mesh or pits. The sieve cell mainly occurs in gymnosperms, as well as in some dicots. From the fusion of several cells, usually elongated and always containing cytoplasm, another important and much more common element of the phloem, the **sieve tube**, is formed. The side walls of its constituent cells, the sieve tube members, are made of cellulose, usually thin, but their cross walls, or sometimes both, are partially absorbed in a sieve or fibrous manner. The perforations occur within smaller or larger fibre fields. The cell wall part containing the perforations is called a sieve plate. The tubular cytoplasm of the sieve tube members is in direct contact with each other through the holes of the sieve plate.

They usually function for one, sometimes several years, then the sieve plates are covered and sealed by a thinner or thicker layer of callus.

In addition to the sieve tubes, the phloem has very characteristic elements, the **companion cells** with abundant cytoplasm, which store reserve nutrients. The thin-walled phloem parenchyma cells, which are similar to the xylem parenchyma, also perform a similar function. The listed elements are complemented in most plants by supporting, thicker-walled, narrow-cavity, cytoplasm-free, often very elongated phloem fibres. In addition to their position and significant length, they differ from the xylem fibres in that their cell walls are not, or only slightly, lignified.

The Tissue Structure of Plant Organs 10

10.1 Root

The root is a radially symmetrical organ for anchoring the plant in the soil and storing inorganic nutrients (water and mineral salts) and organic matter. Roots are present in most shoot plants, only a few species are rootless (e.g. *Utricularia*, *Ceratophyllum*, and *Salvinia*).

The root, which develops from a radicle, never bears leaves and, unlike the shoot, has no stomas in its dermal tissue. The apical part of the root is protected by a characteristic formation, the root cap (Fig. 10.1).

The outer cell wall of the root cap is sloughed off, which makes it much easier for the growing root to advance in the soil. As the root progresses, the cells of the root cap are constantly breaking off and disintegrating, forming a slimy, gelatinous layer at the root tip. The detached cells are replaced from the inside by regenerating cells from the apical meristem.

The cells in the root cap sense the Earth's gravitational direction through the starch granules inside them, and therefore play an important role in the root growing towards the centre of the Earth. The root is formed from the promeristematic region.

The cells of the apical meristem can be observed behind the sloughing root cap, from which the rhizodermis, the primary cortex, and the stele containing the vascular tissue develop above the root apex.

The apical meristematic cells are rounded, large nucleated, and have a full plasma. The cells of the root apex, which is covered by the root cap, continuously develop the root by a series of mitosis. This part is the region of cell division.

In the elongation region of the root, the tissues of the dermis, the cortex, and the stele begin to separate and form the functional organ. The elongation area is followed by the maturation region, which can be recognised by the appearance of root hairs. Root hairs (Fig. 10.2) are unicellular formations of rhizodermis that form continuously after the elongation region and die towards the end of the maturation area.

G. Kriska, *Teaching Biology*, https://doi.org/10.1007/978-3-031-84085-2_10

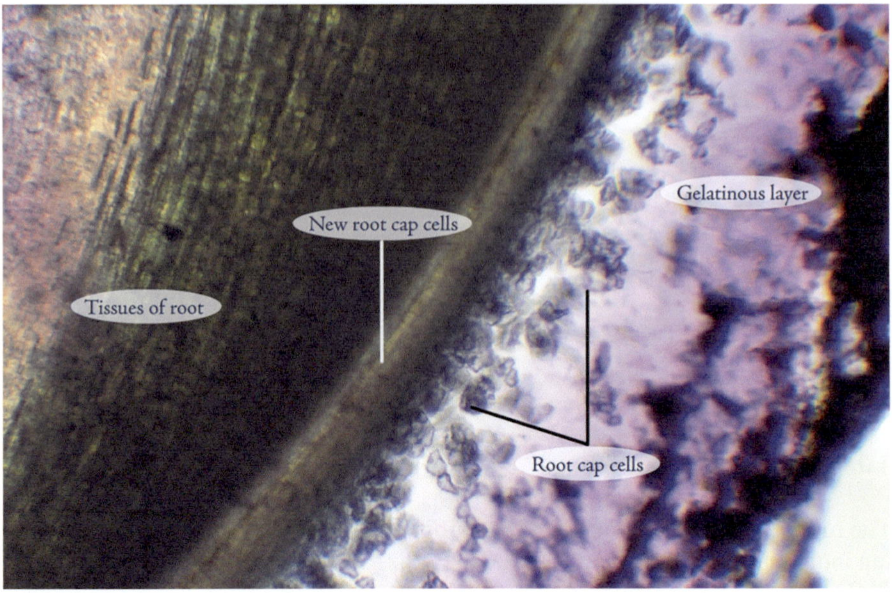

Fig. 10.1 Longitudinal, radial section of a maize (*Zea mays*) aerial root stained with toluidine blue

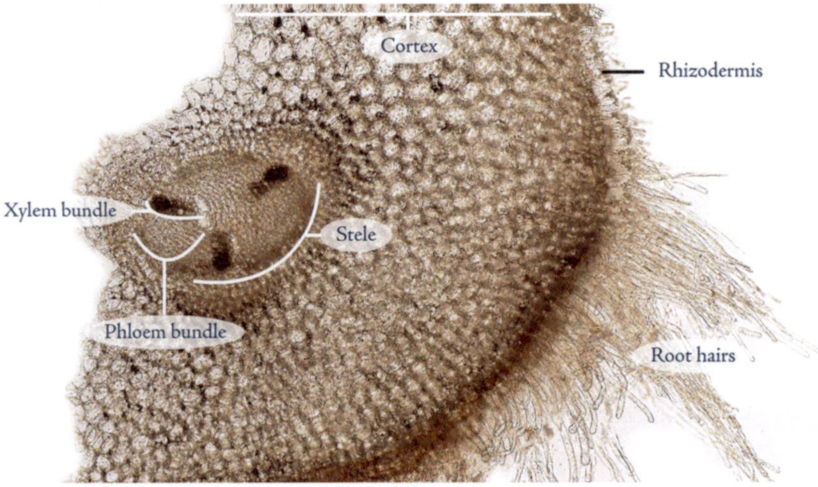

Fig. 10.2 Cross-section of a young pea (*Pisum sativum*) root

Thus, as the root apex progresses, the maturation zone is in contact with new soil parts from which it can take up nutrients. The nutrients taken up by the root hairs, water and water-soluble mineral salts, are transported through the root's cortical tissue to the vascular tissues, from where they are transported to the photosynthesis site, the shoot.

The next part is the branching region, where the lateral roots start to form. Further away from the root apex, in dicots, is the secondary growth zone, which transmits ingested nutrients, anchors and consolidates the plant, and stores nutrients. Beneath the single-cell layer of dermal tissue, the rhizodermis, is the primary cortex, which consists of parenchymatous tissue richly interconnected by intercellular passages. Above the maturation region, the outermost cell of the cortex, the exodermis, takes over the protective role as the rhizodermis is sloughed off.

The innermost cell of the cortex is the endodermis (Figs. 10.3 and 10.4), whose cell walls are thickened and are crucial in regulating water uptake.

Between cells with highly thickened walls, usually in the direction of the simple xylem bundles, a single cell with a thin wall remains. These are the so-called **passage cells**, which play an important role in the plant's nutrient cycling. In addition to the root, endodermis may also be present in underground modified stems, for example in the rhizome of lily of the valley (*Convallaria majalis*).

In general, the anatomical structure of all young roots is characterised by a relatively small diameter central **stele** in the middle of the organ, usually surrounded by a well-developed primary **cortex** (Fig. 10.2).

Below the endodermis is the outermost cell of the stele, the pericycle, which may play an important role in the further organisation of the root (Fig. 10.4).

As the root's vascular bundles form, the stele's meristem gives rise to separate phloem bundles, followed by the xylem bundles (Figs. 10.2, 10.3, and 10.4).

The first to appear are the **protophloem**, consisting of the narrower phloem elements, and the **protoxylem**, consisting of the narrower xylem elements (Figure 10.4). Further inwards, towards the centre of the stele, the **metaphloem** and **metaxylem**, consisting of the broader phloem and xylem elements, develop. This structure is

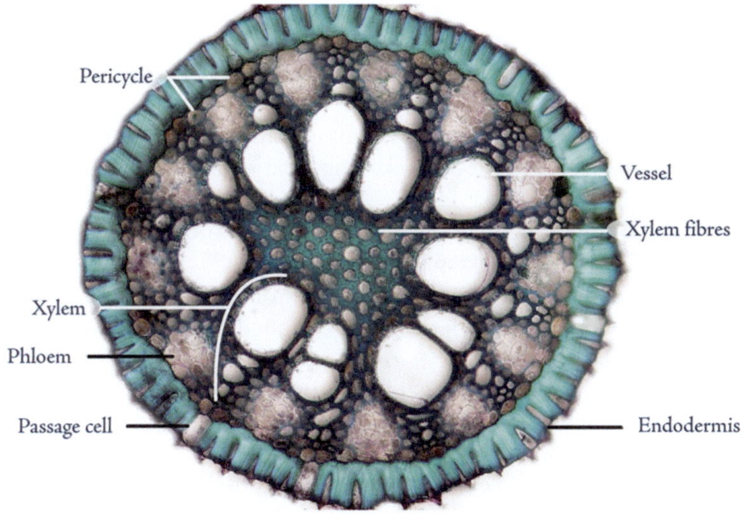

Fig. 10.3 Single-layered pericycle in the stele of the *Iris* root

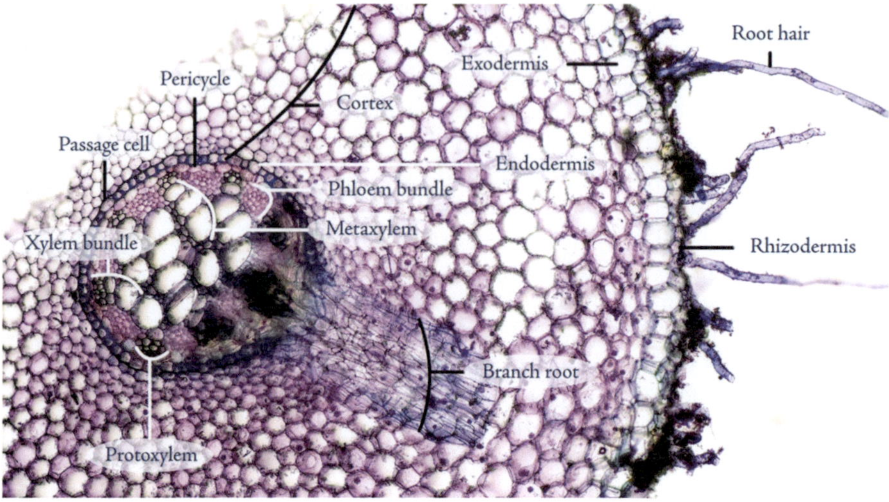

Fig. 10.4 Lily (*Lilium*) root cross-section

typical of **exarch xylem**. The xylem and phloem bundles are alternately arranged in the stele.

The central part of the stele is filled with **pith** tissue, the cells of which often form intestinal sclerenchyma, which contributes to the strengthening of the root (Fig. 10.3). Some of the roots have a xylem bundle in the centre, to which the other xylem bundles are radially connected.

The number of xylem bundles is higher in monocots, seven to nine or more, and lower in dicots, two to six (Fig. 10.5).

The many vascular bundles of monocots can usually only be arranged in a larger circle, so they may develop more extensive pith tissue within the root (Fig. 10.3) than dicots. The roots of monocots are not thickening secondarily, they have evenly thick, shoot-derived lateral roots.

In the roots of dicots, secondary meristems ensure further growth of the vascular tissues, resulting in secondary root thickening. The most common thickening process is illustrated in Fig. 10.6.

In the secondary growing root of a dicot, there is no endodermis because it is shed with the cortex during thickening.

The outermost cell layer of the stele is a ring of meristem, the pericycle, which plays a role in the formation of lateral roots, secondary root thickening, and the formation of secondary cortex.

The first step of secondary thickening is the formation of a wavy cambium (green) (Fig. 10.6, drawing 1). The wavy cambium is formed by pericycle parts lying outside the xylem bundles and intestinal parenchyma cells lining the inside of the phloem bundles. In the resulting structure, the primary phloem bundles (light blue) are located on the outer side of the cambium ring, while the primary xylem bundles (dark blue) are located on the inner side. The wavy cambium in a

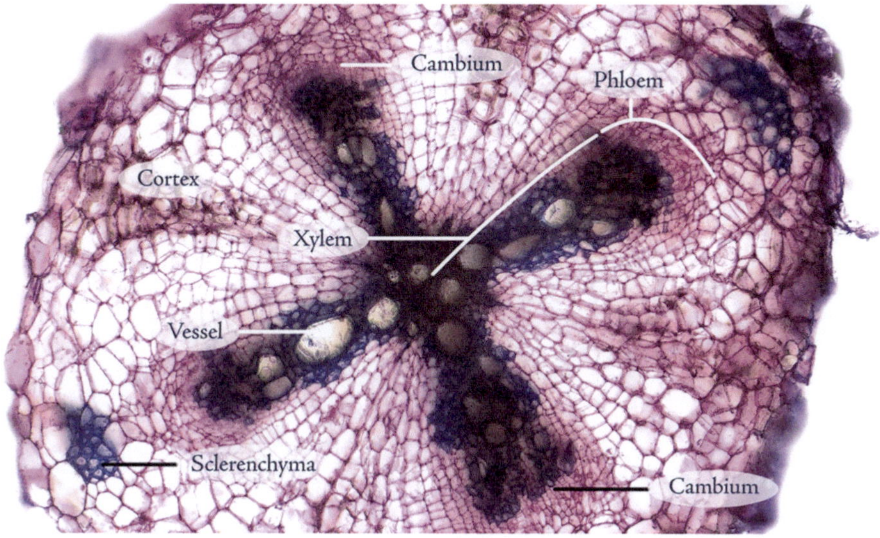

Fig. 10.5 Cross-section of the secondary root of a cucumber (*Cucumis sativus*)

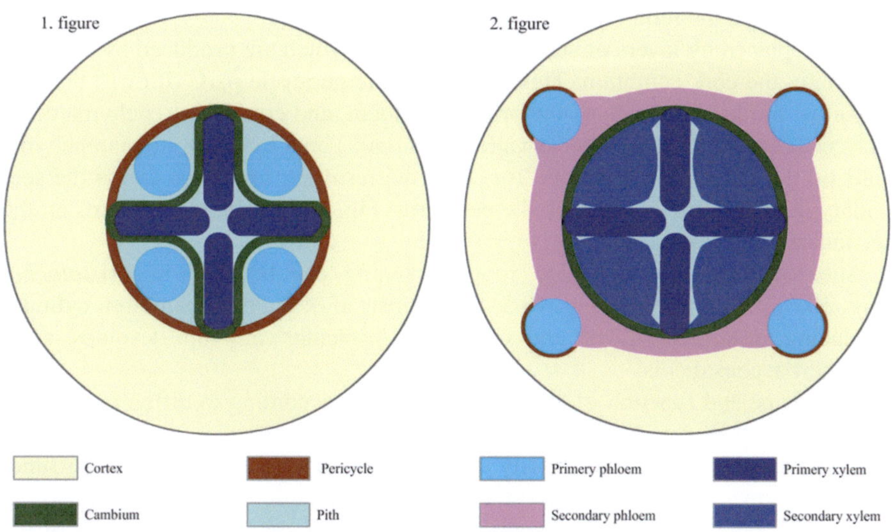

Fig. 10.6 Steps in the secondary growth of the dicot root

continuous ring or bundle structure forms xylem elements (secondary xylem, mid blue) inwards and phloem elements (secondary phloem, pink) outwards. Due to the uneven formation of vascular elements, the initially wavy cambium later becomes circular (Fig. 10.6, drawing 2). As the root thickens, the primary xylem elements are still recognisable in the central part of the stele, while the primary phloem bundles are pushed outwards by the secondary phloem (Fig. 10.6, drawing 2 and Fig. 10.7).

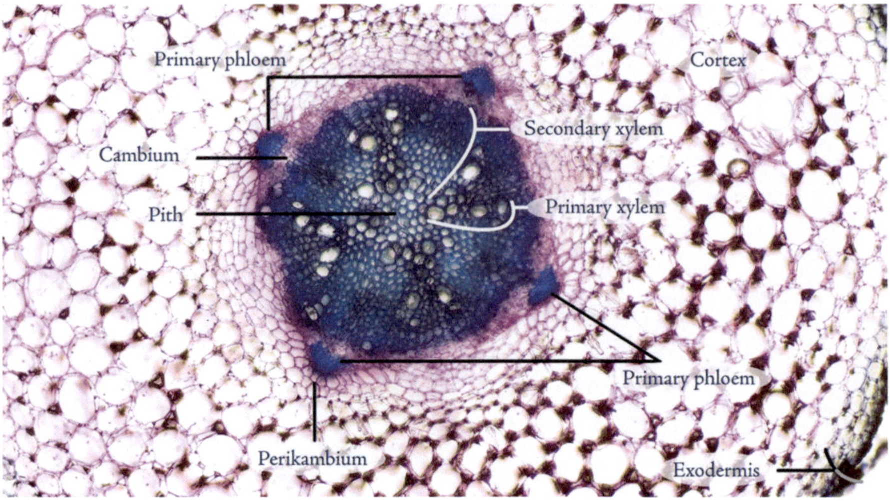

Fig. 10.7 Cross-section of a bean (*Phaseolus vulgaris*) root that thickens secondarily

After a while, the cortex is cracked because of the forming vascular tissues, and breaks off from the root.

It is replaced by layers of suberin walled cells, which are produced by a ring of meristem, the cork cambium. This formed by the pericycle parts (brown) that has remained on the outer side of the primary phloem, and cortical parenchyma cells. The cork cambium forms suberin walled cells (cork) on the outside and parenchyma cells on the inside (phelloderm). Together, the resulting formation forms the secondary dermal tissue of the root, the periderm (Fig. 10.8), which also forms on the secondarily thickening stem.

Specific forms of secondary root thickening can be seen in *Aristolochia* (Fig. 10.8) and cucumber (Fig. 10.5). Here, instead of a closed cambium cylinder producing an unified phloem and xylem, the fascicular cambium develops, as in most herbaceous dicots.

The shape and function of the roots can change according to different environmental factors, leading to modified roots.

The changes in shape are mostly due to the emergence of the following functions: nutrient storage, plant support, clinging, aeration, defence, and parasitism.

Aerial roots are adventitious roots arising from the stem nodes or stems of the shoot, which can be classified into different types according to their function (Fig. 10.9). True aerial roots are those that do not penetrate the soil but bind air humidity and rainwater. This is the case for tree orchids.

Clinging aerial roots, such as the short roots that form on the stem members of ivy, merely anchor the shoot of the emerging plant (Fig. 10.9).

The prop roots are the roots from the lower stem nodes of the shoot, as we have seen earlier in the case of maize (Fig. 10.10).

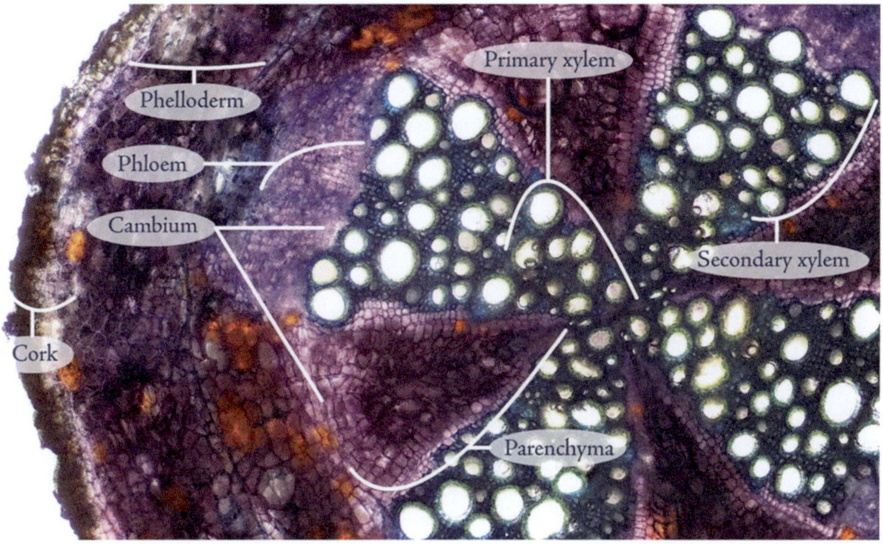

Fig. 10.8 Cross-section of the birthwort (*Aristolochia clematitis*) root

Fig. 10.9 Ivy (*Hedera helix*) aerial rootlets

Fig. 10.10 Maize (*Zea mays*) prop roots

Philodendron's aerial roots are able to bind rainwater on their surface and then transport water-soluble salts when they enter the soil (Figs. 10.11, 10.12, and 10.13).

The characteristic root modifications of legumes (Fabaceae) are the root nodules. Spherical formations formed by cell proliferation in the uptake zone of the root, in which nitrogen-fixing bacteria live in symbiosis with the plant (Fig. 10.14).

The bacteria reduce nitrogen molecules in the air into ammonia, which can then be used by the host plant. Therefore, the nitrogen metabolism of the plant is no longer dependent on the nitrogen content of the soil, while the bacteria obtain nutrients from the plant. The majority of the nitrogen fixed is incorporated into the body of the legume host plant and about a tenth of it is released into the soil. Also essential to the biosphere's nitrogen cycling process, it is estimated that around 35 million tonnes of atmospheric nitrogen are fixed annually by nitrogen-fixing bacteria living in the root nodules of legumes.

At the junction of the root nodules, a cross-section of the root clearly shows the characteristic structure of the root nodule (Figs. 10.15 and 10.16).

Nitrogen-fixing bacteria live in a membrane-enclosed space surrounded by a yellowish-red ring of leghaemoglobin. Leghaemoglobin, which makes up 25–30% of the proteins in the root nodule, can bind oxygen faster and more strongly than the haemoglobin in our blood. This places the bacteria in anaerobic conditions, where their nitrogen-fixing nitrogenase enzyme can reduce atmospheric nitrogen to ammonia.

Taproot is the main root of a primary root system, growing vertically downward. Most dicotyledonous plants produce taproots, and some, such as the edible roots of

Fig. 10.11 *Philodendron* aerial roots

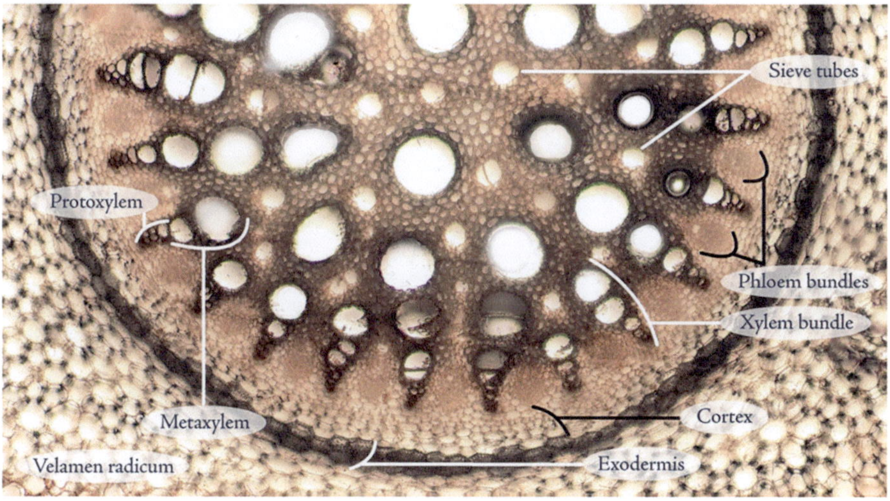

Fig. 10.12 Cross-section of the aerial root of *Philodendron*

carrots, are specialised for food storage. The carrot owes its colour to the colouring substance carotene, which is stored in it and accumulates in large quantities, especially in the outer layers of the carrot. In carrots can be seen four distinct parts in

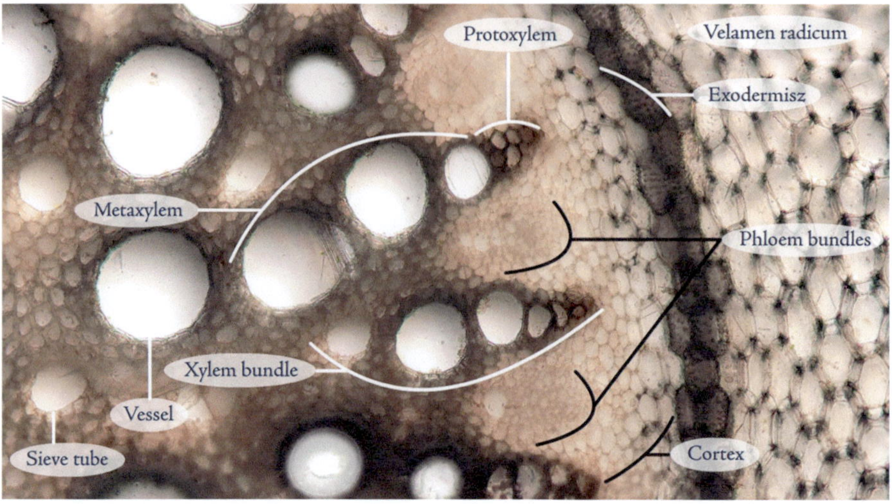

Fig. 10.13 Cross-section of the aerial root of *Philodendron*

Fig. 10.14 Root section of red clover (*Trifolium pratense*) with root nodules

cross-sections. The innermost central part (core) is rapidly hardens and contains less pigment and nutrients.

Next is the part called the flesh (cortex), which is darker in colour and rich in carotenes and nutrients. Between these two layers is a thin ring of meristem that plays a role in thickening the carrot body. Several concentric cambium rings of

Fig. 10.15 Cross-section of pea (*Phaseolus vulgaris*) root and associated root nodule

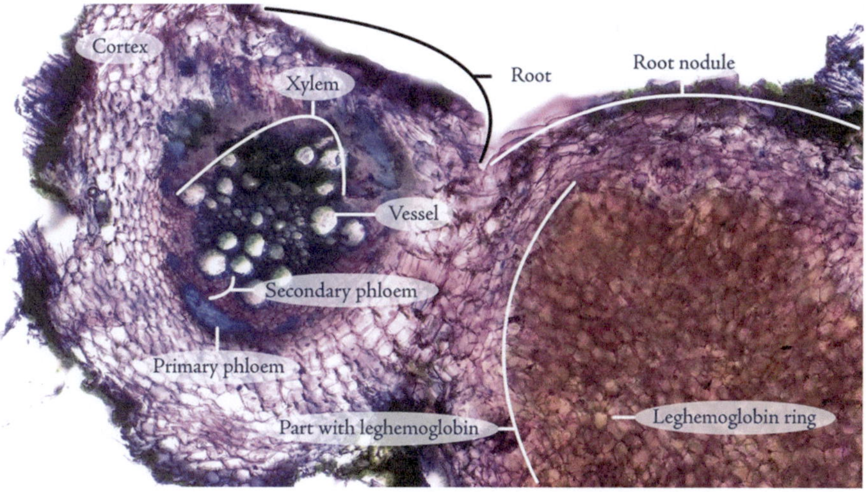

Fig. 10.16 Pea (*Phaseolus vulgaris*) root–root nodule relationship structure

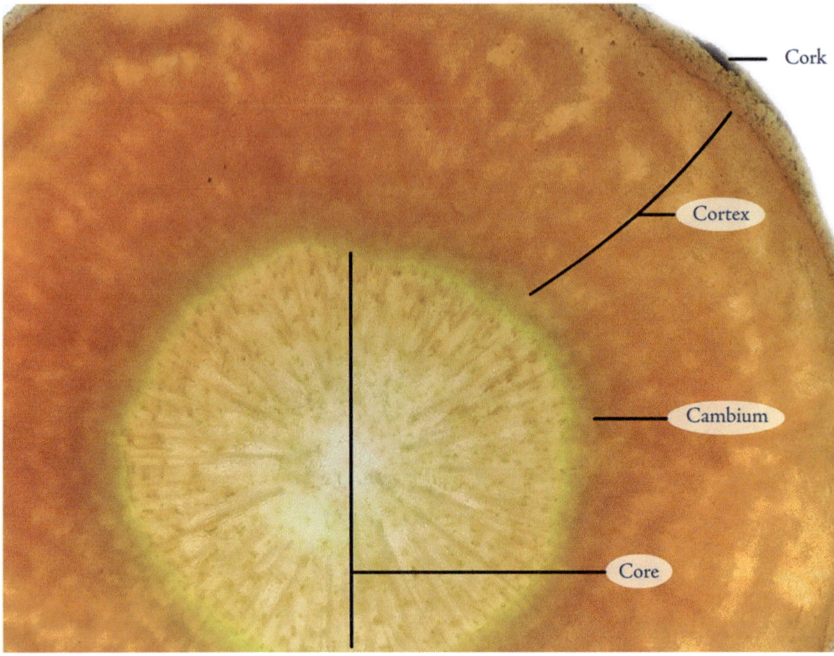

Fig. 10.17 Carrot (*Daucus carota* subsp. *sativus*) cross-section

tissue may form in the carrot body. The outside of the carrot is covered with cork (Figs. 10.17 and 10.18).

10.2 Stem

The stem tissue develops at a certain distance from the shoot apex. In the shoot apex, the peripheral meristem forming the cortex and the intestinal meristem forming the pith are separated by a procambium roller forming the vascular tissues. As the shoot grows, the cells that form in the shoot apex are transported to the site of tissue formation, where the cortical, pith, and vascular tissues begin to separate. The dermal tissue of the shoot, the epidermis, is composed of tightly closed epidermal cells, usually without a chloroplasts, whose outer surface is often covered by a protective layer of a waxy substance (**cutin**), the cuticle (Fig. 10.19).

Like the leaves, the stem also has stomas through which respiratory (oxygen uptake and carbon dioxide release) and photosynthetic (carbon dioxide uptake and oxygen release) gas exchange takes place. The stomas and the hairs of the epidermis also regulate water loss from the stem, evaporation.

In addition to the covering hairs, glandular hairs may also appear on the stem, which have a secretory function.

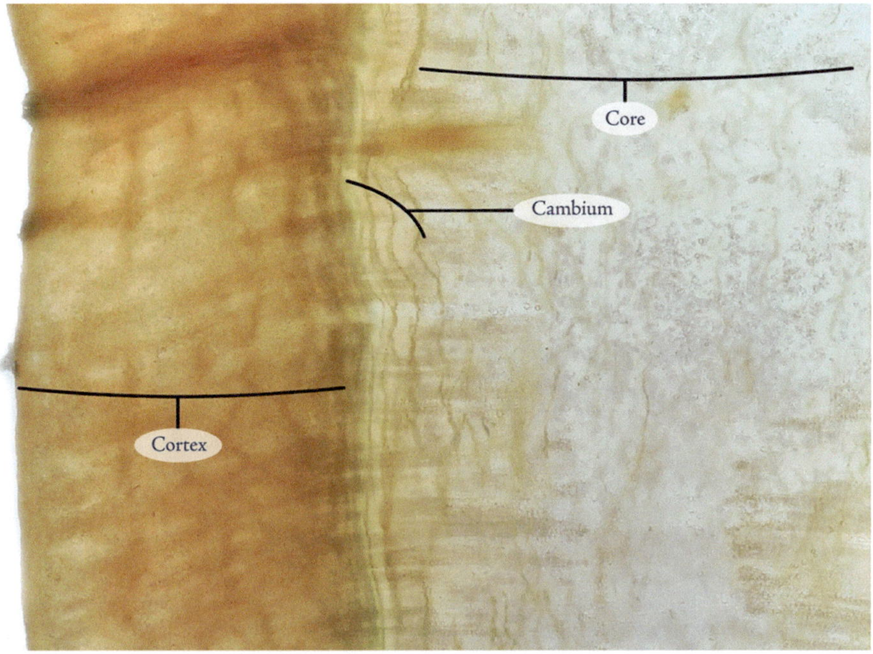

Fig. 10.18 Radial and longitudinal section of carrot (*Daucus carota* subsp. *sativus*)

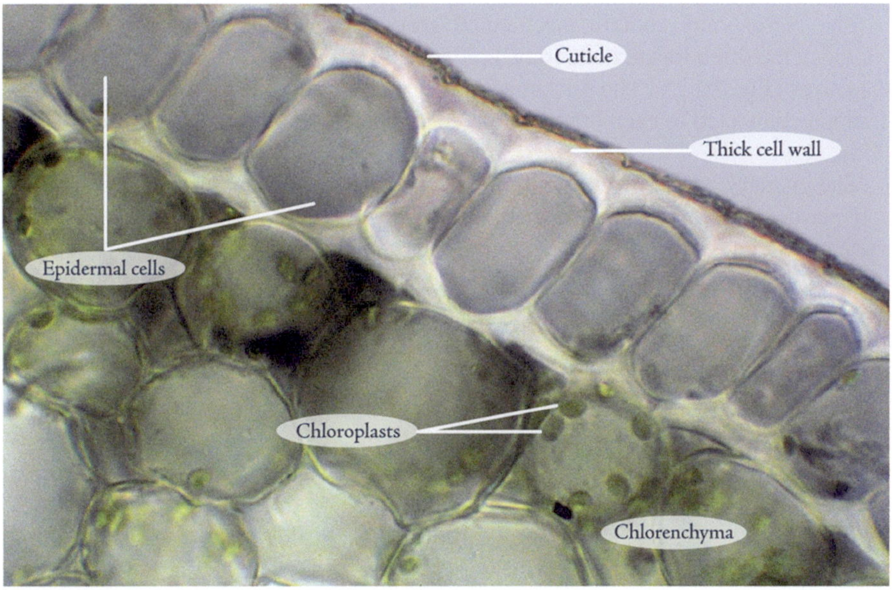

Fig. 10.19 Onion (*Allium cepa*) epidermis and chlorenchyma

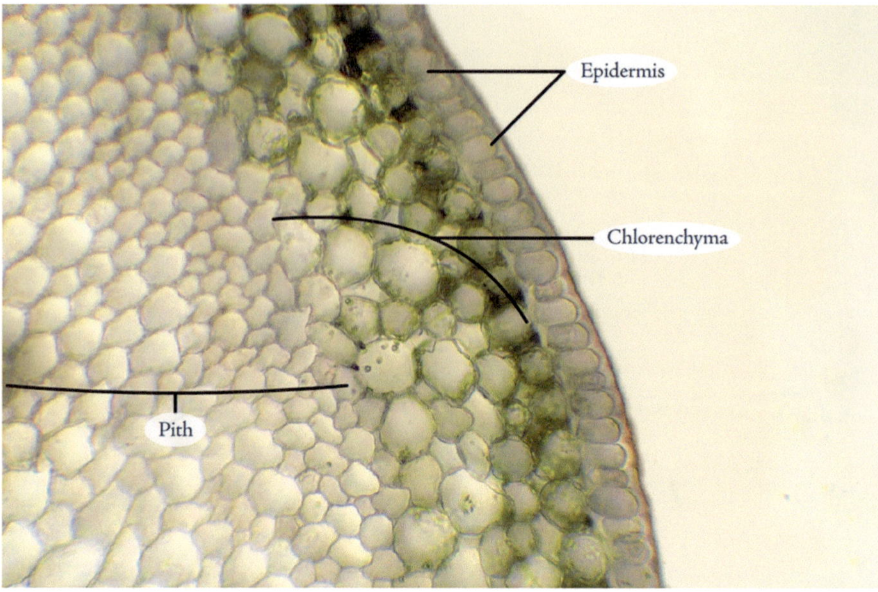

Fig. 10.20 Onion (*Allium cepa*) stem cross-section detail

The primary cortex of the stem is made up of ground tissue with a lot of intercellular ducts. The subepidermal cells in young stems contain chloroplasts (assimilatory parenchyma and chlorenchyma) (Fig. 10.20).

The cortex may contain mechanical ground tissues consisting of thickened-walled cells, often located under the epidermis (hypodermis) or on the inner side of the cortex (Fig. 10.25). The thickened-walled fibres sometimes form a T or a double T shape connected to each other, which contribute significantly to the solidification of the stem.

The primary cortex of the stem, unlike the root, is not separated by endodermis and pericycle from the central part (stele and vascular cylinder) containing the transport elements.

The central part of the young stem is filled with ground tissue consisting of large parenchyma cells. During the development of the stem with vascular bundles, the cortex and the intestinal tissue are in direct contact, forming a single formation. In the case of hollow stems, as the stem thickens, the pith is torn away and a pith cavity forms in its place. In the stem tuber, the pith becomes a storage parenchyma, whereas in the stem of water plants it becomes aerenchyma.

The vascular tissues of the stem are formed from the procambium of the shoot apex. In the herbaceous dicots, vessel bundles develop from procambium parts that are mostly divided into bundles (Fig. 10.23). In the older secondarily thickening stem, the bundle structure may be significantly modified. In the majority of woody stems, a single procambium cylinder produces the vascular tissues.

In some herbaceous dicots (e.g. common mallow (*Malva sylvestris*)), the primary cambium forming a cylindrical cambium, forms phloem outwards and xylem inwards (Fig. 10.21).

In the cross-section of the monocot stem, the vessel bundles are distributed in a scattered pattern (Fig. 10.22), whereas in the dicots they are arranged in a circular pattern (Fig. 10.23).

The scattered arrangement of vessel bundles in the cross-section of the monocot stem can be explained by the different course of the bundles entering the stem from the leaf sheath. The vessel bundles at the edges of the leaf sheath, so the veins, enter the stem immediately downwards, whereas the central veins of the leaf sheath, having penetrated deep into the stem, only then descend. As a result, in pithy-stalked monocots such as maize, we also see a scattered arrangement of the vessel bundles in the cross-sectional view of the stem.

In monocots and some dicots, the meristem that makes up the vessel bundles is entirely converted into vascular tissue cells, so there is no meristem (cambium), between the xylem and the phloem. This type of bundle is called a closed vascular bundle (Fig. 10.27).

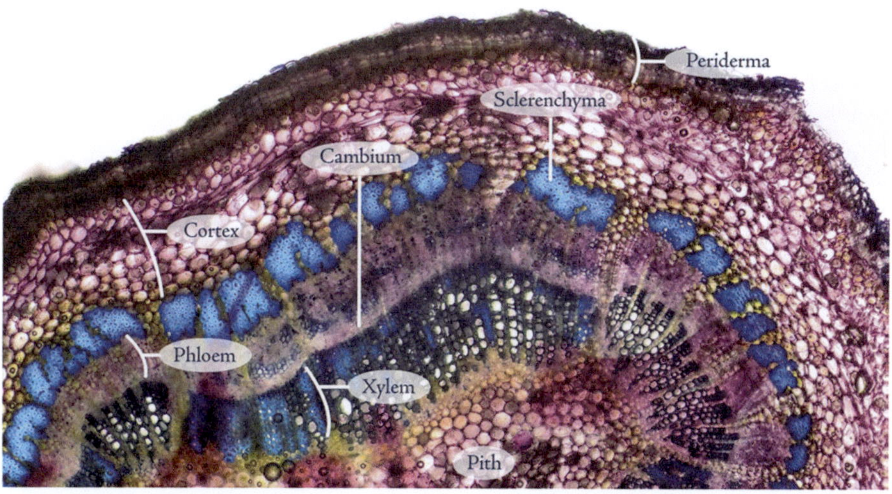

Fig. 10.21 Stem cross-section of common mallow (*Malva sylvestris*)

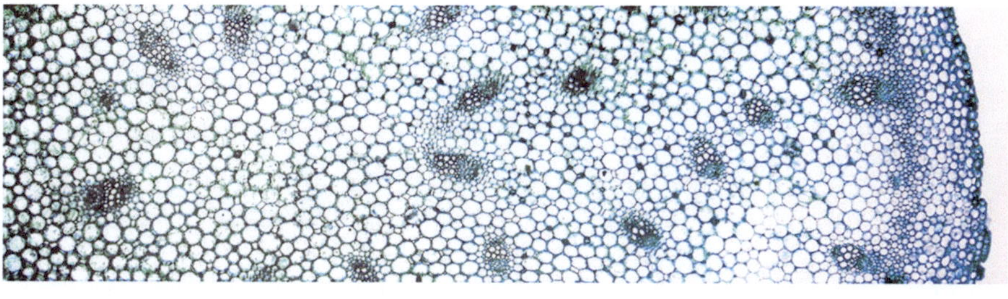

Fig. 10.22 Scattered arrangement of vessel bundles in the stem of a tulip (*Tulipa gesneriana*)

The narrow cavity xylem and phloem elements (protoxylem and protophloem) that are primarily formed during bundle formation are soon torn away by elongation of the stem and replaced by wider cavity vascular elements (metaxylem and metaphloem) corresponding to the final length of the stem.

In the case of the stem, the inward facing part of the xylem is the protoxylem, while the outward facing part is the metaxylem. This structure is a characteristic of **endarch xylem**.

The formation of vessel bundles in dicots is similar to that in monocots, except that they retain a cambium, the meristem between the phloem and the xylem, which is capable of producing new vascular elements. This type of vessel bundle called an open vessel bundle. In the xylem of the vessel bundles, the mineral saline solution taken up by the root is transported to the shoot, while the phloem part transports organic matters (Fig. 10.23).

Among the dicots' open vessel bundles are those with two phloem portions that hold the xylem between them. In these bundles, active cambium is found only between the outer phloem and the xylem. Such so-called bicollateral bundles are typical of the cucurbits (Cucurbitaceae) and nightshades (Solanaceae) families (Fig. 10.24).

In the stem and rhizome of ferns, the vascular bundles are surrounded by a single cell line endodermis and starch sheath. In the vascular bundles, the phloem elements encircle the centrally positioned xylem (Fig. 10.25).

The cortex of an ageing soft stem undergoes characteristic histological changes even if the stem is not capable of secondary thickening. These changes are mainly due to the appearance of the mechanical tissues, the sclerenchyma and the collenchyma. The cells of the sclerenchyma are thick-walled, died at maturity and have only a mechanical function. The walls of the cells of the collenchyma are also thickened, but collenchyma cells are living and containing cytoplasm.

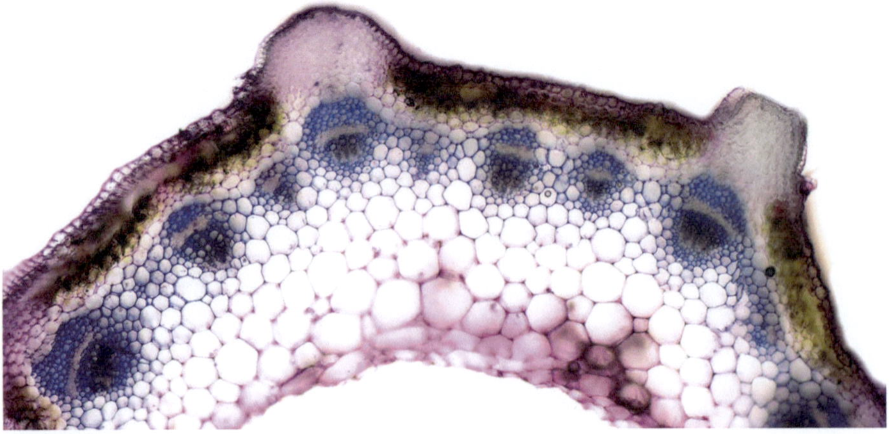

Fig. 10.23 Vessel bundles of spring ragwort (*Senecio leucanthemifolius*) standing in a circle

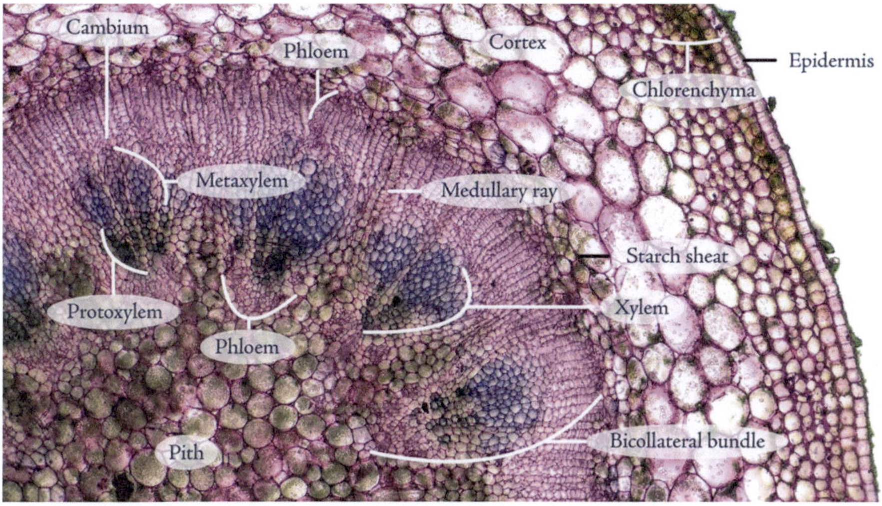

Fig. 10.24 Bicollateral bundles in the cross-section of the jimsonweed (*Datura stramonium*)

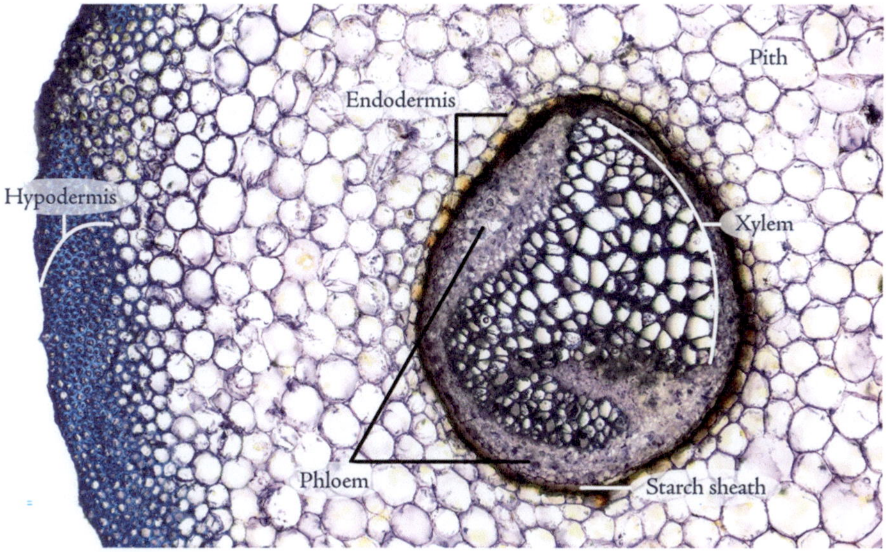

Fig. 10.25 Cross-section of male fern (*Dryopteris filix-mas*) rhizome

In the case of the angular collenchyma, the wall thickening occurs at the corners of the cell, while in the case of the lamellar collenchyma, the thickening extends to the cell walls parallel to the surface of the plant part. In dicots, a lamellar collenchyma often develops beneath the unicellular epidermis, forming a single supporting cylinder called the hypodermis (Fig. 10.26).

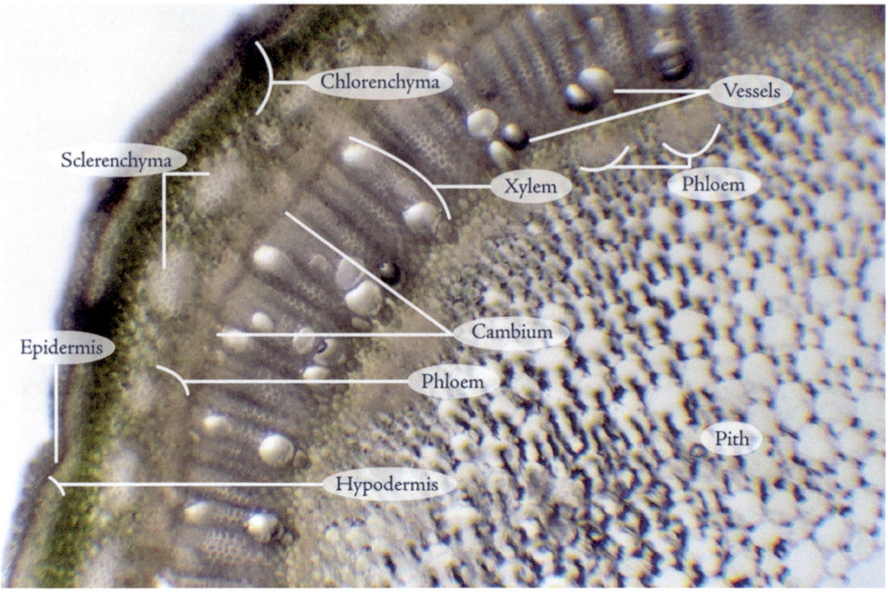

Fig. 10.26 Hypodermis below the epidermis in a cross-section of the trumpet vine (*Campsis radicans*) stem

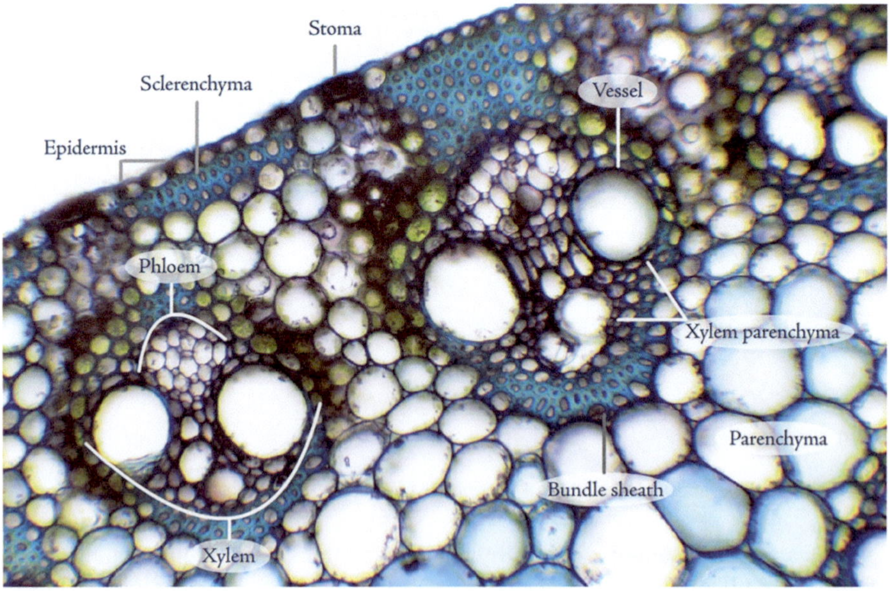

Fig. 10.27 Hypodermis in maize (*Zea mays*) stem is made up of sclerenchyma

In monocots, the hypodermis beneath the epidermis is composed of sclerenchyma tissue (Fig. 10.27).

Beneath the hypodermis are the parenchymatous (thin-walled, rounded) cells of the cortex, followed by a ring of starchy cells, often called the starch sheath, which often appears at the cortical boundary and helps to separate the primary cortex and the stele within it (Fig. 10.24).

The thickening of plant stems is basically due to two processes. The primary thickening is due to the tissues produced by the divisions of the shoot apex and their elongation, while the secondary thickening is due to the activity of the divisions that develop later. In monocots, secondary thickening of the stem occurs only in a few cases, for example in the dragon trees (*Dracaena draco*) of the family of asparagus. In these plants, secondary thickening is caused by the secondary division of the cortex (Fig. 10.28).

In the central part of the stem of the dragon tree, the vascular bundles are scattered in the pith. The vascular bundles are closed, contain no cambium, and have the xylem on the outside, surrounding the phloem. Secondary thickening of the stem is made possible by a ring of meristem formed by cortical parenchyma cells, which creates a circle of inwardly closed vascular bundles together with the pith parenchyma cells separating them. The formation of the first ring of meristem is later followed by others in the outer layers of the cortex, which create further circles of vascular bundles. Thus the characteristic tissue pattern of the stem cross-section of the dragon tree is formed, characterised by scattered primary vascular bundles in the centre and secondary bundles in a circle outside the centre. In the microscopic section, the central stem part with the closed vascular bundles and the cortex, whose inner cell line has been transformed into a meristem, is clearly distinguished (Fig. 10.29).

In dicots, there are several forms of secondary growth of the stem. Each type is named after a plant for which the particular thickening is characteristic.

During the secondary growth of the birthwort the cambium ring is formed by the closure of parenchymal cells, that have recovered their ability to divide by dedifferentiation, and the fascicular cambiums. The cambium between the vascular bundles produces only parenchymal cells. The cambium ring remains functional for many years, so that the microscopic examination of a multi-year stem cross-section reveals the annual-ring structure of the xylem structure within the continuous growth, ever larger vascular bundles (Fig. 10.30).

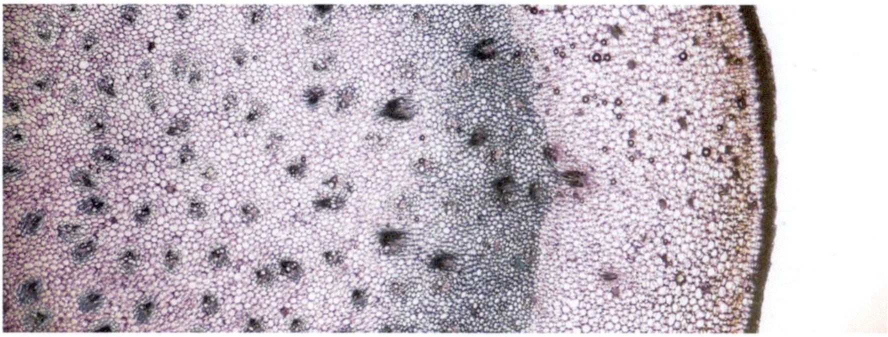

Fig. 10.28 Cross-section of the dragon tree (*Dracaena draco*) stem

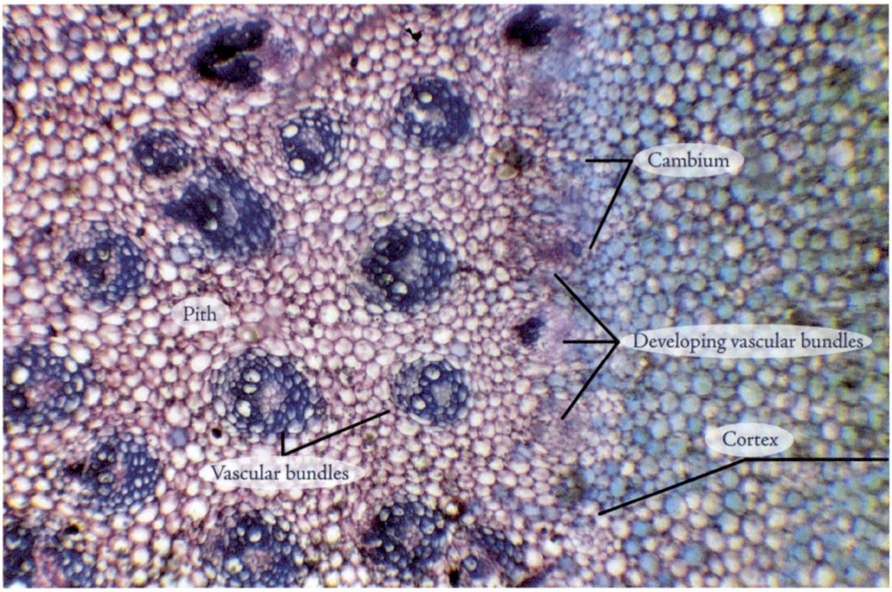

Fig. 10.29 Primary and forming secondary vascular bundles in the stem of dragon tree (*Dracaena draco*)

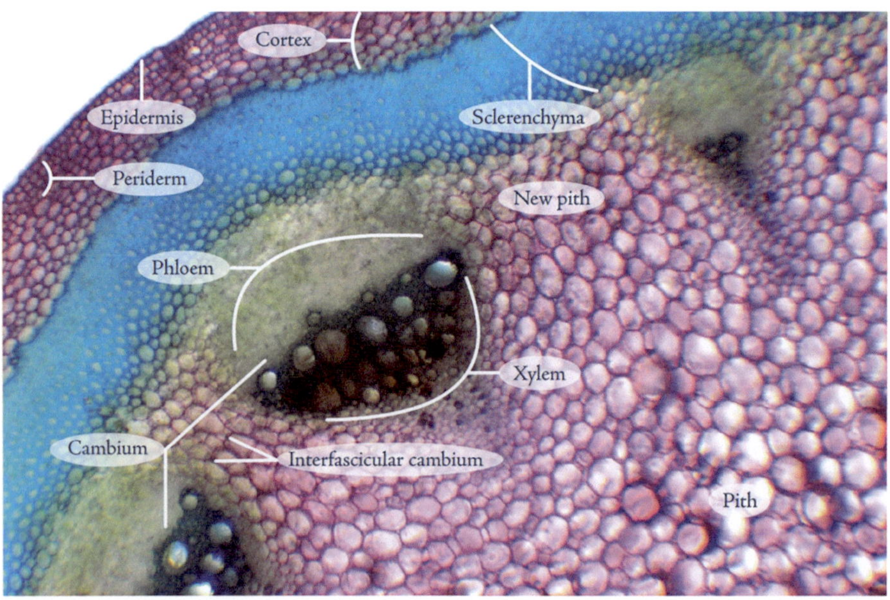

Fig. 10.30 Cross-section of a birthwort (*Aristolochia clematitis*) stem

In the case of secondary growth of the sunflower, as in the birthwort's type, the cambium ring is formed by the fascicular cambiums and parenchymal cells between the fascicular cambiums. The closed cambium ring creates new (smaller) secondary vascular bundles between the initially existing vascular bundles. As the stem thickens, the original bundles continue to grow, so that several bundles of different sizes can be seen in the stem (Fig. 10.31).

During secondary stem growth of the ricinus type, the cambium of the circularly arranged vascular bundles and the parenchyma cells located in their continuation between them also form a closed cambium ring, which produces a continuous xylem part inwards and a continuous part of the phloem outwards. The xylem remnants of the primary bundles form a pith corona towards the pith, while the phloem remnants form a phloem corona towards the cortex. The latter are more difficult to detect in light microscopic sections because they are compressed as the stem thickens (Fig. 10.32).

Linden-type secondary growth is mostly characteristic of trees. In these plants, the meristem (procambium), which produces the primary vascular elements, does not divide into bundles but forms a continuous cylinder and develops a single phloem and xylem ring from the start (Fig. 10.33). In addition to the phloem, the stem contains the primary cortex, consisting of parenchyma cells, and the primary dermal tissue, the epidermis, on the outside. As the stem thickens, the outer layer, the epidermis is eventually shed and replaced by periderm and then rhytidome.

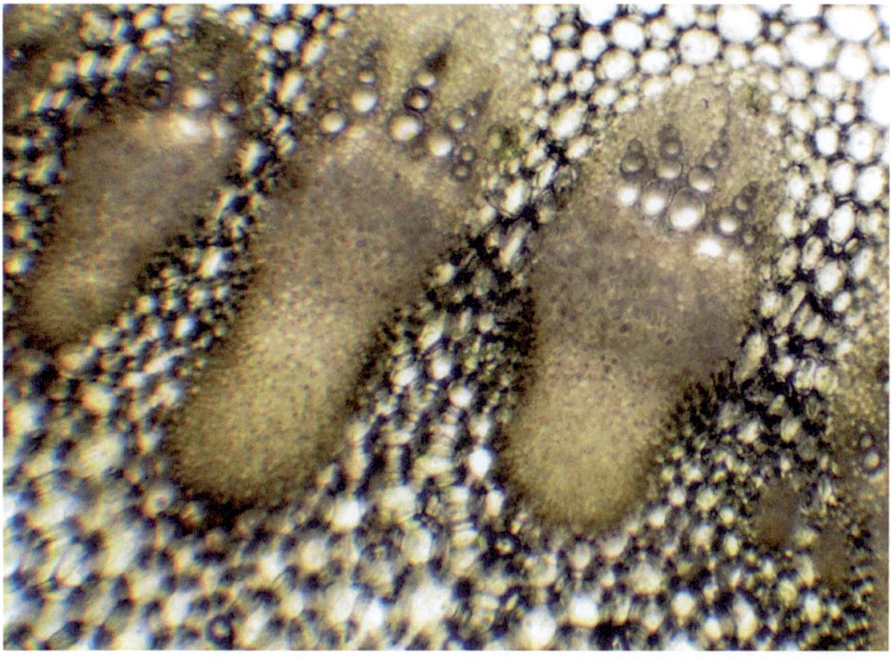

Fig. 10.31 Different sizes of vascular bundles in the sunflower (*Helianthus annuus*) stem

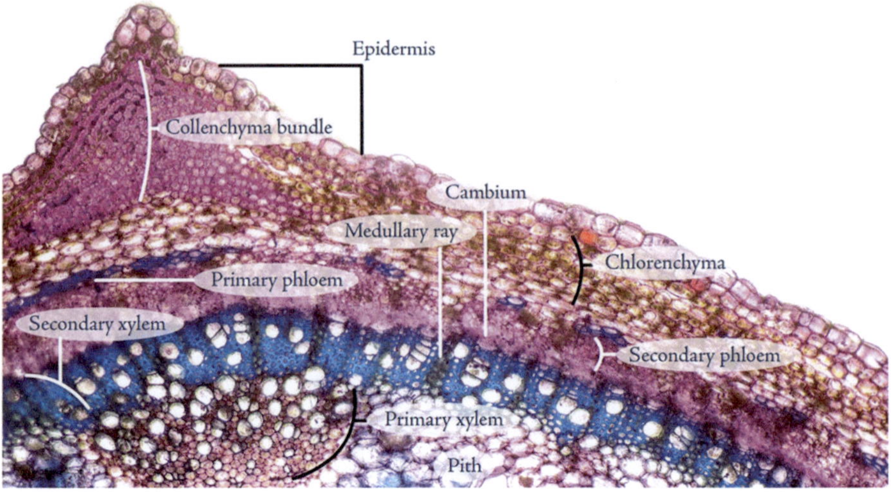

Fig. 10.32 Cross-section of a young stem of black elder (*Sambucus nigra*) showing ricinus-type growth

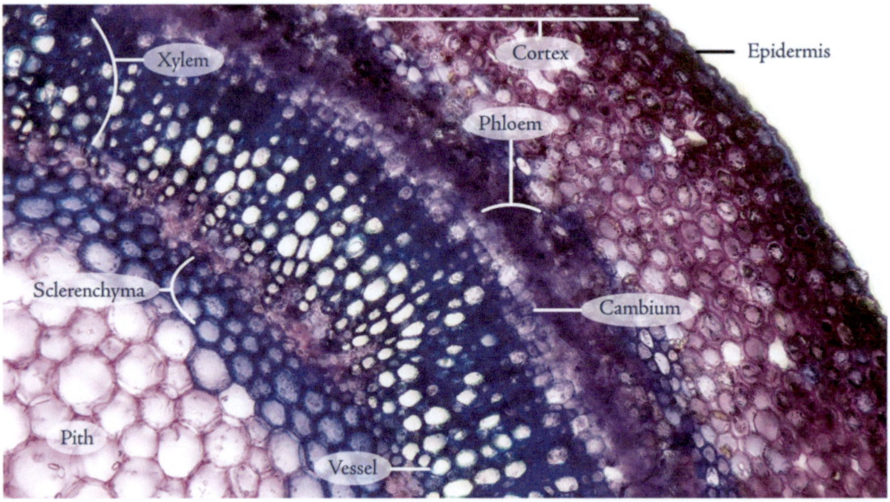

Fig. 10.33 Cross-section of the stem of *Forsythia* showing linden-type thickening

Trees can live for centuries while their trunks grow with secondary thickening of the ricinus or linden type. The largest part of the trunk is the centrally positioned wood, which contains all the xylem elements produced by the cambium during the life of the plant. However, only the ones closer to the periphery of the wood are involved in the transport of nutrients.

The phloem is located outside the cambium, so the circumferential expansion of the trunk causes it to mostly break off, and only the newly formed phloem elements close to the cambium contribute to the transport of material. The dead, fragmented older parts are mostly important in the formation of the rhytidome. The heavily striated rhytidome does not appear on the surface of the trunk in all species, because the cracks formed by the splitting of the bark may be filled by parenchyma cells in the form of intestinal cones (e.g. linden) (Fig. 10.34).

In addition to the xylem and phloem elements, the cambium also produces longitudinal and radial ground tissue cells, which are interconnected to form a network system in the phloem and the xylem. Only the phloem and the xylem, close to the cambium are always involved in the transport of material. But what will be the fate of the previously formed xylem and phloem? The older xylem is positioned inside the trunk, with the oldest xylem (**heartwood**) on the inside and the newer xylem (**sapwood**) on the outside. The inside of the trunk is a mass of nonfunctional xylem with a ring structure, which is the heartwood that has formed over the years and is no longer transporting material. The formation of lighter and darker concentric circles in the dead wood, the **annual rings**, is due to the alternation of the seasons. In spring, after bud break, the plant's water demand increases significantly and the cambium ring forms large cavity transport elements, the lighter coloured layer of which is called the **earlywood**. At the end of summer, after flowering and fruit ripening, the water demand of the tree decreases and narrow cavity transport elements are formed, which give the darker coloured **latewood** (Fig. 10.35).

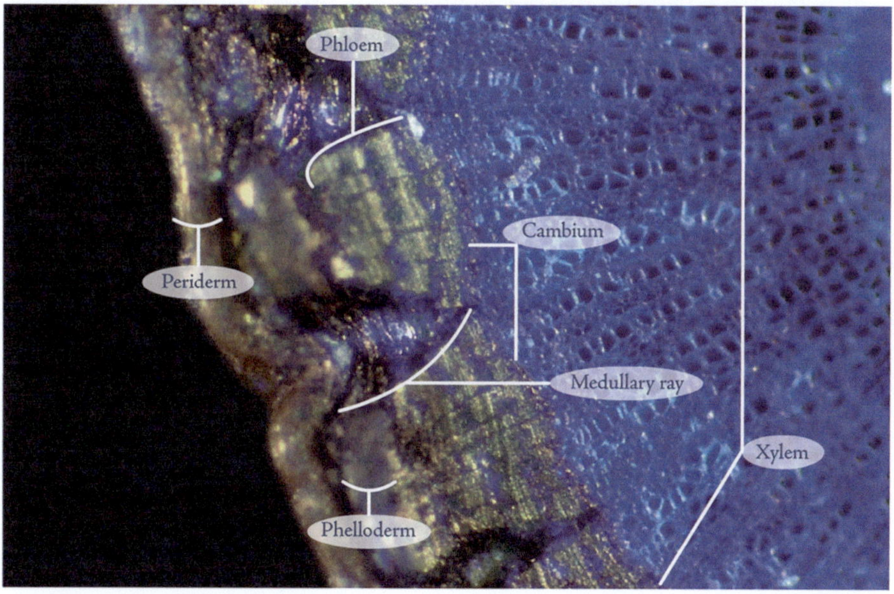

Fig. 10.34 Cross-section of a linden (*Tilia* sp.) branch

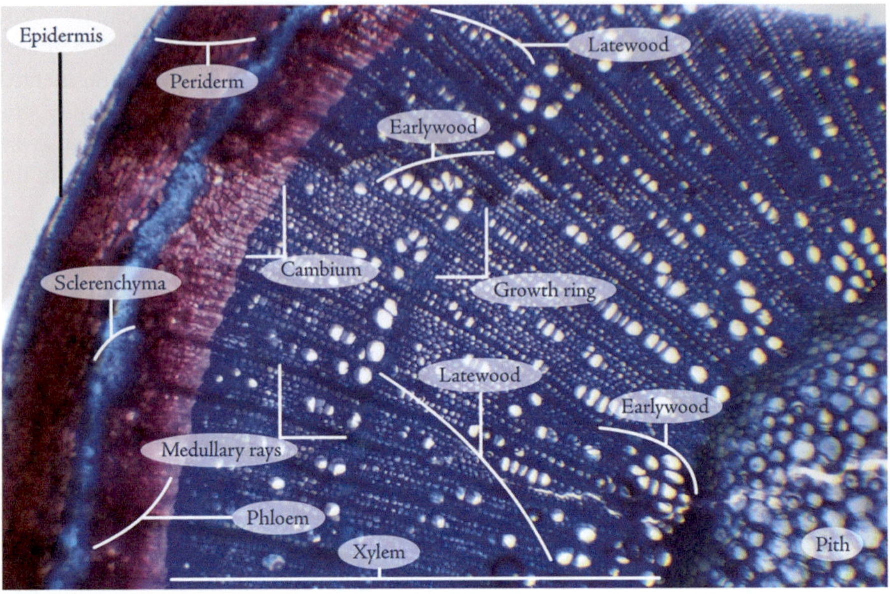

Fig. 10.35 Cross-section of a box elder (*Acer negundo*) branch

You can also estimate the age of a tree by using annual rings, but remember that as you move up the trunk there are fewer annual rings and the main branch at the top of the shoot axis is only one year old. In some cases, two annual rings can be formed in the tree trunk in a single year. If, for example, severe frost damage or caterpillar chewing destroys the entire canopy, new canopy will develop from the dormant buds and two annual rings will form in one year.

The formation of a new annual ring can also be caused by a second flowering during the long autumn, which again causes the production of large cavity transport elements and the appearance of earlywood.

When trees reach the age of nearly twenty years, the annual-ring structure of the xylem starts to senesce, resulting in a significant reduction in the water content of the wood and the accumulation of resins, alkaloids, dyes, and tannins in the dead cells, which make the wood resistant to fungi and other damaging agents. The resulting heartwood is darker than the still functional and more water-rich sapwood adjacent to the cambium. Poplar and willow trees do not form heartwood, so their trunks are more likely to be decayed. Trees are also classified as hardwoods and softwoods. The cambium, in addition to the transport elements and parenchyma cells, also produces thick-walled fibres, which are important for the consolidation of the trunk.

Hardwoods (e.g. acacia, walnut, and beech) contain much more solidifying fibre and their cell walls contain more lignin than cellulose, which makes the cell wall more resistant. The wood often develops knots, making it more difficult to split and saw firewood. These develop when the broken stumps of the side branches are

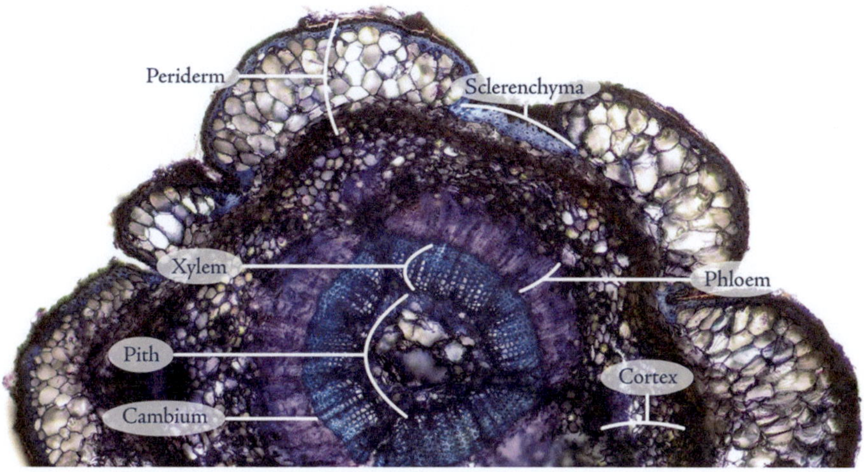

Fig. 10.36 Cross-section of a young pine (*Pinus* sp.) branch

surrounded by the thickening trunk. Knots are particularly common in the trunks of pine trees because in closed pine forests the lower branches receive so little light that they eventually die and their stumps are overgrown by the trunk.

In the cross-section of the xylem, the transport elements, the narrower tracheids, and the wide cavity vessels are clearly visible by light microscopy. These are arranged in bands according to the annual-ring structure of the wood. Between them, the medullary rays, consisting of parenchyma cells, run radially, the longer ones (primary medullary rays) running from the pith to the cortex, while the shorter ones (secondary medullary rays) only span a few annual rings.

The xylem structure of the gymnosperms differs from the structure outlined above. They have only tracheids, and the wide cavity vessels do not appear in the xylem (Fig. 10.36).

The narrow cavity tracheids have thick walls and therefore, in addition to transporting material, they also consolidate the trunk. This explains why, unlike in the case of angiosperm woody stems, no thick-walled consolidating fibres appear in the xylem of gymnosperm woody stems. Wood of angiosperms contains all of the xylem elements (vessel, tracheid, fibre, and xylem parenchyma) and it is called heteroxylous wood. However, gymnosperm wood is homoxylous, because vessels and fibres are absent.

Gymnosperm trees do not have vessels capable of transporting large quantities of water, so unlike angiosperm trees, not only the annual ring near the cambium is involved in water transport, but also the inner ones, which together are capable of transporting larger quantities of water. The location of the vessels in the tree trunk can be of two types. In **diffuse porous woods**, they are evenly distributed throughout the annual ring (e.g. linden, horse chestnut and box elder) (Fig. 10.35), whereas in **ring-porous woods** they occur only in the early part of the annual ring (e.g. common ash and black locust) (Fig. 10.37).

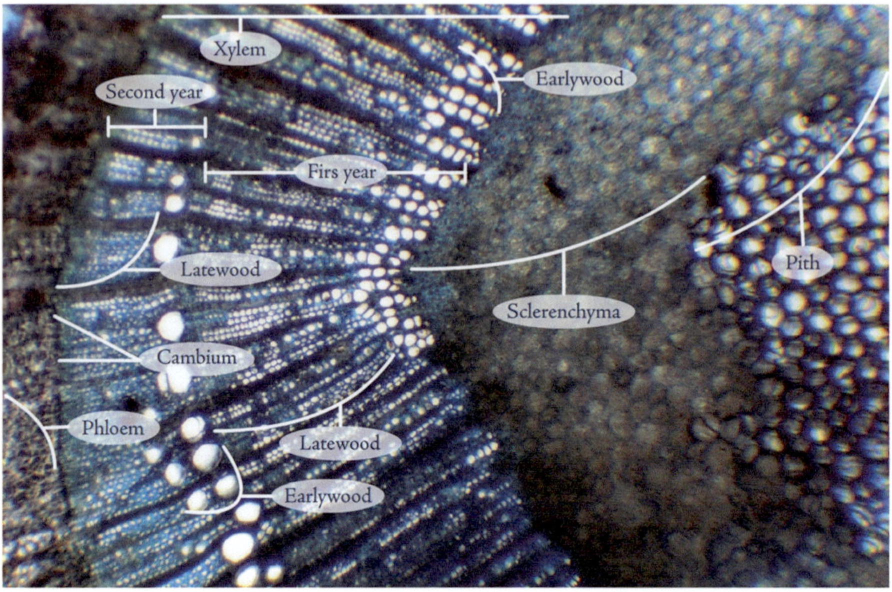

Fig. 10.37 Cross-section of a common ash (*Fraxinus excelsior*) branch

For this reason, early- and latewood are sharply separated in the latter trees. This feature allows a clear identification of the annual-ring boundaries.

10.3 Leaf

The leaf surface is mostly covered with a single layer of epidermis, which in some species may be multilayered (e.g. *Ficus*). The epidermal cells of the epidermis are tightly closed, with a thick, characteristic pattern of cuticle and waxy outer cell walls. The preparation of imprints is an excellent technique for examining the surface of epidermis by light microscopy (Fig. 10.38).

Epidermal cells often contain leucoplasts (e.g. violets, snake plant (*Sansavieria*)), but only rarely chloroplast (e.g. male fern, Canadian water weed, and impatiens).

Plants need water, minerals, and carbon dioxide for photosynthesis. Water and its dissolved mineral salts are taken up from the soil by the roots, while carbon dioxide is supplied from the air to the photosynthesising cells through stomata.

In addition to absorbing carbon dioxide, the stomata are important not only for the release of oxygen produced during photosynthesis, but also for the evaporation of water from plant known as transpiration. This prevents the leaf from overheating by cooling it, while transpiration ensures a constant supply of water from the root. In the stoma, two bean-shaped cells (dicots) or two cells with bulbous end (grasses) cells delimit the pore leading to the air cavity inside the leaf. From this, the air passages can emerge, introducing carbon dioxide into the leaf and, upon transpiration,

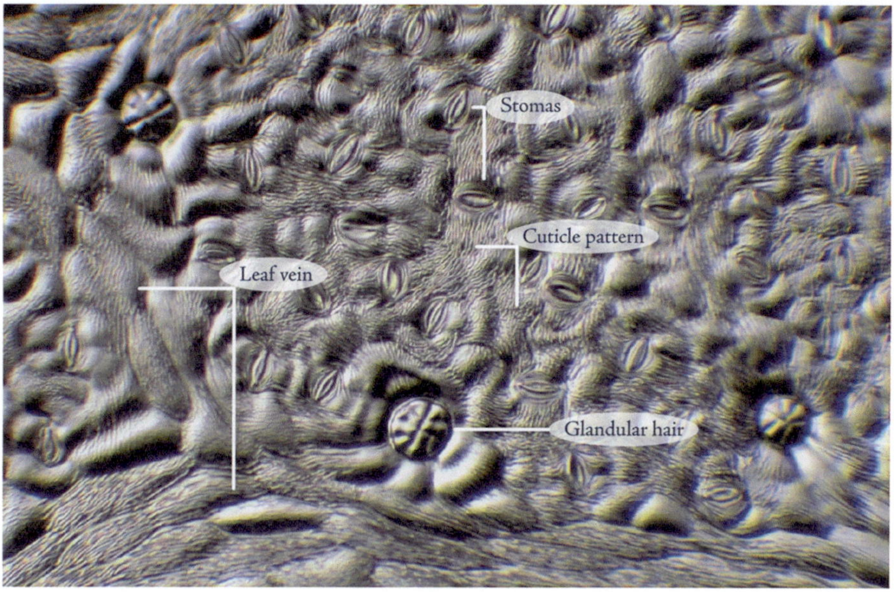

Fig. 10.38 Oblique illumination image of the lilac's (*Syringa vulgaris*) leaf imprint

releasing the water vapour into the environment. Around the guard cells, subsidiary cells with a shape different from other epidermal cells can often be seen. The number and arrangement of these cells is an inherited trait and is therefore an important feature in plant systematics. The unit of structure and function of pores, guard cells, and subsidiary cells is called the stoma.

Several factors influence the functioning of the stomata. One of these is the adequate water supply to the plant, which results in the uptake of water by the guard cells, which then stretch and open the pore between them, allowing the plant to release water by transpiration. In periods of water deficit, the guard cells lose water, the pore closes, and the amount of water released by transpiration decreases. The two states of guard cells can be artificially created by using solutions of different concentrations.

In wet and humid weather, the plant may not be able to release enough water by transpiration to allow the root to take up nutrients, in which case the leaf can get rid of the excess water by guttation. During guttation, the excess water is dripped out of the leaf from the leaf's tips. In nature, it is usually not the abundance of water but the lack of water that causes problems for plants, which plants protect themselves against by reducing evapotranspiration water loss.

Especially in plants living in arid areas, the leaves are densely covered with plant hairs. These cover hairs protect the leaves from overheating and excessive water loss (Fig. 10.39). Plant hairs also include glandular hairs and stinging hairs.

Most of the leaf tissue is composed of the **mesophyll** between the upper and lower epidermis, which is made up of loosely arranged elongated cells (**spongy**

Fig. 10.39 The dense cover of hairs on lower surface of raspberry (*Rubus caesius*) leaf

parenchyma and spongy chlorenchyma) or is often separated into two layers, a **palisade parenchyma** and a spongy parenchyma. The palisade parenchyma is made up of many elongated cells with many chloroplasts, which play a primary role in photosynthesis. The intercellular ducts of the spongy parenchyma are connected to the pores of the stomas, so that in addition to photosynthesis, it also plays an important role in gas exchange and transpiration (Figs. 10.40 and 10.41). The transport veins in the leaf are provided by the vascular bundles running in the leaf.

In the stem the xylem of vascular bundles is positioned inwards, so when bundles enter the leaf as leaf veins, the xylem faces the upper surface of the leaf and the phloem faces the lower surface. The organic matter produced during photosynthesis of the leaf is transported to the rest of the plant, so insect pests (e.g. aphids) that feed on the sugary solution transported in the phloem always make their home on the lower surface of the leaf.

Both the upper and the lower surfaces of the oleander leaf are covered by a thick cuticle-coated epidermis, under which is a two-layered hypodermis of thick-walled cells. Underneath the hypodermis is the palisade parenchyma, 2–4 cell layers thick, which is connected to the spongy parenchyma by collecting cells. The spongy parenchyma surrounds the vascular bundles, which are surrounded by a bundle sheath. It is sclerenchymatised in older leaves and forms mechanical supports attached to the lower and upper hypodermis. Attached to the spongy parenchyma at the lower surface is a layer of palisade parenchyma cells. The pores are located in stomata covered with hairs, which reduces the transpirational water loss of the plant.

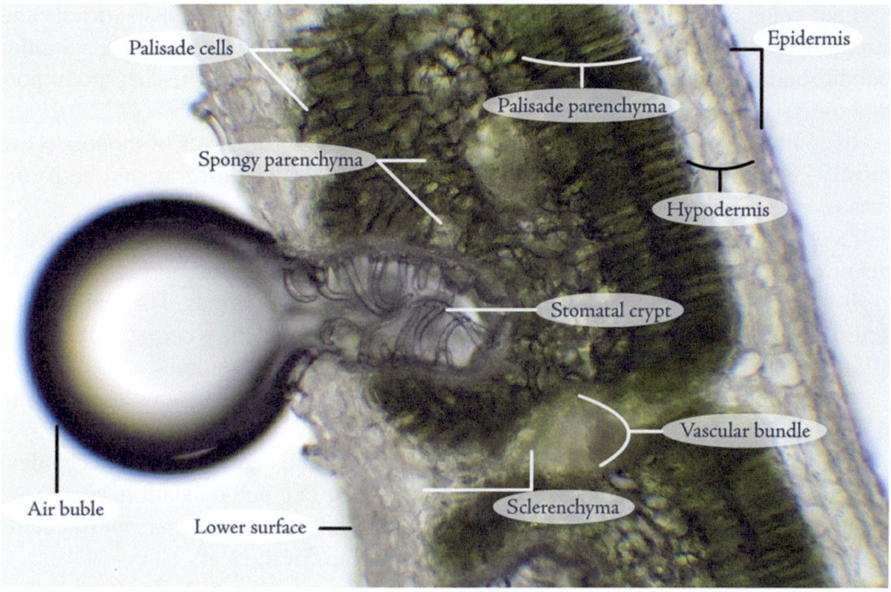

Fig. 10.40 Oleander (*Nerium oleander*) leaf cross-section

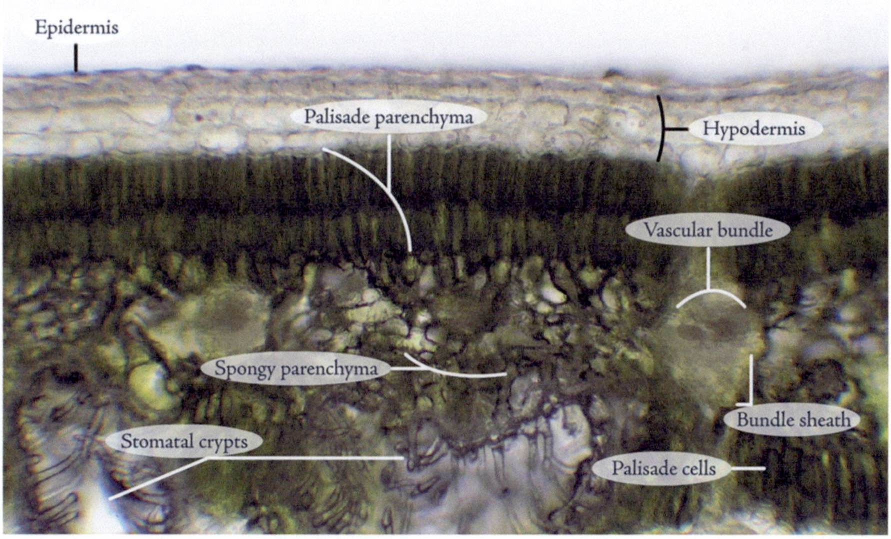

Fig. 10.41 Oleander (*Nerium oleander*) leaf cross-section detail

Leaf veins is not only important for the transport of material, it also stiffens and strengthens the leaf. In this, the sclerenchymatous sheath surrounding the vascular bundle, and the subepidermal and the leaf marginal strengthening tissues are important too.

In contrast to the dicots, the upper and lower parts of the leaves of monocots are mostly very similar, the mesophyll is also considered to be uniform, and both surfaces of the leaf contain stomas (Fig. 10.42).

Occasionally, however, a heterogeneous leaf may also develop in monocots, resulting in a separation of the spongy and palisade parenchyma.

Both on the upper and lower epidermis of the reed have stomas. The leaf is characterised by the formation of bulliform (motor) cells which, when stretched out, allow the leaf blade to expand and, when shrunk, allow it to contract.

The mesophyll is homogeneous, with double-sheathed bundles running through it.

The inner bundle sheath is partially sclerenchymatous and forming mechanical supports attached to the upper and lower epidermis. The outer sheath is composed of large parenchymatous cells. In the xylem, dilated vessels and narrow tracheids can be recognised. The phloem faces to lower surface of the leaf.

Maize photosynthesis is of the so-called C4 type, which is reflected in the characteristic structure of its leaf cross-section (Fig. 10.43). **C4 photosynthesis** enables maize to photosynthesise efficiently even under drier conditions.

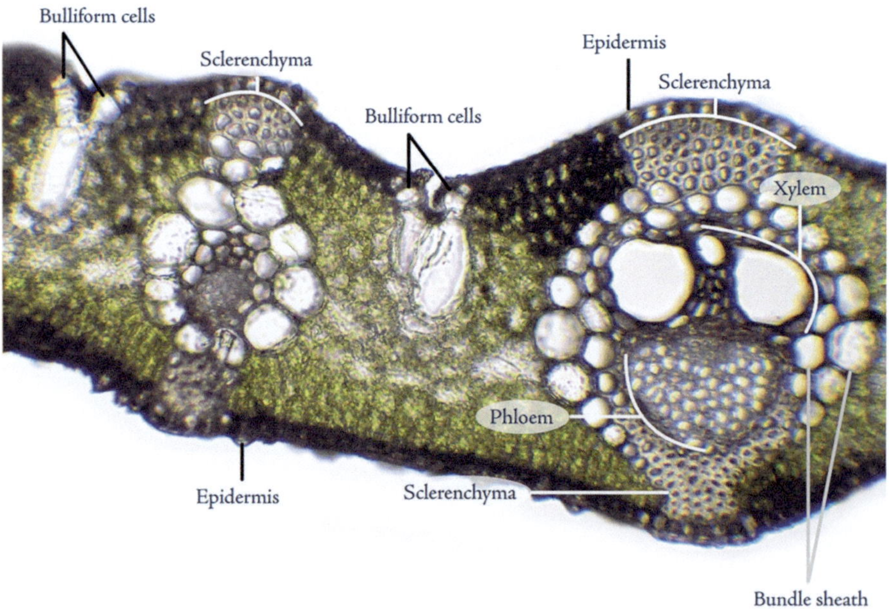

Fig. 10.42 Cross-section of reed (*Phragmites australis*) leaves

Fig. 10.43 Light microscope image of a cross-section of a maize (*Zea mays*) vascular bundle

In maize leaves, the bundle sheath cells surrounding the vascular bundle contain different chloroplasts (lacking the granums) than the mesophyll cells, which are radially positioned from them. As a consequence, the bundle sheath parenchyma cells do not contain photosystem II and therefore do not able to split water. Water splitting takes place in the mesophyll cells surrounding the bundle sheath parenchyma, while NADPH and oxygen are produced. Here malic acid is formed from carbon dioxide using NADPH. The malic acid enters the sheath parenchyma cells where it is converted to pyruvic acid by the formation of carbon dioxide and NADPH, which returns to the mesophyll parenchyma cell so that malic acid production can begin again. The ATP produced in the bundle sheath parenchyma cells (photosystem I) and the carbon dioxide and NADPH produced from malic acid provide all the material for the dark reactions of photosynthesis (Calvin cycle).

The mesophyll parenchyma cells supply large amounts of malic acid to the sheath parenchyma cells, which, together with the reduced oxygen content (chloroplasts do not have a functioning photosystem II and therefore no oxygen production), contributes significantly to the high efficiency of the enzyme ribulose-1,5-biphosphate carboxylase-oxygenase (the enzyme's carbon dioxide fixation is inhibited by oxygen). The efficient use of carbon dioxide allows the stomas to remain closed for longer periods of time, which makes the plant more drought tolerant due to less transpirational water loss. This type of photosynthesis is called C4 photosynthesis, after the four carbon atoms of malic acid.

A good guide to identifying the upper and lower surface of a leaf is usually the fact that the xylem part of the vascular bundles entering the leaf runs close to the

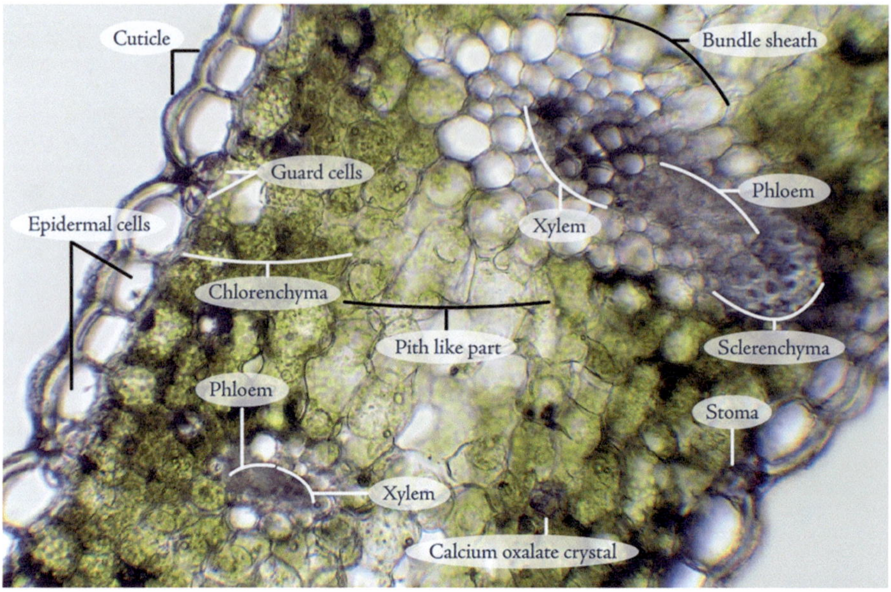

Fig. 10.44 Cross-section of *Iris* leaf

upper surface of the leaf, while the phloem part runs close to the lower surface. However, in a proportion of monocots the leaf develops into unifacial (Fig. 10.44).

Both sides of this type of leaf are formed from the lower part of original leaf, so that on the new surfaces the xylems of the vessel bundles face inwards. The *Iris*'s is a typical example of this leaf development. The characteristic structure is formed by the original leaf folding towards the upper surface at the midline and then the leaf halves fusing towards the tip. Thus, the two surfaces of the new leaf develop from the lower surfaces of the original leaf.

Pine needles are organs with a characteristic shape and tissue structure, which carry the characteristics of drought tolerance. The elongated, corrugated-walled epidermal cells, which are thick-walled and covered with cuticle, are clearly visible on the imprint of scots pine needle. The stomas, sunk into the epidermis, are arranged in parallel rows (Fig. 10.45).

Underneath the epidermis, the effect of toluidine blue staining reveals the hypodermis, which is coloured light blue (Fig. 10.46).

Underneath is the bright green chlorenchyma, with the resin ducts running into it. The central part is surrounded by endodermis.

The central part shows the cross-section of the vascular bundle. The cross-section clearly shows the xylem, the phloem, and the cambium. The central part is filled with **transfusion tissue**, which stores water and provides lateral flow (Fig. 10.47).

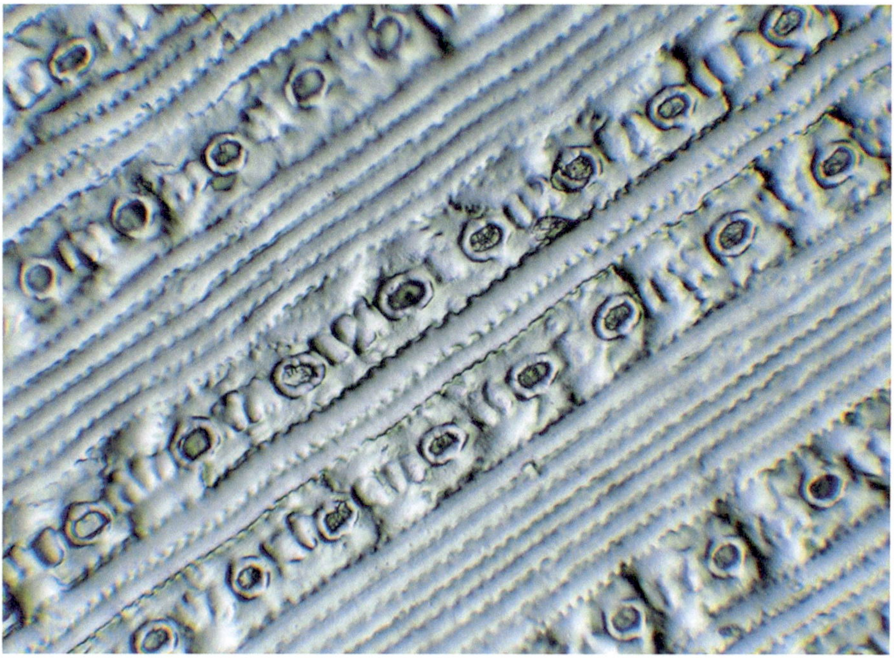

Fig. 10.45 Imprint the epidermis of scots pine (*Pinus sylvestris*) needle

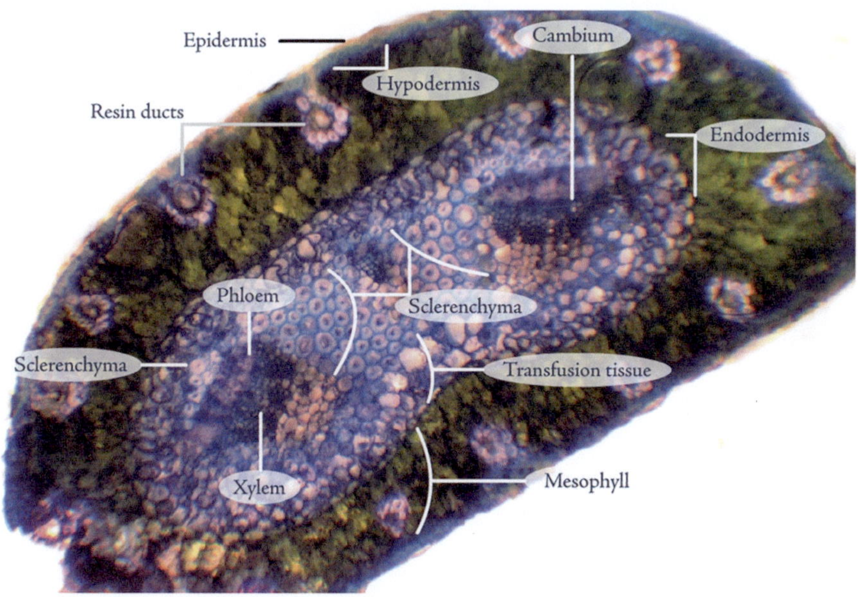

Fig. 10.46 Cross-section of scots pine (*Pinus sylvestris*) needle stained with toluidine blue

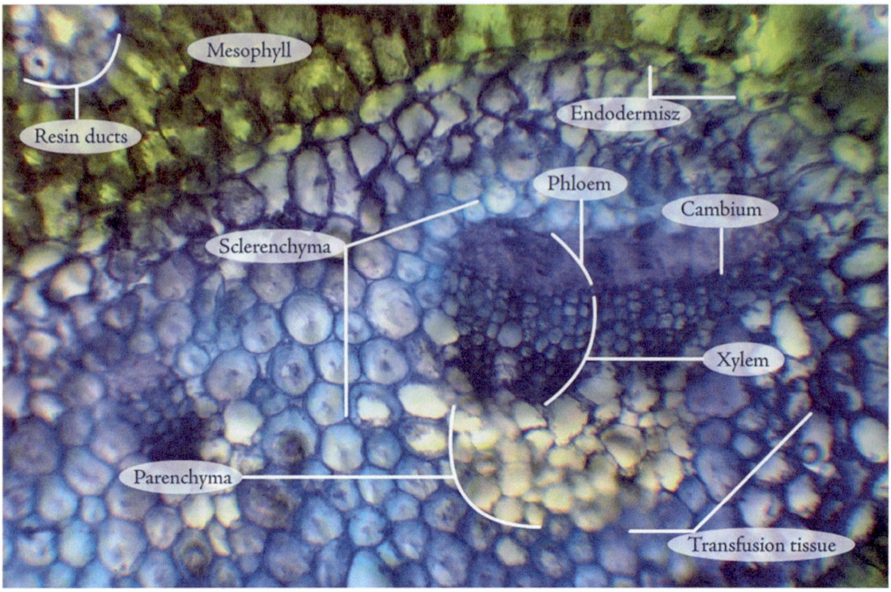

Fig. 10.47 High magnification detail of the central part of the needle leaf

Organ Histological Examinations 11

11.1 Root

11.1.1 Primary Growth of Root

11.1.1.1 Lily (*Lilium* sp.)

The shoot-root of the lily (Fig. 11.1) is covered on the outside by a rhizodermis with root hairs. The cell line underneath is parched to form a protective layer, the exodermis, which later takes over the role of the degraded rhizodermis.

Then comes the very wide cortical parenchyma further inwards, the cells of which become smaller and smaller as you move inwards. The innermost cell line of the cortex is the endodermis, composed of cells with thickened walls. In the endodermis, thin-walled permeable cells, passage cells are observed.

Inside the stem is the stele with eight xylem and eight phloem bundles (Figs. 11.2 and 11.3). The outer, narrow but thick-walled elements of the xylem bundles form the protoxylem. These are joined inwards by larger hollow vessels which converge at the centre. This part forms the metaxylem.

Between the xylem bundles are colourless, white shiny glowing phloem, which consist of sieve tubes and companion cells. There is no direct contact between the xylem and the phloem bundles, which are separated by a layer of parenchyma. The alternating xylem and phloem bundles are surrounded externally by pericycle. The roots of other bulbous plants have a similar structure, but the number of xylem bundles may be different (Figs. 11.2 and 11.3).

During lateral root formation, a cluster of meristem is formed from the pericycle opposite one of the xylem bundle to form the lateral root tissues (Figs. 11.4 and 11.5).

At the affected site, the endodermal cells regain their ability to divide and, enveloping the developing lateral root, keep pace with its growth for a while and then rupture.

The xylem bundles contain large vessels, whose different cell wall thicknesses can be studied in longitudinal sections (Figs. 11.6 and 11.7).

G. Kriska, *Teaching Biology*, https://doi.org/10.1007/978-3-031-84085-2_11

Fig. 11.1 Flowering lily

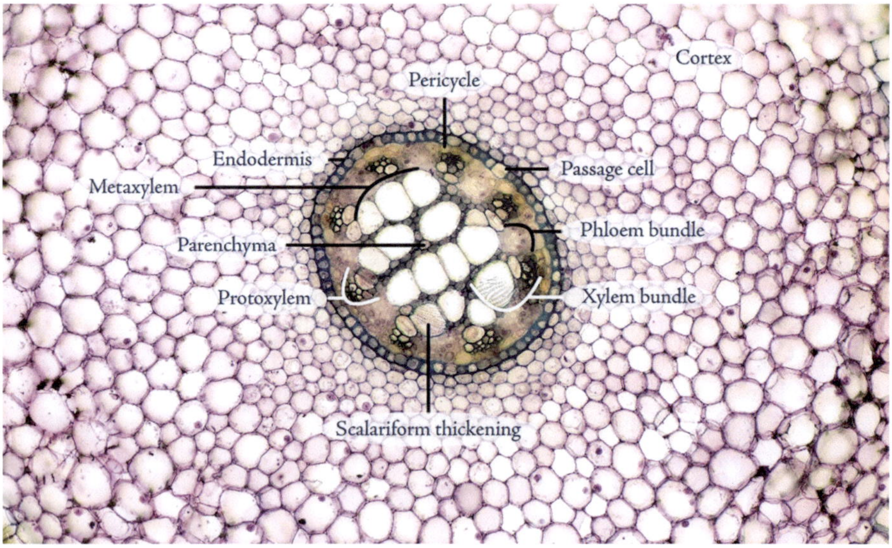

Fig. 11.2 The stele of the lily (*Lilium* sp.) root

Make a section of the bulb of the lily and stain it with Lugol's solution (diluted iodine-potassium iodide solution). In the section, we can observe the unilayered epidermis covering the outer surface and the exodermis underneath, which is also

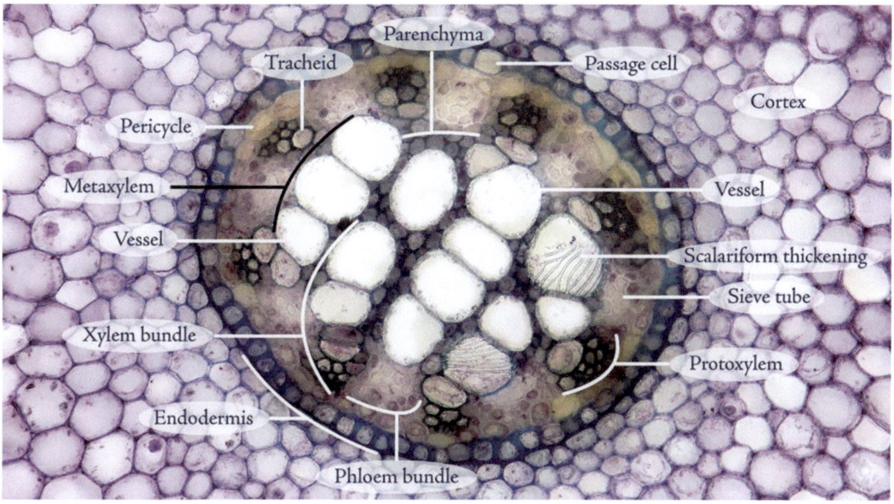

Fig. 11.3 High magnification image of the stele of a lily (*Lilium* sp.) root

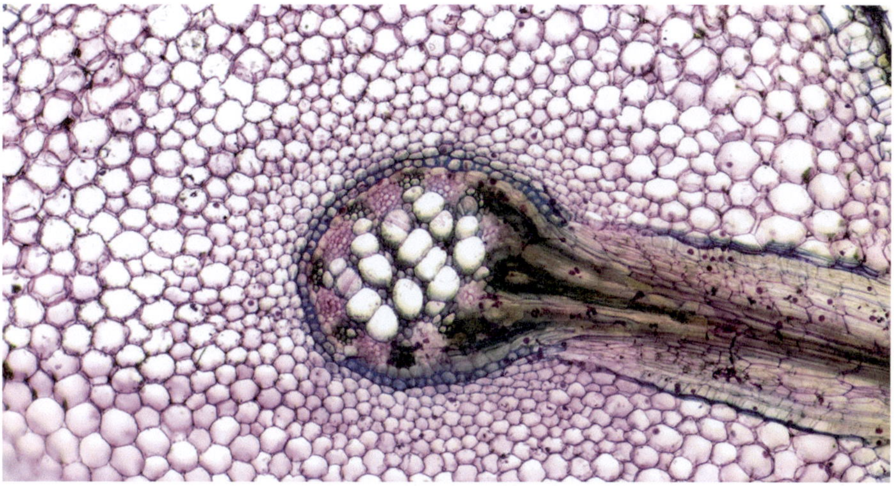

Fig. 11.4 Lily (*Lilium* sp.) lateral root formation

unilayered. The storage parenchyma filling the bulb contains purple starch granules (Figs. 11.8 and 11.9).

11.1.1.2 *Iris* sp.

In monocots, the rootstock (rhizome) develops side roots of uniform thickness. These are thick enough in the case of the Iris to make thin cross-sections (Fig. 11.10).

When making the sections, it is often the case that the stele, surrounded by endodermis, is cut out from the looser structured cortex.

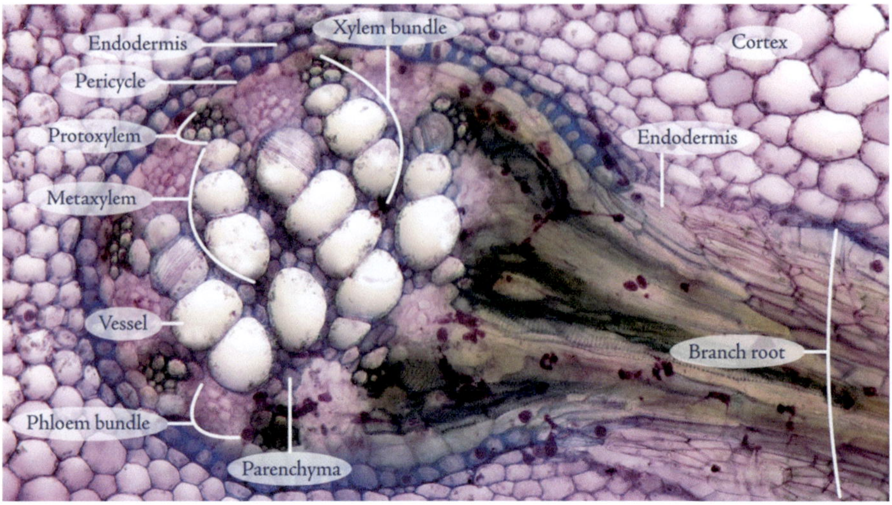

Fig. 11.5 Lily (*Lilium* sp.) root with lateral root

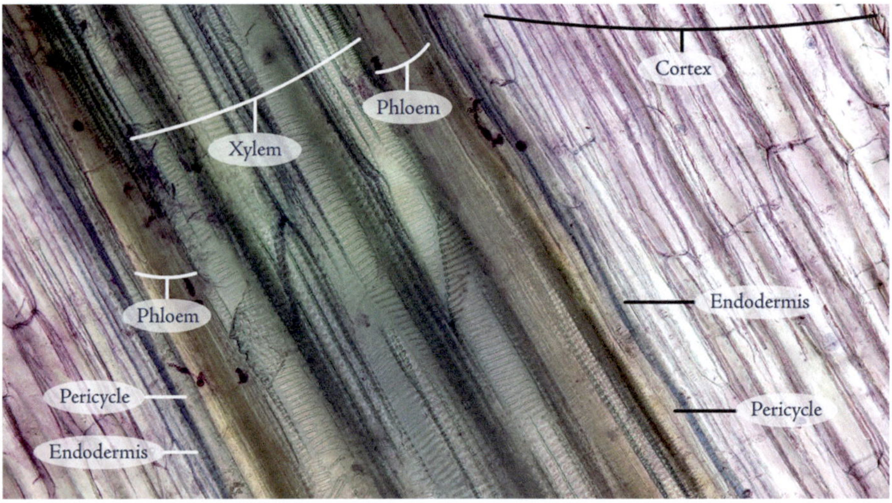

Fig. 11.6 Longitudinal section of a xylem bundle

Staining the sections with toluidine blue reveals the structure typical of monocots (Figs. 11.11, 11.12, and 11.13). In the central part of the section, the vascular cylinder (stele) can be observed, surrounded on the outside by the endodermis, the innermost cell line of the cortex. This cell layer, which surrounds the stele like a cloak, plays an important role in the transport of solutions absorbed from the soil into the vascular tissue.

The endodermis prevents the flow of material between the gaps in the cell walls and between the cell wall and the cell membrane.

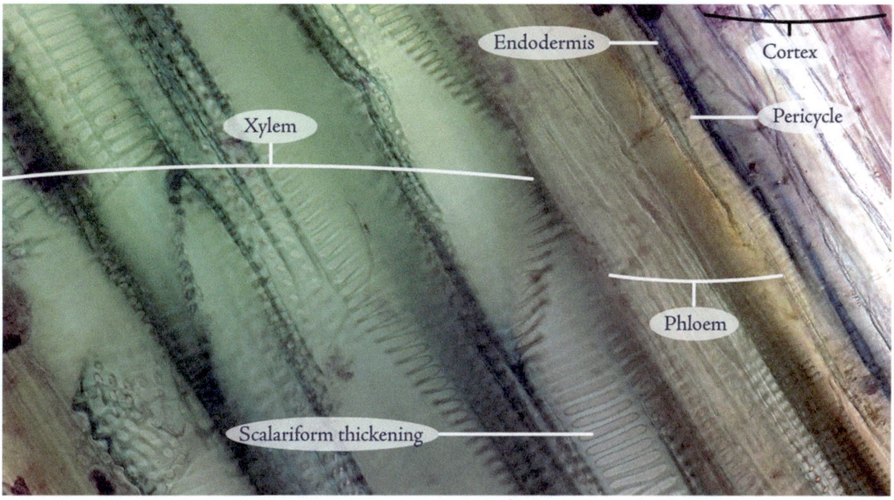

Fig. 11.7 High magnification image of a longitudinal section of a xylem bundle

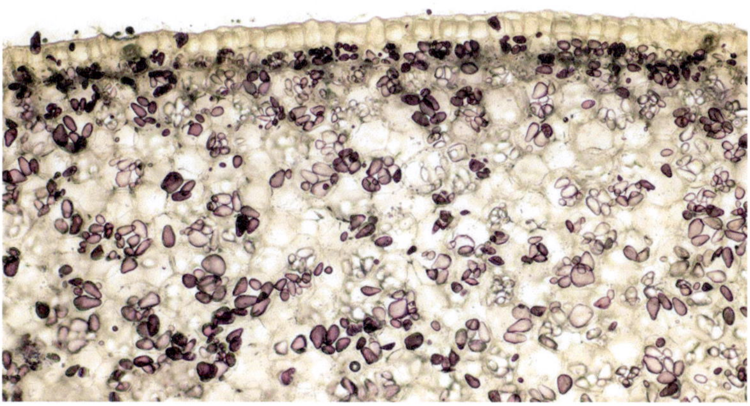

Fig. 11.8 Section of the lily (*Lilium* sp.) bulb

The solutions can only pass through the endodermis if their molecules enter the plasma of the endodermis cells, which provides the plant with controlled uptake. Sections taken further from the root tip show a tertiary endodermis with a supporting function, characterised by a "U" shaped thickening of cells formed by layers of cellulose and lignin. Between the thick-walled endodermis cells, passage cells for substance transfer are retained. The passage cells are always preserved above the xylem bundles.

In monocots, such as the Iris root, no new vascular tissues are organised. The vascular cylinder shows the general structure (polyarch stele) typical of monocots. Under the endodermis is a single-layered pericycle.

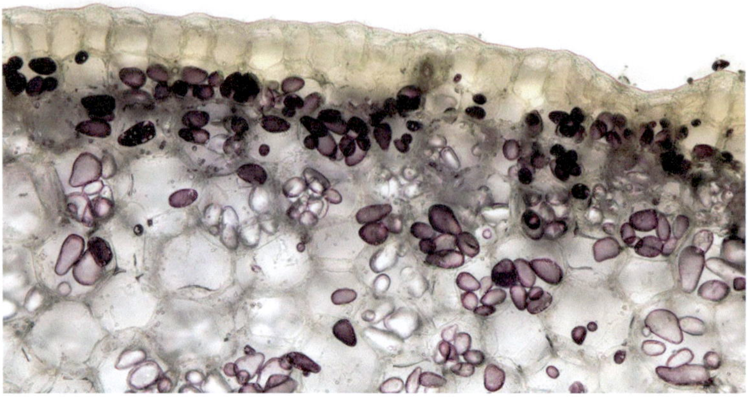

Fig. 11.9 Starch particles stained in the lily (*Lilium* sp.) bulb

Fig. 11.10 Flowering *Iris* sp.

This meristem is not functional here, but in dicots it allows the secondary growth of the root, the formation of the lateral root and the periderm.

In the vascular cylinder, simple xylem bundles alternate with simple phloem bundles. The xylem bundles are radial, while the phloem bundles are rounded and fibreless (soft phloem). The number of bundles is more than 10, for example 16 bundles can be observed in Fig. 11.11. The central part of the stele is filled with sclerenchyma.

This creates the cable structure typical of roots, which increases the tensile strength of the organ.

Iris species spread by a rhizome close to the soil surface, so it is easy to dig up a piece for microscopic examination.

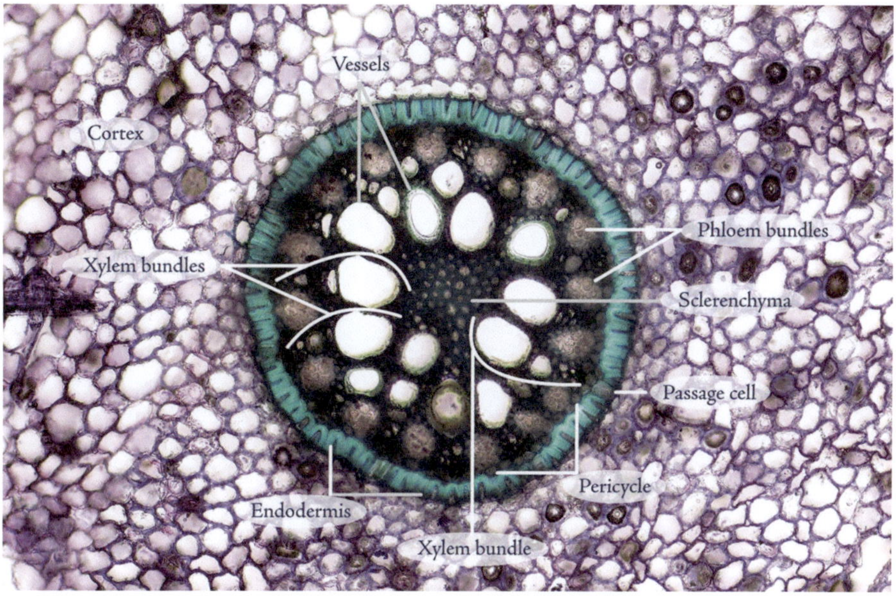

Fig. 11.11 Light microscopy image of the central part of the *Iris* lateral root cross-section. Toluidine blue staining

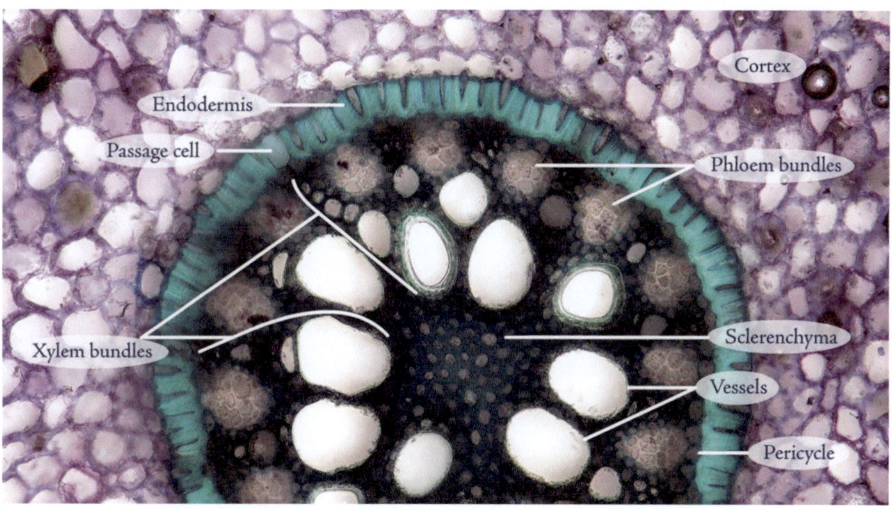

Fig. 11.12 High magnification image of the central part of the *Iris* lateral root cross-section

In the cross-section of the rhizome, it can be observed that after the rhizodermis has been disappeared, its protective role is taken over by the exodermis, the outer layer of cells of the cortex. The inner cortex is composed of parenchymatous cells containing starch granules (Figs. 11.14 and 11.15).

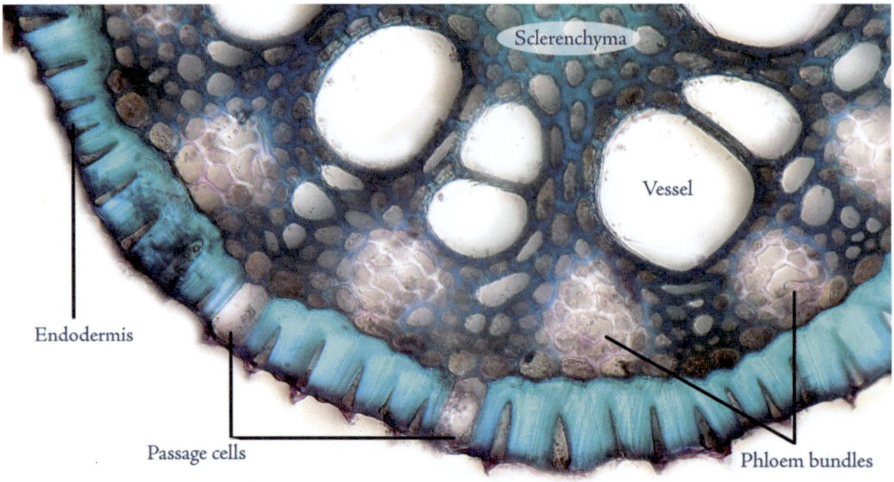

Fig. 11.13 Detail of the vascular cylinder delimited by the endodermis, torn from the cortex

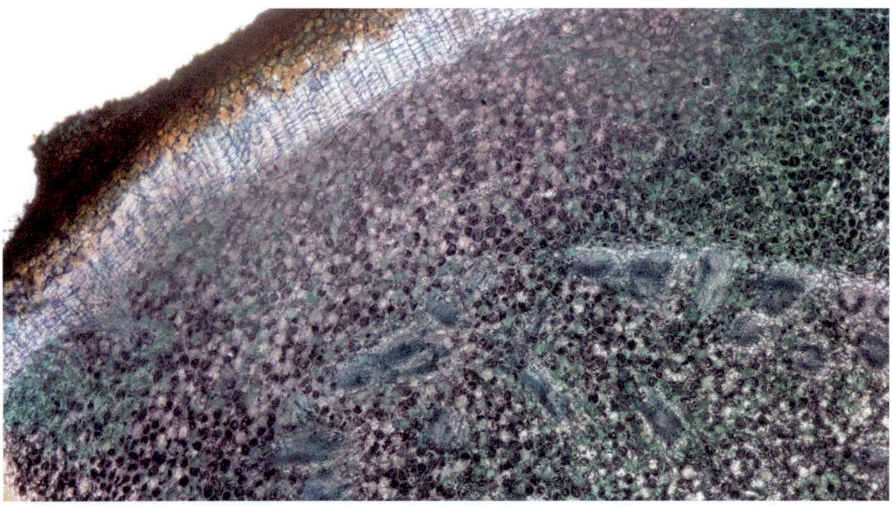

Fig. 11.14 Cross-section of *Iris* rhizome

By staining the sections with Lugol's solution, we can observe the location of starch granules in the cortical parenchyma cells (Fig. 11.16).

The cross-sections of the vascular bundles inside the rhizome are clearly visible in the centre (Fig. 11.17).

11.1.1.3 Lesser Celandine (*Ficaria verna*)

Lesser celandine is a herbaceous perennial plant. Its flowers open from late March to May (Fig. 11.18). By early June, the above-ground parts of the plant die and

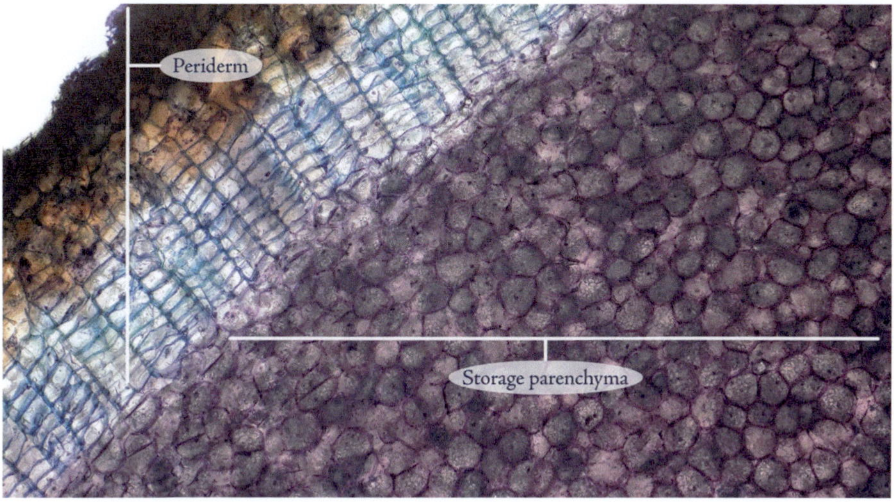

Fig. 11.15 High magnification image of a cross-section of the *Iris* rhizome

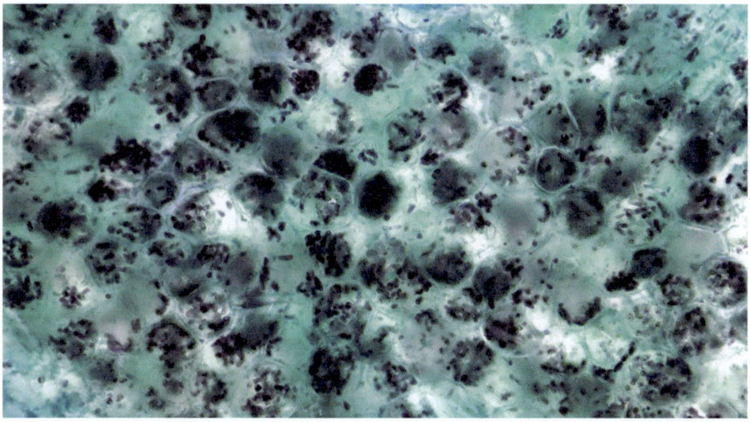

Fig. 11.16 Light microscopy image of a Lugol's solution-stained rhizome

continue to live in the soil in tiny root tubers until the following spring (Fig. 11.19). Because of its short above-ground life cycle, it needs to photosynthesise vigorously to accumulate sufficient nutrients, starch, in its root tubers to survive. For this reason, in spring its leaves almost completely cover the soil surface in its habitat.

The root of the lesser celandine is very thin and not a very suitable object for light microscopy, but cross-sections of the root tubers can be easily made (Figs. 11.11 and 11.12).

These sections illustrate the root vascular tissue in addition to the large amount of starch-storing ground tissue.

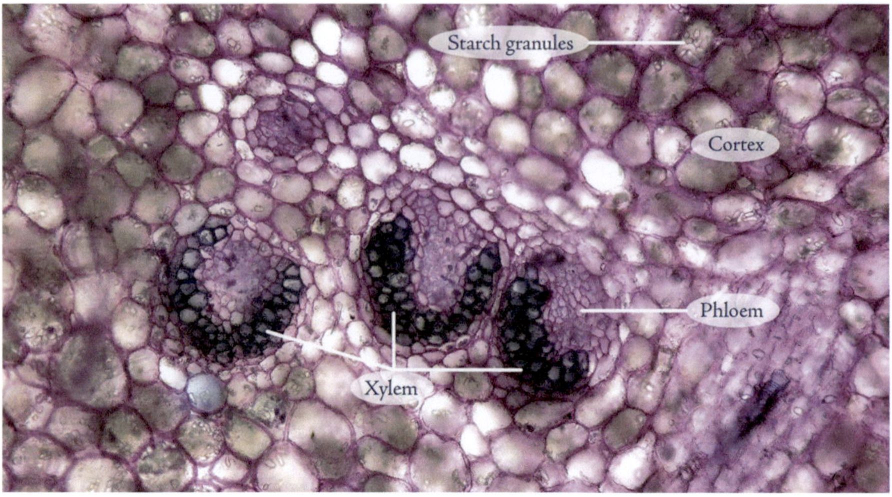

Fig. 11.17 Cross-sections of the vascular bundles inside the rhizome

Fig. 11.18 Lesser celandine (*Ficaria verna*)

When the sections are covered, large amounts of starch particles are released, which should be stained with Lugol's solution. After just one to two drops, the purple and over-stained black particles will appear. The cortical parenchyma of the root tuber, which reaches a considerable thickness, is filled with cells packed with starch granules (Figs. 11.20 and 11.21).

A section of the root tuber also shows the root structure of the lesser celandine, an ancient dicotyledonous plant.

Fig. 11.19 Root tubers of lesser celandine (*Ficaria verna*)

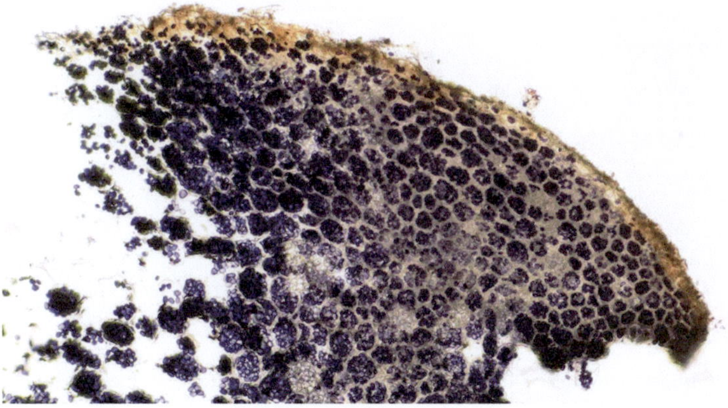

Fig. 11.20 Lugol's solution-stained starch granules in the cortical parenchyma cells of the root tuber

Moving inwards in the section from the outside can be seen the rhizodermis covered with root hairs, the starch-storing cortex parenchyma and the vascular cylinder (Fig. 11.22).

To examine this, however, it is definitely worthwhile to perform a toluidine blue staining. After 10–20 s of toluidine blue staining, the vascular cylinder and its parts are clearly visible in the tuber section (Figs. 11.23, 11.24 and 11.25).

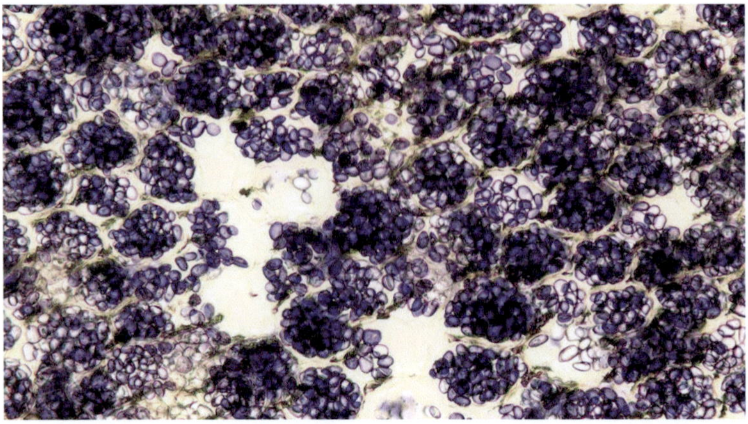

Fig. 11.21 Cortical parenchyma cells filled with stained starch granules

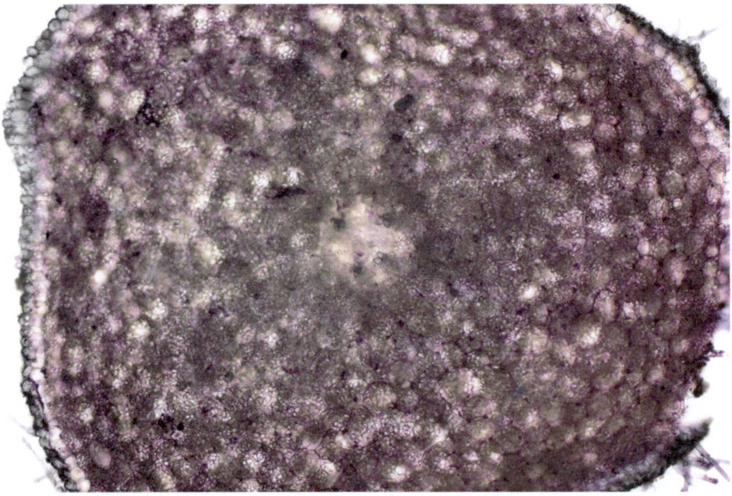

Fig. 11.22 It is difficult to see the vascular cylinder in a section of the root tuber at low magnification

The innermost cell layer of the cortex is the endodermis. The layer of cells surrounding the stele in the form of a mantle plays an important role in the transport of solutions absorbed from the soil to the vascular tissue. The outer cell layer of the central cylinder adjacent to the endodermis is the pericycle. In lesser celandine, this cell layer loses its functionality and the root does not thicken secondarily. In the stele, the vascular tissues are organised in simple, alternating bundles of xylem and phloem, separated by parenchymatous cells. In the xylem bundles, the smaller xylem elements that are formed first are arranged towards the periphery, while the larger ones that are formed later are arranged towards the centre. Pith consisting of a few large parenchyma cells is observed in centre. Lesser celandine is usually

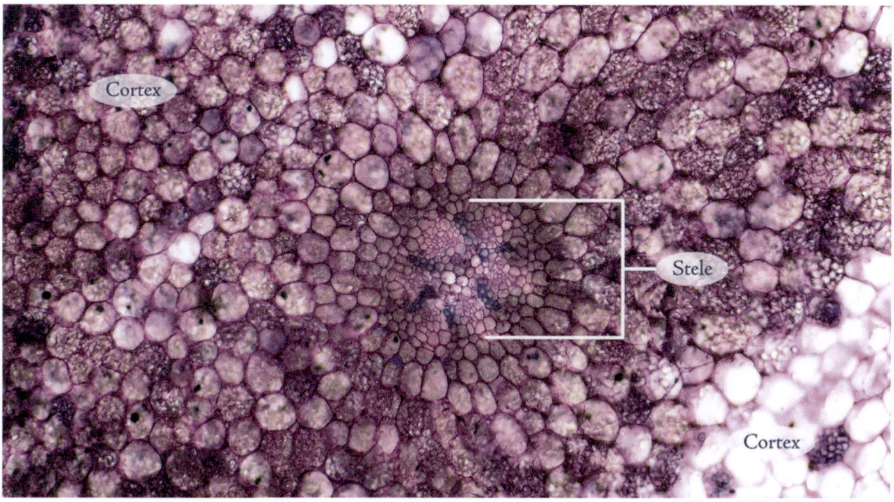

Fig. 11.23 Light microscope image of a root tuber section stained with toluidine blue

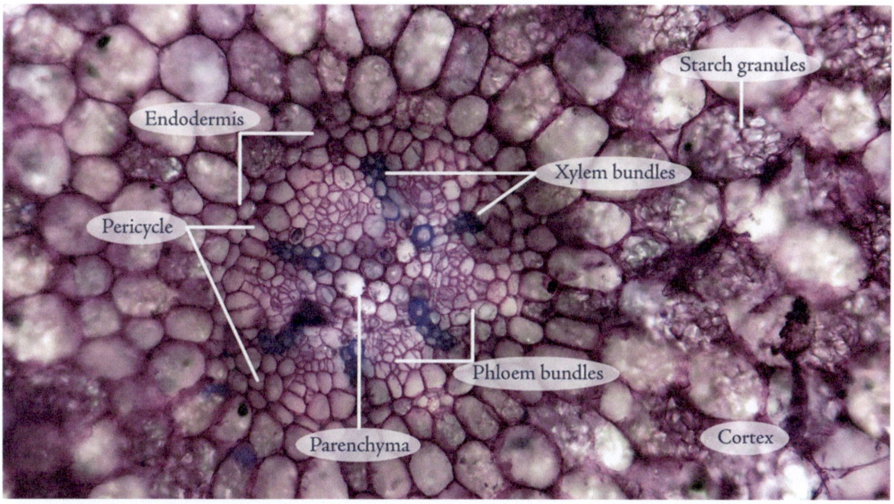

Fig. 11.24 Higher magnification image of a root tuber section stained with toluidine blue

characterised by a 3–4 xylem bundles structure, but here we see an example of a stele with 6 xylem bundles.

11.1.1.4 Dragon Tree (*Dracaena draco*)

Make a cross-section from a dragon tree (Fig. 11.26) root with a razor blade, then stain the cross-section with toluidine blue. Cover the preparation and examine it by a light microscope.

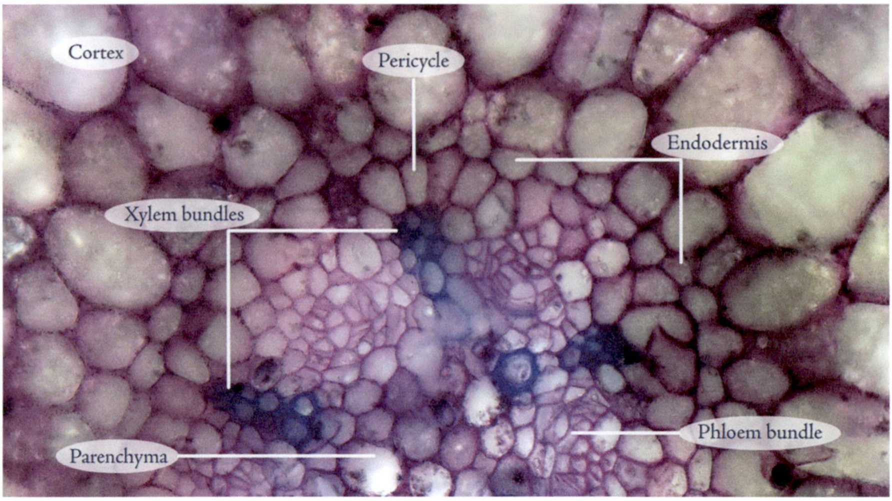

Fig. 11.25 High magnification light microscopy image of a root tuber section

Fig. 11.26 Dragon tree (*Dracaena draco*)

In the section, the U-shaped thickened cell walls are clearly visible at the endodermis, the innermost cell layer of the cortex. The outer cell of the stele, the pericycle, is easily identified. Further inside, the xylem and phloem bundles are alternately located.

The protoxylem and the later developing metaxylem can be clearly distinguished in the xylem bundles. The central part of the root is filled with pith with scattered

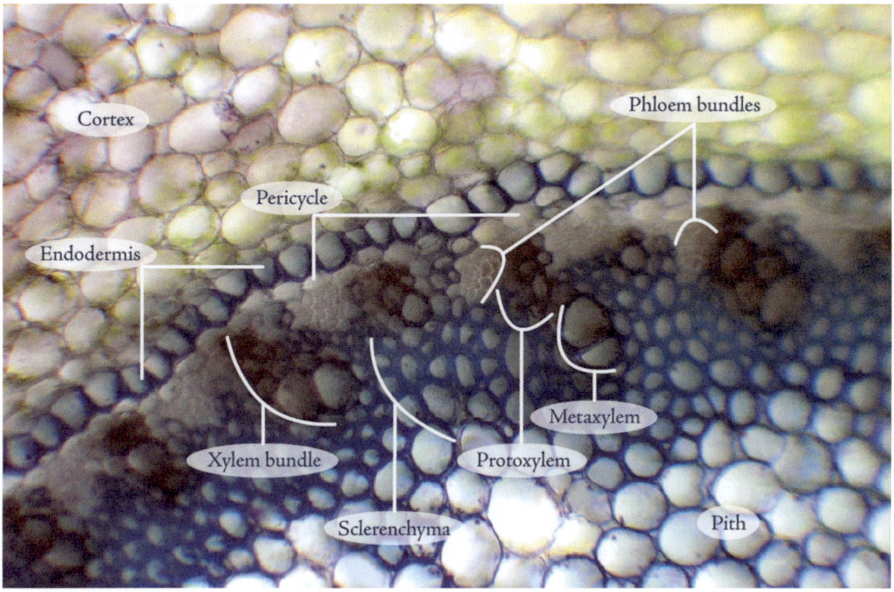

Fig. 11.27 Root cross-section of a dragon tree (*Dracaena draco*) stained in toluidine blue

vessels. These are surrounded by thick-walled parenchyma cells, forming rings of vessels (Figs. 11.27, 11.28, 11.29, and 11.30).

11.1.2 Secondary Growth of Dicot Root

11.1.2.1 Bean (*Phaseolus vulgaris*)

Make a thin section of the developing root of a 4- to 5-day-old bean sprout (Fig. 11.31) in the region of maturation where the lateral roots appear.

Place the section in a toluidine blue solution for 10–15 s, then cover and examine it under a light microscope.

The different types of tissue are clearly distinguished in the section (Figs. 11.32, 11.33, 11.34, 11.35, and 11.36). Under the epidermis, the pink-stained cells of the cortical parenchyma are visible. The innermost cell of the cortex is the endodermis, followed by the outermost cell of the stele, the pericycle. The lateral roots are formed from this meristem. The lateral roots are formed in the direction of the four primary xylems, which are radial and perpendicular to each other.

The narrower, outer elements of the primary xylem form the protoxylem, which is the first to form during root development, while the metaxylem, which consists of wider elements, is located further in.

In addition to the four primary xylems, elements of secondary xylem from the cambium fill the stele, while in the innermost part, the centrally positioned pith is observed. The cambium develops outwardly into the secondary phloem, with the insular primary phloem visible on the outer side. Recognition is aided by the

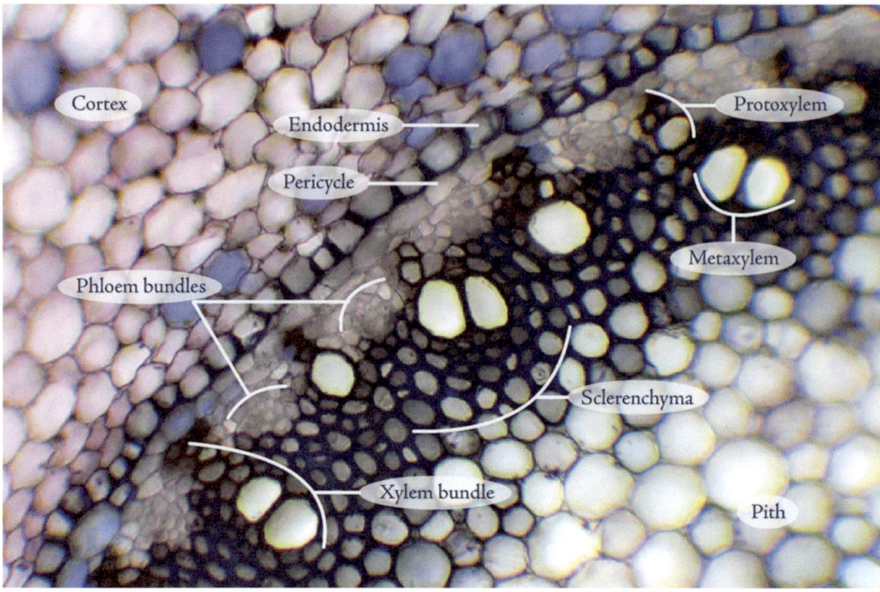

Fig. 11.28 Root cross-section of a dragon tree (*Dracaena* sp.) stained in toluidine blue

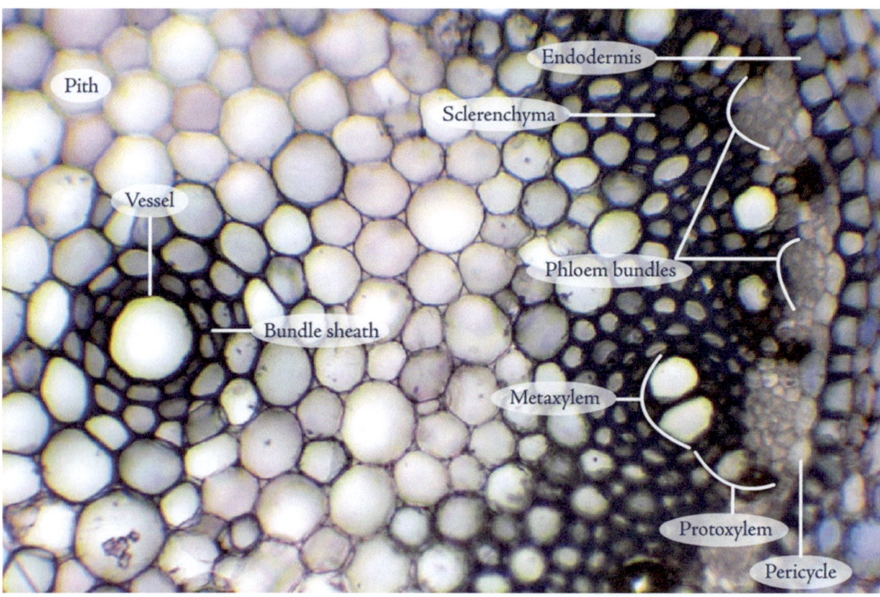

Fig. 11.29 Dragon tree (*Dracaena draco*) root cross-section with internal vessel

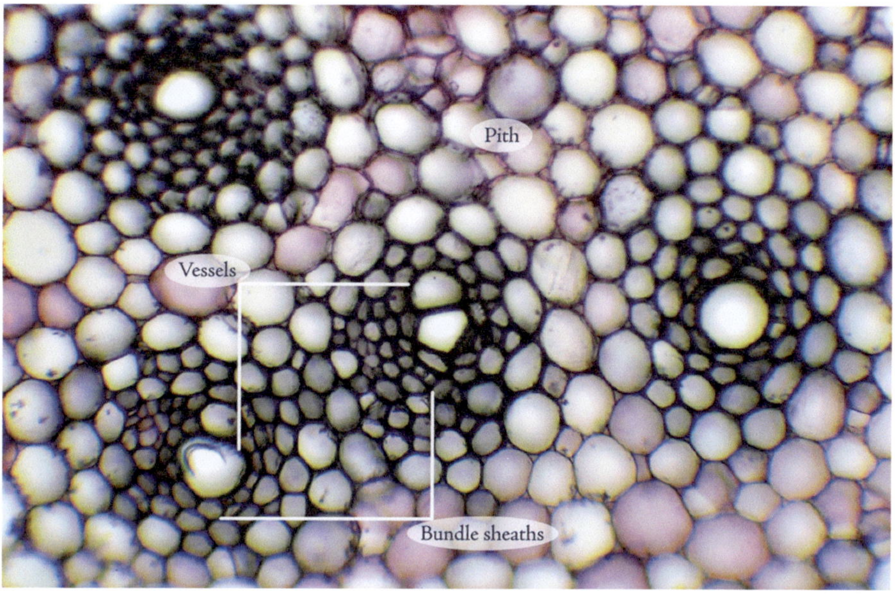

Fig. 11.30 Scattered vessels in the pith of the dragon tree (*Dracaena draco*)

Fig. 11.31 5-day bean (*Phaseolus vulgaris*) sprout

staining of the primary (blue) and secondary (pink) phloem. The pericycle, which will later be involved in the formation of the periderm, is visible above the primary phloem. The cambium is several rows of cells wide on the primary phloem, tapering to a single row of cells towards the xylem bundles. Its function is the most intense on the side of the primary cambium: it creates inwardly dilated vessels. The cells of the pith are enlarged.

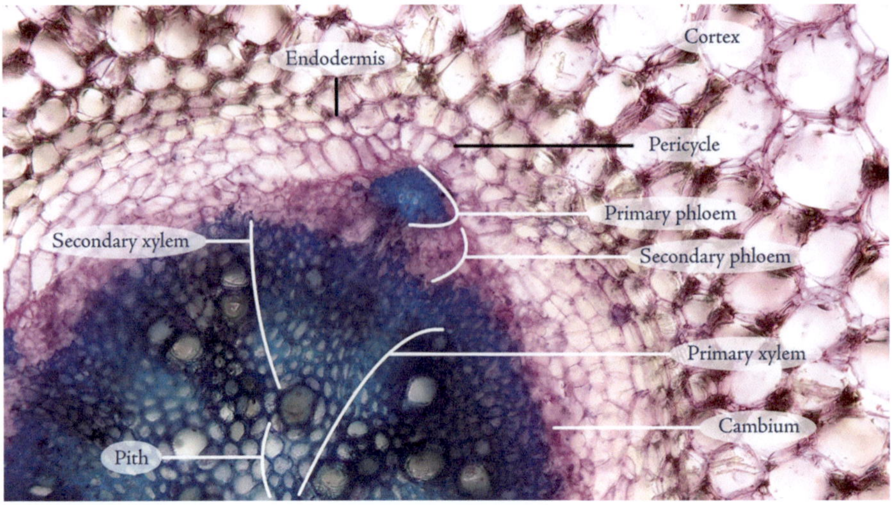

Fig. 11.32 Cross-section of a bean (*Phaseolus vulgaris*) root with secondary growth, stained with toluidine blue

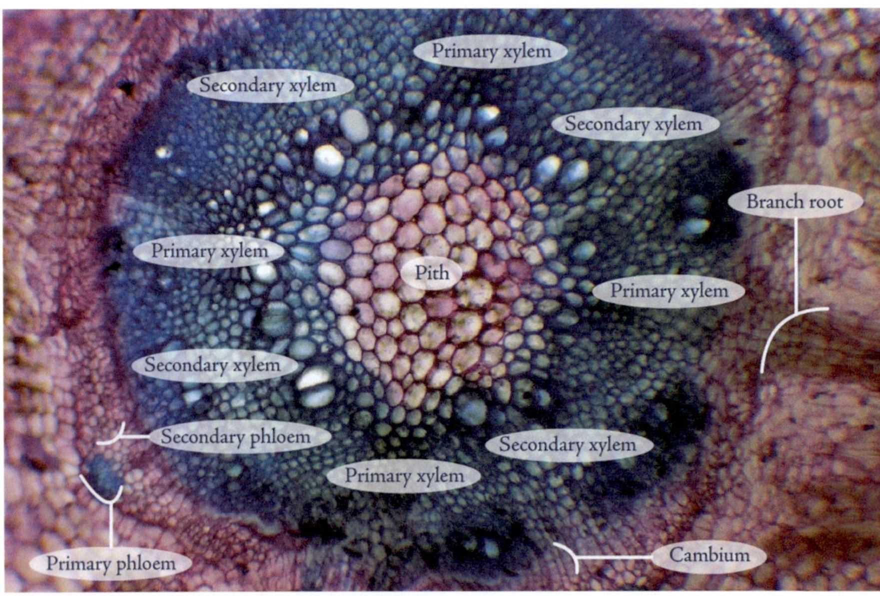

Fig. 11.33 Central part of a secondarily thickening bean (*Phaseolus vulgaris*) root

Outward from the cambium, dilated sieve tubes, plasma-containing companion cells, and a phloem parenchyma are formed.

As the cambium continues to function, it develops a coherent system of vascular tissues, which is divided in places by 3–9 cell rows of wide medullary rays. The xylem is built up by vessels, rows of tracheids and mainly by xylem fibres. The

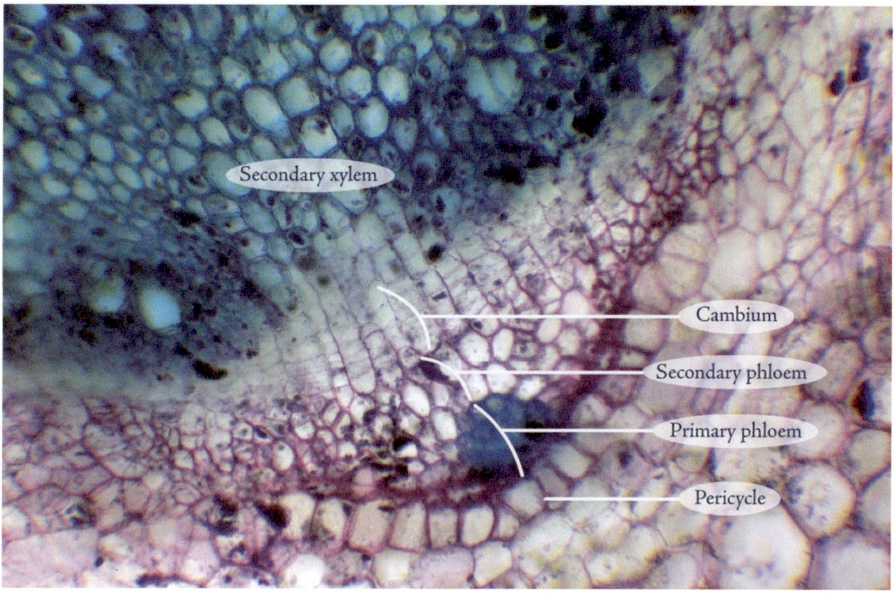

Fig. 11.34 High magnification image of the vascular cylinder detail

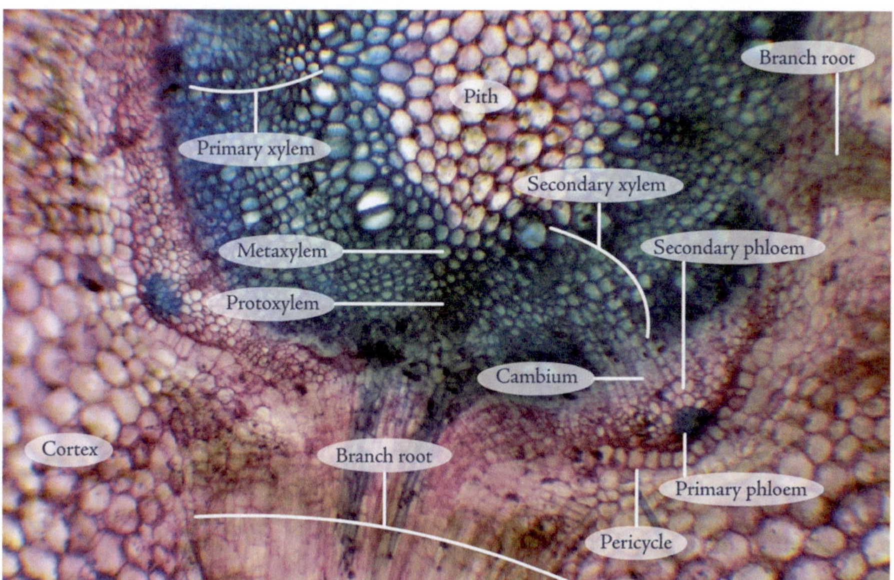

Fig. 11.35 Lateral root formation in a secondary thickened bean (*Phaseolus vulgaris*) root

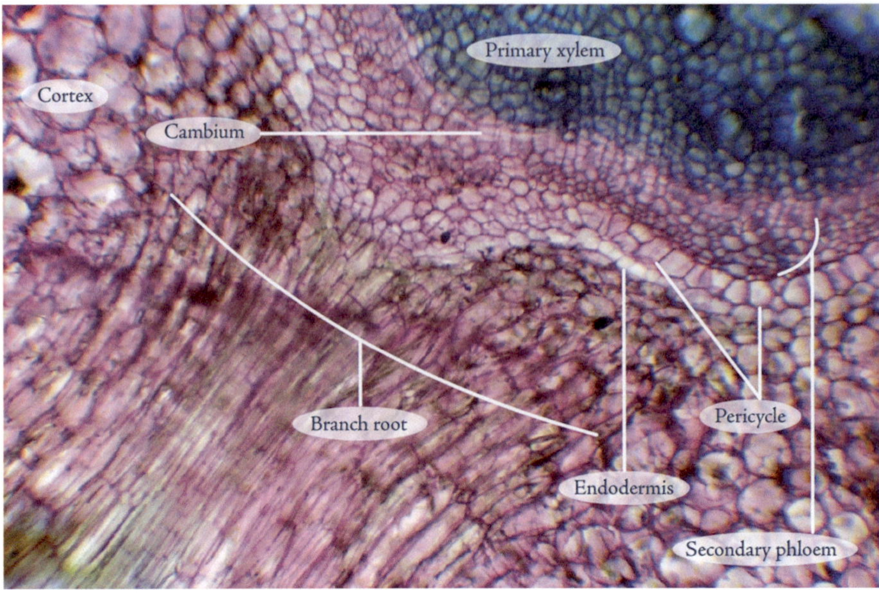

Fig. 11.36 High magnification image of the site of lateral root formation

Fig. 11.37 Pea (*Pisum sativum*) root with root nodules

phloem grows at a slower rate and covers the cambium as a narrow tissue. Within the rhizodermis, which is still present but already ruptured, the cells of the primary cortex begin to cover and protect the root as exodermis.

11.1.2.2 Pea (*Pisum sativum*)

The pea has a deep root system with root nodules on the lateral roots, in which nitrogen-collecting bacteria live (Fig. 11.37). The cross-section taken above the root apex in the region of maturation shows the characteristic histological structure without staining.

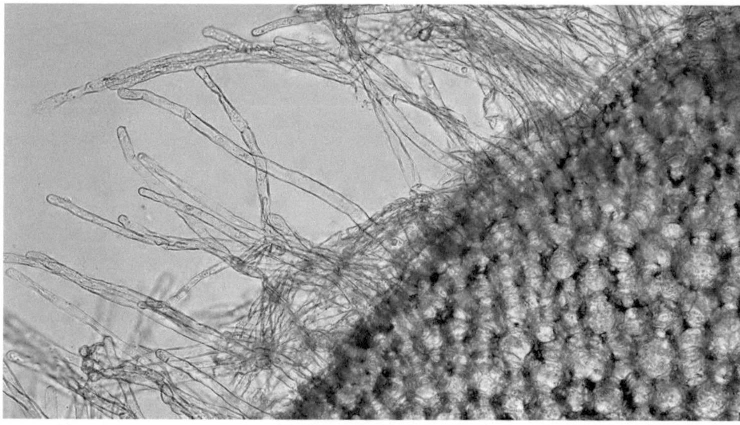

Fig. 11.38 Cross-section of pea (*Pisum sativum*) root with cortex and root hairs

Note the rhizodermis, which forms unicellular root hairs, and the thick cortical parenchyma beneath, formed by rounded cells filled with starch granules, often with an angular cell wall thickening (Fig. 11.38).

The innermost cell layer of the cortex is the endodermis, beneath which the outer cell layer of the stele is the pericycle. The stele contains alternating xylem and phloem bundles, 3 to 3 or 4 to 4 (Fig. 11.39).

In the section taken above the region of maturation, the structure of the secondary thickened root can be examined by light microscopy. The structure of the cross-section is clearly visible due to the staining with toluidine blue (Figs. 11.40 and 11.41).

The first step in secondary growth is the formation of a wavy cambium. This is formed by the pericycle parts lying outside the xylem bundles and the parenchyma cells edging the phloem bundles from inside. In the resulting structure, the primary phloem bundles (light blue) are located on the outer side of the cambium ring, while the primary xylem bundles (dark blue) are located on the inner side. The wavy cambium in a continuous ring or lobe structure forms xylem elements (secondary xylem) inwards and phloem elements (secondary phloem) outwards. Due to the uneven formation of xylem elements in the cambium, the originally wavy cambium later becomes circular.

As the root thickens, the primary xylem bundles can still be recognised in the central part of the stele, while the primary phloem is pushed outwards by the secondary phloem.

11.1.2.3 Birthwort (*Aristolochia clematitis*)

Make a thin cross-section of the root of a birthwort (Fig. 11.42) and examine it with a light microscope, stained with toluidine blue.

In the vascular cylinder, xylem bundles, containing wide vessels and purplish-coloured phloem bundles alternate with each other to form a spectacular cross-shaped pattern. The secondary thickening of the root is provided by the pericycle,

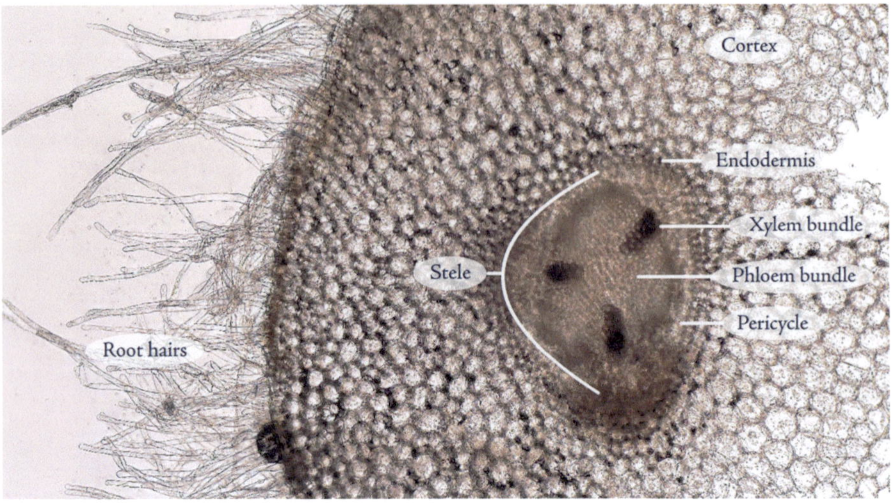

Fig. 11.39 Cross-section of pea (*Pisum sativum*) root

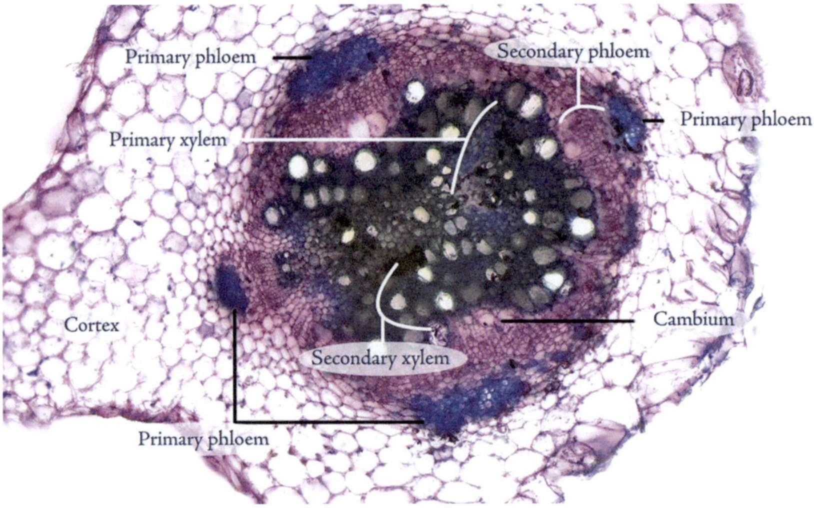

Fig. 11.40 Cross-section of a secondarily thickened pea (*Pisum sativum*) root

the part of which, outward from the xylem bundles, escapes the inhibitory influence of the apical meristem and develops xylem inward and phloem outward, thus forming large vascular bundles, similar to what we saw in the stem. The phloem bundles stop growing, and the growing xylem parts compress them more and more. The part of the pericycle above the phloem bundles thickens the cortex outwards. The rhizodermis, which is torn off during thickening, is replaced by a secondary dermal tissue, the periderm (Figs. 11.43, 11.44, and 11.45).

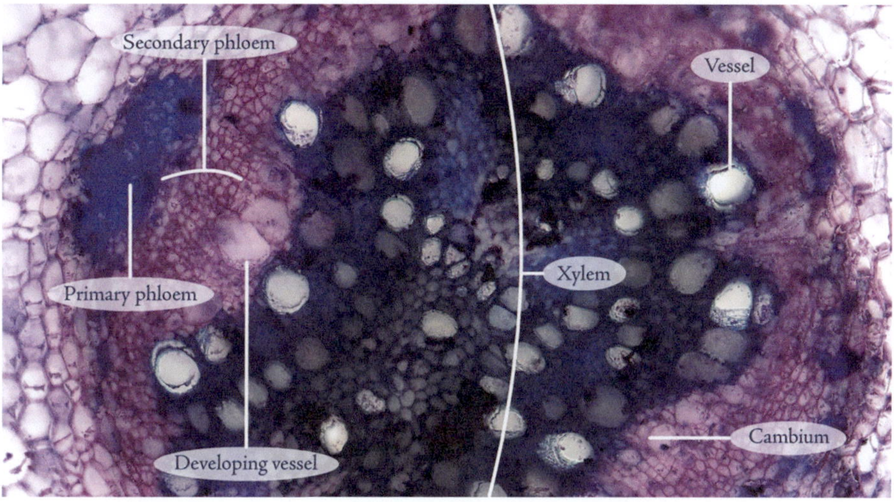

Fig. 11.41 Central part of a secondarily thickened pea (*Pisum sativum*) root

Fig. 11.42 Flowering birthwort (*Aristolochia clematitis*)

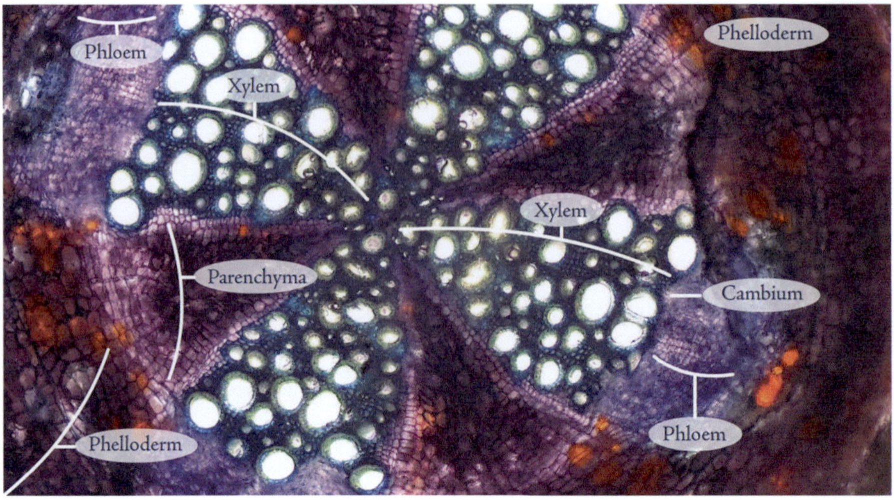

Fig. 11.43 Central part of the root of a birthwort (*Aristolochia clematitis*)

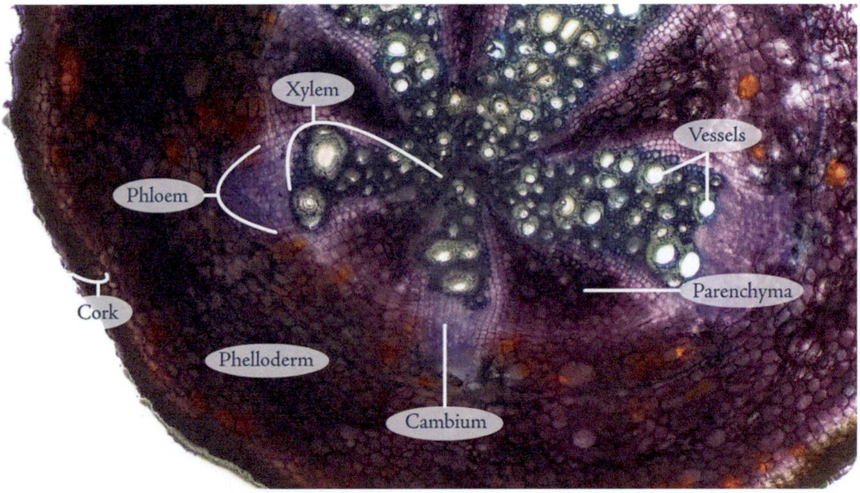

Fig. 11.44 Cross-section of the root of a birthwort (*Aristolochia clematitis*)

11.1.2.4 White Campion (*Silene latifolia*)

A section of the older root of the white campion (Fig. 11.46) shows the tissue structure of the root as it thickens secondarily.

The root is now covered with periderm on the outside, because the rhizodermis and the underlying primary cortex have been worn away during the thickening of the organ. Since the cork cambium, which forms the periderm, is not formed in the primary cortex but deeper, immediately above the phloem, the parenchymatous part

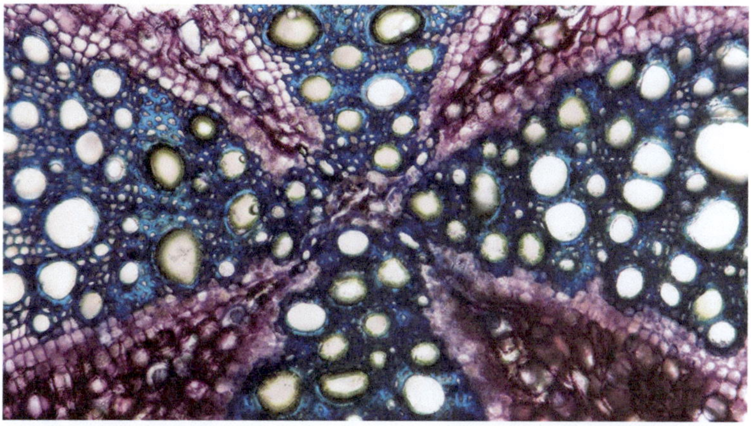

Fig. 11.45 High magnification image of the central part of the root cross-section of a birthwort (*Aristolochia clematitis*)

Fig. 11.46 White campion (*Silene latifolia*) flower

between the cork and the phloem is also formed by the cork cambium and corresponds to the phelloderm.

The phloem and xylem form a single cylindrical layer, made by the cambium ring between them. The xylem contains dilated vessels, narrower tracheids, xylem fibres, and xylem parenchyma. The wood parenchyma forms lighter intestinal rays, which continue into the phloem. The xylem completely fills the central part of the root, no pith is formed (Figs. 11.47, 11.48, and 11.49).

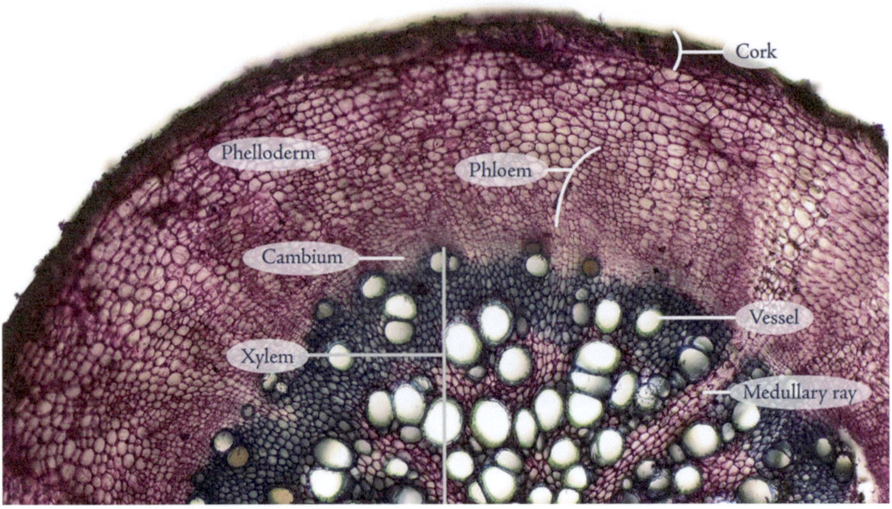

Fig. 11.47 Cross-section of an older white campion (*Silene latifolia*) root

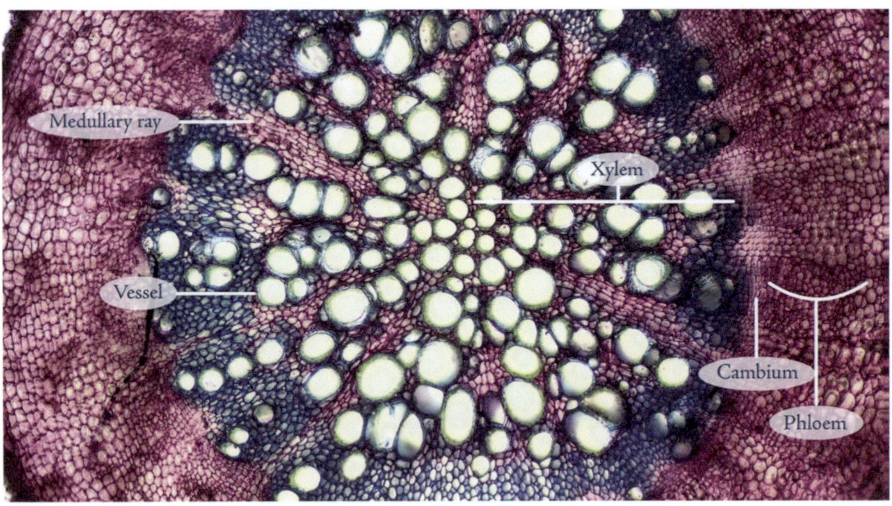

Fig. 11.48 Central part of the old root of white campion (*Silene latifolia*)

11.1.2.5 Wild Pansy (*Viola tricolor*)

The thicker roots of the wild pansy (Fig. 11.50) are easy to make sections from. These roots already show secondary thickening. Their characteristic tissue structure can be studied in cross-sections stained with toluidine blue.

The central part of the root cross-section is filled by the primary xylem, in which the wide cavity vessels are clearly visible, as are the cross-sections of the narrower cavity tracheids and the pink-stained xylem parenchyma cells. Moving outwards,

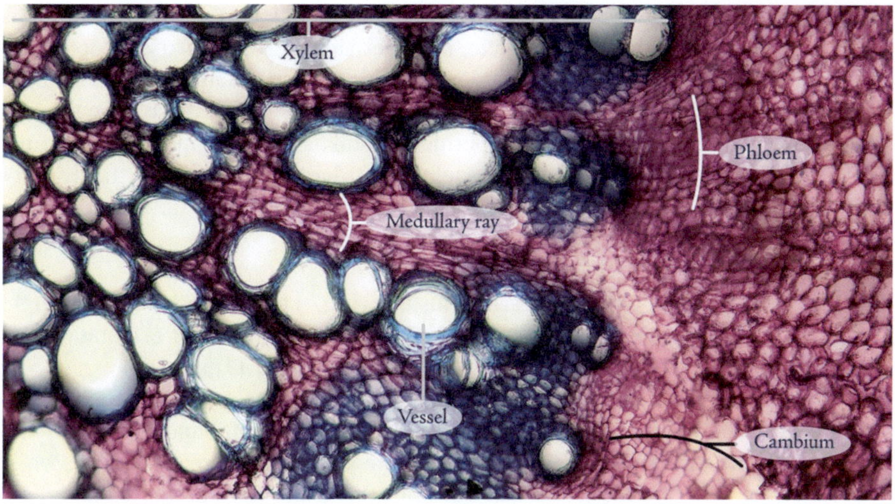

Fig. 11.49 High magnification image of the old root of a white campion (*Silene latifolia*)

Fig. 11.50 Wild pansy (*Viola tricolor*)

the radially thickening secondary xylem is visible with a blue staining. Here, too, cross-sections of vessels, tracheids, and thick-walled xylem fibres are clearly visible. At the outer boundary of the xylem is the few-celled cambium, which develops the xylem inwards and the pinkish-coloured phloem outwards. Further out, a transformation of the primary cortex is observed. The unicellular rhizodermis that originally covered the root has been shed, so that the outer layer is now formed by dead cells of the primary cortex. Beneath them, a ring of meristem, the cork cambium, is now observed, formed by the pericycle, on outer side of the primary phloem bundles, and cortical parenchyma cells. Cork cambium forms flattened cells of cork outwards and parenchyma cells (phelloderm) inwards. Together, these form the secondary dermal tissue of the root, the periderm. This is only the initial stage of the transformation, so the part outside the stele is still largely made up of the primary cortex (Figs. 11.51, 11.52, and 11.53).

11.1.2.6 Garlic Mustard (*Alliaria petiolata*)

The structure of the root of the garlic mustard (Fig. 11.54) can be seen by light microscopy without staining (Figs. 11.55 and 11.56), but for easier orientation it is worth using toluidine blue staining (Figs. 11.57, 11.58, and 11.59).

In the central part of the root, an intestinal parenchyma can be observed, followed outwards by the darker primary xylem. Here the xylem bundles are still visible. In addition, the bright blue secondary xylem and the phloem are visible, while in between the two, the meristem that develops them, the few cellular cambium, is visible.

Outside the phloem we find the secondary dermal tissue, the periderm and the cortex. The former takes the place of the unilayered rhizodermis and cortex, which are torn away by root thickening (Figs. 11.57–11.59).

11.1.2.7 Common Purslane (*Portulaca oleracea*)

Common purslane is native to Asia, but is widespread in gardens and weed communities worldwide (Fig. 11.60). Make a cross-section of the thinner root of this annual plant and examine it with a light microscope, stained with toluidine blue.

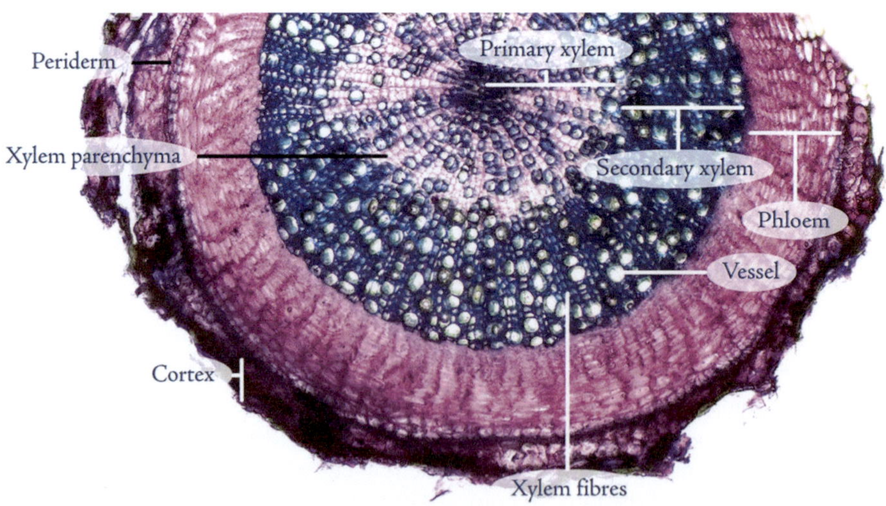

Fig. 11.51 Root cross-section of wild pansy (*Viola tricolor*) stained with toluidine blue

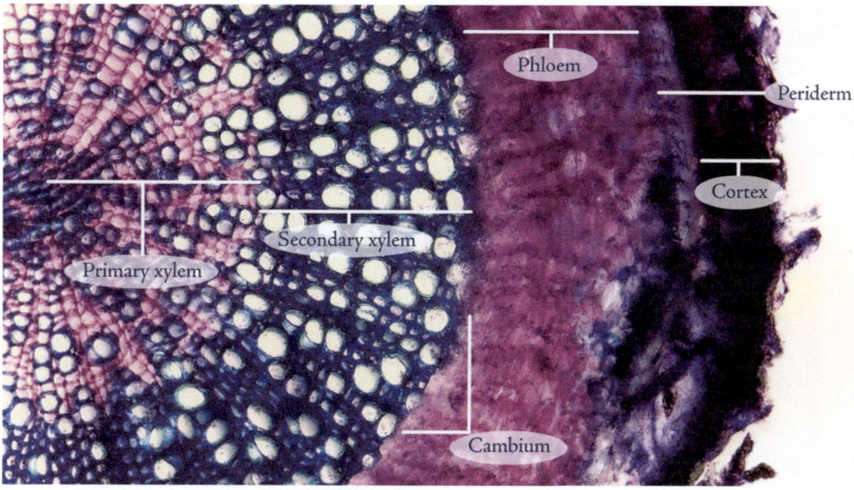

Fig. 11.52 Higher magnification image of the root cross-section of the wild pansy (*Viola tricolor*)

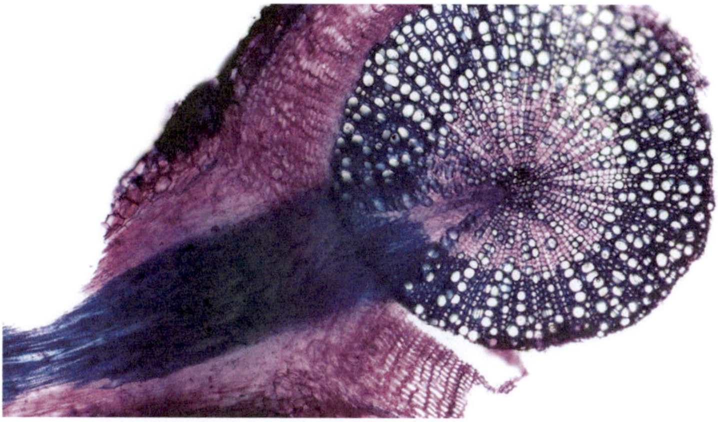

Fig. 11.53 Side root formation

In the central part of the root, you can see primary xylem bundles, which consist of a few vessels and tracheids. Moving outwards from these, the heterogeneous secondary xylem appears (Fig. 11.61).

In it, clusters of vessels and tracheids, xylem parenchyma cells and thick-walled xylem fibres are clearly visible. The closed multicellular cambium ring thickens the xylem inwards, while outwards it forms the phloem groups, separated by funnel-like widening medullary rays. Outside the phloem is the phelloderm, which is derived from the cork cambium of the periderm.

Fig. 11.54 Garlic mustard (*Alliaria petiolata*)

Fig. 11.55 Older root cross-section of garlic mustard (*Alliaria petiolata*)

In secondarily thickening roots, the cork cambium, which forms the periderm, develops above the phloem, so that the primary cortex is torn off and replaced by the periderm. The central part of the thicker, older root is filled with a starchy pith (Figs. 11.62 and 11.63).

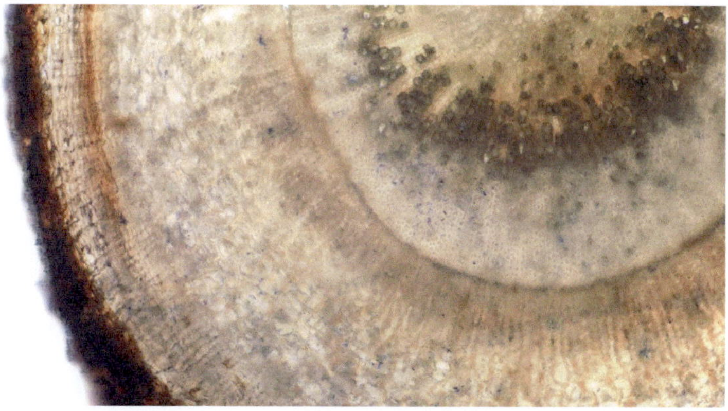

Fig. 11.56 Higher magnification image of the root cross-section of garlic mustard (*Alliaria petiolata*)

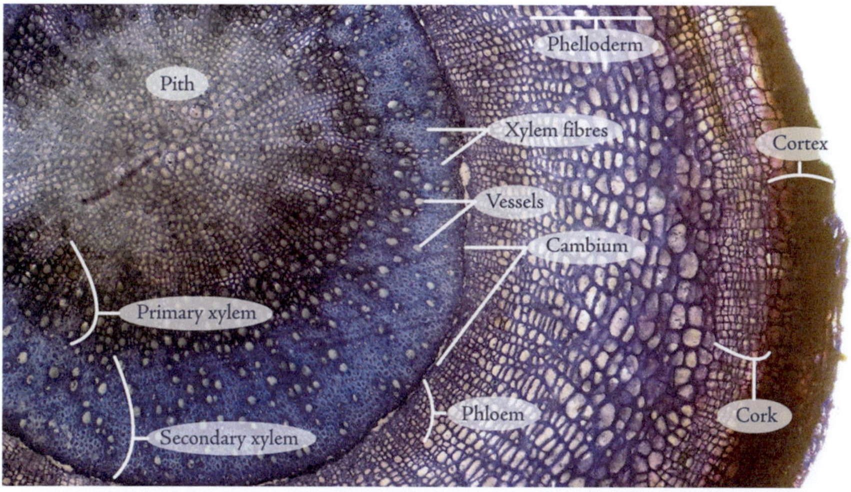

Fig. 11.57 Root cross-section of a garlic mustard (*Alliaria petiolata*) stained with toluidine blue

11.1.2.8 Horseshoe Geranium (*Pelargonium zonale*)

In a section of an older root of horseshoe geranium, you can observe the tissue structure of the secondarily thickened root. The root is already covered with periderm on the outside, the rhizodermis and the underlying primary cortex have been detached during growth. The cork is very thick, with the single-layered cork cambium underneath. Underneath, the cylindrical layers of the phloem, cambium, and xylem are visible (Figs. 11.64 and 11.65).

Sclerenchymal parts appear in the phloem. The central part of the root is filled by the primary xylem. Within the xylem, groups of vessels and xylem fibres are clearly

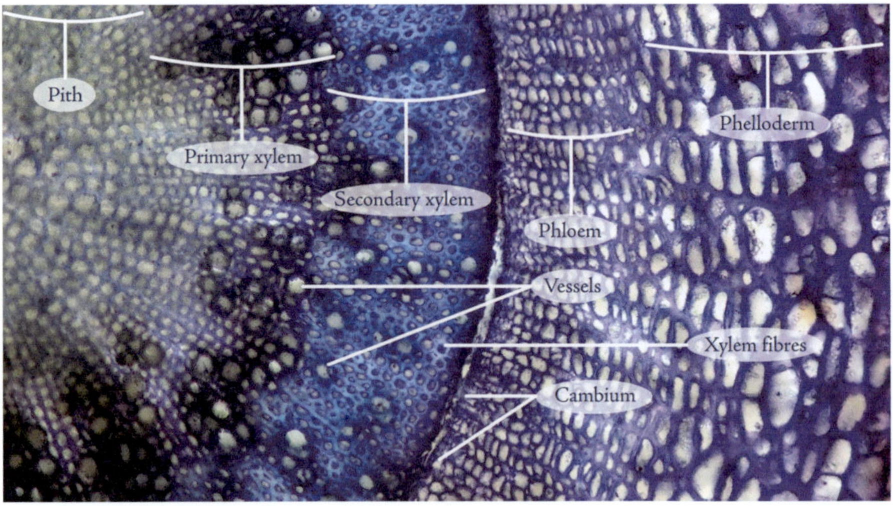

Fig. 11.58 Root structure of garlic mustard (*Alliaria petiolata*)

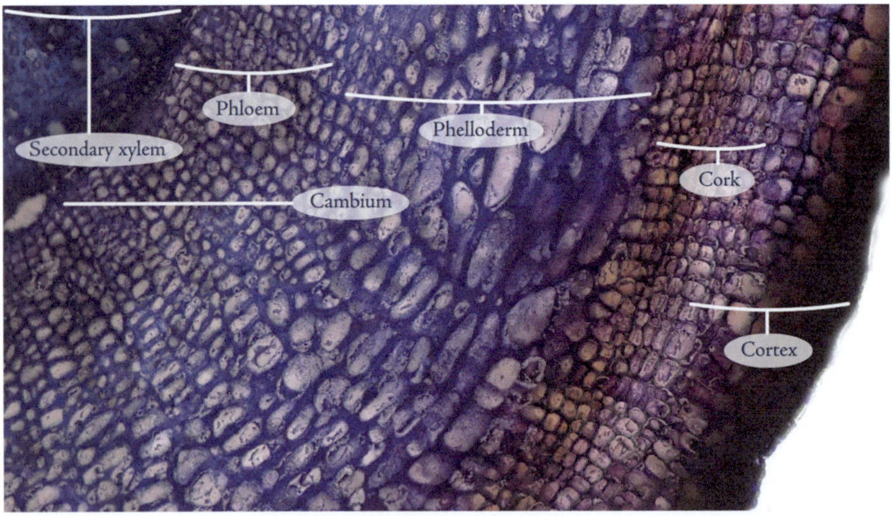

Fig. 11.59 Higher magnification image of the root cross-section of garlic mustard (*Alliaria petiolata*)

visible (Figs. 11.64 and 11.65). Further up, there are two to three cellular layers of large parenchyma cells (phelloderm) containing many starch granules.

11.1.2.9 Petunia (*Petunia* sp.)

In the toluidine blue-stained root cross-section of petunia (Fig. 11.66), the broad periderm consisting of a cork cambium and a cork above the phloem is clearly

Fig. 11.60 Common purslane (*Portulaca oleracea*)

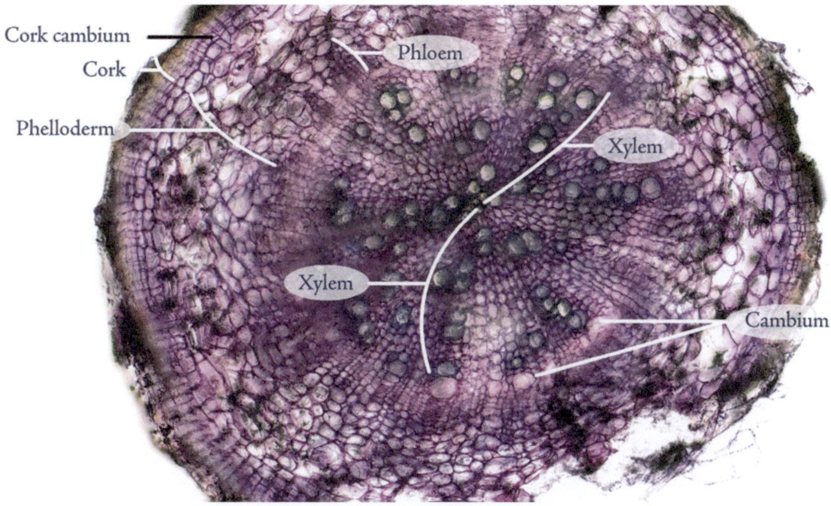

Fig. 11.61 Cross-section of the thin root of a common purslane (*Portulaca oleracea*). Preparation stained with toluidine blue

visible. Beneath the ring of phloem is the cylindrical cambium, which develops the phloem towards the outside and the xylem, which completely fills the central part of the root, towards the inside. In the xylem are clearly visible the large cavity vessels, the narrow cavity tracheids, and the xylem fibres with thickened cell walls (Fig. 11.67).

Fig. 11.62 Cross-section of the root of an older common purslane (*Portulaca oleracea*) stained with toluidine blue

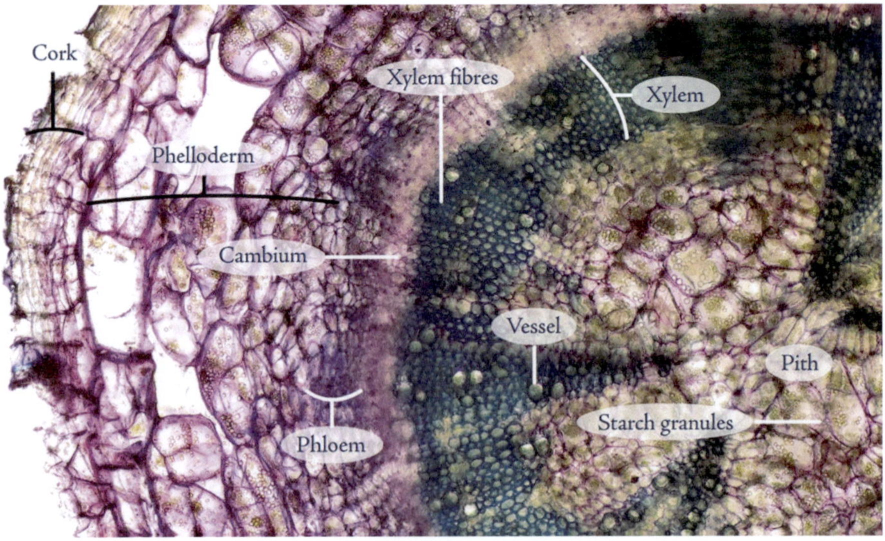

Fig. 11.63 Cross-section of the root of a common purslane (*Portulaca oleracea*)

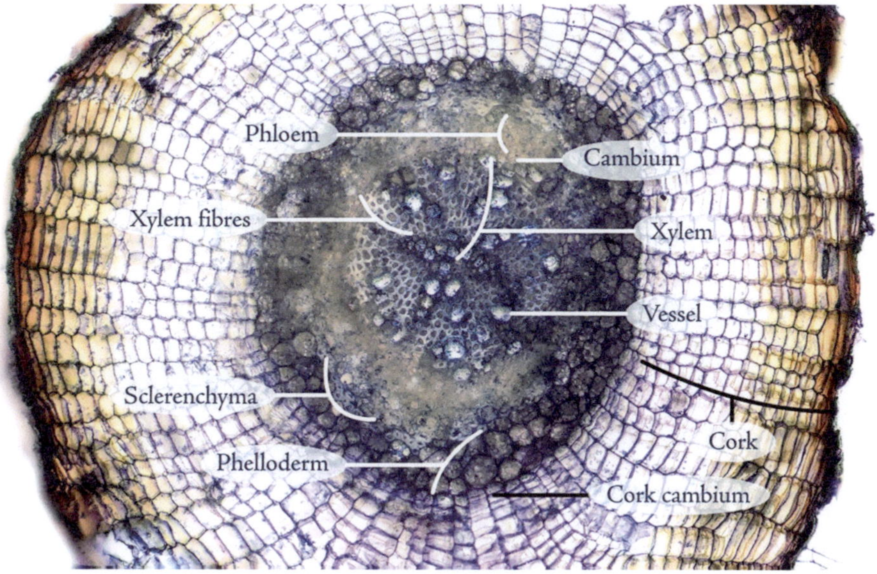

Fig. 11.64 Cross-section of an older root of horseshoe geranium (*Pelargonium zonale*)

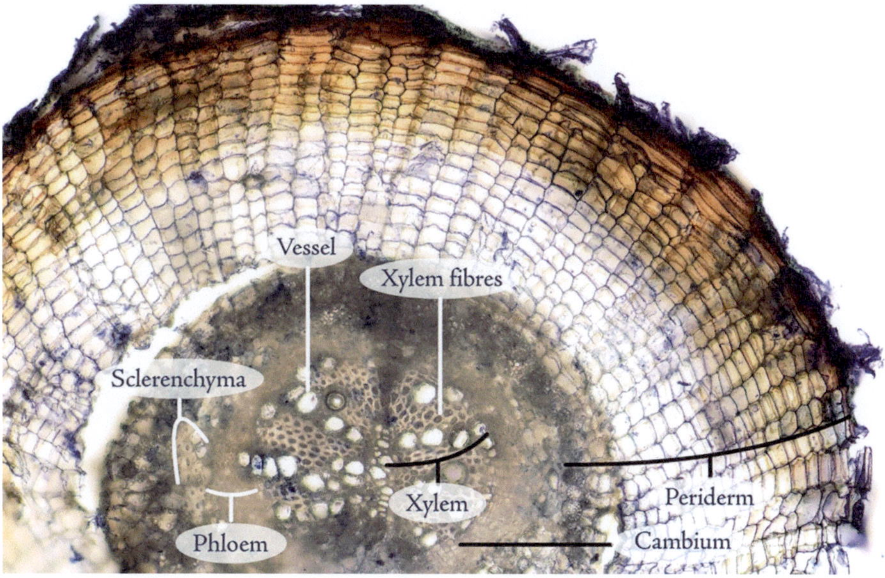

Fig. 11.65 Cross-section of an older root of horseshoe geranium (*Pelargonium zonale*)

11.1.2.10 Common Poppy (*Papaver rhoeas*)

The central part of the older root of the poppy (Fig. 11.68) is composed of xylem, in which the cross-sections of the dilated vessels and the narrower tracheids are clearly visible. Between them, pinkish xylem parenchyma and dark blue-stained

Fig. 11.66 Petunia (*Petunia* sp.)

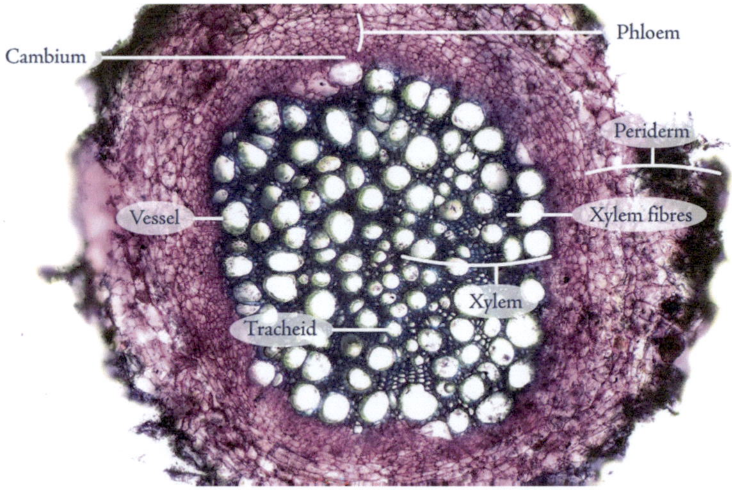

Fig. 11.67 Petunia's root cross-section with toluidine blue staining

xylem fibre bundles can be observed. In some places, the xylem parenchyma forms medullary rays. These extend to the periderm (Figs. 11.69 and 11.70).

11.1.2.11 Golden Deadnettle (*Lamium galeobdolon*)

A cross-section of the root of the golden deadnettle (Fig. 11.71) can be easily made, and when stained with toluidine blue it is a good subject for histological examination.

Figures 11.72 and 11.73 show a cross-section of a young root. Here the root is still covered by rhizodermis, under which a broad layer of cortex parenchyma can be observed. However, the pericycle has already formed from the innermost row of

Fig. 11.68 Field of poppies (*Papaver rhoeas*)

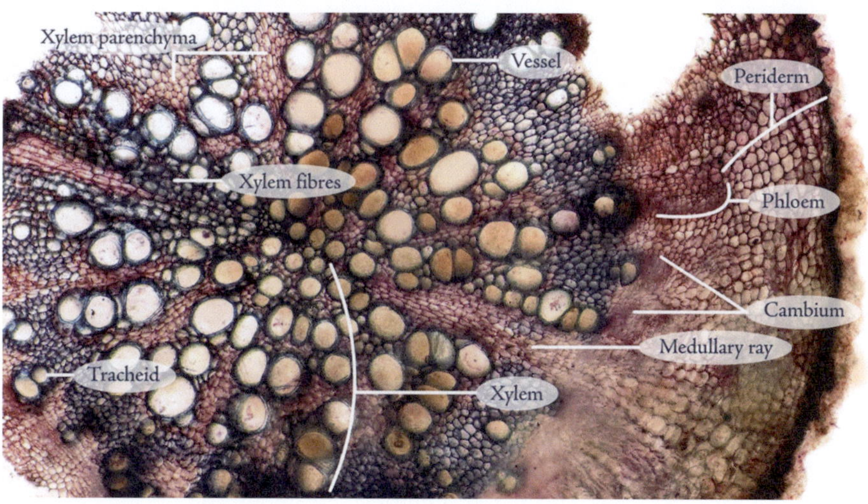

Fig. 11.69 Cross-section of an older poppy (*Papaver rhoeas*) root stained with toluidine blue

cells. Later, this gives rise to the secondary dermal tissue, the periderm, while the cortical parenchyma and the rhizodermis are detached during the secondary thickening of the root.

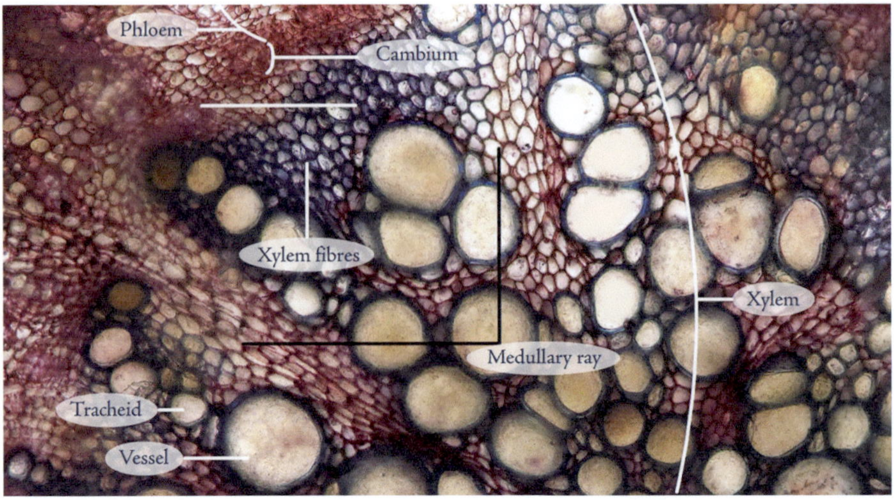

Fig. 11.70 Central part of poppy (*Papaver rhoeas*) root

Fig. 11.71 Golden deadnettle (*Lamium galeobdolon*)

Below the pericycle, the phloem, the cylindrical cambium, and the blue-staining xylem can be observed.

The middle of the root is filled with pith parenchyma. The secondary xylem by its dilated vessels is clearly distinguishable from the inward-facing primary xylem, which are mainly composed of xylem fibres and tracheids. The bundles of the xylem are separated by medullary rays.

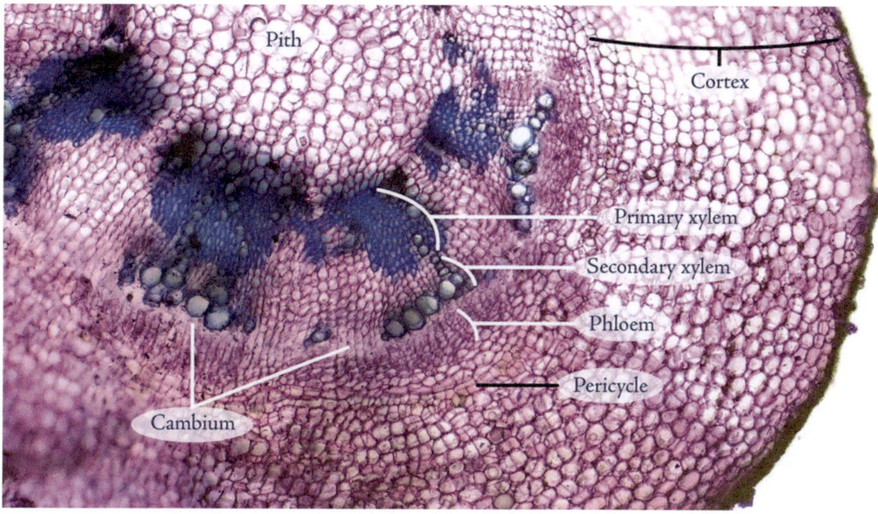

Fig. 11.72 Root cross-section of a golden deadnettle (*Lamium galeobdolon*) stained with toluidine blue

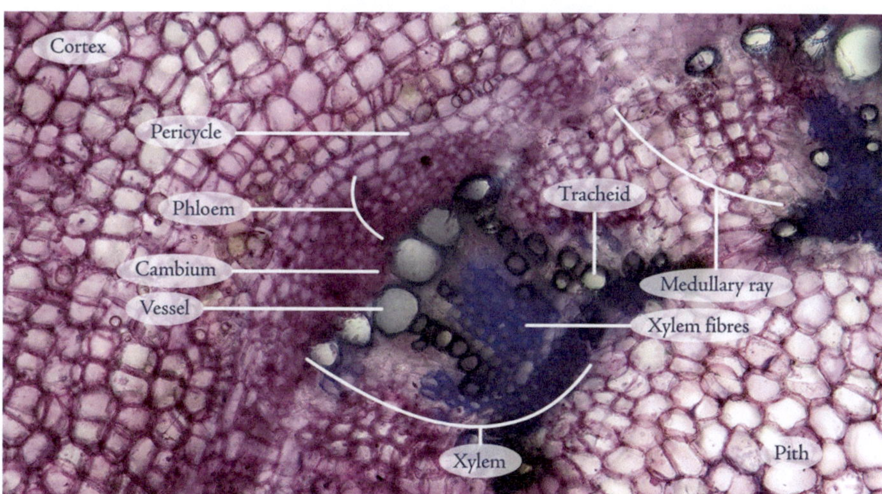

Fig. 11.73 High magnification image of a cross-section of the root cross-section of the golden deadnettle (*Lamium galeobdolon*)

11.1.2.12 Crownvetch (*Securigera varia*)

Figures 11.74 and 11.75 show a cross-section of the older, secondarily thickened root of the crownvetch (Fig. 11.76), stained with toluidine blue.

The root is covered with a thick periderm, and underneath is a purple layer of phloem. In the phloem, blue-coloured sclerenchyma islands are visible.

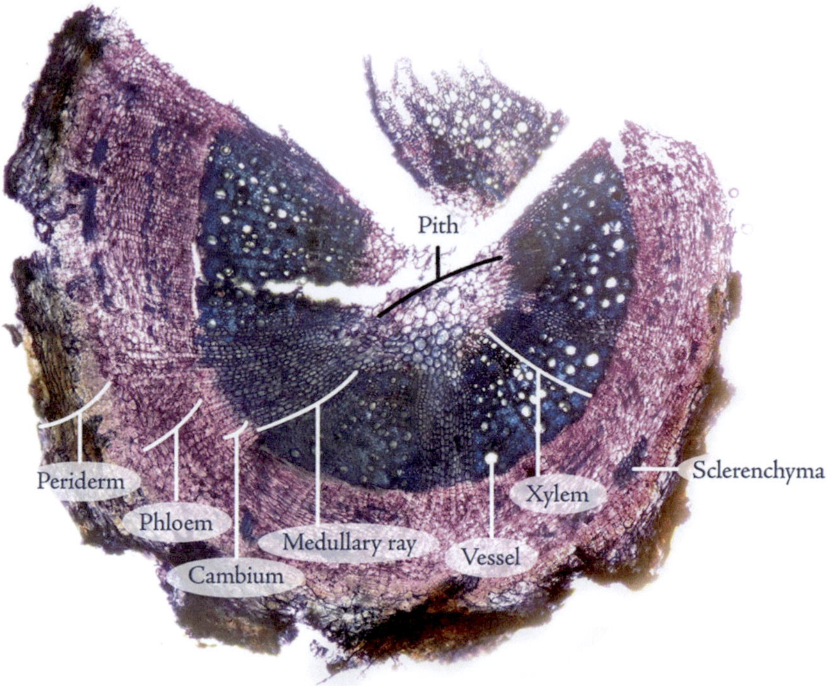

Fig. 11.74 Root cross-section of a crownvetch (*Securigera varia*) stained with toluidine blue

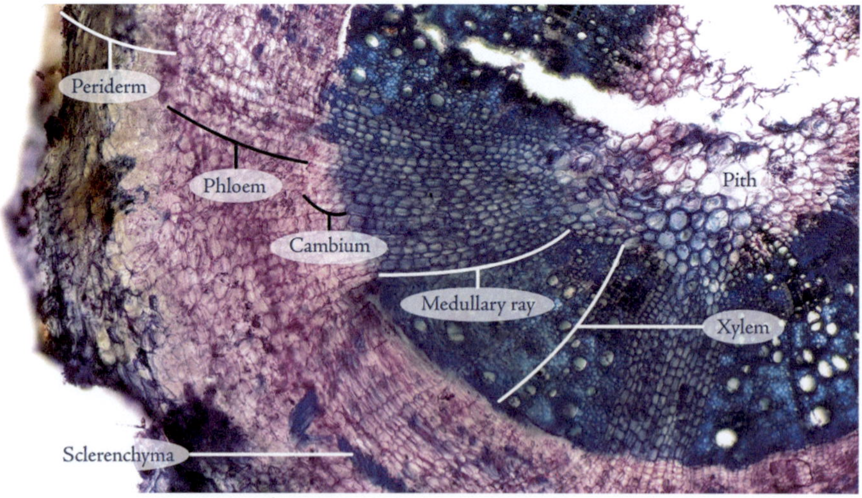

Fig. 11.75 Higher magnification image of the root cross-section

Fig. 11.76 Crownvetch (*Securigera varia*)

Further in, there is a thick layer of the lighter cambium, followed by a thick layer of coherent blue-stained xylem. In the xylem, radial medullary rays of xylem parenchyma can be discerned at higher magnification. In the xylem, the round cross-sections of the dilated vessels and the narrower tracheids are clearly identifiable. The inside of the root is filled with pith.

11.1.2.13 Eastern Groundsel (*Senecio leucanthemifolius* subsp. *vernalis*)

Eastern groundsel is a plant of the Asteraceae family (Fig. 11.77). From the eighteenth century onwards, it invaded central and northern Europe from eastern and south-eastern Europe, where it became a mass weed of disturbed fields. Flowering from April to May, the plant is most common at the railway lines, and its yellow carpet of flowers can also be seen in many urban areas. It is toxic to horses and cattle because of its pyrrolizidine alkaloid content. The annual or biennial plant can grow up to 40 cm tall. The thicker roots of eastern groundsel are easy to make sections from.

These roots already show secondary thickening. Their characteristic tissue structure can be studied in cross-sections stained with toluidine blue.

The central part of the younger roots is filled with the light-coloured pith, in which the primary xylem with darker staining can be recognised.

Here you can still recognise the xylem bundles. Moving outwards, the blue-tinged secondary xylem thickens radially. The cross-sections of the dilated vessels and xylem parenchyma cells are clearly visible. At the outer boundary of the xylem is the few-celled cambium, which develops the xylem inwards and the pinkish-coloured phloem outwards. Further out, the detached primary cortical parenchyma follows.

Fig. 11.77 Eastern groundsel (*Senecio leucanthemifolius* subsp. *vernalis*)

The rhizodermis is already detached, replaced by exodermis covering the root, which is made up of dead cortical cells. When a lateral root is formed, a group of meristem develops from the pericycle opposite the xylem bundles, and gradually gives rise to the tissues of the lateral root. The lateral root breaks through the primary cortex of the parent root and penetrates the soil. The young lateral root is similar in structure to the primary root (Figs. 11.78, 11.79, 11.80, and 11.81).

In the older roots, the central parenchyma develops into a mechanical tissue, forming a cable structure that increases the tensile strength of the organ (Figs. 11.82, 11.83, and 11.84).

The darkening of the sclerenchyma makes the primary xylem clearly distinguishable from the secondary xylem. In the secondary xylem, the xylem parenchyma is replaced by dead wood fibres with thick walls. In older roots, the cortex is also characteristically modified. After a while, the vascular tissues crack the cortex, which breaks away from the root. It is replaced by cork, which are produced by a ring of meristem, the cork cambium, formed by the pericycle and the cortical parenchyma cells.

The cork cambium forms cork cells on the outside and phelloderm cells on the inside. Together, the resulting formation forms the secondary dermal tissue of the root, the periderm (Fig. 11.82).

Moving upwards with the incision planes towards the stem, the tissue structure of the organ shows a peculiar pattern (Fig. 11.85). Like the stem, vascular bundles appear circularly surrounded by cortical and pith parenchyma cells filled with starch granules (Fig. 11.86).

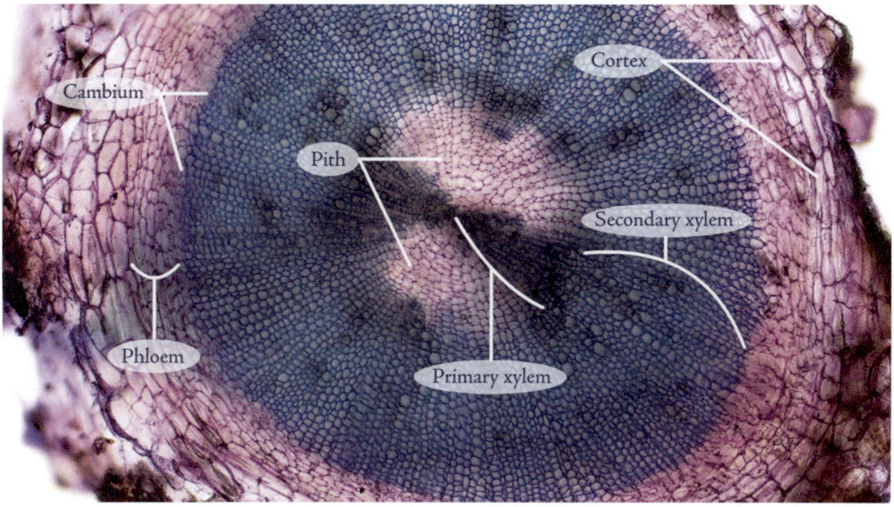

Fig. 11.78 Tissue structure of the secondary thickened root of eastern groundsel (*Senecio leucanthemifolius* subsp. *vernalis*)

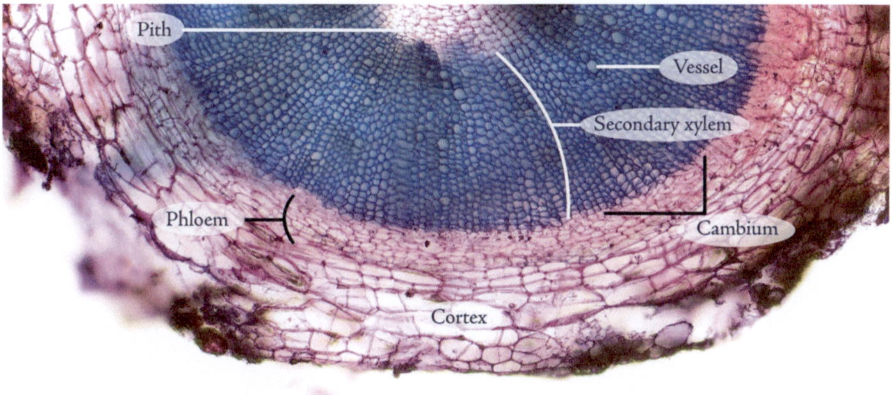

Fig. 11.79 Staining with toluidine clearly distinguishes the different tissues

11.1.2.14 Viper's Bugloss (*Echium vulgare*)

Spreading up to a metre tall, viper's bugloss is a common plant in weed communities (Fig. 11.87). The biennial plant develops only leaves in its first year. The blue flower has a long style, which makes it look like a snake with long, forked tongue out. In the Middle Ages, it was recommended for snake bites, for example.

The stem, which has a higher mucilage content, is difficult to section, but the root is a suitable subject for histological examination.

The older, secondarily thickening root is covered on the outside by a thick periderm, under which a thick layer of phloem is also observed.

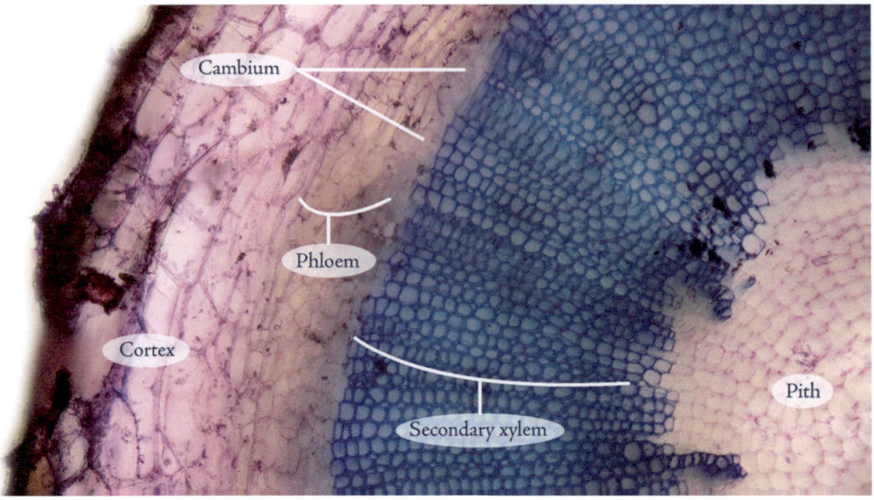

Fig. 11.80 High magnification image of the root of eastern groundsel (*Senecio leucanthemifolius* subsp. *vernalis*)

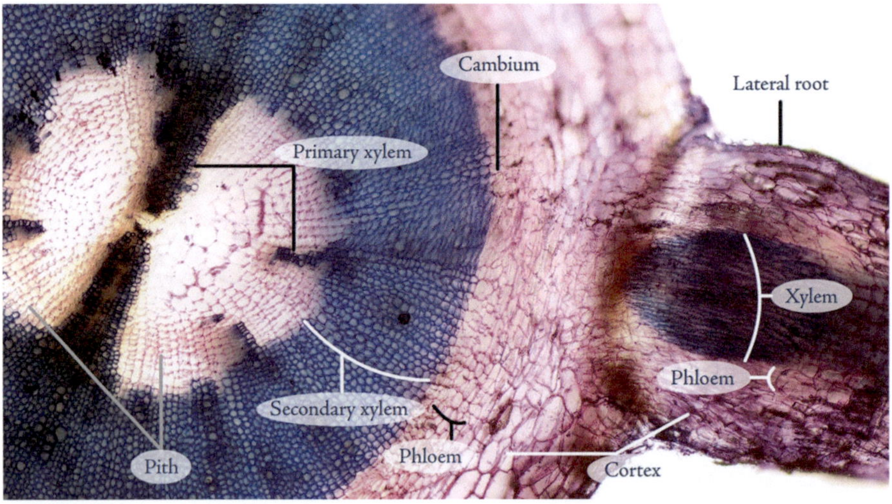

Fig. 11.81 Lateral root formation in eastern groundsel (*Senecio leucanthemifolius* subsp. *vernalis*)

Between the purplish phloem and the dark blue xylem is the cylindrical cambium. This meristem develops the phloem outwards and the xylem inwards.

In the wood tissue, large islands of xylem parenchyma can be observed, which are stained pale purple. In the xylem, the circular cross-sections of the dilated vessels and the narrower tracheids are clearly distinguishable (Figs. 11.88 and 11.89).

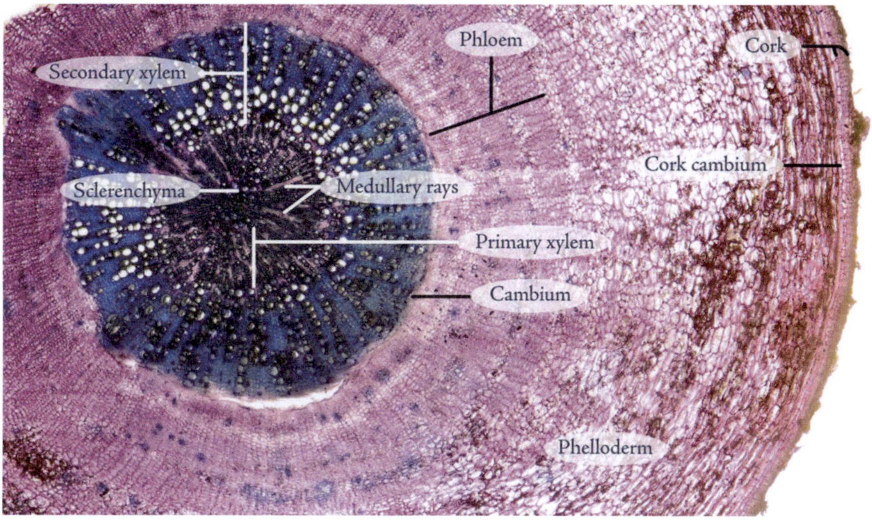

Fig. 11.82 Tissue structure of older root of eastern groundsel (*Senecio leucanthemifolius* subsp. *vernalis*)

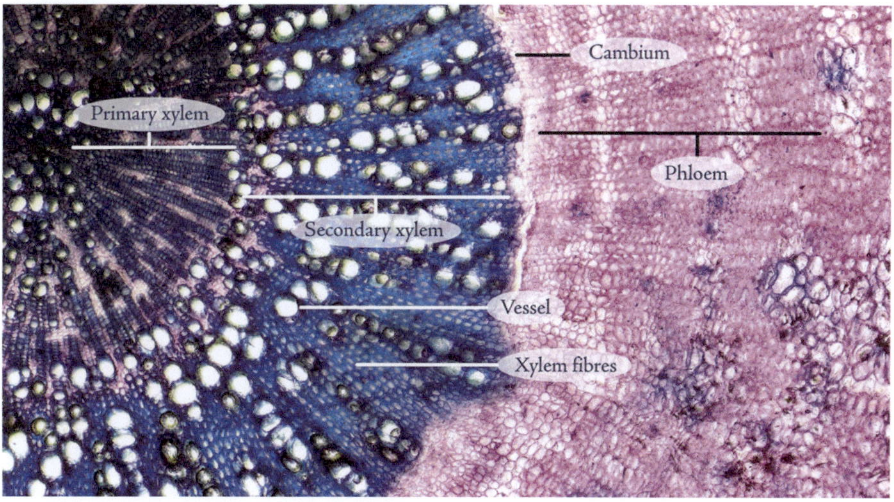

Fig. 11.83 Higher magnification image of the cross-section of the older root

11.1.2.15 Cucumber (*Cucumis sativus*)

Cucumbers (Fig. 11.90) are a member of the gourd family and their fruits are eaten. Originating in South Asia and is a widespread crop throughout the world. Its older root, which is already undergoing secondary thickening, has a peculiar histological structure.

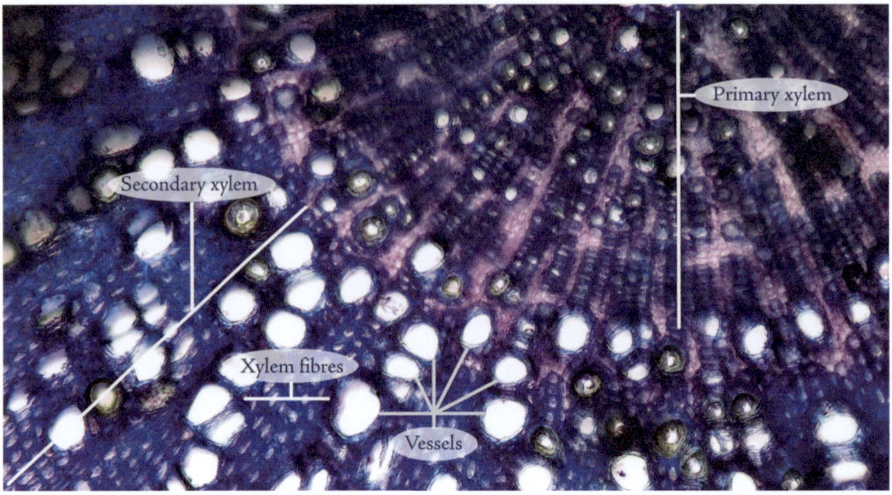

Fig. 11.84 High magnification image of the primary and secondary xylem of an eastern groundsel (*Senecio leucanthemifolius* subsp. *vernalis*) root

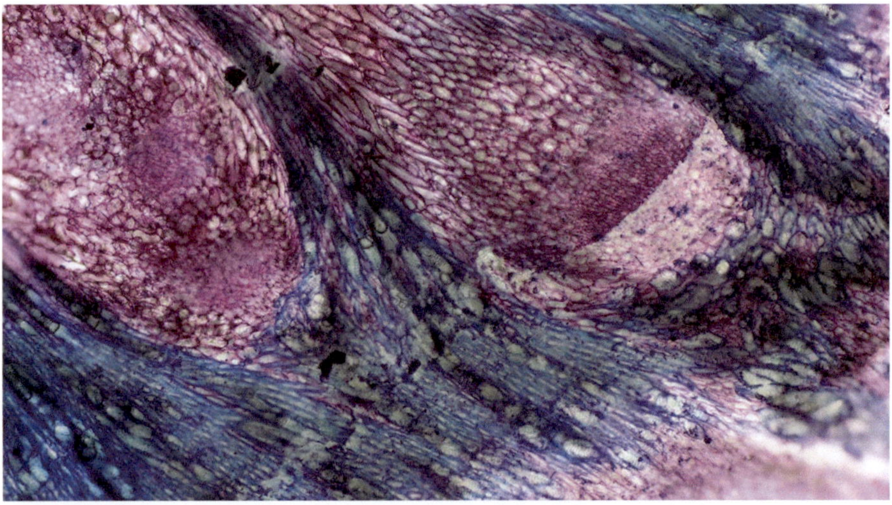

Fig. 11.85 Tissue structure of the root-stem transition of spring ragwort (*Senecio leucanthemifolius* subsp. *vernalis*)

The root is covered with a thick layer of periderm, indicating that the rhizodermis and cortical parenchyma originally covering the root have been detached from the organ during secondary thickening. Nevertheless, we do not see a broad, cylindrical of xylem in the toluidine-stained root cross-section as in the older roots presented earlier. Instead, at first glance, the histological image still shows the four xylem bundle structure typical of young roots. However, on closer examination of

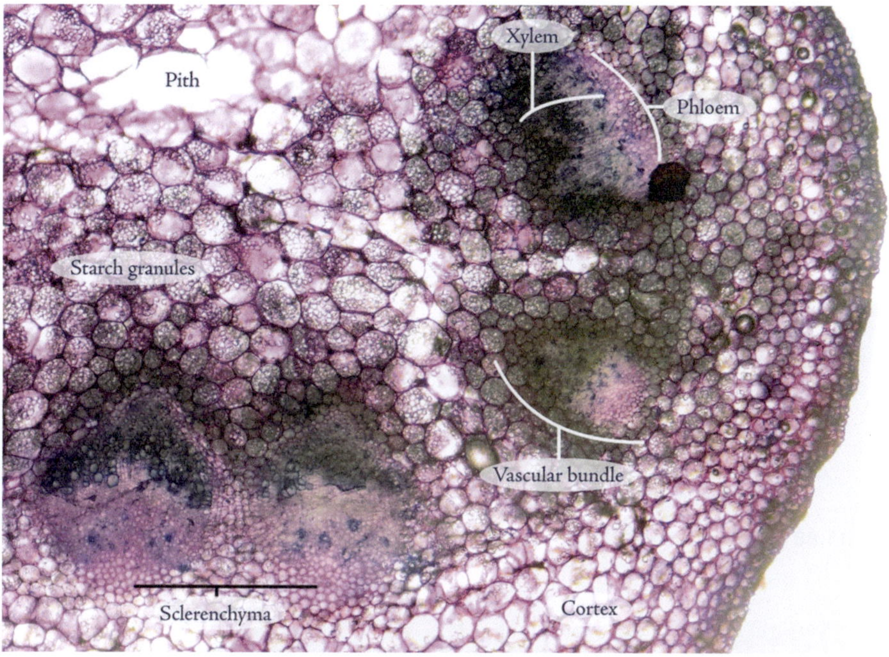

Fig. 11.86 Tissue structure of the lower part of the stem

Fig. 11.87 Viper's bugloss (*Echium vulgare*)

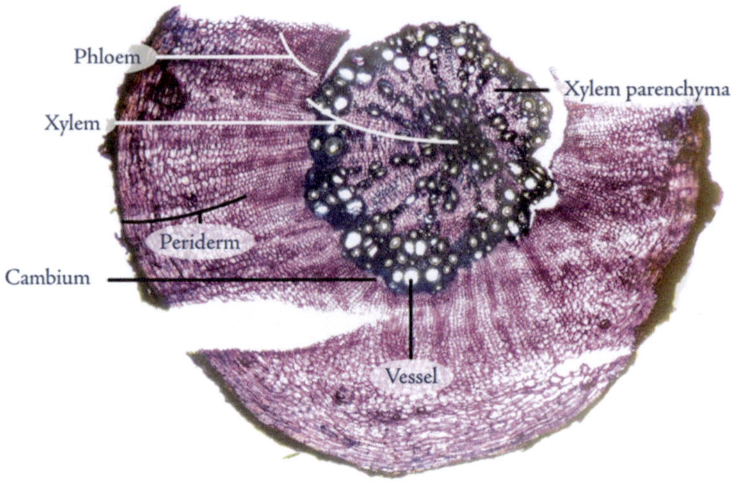

Fig. 11.88 Cross-section of the root of a viper's bugloss (*Echium vulgare*) stained with toluidine blue

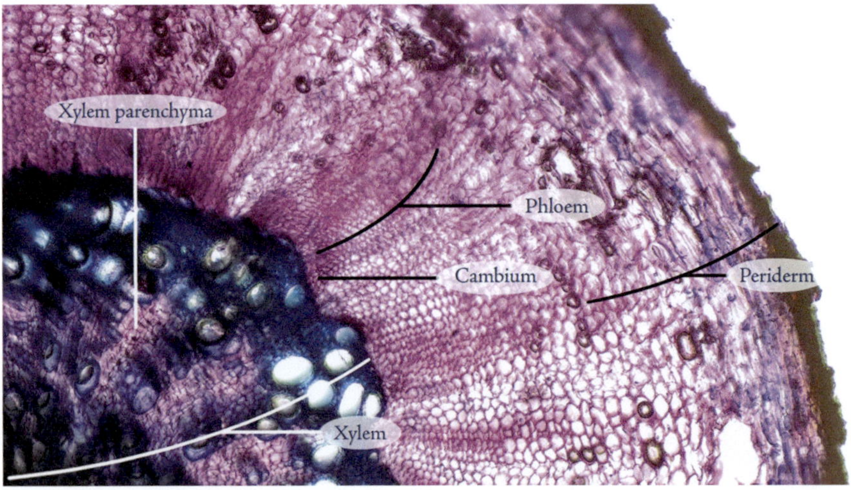

Fig. 11.89 Higher magnification image of the root cross-section

the section, it can be seen that, unlike in the young roots, the phloem is not located between xylem bundles in the form of simple phloem bundles. Phloem is located outwards from the secondary xylem and between them there is cambium. Thus, the cucumber root has developed vascular bundles with a cambium in the direction of the original xylem bundles (Figs. 11.91 and 11.92).

Fig. 11.90 Cucumber (*Cucumis sativus*)

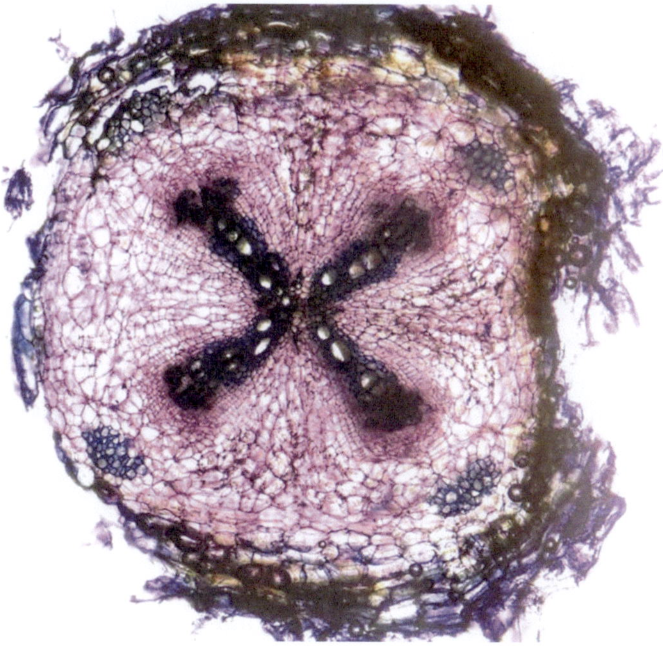

Fig. 11.91 Cucumber (*Cucumis sativus*) root cross-section stained with toluidine blue

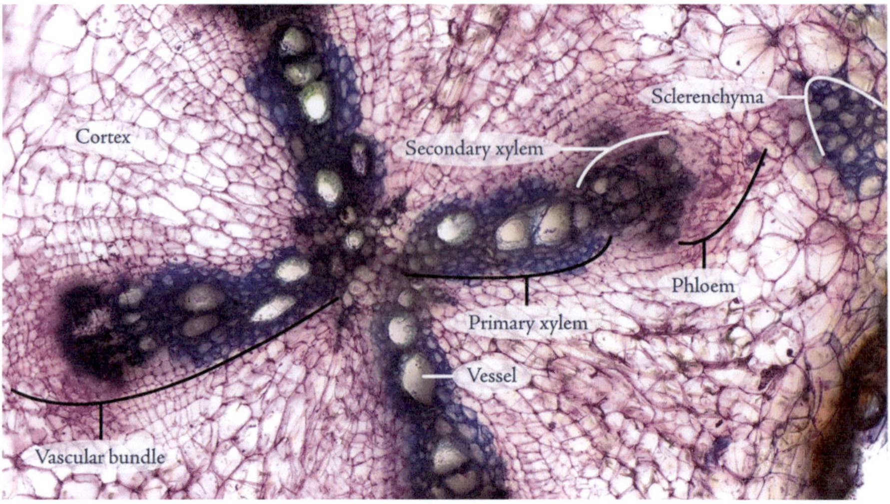

Fig. 11.92 Cucumber (*Cucumis sativus*) root tissue structure

Fig. 11.93 Sainfoin milk vetch (*Astragalus onobrychis*)

11.1.2.16 Sainfoin Milk Vetch (*Astragalus onobrychis*)

Sainfoin milk vetch is common throughout the grassy slopes (Fig. 11.93). In the cross-section of the older root, the wide periderm layer, which replaces the rhizodermis and cortex parenchyma that originally covered the root during secondary thickening, is clearly visible under the light microscope.

The cylindrical cambium produces the phloem outwards and the xylem inwards. In the latter, the round cross-sections of the vessels are clearly visible. The xylem is traversed radially by broad medullary rays of xylem parenchyma. They also partially penetrate the phloem.

The central part of the root is filled with a pale purple coloured pith parenchyma, separated from the xylem by a layer of sclerenchyma (Figs. 11.94 and 11.95).

11.1.3 Rhizome

11.1.3.1 Male Fern (*Dryopteris filix-mas*)

The male fern (Fig. 11.96) is a large perennial fern. Make a thin section of the fern's rhizome with a razor blade and stain it with toluidine blue. Cover the preparation in a drop of water and examine it with a light microscope.

In the cross-section of the rhizome, the hypodermis, formed from thickened-walled, sclerenchymatised ground tissue cells, is located beneath the epidermis (Fig. 11.97), adjacent to thinner-walled parenchymatous cells inwardly. Embedded in the pith are the vascular bundles, which are bordered by the unicellular endodermis and the starch sheath (Figs. 11.98, 11.99, and 11.100).

The centrally positioned xylem is completely surrounded by a continuous band of phloem, more specifically by sieve cells with a wider cavity and phloem parenchyma cells that store reserve material. The sieve tubes, companion cells, and phloem fibres are absent, as in the vascular system of the ferns generally.

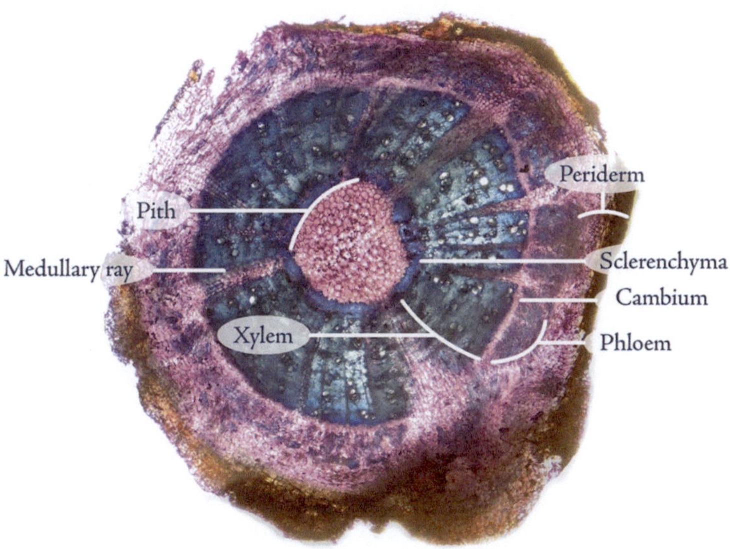

Fig. 11.94 Root cross-section of sainfoin milk vetch (*Astragalus onobrychis*) stained with toluidine blue

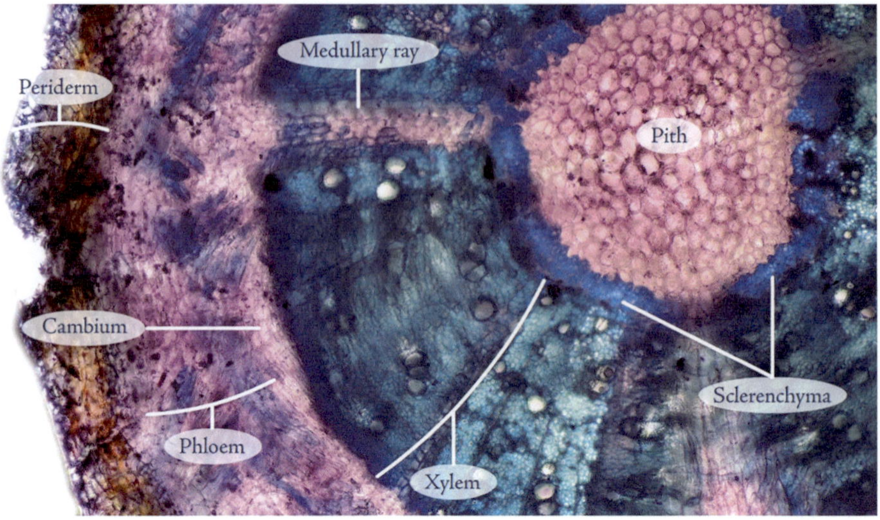

Fig. 11.95 High magnification image of a root cross-section

Fig. 11.96 Male fern (*Dryopteris filix-mas*)

The xylem is made up largely of large, thick-walled vessels and smaller tracheids, which are closely spaced. Between the two tissues there is no cambium, and the vascular bundle is therefore closed. This type of complex bundle has a vasocentric (amphicribral) structure.

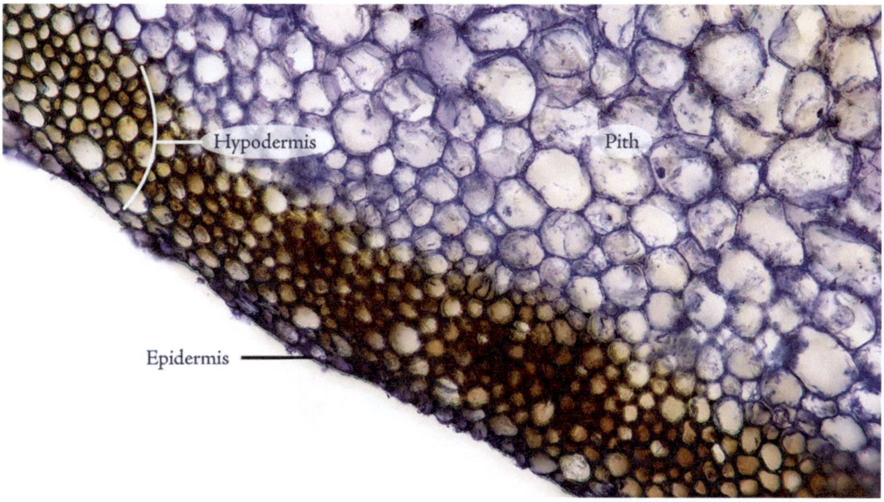

Fig. 11.97 Cross-section of male fern (*Dryopteris filix-mas*) rhizome

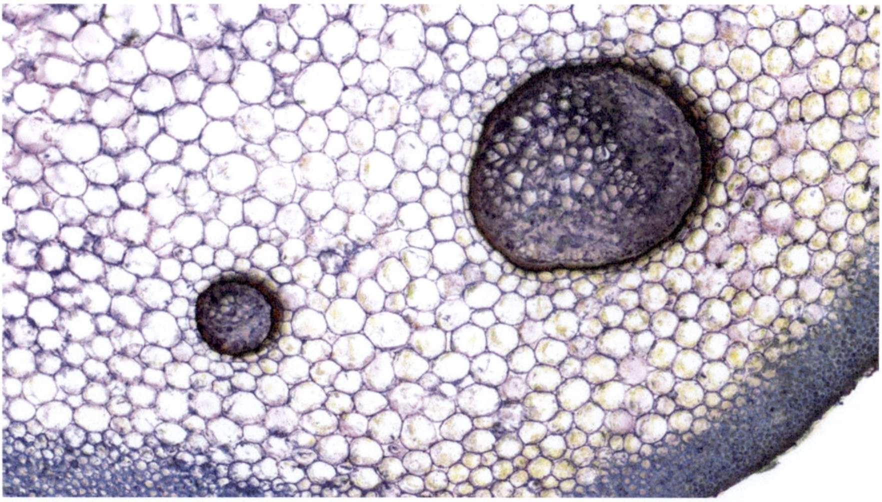

Fig. 11.98 Two vasocentric (amphicribral) vascular bundles in the rhizome

11.1.3.2 Lily of the Valley (*Convallaria majalis*)

The lily of the valley is a species of the asparagus family, native to temperate regions of Europe and Asia. It is found mainly in oak woodlands, but is widely planted in gardens. Make thin cross-sectional sections of the rhizome of the plant (Fig. 11.101) around the middle of the stalk and stain with toluidine blue.

The primary cortical parenchyma beneath the monolayer epidermis is rich in intercellular cavities. Its cells contain a large amount of starch, forming a storage parenchyma (Fig. 11.102). The inner boundary of the cortex is formed by the

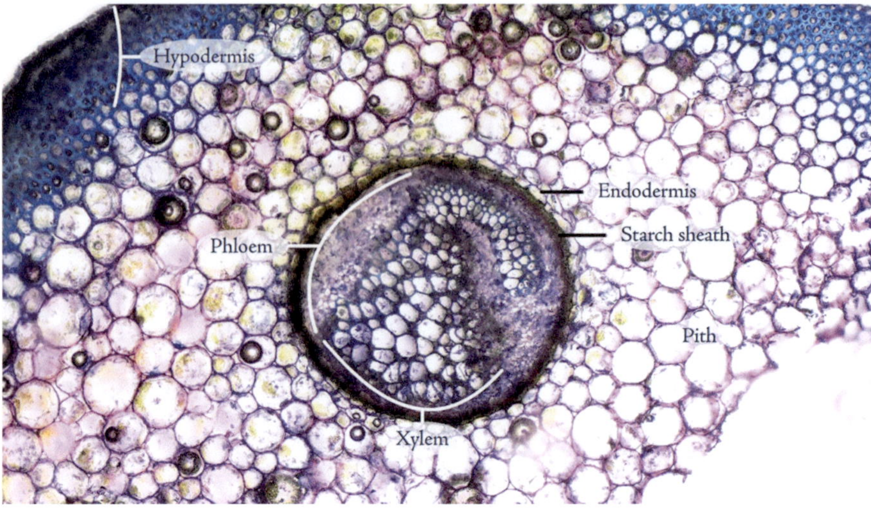

Fig. 11.99 Tissue structure of a male fern (*Dryopteris filix-mas*) rhizome stained with toluidine blue

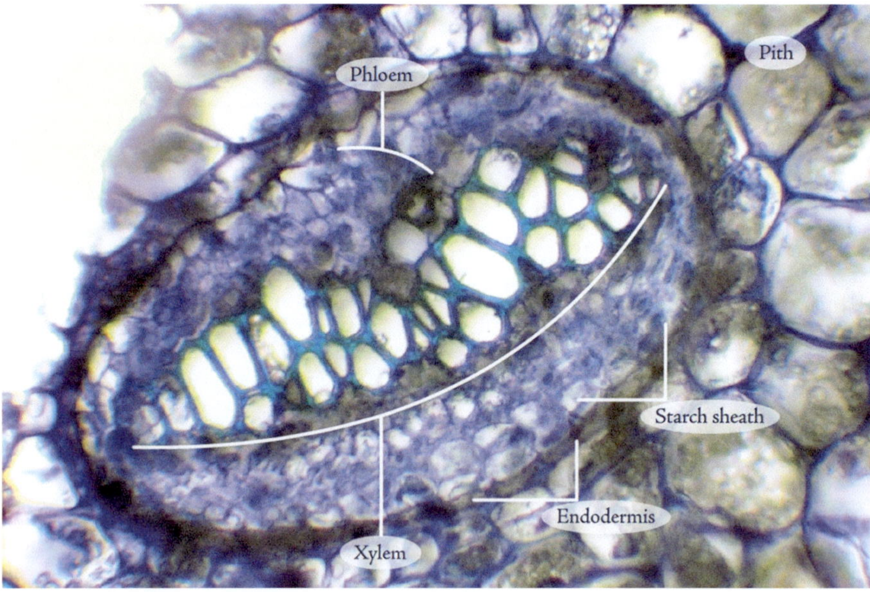

Fig. 11.100 High magnification image of the vascular bundle

single-layered endodermis. Its cell walls are thickened in a U-shape (Figs. 11.103 and 11.104).

In the rhizome of the lily of the valley are vascular bundles. A few of the bundles (6–8) are scattered in the central part of the rhizome, surrounded by the

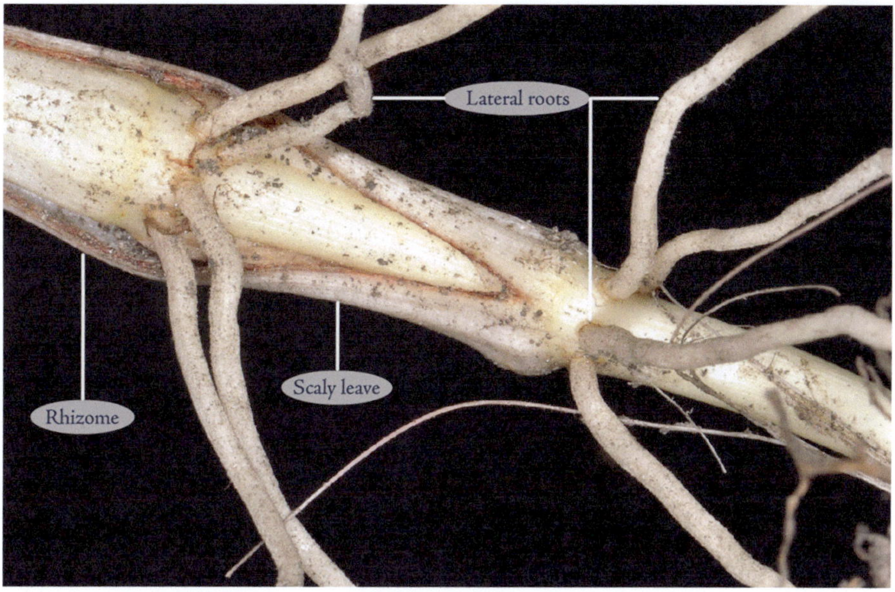

Fig. 11.101 Detail of the rhizome of the lily of the valley (*Convallaria majalis*)

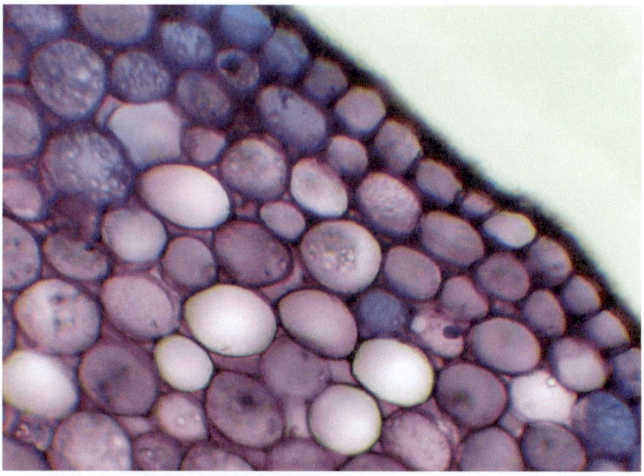

Fig. 11.102 Single-cell epidermis and underlying cortical parenchyma

parenchyma. They have a closed and concentric structure similar to that of ferns, but differ from them in that the xylem elements forming the vascular bundles are located outside and completely surround the centrally positioned phloem. There is also no cambium between the two types of tissue, so that these bundles are also closed. This type of vascular bundle is called leptocentric (amphivasal) (Fig. 11.105).

However, in the rhizome of the lily of the valley, other types of vascular bundles are also found.

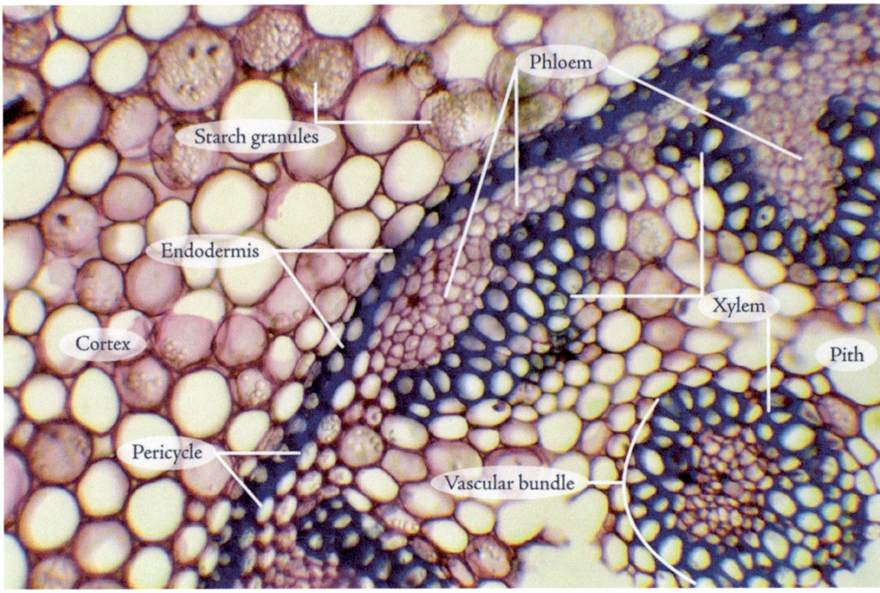

Fig. 11.103 Cross-section of a rhizome of a lily of the valley (*Convallaria majalis*) stained with toluidine blue

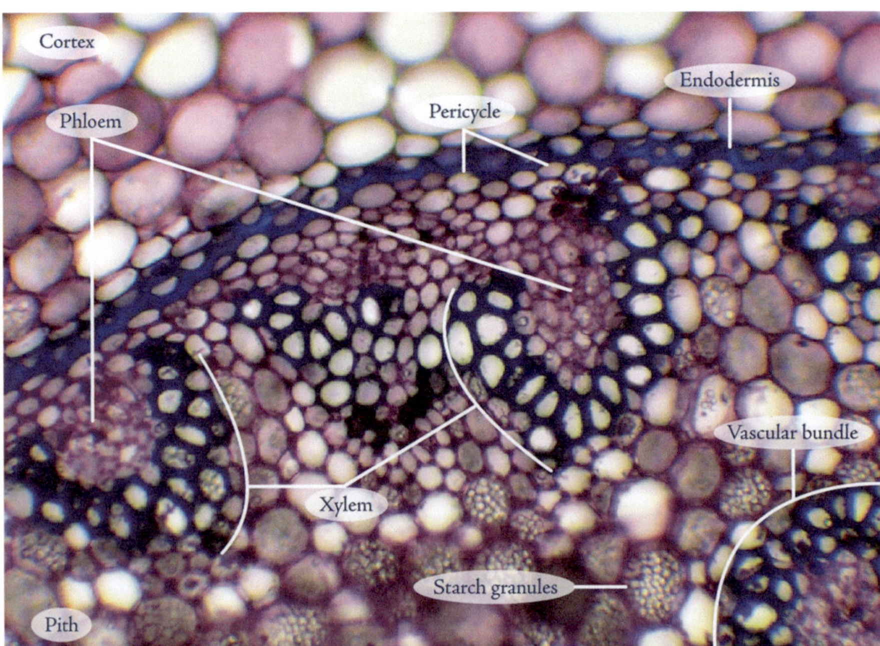

Fig. 11.104 Cross-section of a rhizome of a lily of the valley (*Convallaria majalis*) stained with toluidine blue

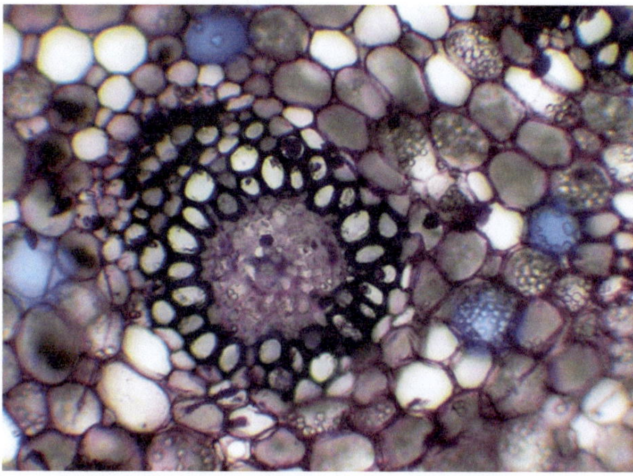

Fig. 11.105 Leptocentric vascular bundles of the lily of the valley (*Convallaria majalis*)

They are located on the inner side of the endodermis, which is made up of U-shaped, thickened-walled cells. Most of them are not concentric, because their xylem does not completely surround the phloem. They do not contain cambium and are therefore closed. The two types of bundles described are generally made up of the same elements in both the xylem and the phloem. In the phloem part of the bundles, the broader cavity of the sieve tubes is visible, together with the very small but, because of their cytoplasmic content, well-staining companion cells. The phloem parenchyma elements and the phloem fibres are absent. In the xylem of the bundle there are round, oval or angular, thick-walled water transport elements, vessels, tracheids, and compressed protoxylem elements, which are only visible at high magnification. The xylem parenchyma is absent. The leptocentric vascular bundles observed here are mostly found in the rhizomes of monocots.

11.1.3.3 Reed (*Phragmites australis*)

Make a cross-section of the rhizome and stain it with toluidine blue! The cross-section is bordered on the outside by a monolayer of epidermis composed of thick-walled cells. Underneath is a layer of three to four rows of cells of the supporting hypodermis. The cortex shows a section of several small vascular bundles protruding into the sheath leaves. These are enclosed by a layer of cells (the bundles sheath) with a thickened wall in a "U" shape. The vascular bundles are formed by a few xylem and phloem elements (Fig. 11.106).

The major part of the cross-section of the stem is the cortex, which is bordered by the clearly visible endodermis and epidermis. The cortex is composed of thin-walled parenchyma cells with small intercellular spaces (Fig. 11.107).

The stele is made up of annular pith, two rows of vascular bundles, and a pith cavity. The ground tissue under the endodermis is composed of thick-walled sclerenchyma fibres (Figs. 11.108 and 11.109).

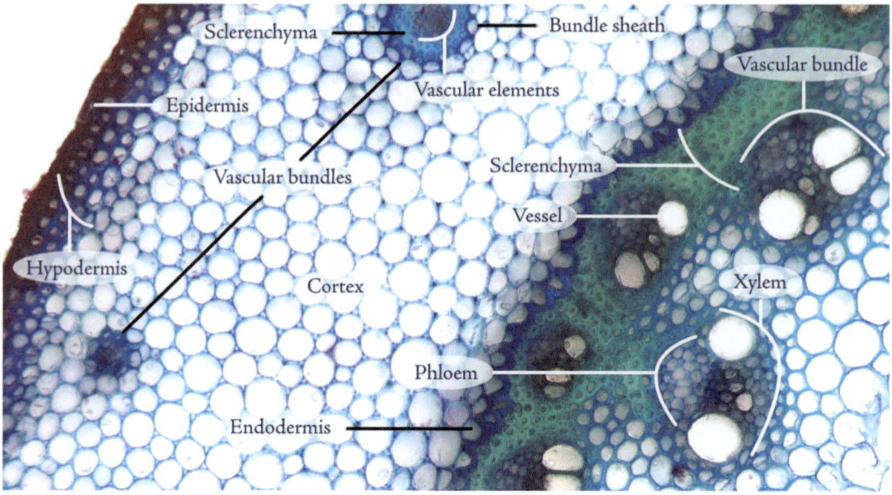

Fig. 11.106 Enlarged image of the rhizome

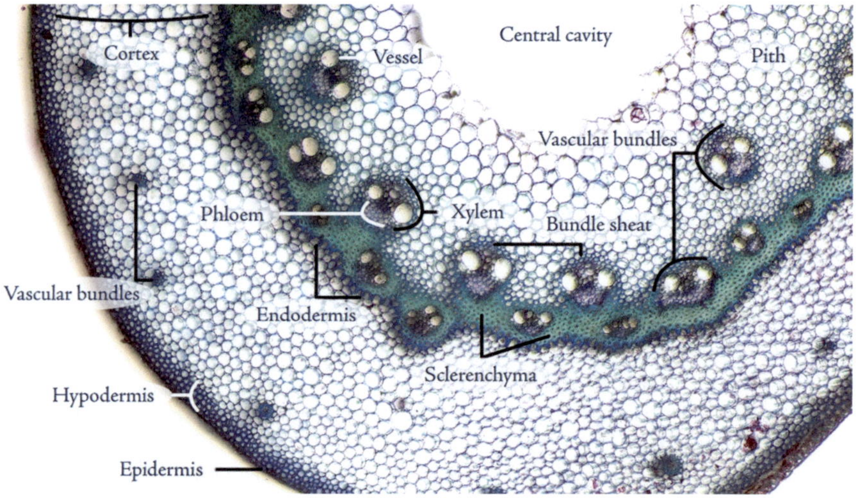

Fig. 11.107 Cross-section of reed (*Phragmites australis*) rhizome

11.1.3.4 Common Nettle (*Urtica dioica*)

Common nettle is a weed and medicinal plant found throughout Europe (Fig. 11.110). It is common in dry, moist woodlands, weed communities, swamps, and streamsides. It prefers nitrogen-rich soils, so it is abundant in planted acacia groves and in disturbed soils. The rhizome, stained with toluidine blue, shows a peculiar structure even at low magnification when examined with a light microscope (Fig. 11.111), so it is worth examining the tissue structure more closely.

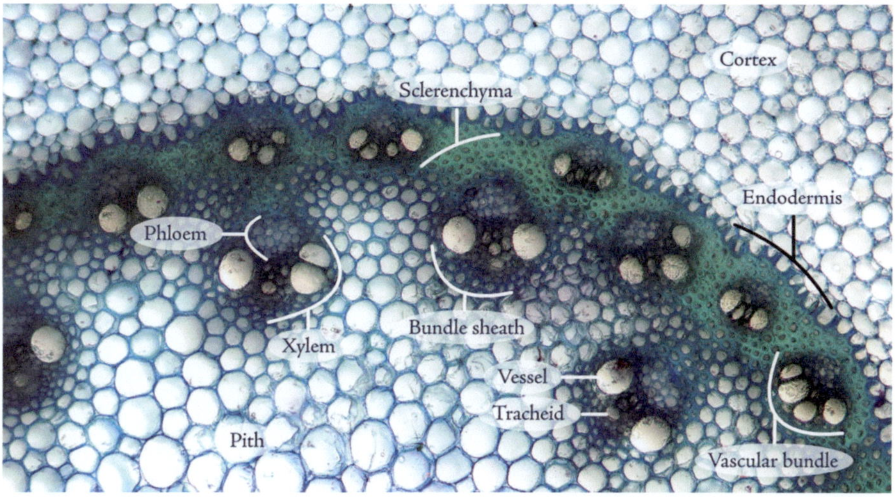

Fig. 11.108 Cross-section of the stele

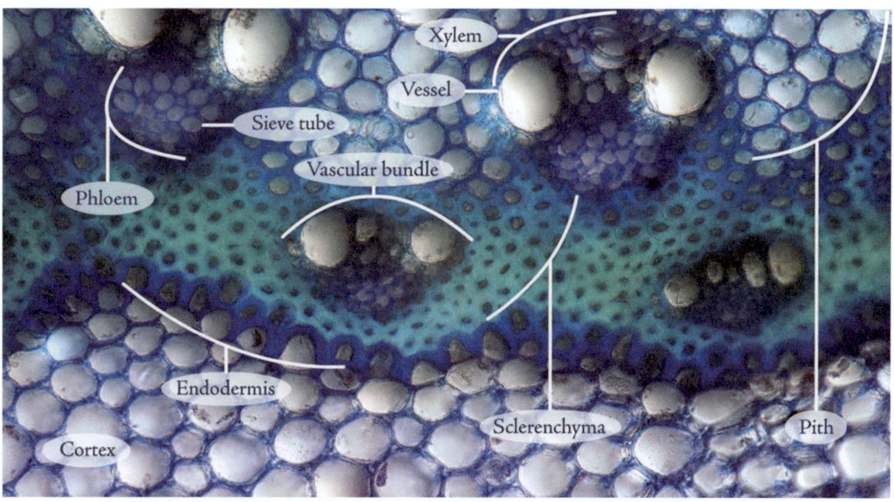

Fig. 11.109 High magnification image of the cortex boundary

A cross-section of the underground rhizome of the common nettle shows a stem-like structure with vascular bundles, which, however, continues to develop. Here, unlike the stem above ground, the secondary thickening of the plant part appears. The secondary thickening of the rhizome causes the epidermis to break away and be replaced by periderm. Its layer can be observed under the degraded cortical parenchyma. Beneath the periderm, blue-staining solidifying elements can be seen, which are remnants of the primary phloem. Secondary phloem is observed developing outwards from the cambium, while secondary xylem is forming inwards. The

Fig. 11.110 Common nettle (*Urtica dioica*)

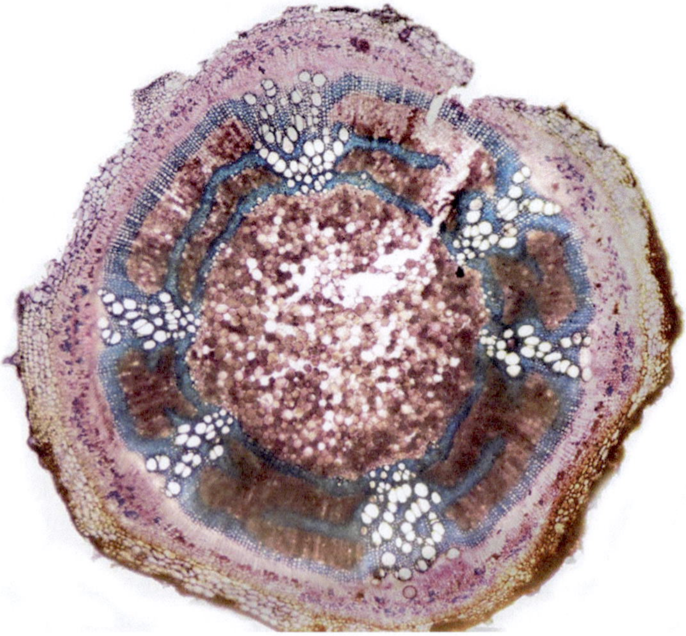

Fig. 11.111 Cross-section of a rhizome stained with toluidine blue of a common nettle (*Urtica dioica*)

dilated vessels tubes are formed from the cambium part of the original vascular bundles, and the xylem formed from the intermediate (interfascicular) cambium parts does not contain vessels only xylem fibres and xylem parenchyma. The unicircular vascular bundles are connected by a lateral sclerenchyma stripe that forms a ring with the inner protoxylem. Outside this ring of sclerenchyma, additional bands of sclerenchyma appear laterally from the vascular bundles' xylem parts, which may connect 1 to 1 vascular bundle, but do not close into a ring. Inside the rhizome, parenchyma cells filled with starch granules can be observed (Figs. 11.112, 11.113, 11.114, and 11.115).

11.1.4 Aerial Root

11.1.4.1 Swiss Cheese Plant (*Monstera deliciosa*)

The Swiss cheese plant (*Monstera deliciosa*), a popular ornamental plant, is a monocot tropical liana with characteristic aerial roots. These are nutrient-transporting aerial roots, which play a role in absorbing rainwater that runs off them and later, once in the ground, take up nutrients dissolved in water.

Make a thin cross-section of the aerial root! The aerial roots have a multi-layered root sheath (velamen radicum). It is the outermost layer of the root that binds water. Beneath this is a layer of thickened-walled cells, followed by a peridermal part. The fourth layer is the collenchyma, consisting of thickened-walled cells. The inner parenchymatous layer of the root sheath is in contact with the thickened-walled cells of the cortex, which form the 2–3 row exodermis (Fig. 11.116).

The cells of the exodermis are hexagonal, with highly thickened walls and a pitted thickening. Within this is the parenchymatous cortex.

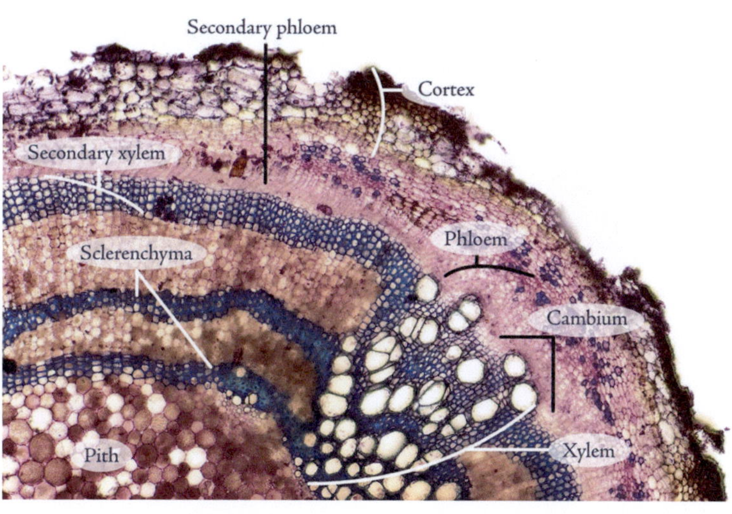

Fig. 11.112 Tissue structure of the rhizome

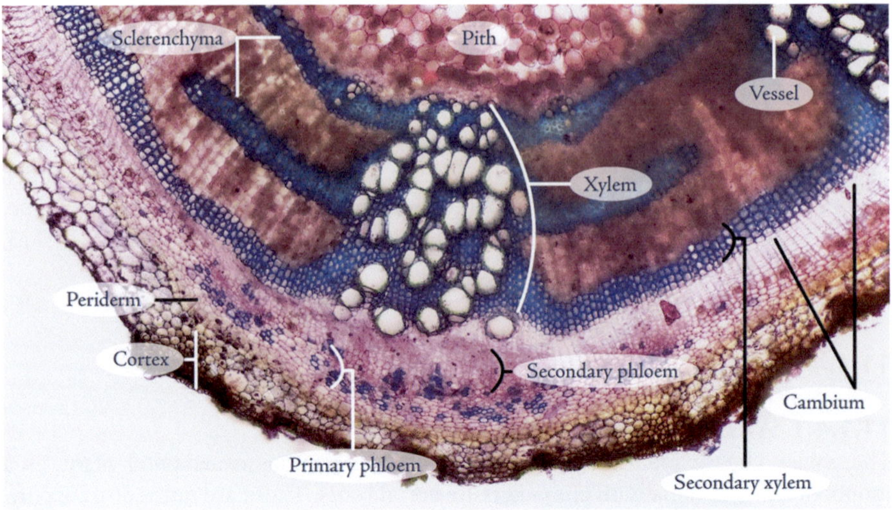

Fig. 11.113 Tissue structure of the rhizome of the common nettle (*Urtica dioica*)

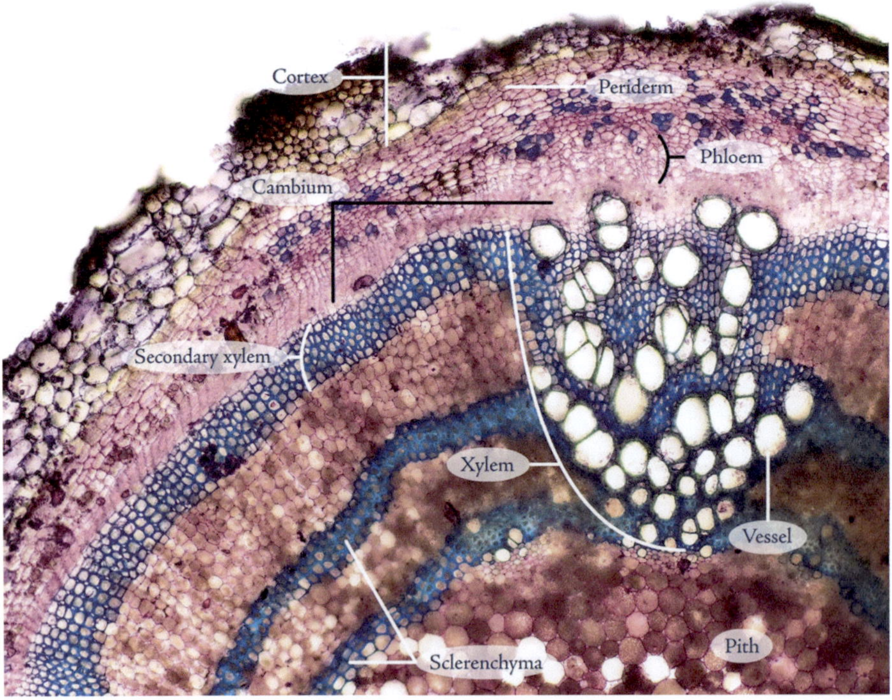

Fig. 11.114 Higher magnification image of the cross-section of the rhizome

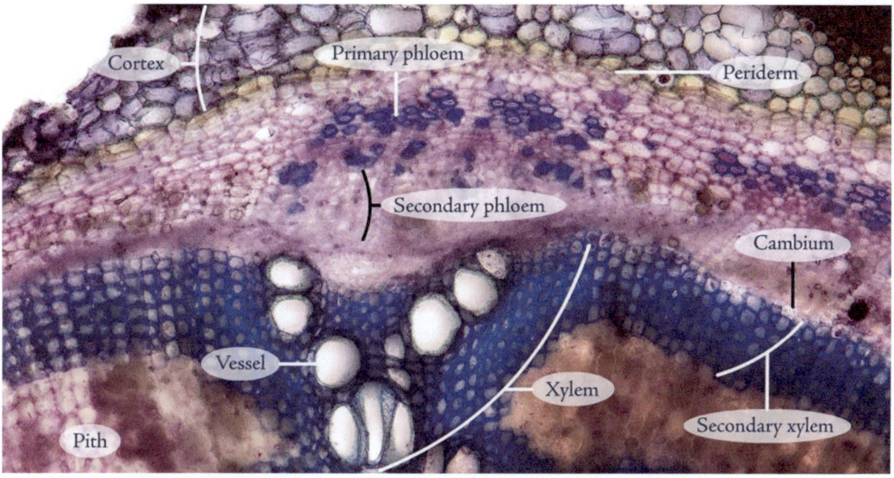

Fig. 11.115 High-magnification image of a vascular bundle environment

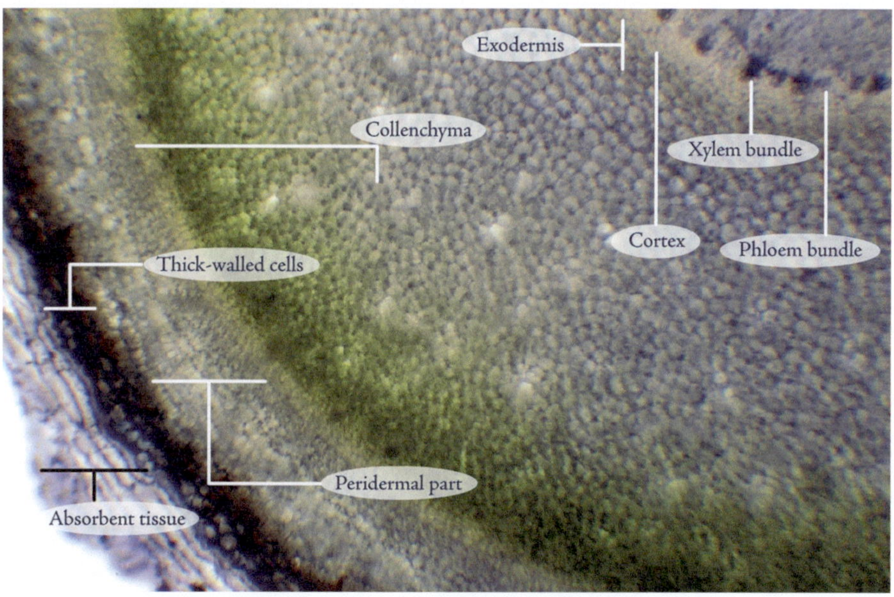

Fig. 11.116 Aerial root parts

The stele occupies about a third of the entire root. In the stele, the xylem and phloem bundles alternate.

Vessels and tracheids are clearly visible in the xylem bundles, and sieve tubes and companion cells in the phloem bundles.

Within the circle of simple bundles, we can see large separate or twin-pore vessels, and sieve tubes surrounded by parenchyma cells, in an irregular (scattered)

arrangement. The pith is composed of sclerenchymatically thickened cells (Figs. 11.117 and 11.118).

11.1.4.2 Maize (*Zea mays*)

Maize is a very important fodder crop. Make a thin cross-section of a maize aerial root (Fig. 11.119) a few centimetres from the root tip. Stain the section with toluidine blue and examine it with a light microscope after covering. The outer edge of the root cross-section shows the monolayer epidermis. Underneath the epidermis is the exodermis, which consists of the outer cell layers of the cortex.

The exodermis replaces the epidermis, which wears away over time, and protects the root from external damage. Beneath the exodermis is the cortex, composed of parenchymatous cells. The cortical part consists of advanced parenchymal cells. They become smaller towards the outside, with thicker walls, and form the solidifying exodermis. The innermost layer of the cortex is the endodermis. The outermost cell of the stele is the pericycle, beneath which the xylem and phloem bundles are arranged radially, alternating. The number of xylem bundles is numerous (40–50). The outer part, below the pericycle, is the protoxylem, which consists of narrow cavities in the xylem elements, while the inner part is the metaxylem, which contains the large vessels. In the caudal part, the narrow cavity tracheids are separated from the wide cavity vessels by parenchyma cells. The phloem bundles are composed of few elements and are located between the protoxylems. The mechanical tissues thus form a triple circle: the hypodermis, the endodermis, and the xylem ring. The central part of the root is filled with pith parenchyma (Figs. 11.120, 11.121, and 11.122).

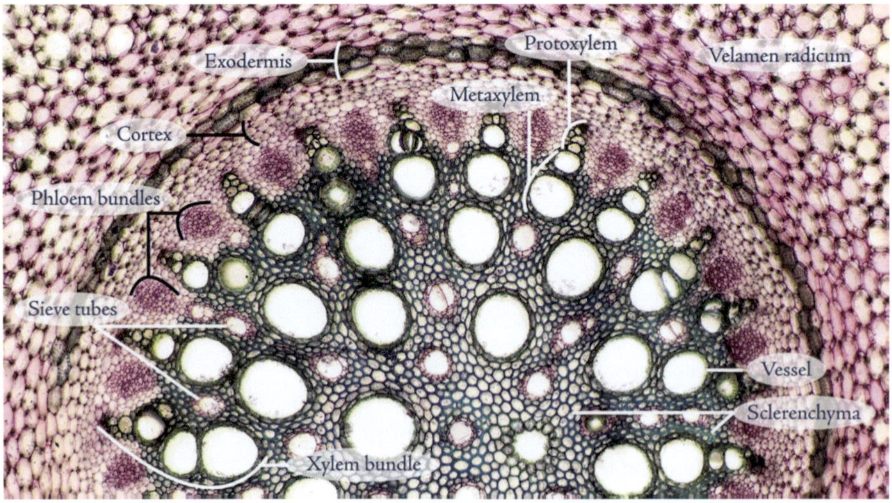

Fig. 11.117 Cross-section of aerial root stained with toluidine blue

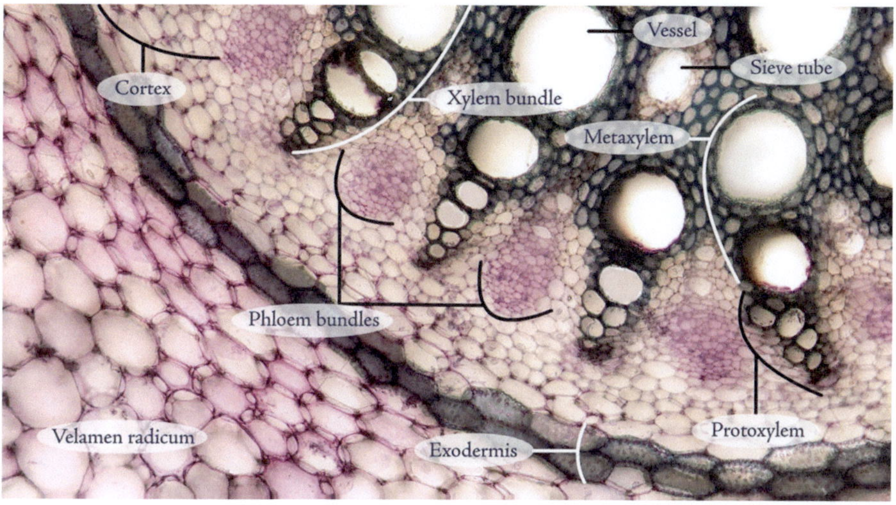

Fig. 11.118 High magnification image of a cross-section of the aerial root

Fig. 11.119 Maize (Zea mays) aerial roots

11.2 Stem

11.2.1 Horsetail and Monocot Stem

11.2.1.1 Maize (*Zea mays*)

Make a cross-section of a thinner piece of maize stem (Fig. 11.123) and examine it with a light microscope, stained with toluidine blue (Fig. 11.124), and native (Figs. 11.125 and 11.126).

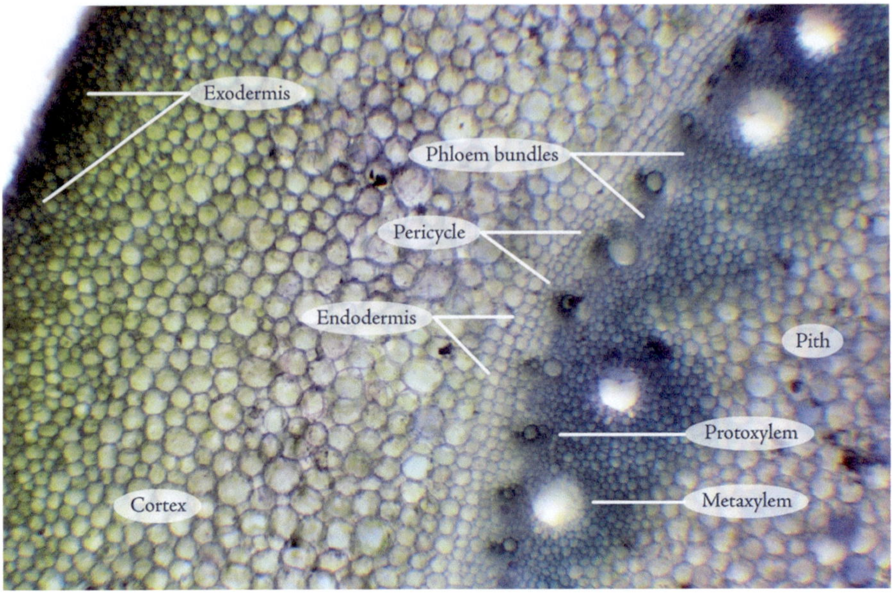

Fig. 11.120 Cross-section of the aerial root stained with toluidine blue

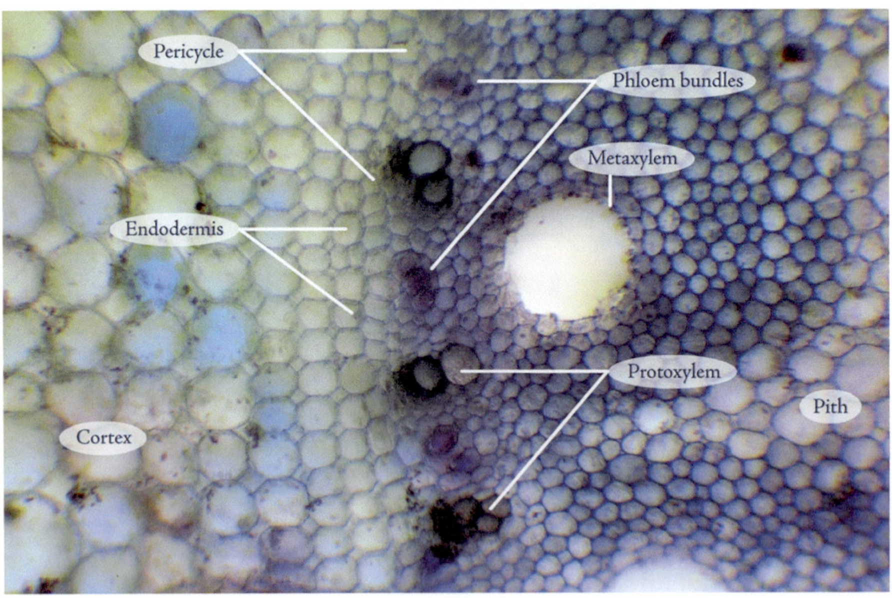

Fig. 11.121 High magnification image of the junction of the cortex and the stele

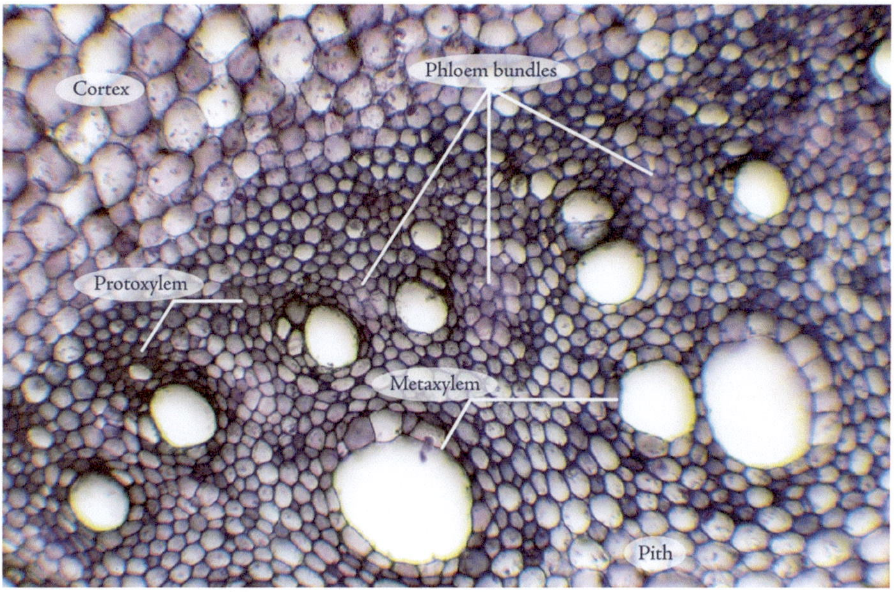

Fig. 11.122 High magnification image of the junction of the cortex and the stele

Fig. 11.123 Maize (*Zea mays*)

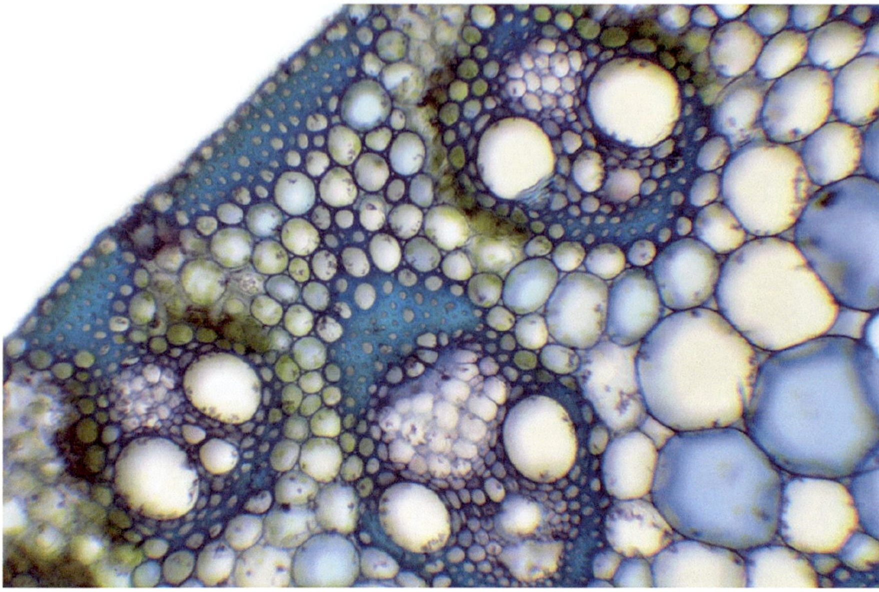

Fig. 11.124 Maize (*Zea mays*) stem cross-section stained with toluidine blue

In the cross-section, a low magnification shows that in the parenchymatous ground tissue is scattered with vascular bundles, more densely near the surface. A layer of sclerenchyma fibres, the bundle sheath, is observed along the vascular bundles (Figs. 11.125 and 11.126).

Two parts of tissue can be observed inside the vascular bundle. The outer part towards the surface of the stem is the phloem, and the inner part towards the centre is the xylem.

In the cross-sectional view of the phloem, the elements show a regular arrangement. The phloem is also missing the phloem parenchyma and the phloem fibres. The wider cavity, thinner-walled, angular elements are the sieve tubes.

In some of the sieve tubes you can see tiny refractive dots, which are the pore of the sieve plates. One or more thin-walled, plasma-filled, highly staining cells of smaller diameter are located in the immediate vicinity of the sieve tubes. These are the companion cells.

On the part of the phloem towards the surface, near the inner boundary of the sheath, we can observe the first phloem elements (protophloem) that were formed and compressed during stem growth.

Even at a low magnification, two or more very large, thick-walled vessels can be seen in the vascular bundles directly in contact with the phloem. Between these wide tracheae, a few more narrowly vessels and tracheids are visible, mostly surrounded by xylem parenchyma.

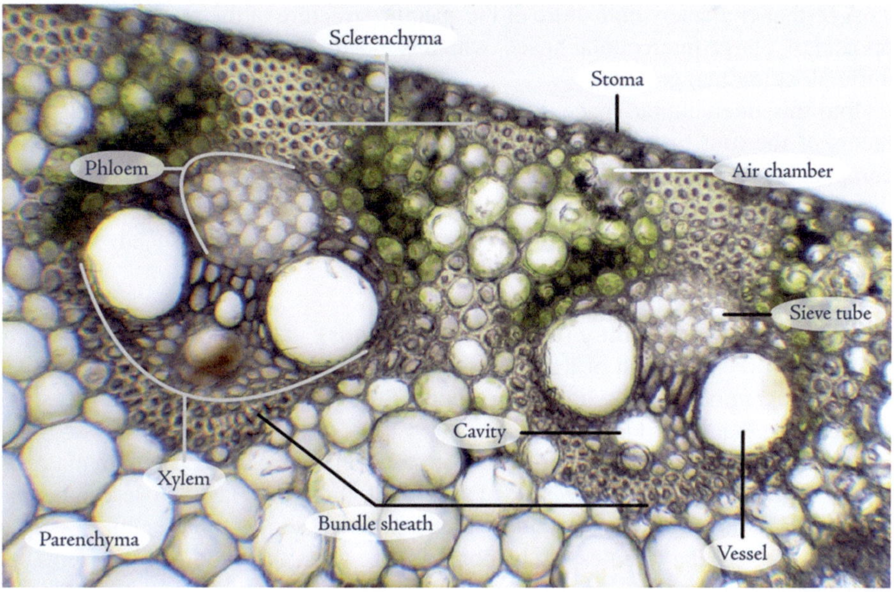

Fig. 11.125 Maize (*Zea mays*) stem cross-section detail

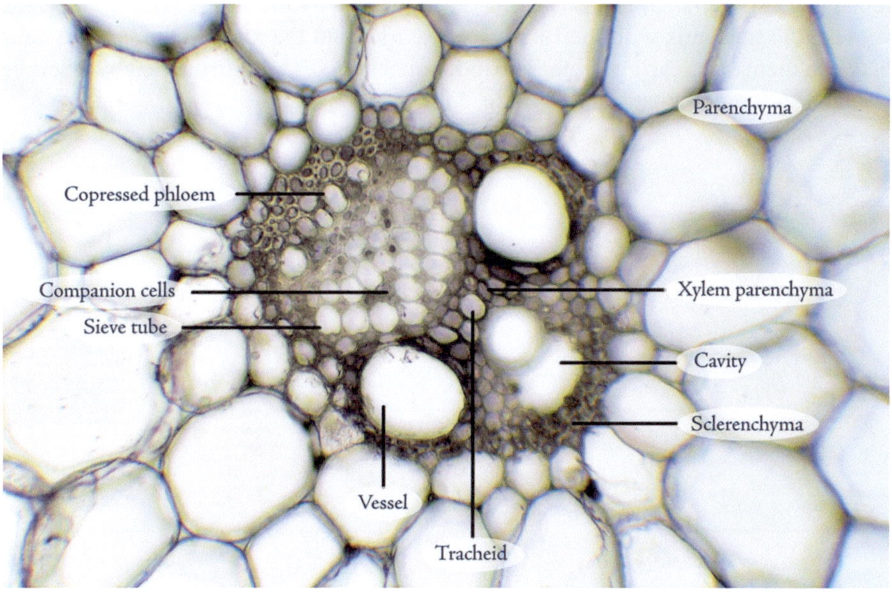

Fig. 11.126 Structure of the maize (*Zea mays*) vascular bundle

A further characteristic feature of the specific structure of the xylem is the development of a large intercellular space, which is mostly the result of tissue separation during longitudinal growth.

Into this intercellular space, the spirally or annularly thickened cell wall fragments of the first to form xylem elements protrude. Maize vascular bundles are complex because they are composed of a phloem and a xylem, collateral because the phloem and xylem parts are in contact with each other only on one side, and closed in the manner typical of monocots, because they do not contain any meristem (cambium).

11.2.1.2 Lily of the Valley (*Convallaria majalis*)

Make a thin cross-section of the stem of a lily of the valley, stain it with toluidine blue, and examine it with a light microscope. So we can study the sections of stem and leaf together (Fig. 11.127).

Beneath the monolayer epidermis of the stem, a layer of small parenchyma cells containing chloroplasts is observed. Further up, there is a layer of larger parenchyma cells including the vascular bundles. In this layer, intercellular spaces between the vascular bundles can be observed. Towards the middle of the stem, the parenchyma cells become smaller and smaller. The cortex and the pith are in direct contact, forming a single formation.

The outward-facing part of the vascular bundles is covered with a light blue sclerenchyma cap. Beneath this is the phloem, followed by the xylem. In the latter, the vessels, the tracheids with their narrower cavities and the xylem parenchyma cells between them are clearly visible. The individual bundles are surrounded by a bundle sheath of sclerenchymatised cells (Fig. 11.128).

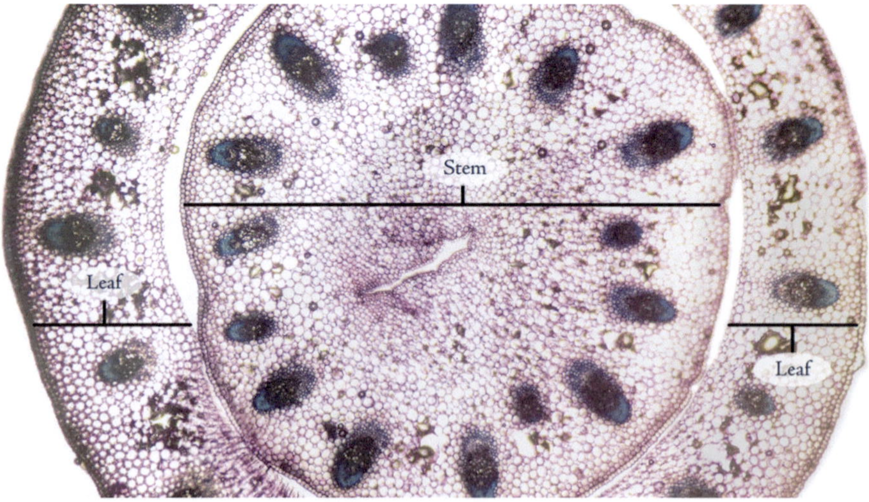

Fig. 11.127 Lily of the valley (*Convallaria majalis*) stem and leaf cross-section

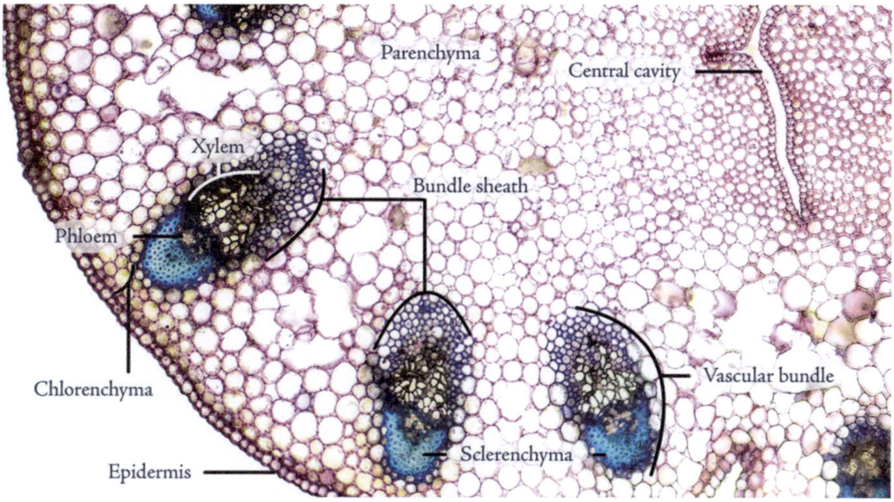

Fig. 11.128 Detail of a cross-section of a stem stained with toluidine blue

11.2.1.3 Horsetail (*Equisetum arvense*)

Make a cross-section of the stem of a horsetail and examine it with a light microscope after staining with toluidine blue.

The strongly ribbed stem is stiffened by sclerenchyma bundles in the prominent ribs. Below these is the chlorenchyma containing chloroplasts, with large vallecular canals in the direction of the furrows. The cortical parenchyma is closed by endodermis, within which the vascular bundles are always in the direction of the ribs and in which water-containing cavities (carinal canals) are formed. The xylem part of the bundles is composed of tracheids and parenchyma, and the phloem part of the bundles of sieve cells and phloem parenchyma (Fig. 11.129).

11.2.1.4 Iris (*Iris* sp.)

The peduncle of the *Iris* sp. is an excellent example of the characteristic tissue structure of monocots. Make as thin cross-sections as possible from the stems of the Iris and examine them with a light microscope after staining with toluidine blue.

The stem is covered by a single layer of epidermis, under which is chlorenchyma. The parenchymatous cells located here contain many chloroplasts. Underneath this is a layer of light blue-coloured sclerenchyma, which strengthens the stem. In the underlying part of the parenchymatous cells are scattered cross-sections of the vascular bundles. The bundles are closed and contain no cambium. Most of vascular bundles have a sclerenchyma part (Figs. 11.130, 11.131, and 11.132).

11.2.1.5 Dragon Tree (*Dracaena draco*)

In monocots, secondary thickening of the stem occurs only in a few cases, for example in dragon trees. In these plants, secondary thickening is caused by the secondary

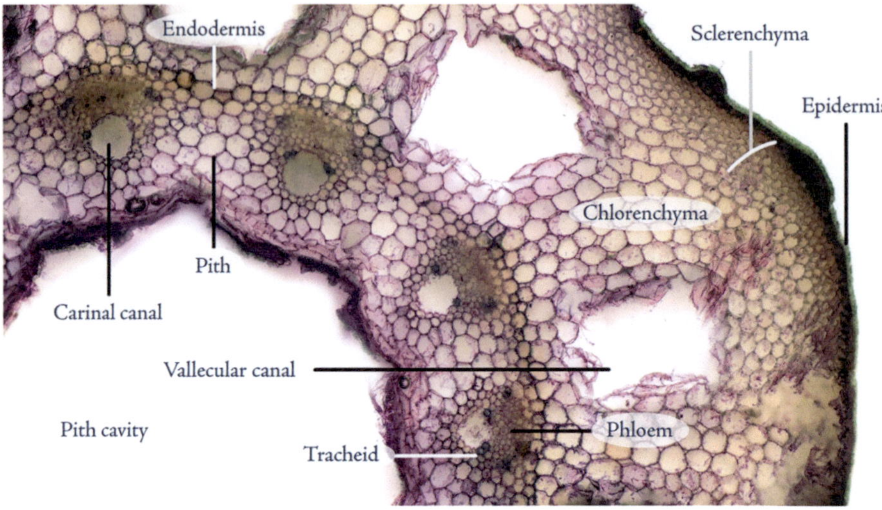

Fig. 11.129 Light microscopy image of the stem of the horsetail (*Equisetum arvense*)

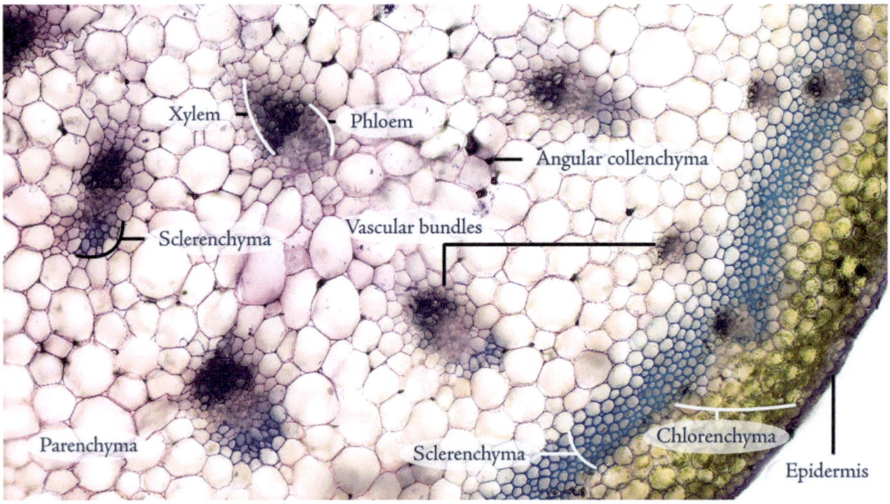

Fig. 11.130 Stem cross-section detail of an *Iris* sp. stained with toluidine blue

meristem of the cortex. Make a thin cross-section of the slender stem of the dragon tree, then stain it with toluidine blue and cover it with a drop of water and examine it with a light microscope.

In the central part of the stem of the dragon tree, the vascular bundles are scattered in the pith. The vascular bundles are closed, contain no cambium, and have the xylem on the outside, surrounding the phloem. Secondary thickening of the stem is made possible by a ring of meristem formed by cortical parenchyma cells, which

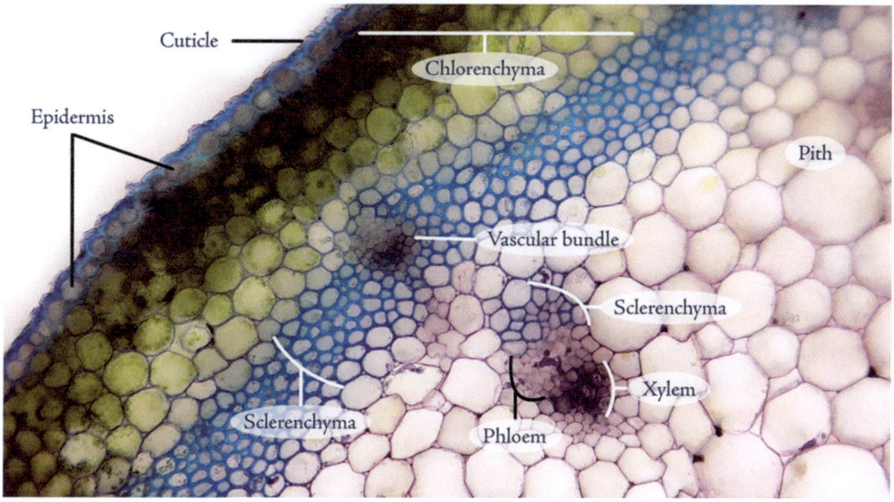

Fig. 11.131 Higher magnification image of a cross-section

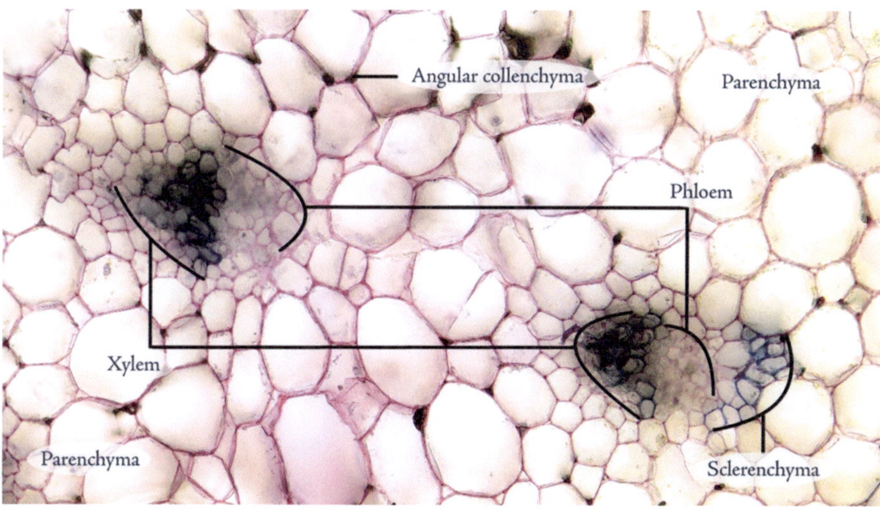

Fig. 11.132 Vascular bundles in the central part of the stem

creates a circle of inwardly closed vascular bundles together with the intestinal parenchyma cells separating them. The formation of the first ring of meristem is later followed by others in the outer layers of the cortex, which create further circles of vascular bundles. Thus, the characteristic tissue pattern of the stem cross-section of the dragon tree is formed, characterised by scattered primary vascular bundles in the centre and secondary bundles in a circle outside the centre. In the microscopic section, the central stem part with the closed vascular bundles and the cortex, whose

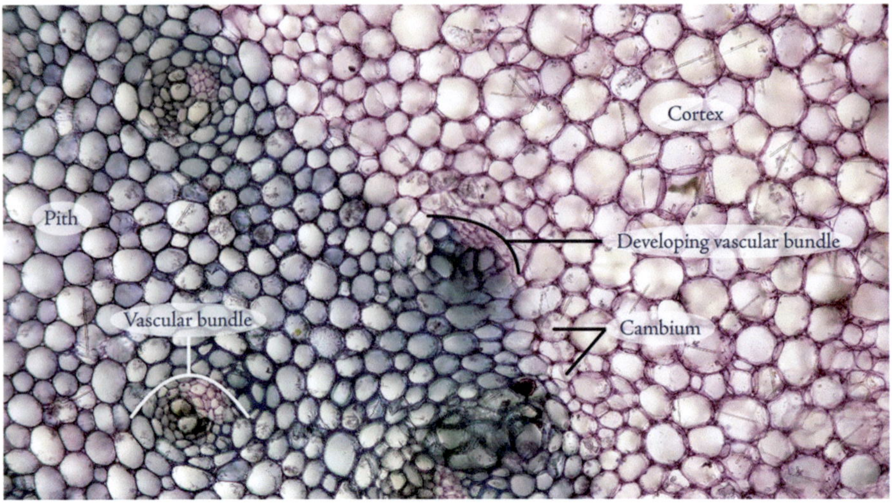

Fig. 11.133 Cross-section of a dragon tree (*Dracaena draco*) stem

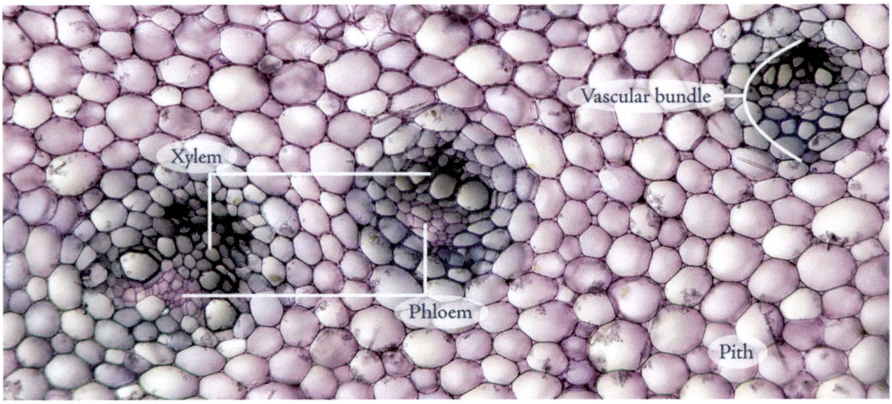

Fig. 11.134 Cross-sections of primary vascular bundles

inner cell line has been transformed into a meristem, are clearly separated (Figs. 11.133, 11.134, and 11.135).

11.2.1.6 Common Bulrush (*Scirpus lacustris*)

Make a cross-section of the stem of a common bulrush and examine it with a light microscope. The figure shows the chlorenchyma, which is composed of palisade parenchyma under the epidermis and contains many chloroplasts.

The parenchyma, which forms the base of the stem, forms large air cavity. In the scattered cross-sections of the vascular bundles, the sheath of the sclerenchyma, the xylem facing inwards towards the stem, and the phloem facing the outer surface are clearly visible.

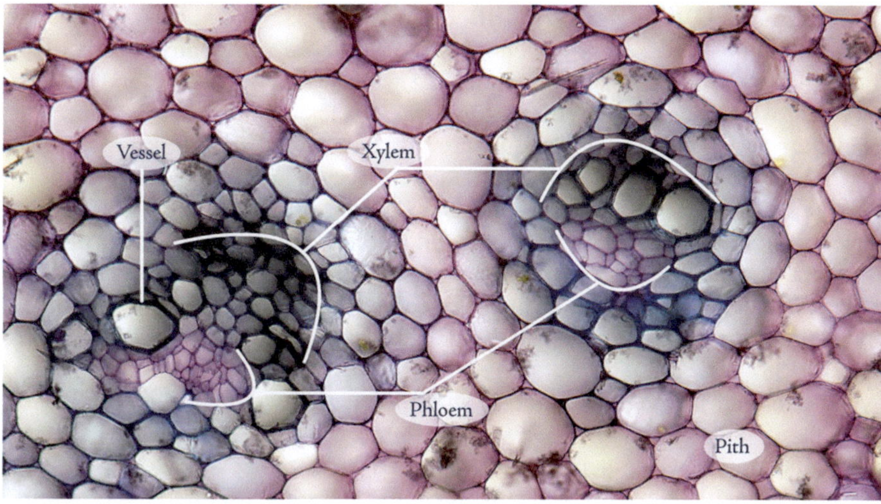

Fig. 11.135 High magnification image of closed vascular bundles

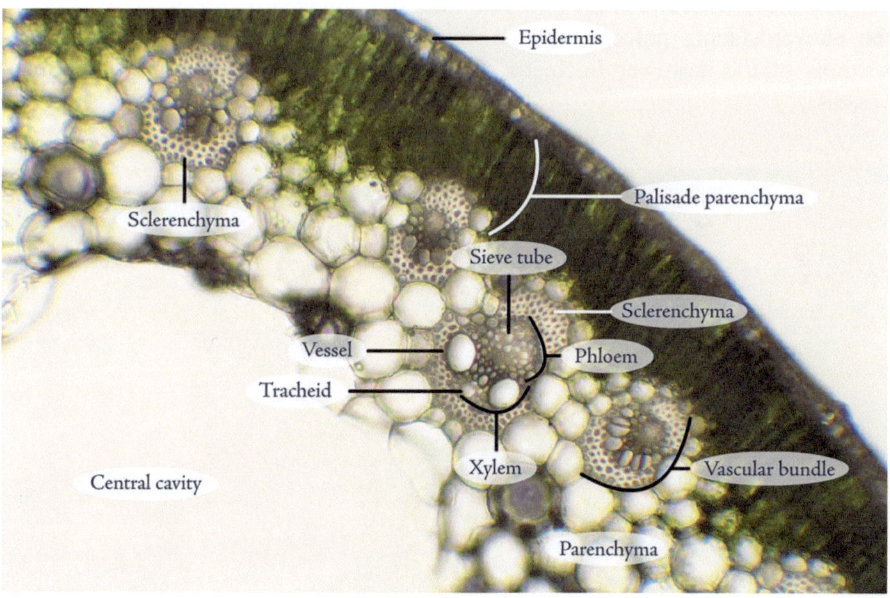

Fig. 11.136 Light microscopy image of a cross-section of the stem of a common bulrush (*Scirpus lacustris*)

In the xylem, sections of the dilated vessels and the tracheids are clearly distinguished (Fig. 11.136).

11.2.1.7 Tulip (*Tulipa gesneriana*)

Most wild tulip species occur in Central Asia, but they are also native to other parts of Asia, North Africa, and Europe. About 125 wild species are known, and the number of cultivated species is 3–4000 (Fig. 11.137). The peduncle of the tulip is an excellent way to study the tissue structure of the monocot stem. It is easy to make very thin sections of a sufficiently thick stem. Place them on a slide in one drop of water and stain them with toluidine blue!

With a light microscope, a small magnification of the cross-section of the stalk shows the cortical parenchyma consisting of rounded cells underneath the unilayered epidermis. Underneath is a ring of sclerenchyma, followed by the pith parenchyma filling the inside of the stem. In the latter, the vascular bundles are scattered (Fig. 11.138).

Under higher magnification, the spectacular stem structure of the tulip stem is revealed in more detail (Figs. 11.139, 11.140, and 11.141). The outer surface of the closely interconnected epidermal cells is covered with a cuticular layer. Underneath, a row of cells with thickened walls (hypodermis) is visible, followed by the cortical parenchyma. Underneath is a ring of sclerenchyma, which supports the stalk as a cylindrical structure. In the pith parenchyma, cross-sections of the vascular bundles can be observed, in which the toluidine-blue staining clearly distinguishes between the outward-facing phloem and the inward-facing xylem. In the xylem, cross-sections of the narrower tracheids can be seen outwards from the wide cavity vessels.

Fig. 11.137 Tulip (*Tulipa gesneriana*)

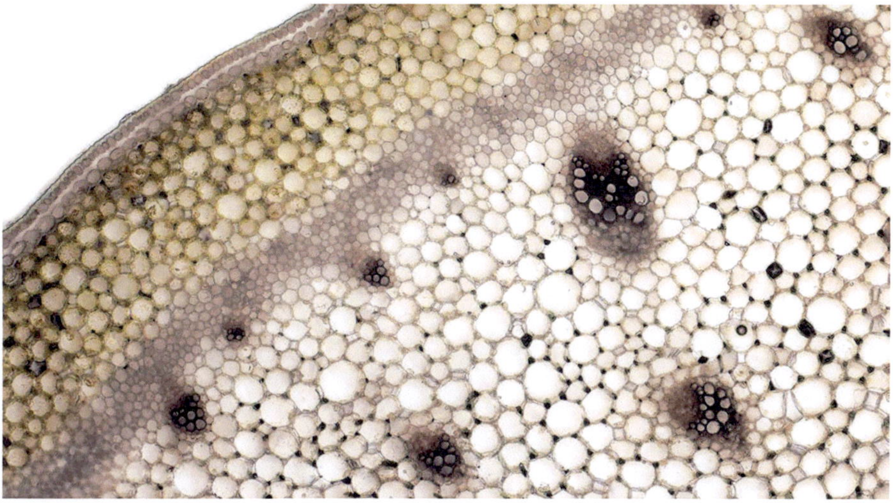

Fig. 11.138 Light microscope image of a tulip (*Tulipa gesneriana*) stem cross-section

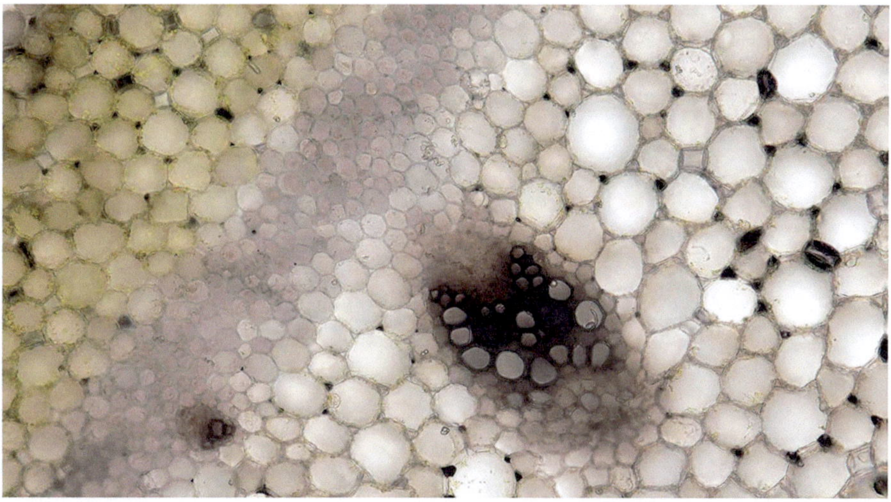

Fig. 11.139 High magnification image of a tulip (*Tulipa gesneriana*) stem cross-section

When examining the longitudinal section, the best way to identify the vascular bundles is to look at the spiral cell wall thickening vessels.

At the higher magnification the elongated elements of the xylem and phloem that form the vascular bundles can be clearly identified. As seen in the cross-section, the tracheids can be observed immediately adjacent to the phloem, while the dilated vessels are observed towards the interior of the pith parenchyma, away from the phloem (Fig. 11.142).

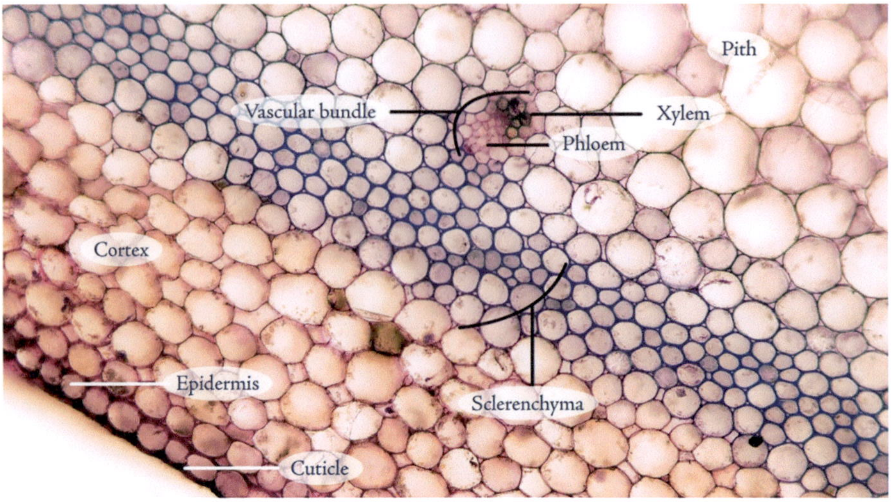

Fig. 11.140 Tulip (*Tulipa gesneriana*) stem cross-section stained with toluidine blue

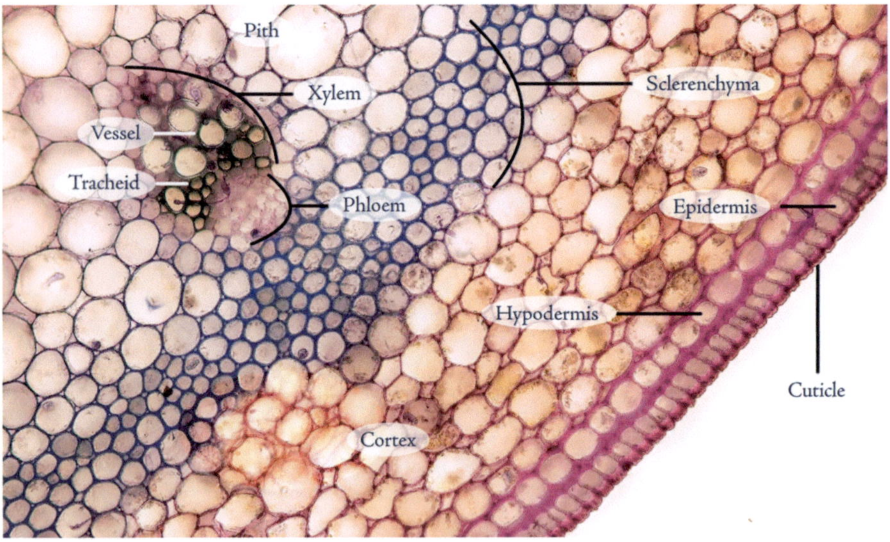

Fig. 11.141 Higher magnification light microscope image of a tulip (*Tulipa gesneriana*) stem cross-section

11.2.2 Primary Growth in Dicot Stem

11.2.2.1 Begonia (*Begonia* sp.)

Begonia species and their hybrids are popular ornamental plants. Its native species are found in tropical and subtropical zones in the Americas, Africa, and Asia (Fig. 11.143).

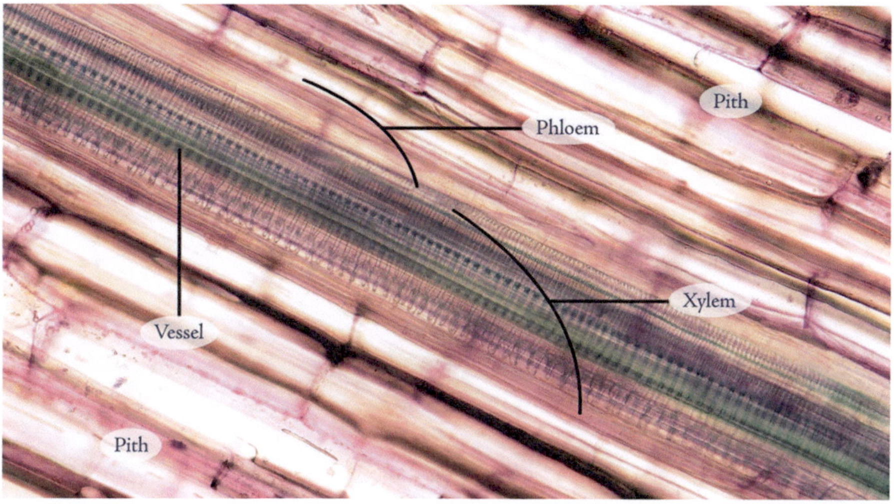

Fig. 11.142 Light microscope image of a longitudinal section of peduncle stained with toluidine blue

Fig. 11.143 Begonia (*Begonia* sp.)

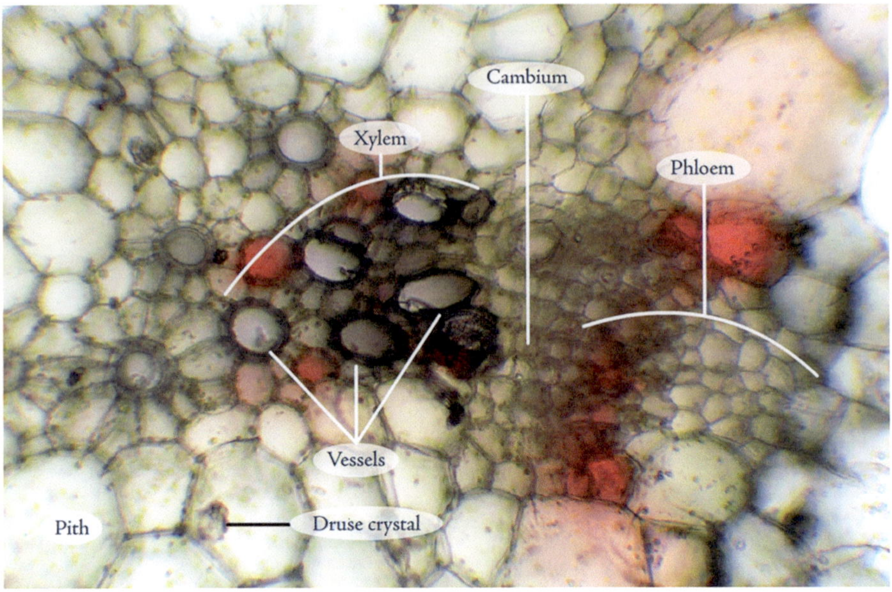

Fig. 11.144 Cross-section of a vascular bundle

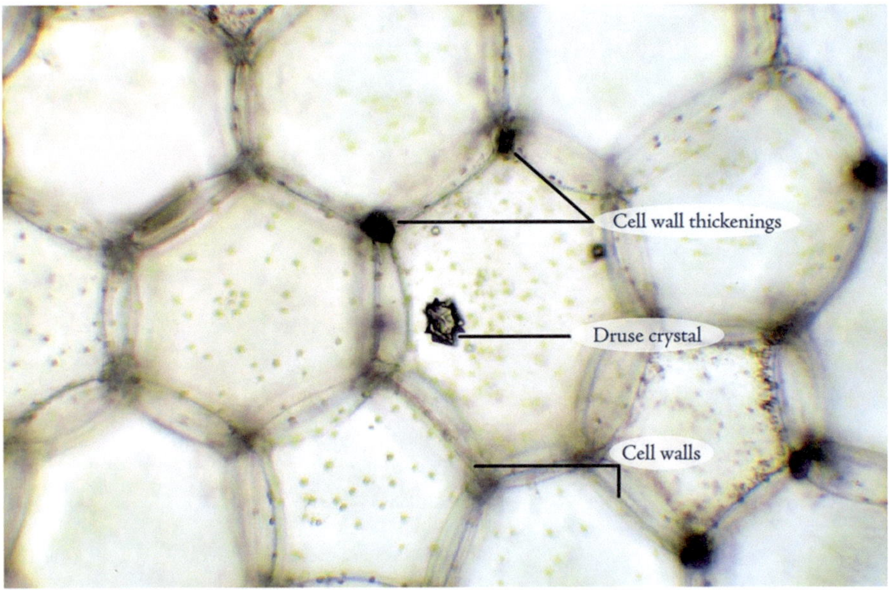

Fig. 11.145 Parenchyma cell with calcium oxalate crystal

Make a thin section of the begonia stem and examine it with a light microscope. The begonia stem has open vascular bundles with cambium that functions between the xylem and the phloem (Fig. 11.144). The parenchyma cells between the vascular bundles may contain druse crystals (calcium oxalate) (Fig. 11.145).

11.2.2.2 Pea (*Pisum sativum*)

The tissue structure of the pea stem (Fig. 11.146) can be revealed by thin cross-sectioning and light microscopy. The toluidine blue-stained preparation shows that the open vascular bundles are arranged in a circle (Fig. 11.147).

Moving inwards from the outside, a few rows of thick chlorenchyma are visible under the monolayer of epidermis, followed by a cortex of thin-walled, rounded parenchyma cells. The outer surface of the vascular bundles is covered by a sclerenchyma cap, followed by pinkish-tinged phloem and then by the lighter cambium.

Inside is the xylem containing the wide vessels, in which inwardly the protoxylem are located.

The pith parenchyma is located in the middle part of the stem, and the central cavity appears when it is torn away (Fig. 11.148).

11.2.2.3 Courgette (*Cucurbita pepo*)

Courgette is a subspecies of the pumpkin species *Cucurbita pepo*. Courgette is available in grocery stores at all times of the year and is therefore a good experimental subject for stem investigating. Make a cross-section of the courgette pedicel, then stain it with toluidine blue and examine it with a light microscope. The elements of the xylem of the bicollateral vascular bundle turn blue, whereas the two parts of the cambium and the parenchyma cells turn pale purple when stained with toluidine blue. Between the cambium and the xylem, which faces the outer surface of the stem, the whitish cells of the cambium are clearly visible (Fig. 11.149).

Fig. 11.146 Pea (*Pisum sativum*) plant

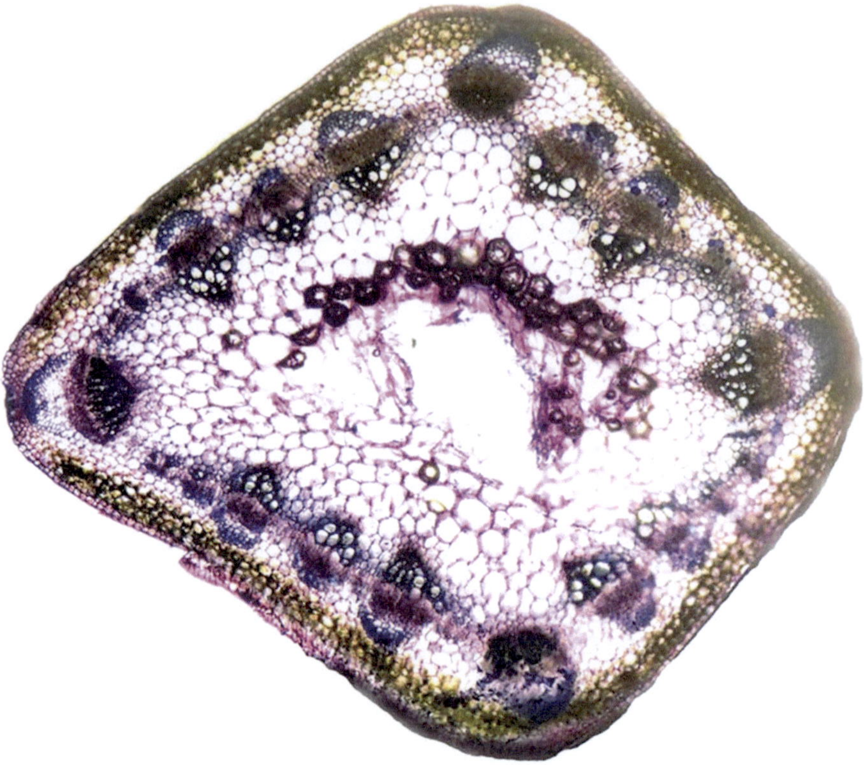

Fig. 11.147 Pea (*Pisum sativum*) stem cross-section stained with toluidine blue

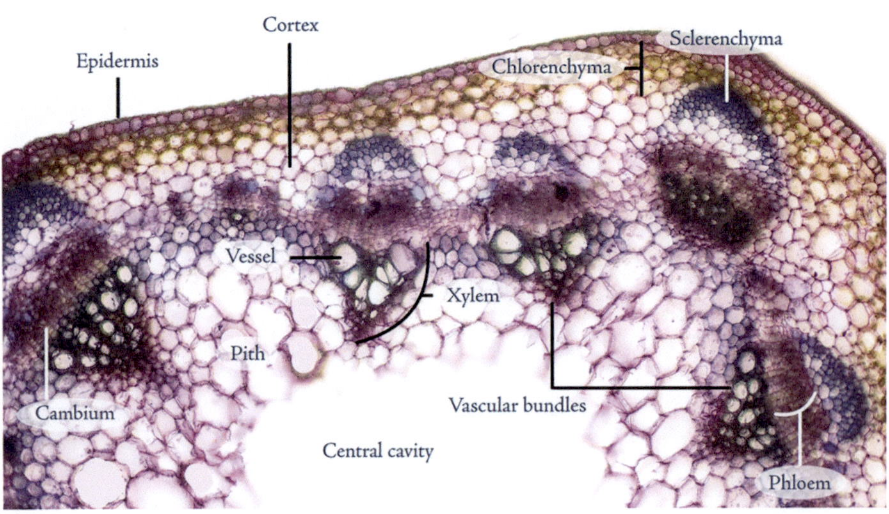

Fig. 11.148 Light microscope image of a cross-section of a pea (*Pisum sativum*) stem

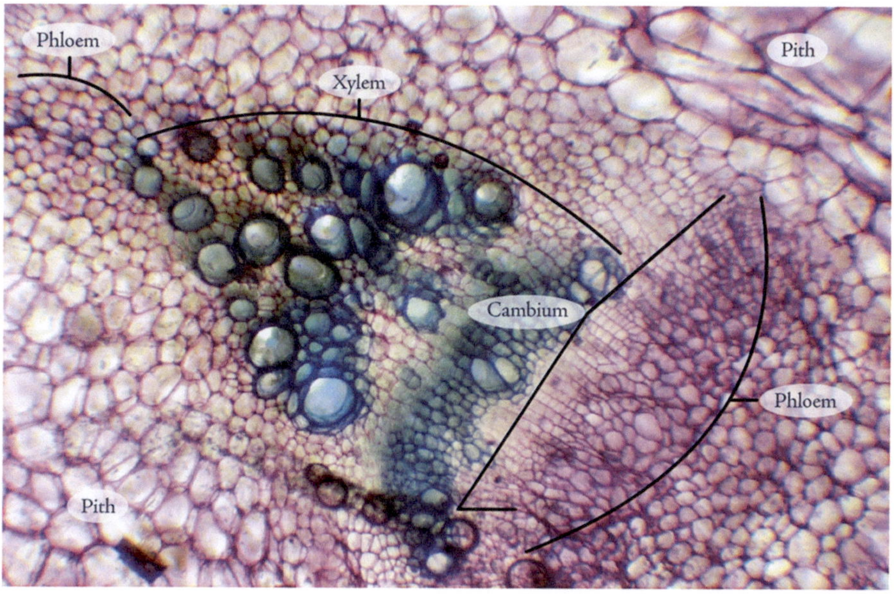

Fig. 11.149 Cross-section of the bicollateral vascular bundle of courgette (*Cucurbita pepo*), stained with toluidine blue

The high magnification image of the vascular bundle shows the phloem facing the outer part of the stem, and inward can be seen the cambium and the xylem. At this magnification, cross-sections of tracheids, dilated vessels, and xylem parenchyma cells can be distinguished in the xylem (Fig. 11.150).

11.2.2.4 Jimsonweed (*Datura stramonium*)

A poisonous plant species of the nightshades (Solanaceae) family, it is native to Central America but can be found in weedy communities in nitrogen-rich soils almost worldwide (Fig. 11.151).

Its stem is covered by a single layer of epidermis, under which are a few cellular chlorenchyma. The layer of parenchyma cells, rich in chloroplasts, is followed by a large layer of parenchymatous cells, and then by the innermost layer of the cortex, a starch sheath containing many starch granules follows.

Bicollateral bundles separated by medullary rays are arranged in a circle in the stele. In the bicollateral bundles, both the outer and inner sides of the xylem are connected by phloem. Cambium attached only to the outer phloem. The darker colour of the xylem clearly distinguishes the inward-facing protoxylem from the later metaxylem. The inside of the stem is filled with a pith parenchyma containing many starch granules (Figs. 11.152 and 11.153).

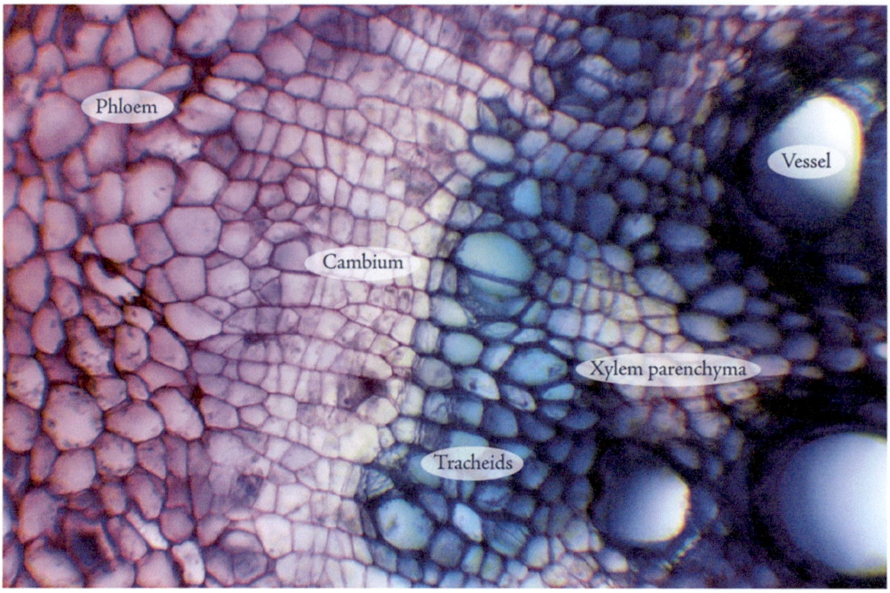

Fig. 11.150 High-magnification image of a courgette (*Cucurbita pepo*) vascular bundle

Fig. 11.151 Jimsonweed (*Datura stramonium*)

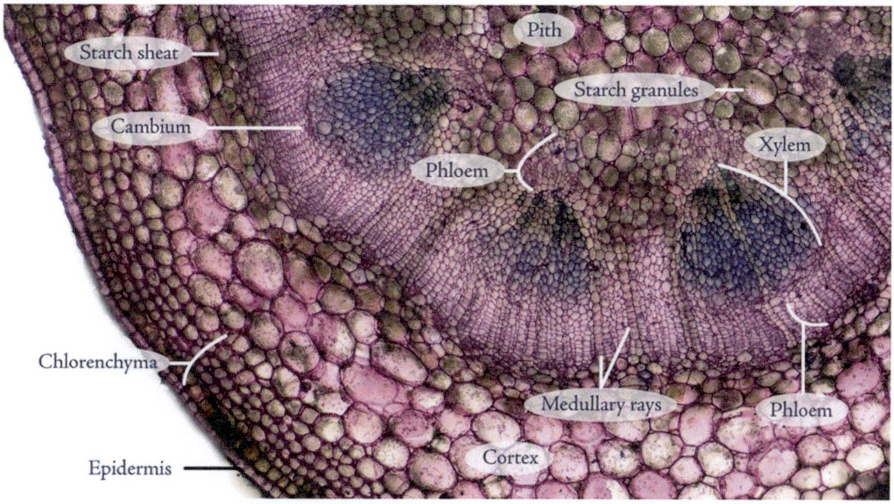

Fig. 11.152 Cross-section of a jimsonweed (*Datura stramonium*) stem stained with toluidine blue

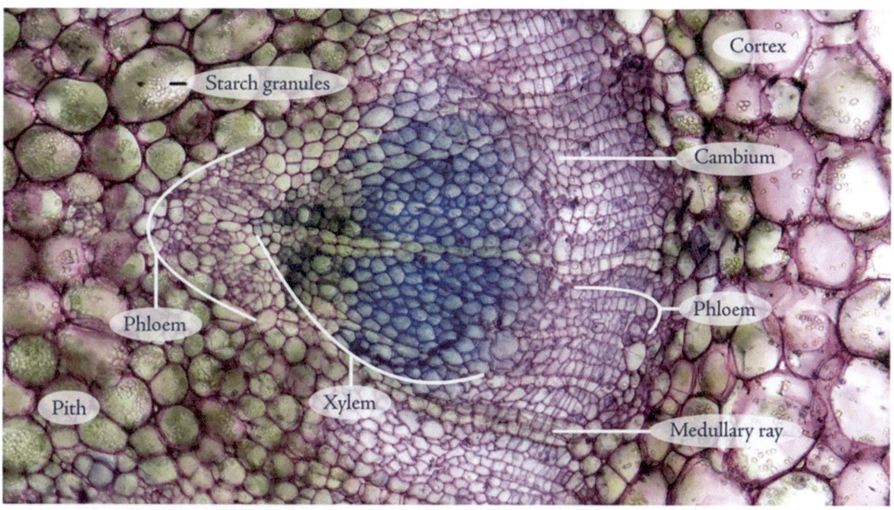

Fig. 11.153 High magnification image of a vascular bundle

11.2.2.5 False Sunflower (*Heliopsis helianthoides*)

The false sunflower is a North American plant that grows up to 1.5 m tall (Fig. 11.154). Its showy yellow flowers make it a popular garden plant. It spreads rapidly once it is established in the wild, as it is drought-tolerant.

Make a cross-section of the soft stem and stain it with toluidine blue! Examined with a light microscope, from the outside inwards, the monolayer epidermis and the green chlorenchyma underneath are clearly visible. Further in, parenchymatous cells form the cortex.

Fig. 11.154 False sunflower (*Heliopsis helianthoides*)

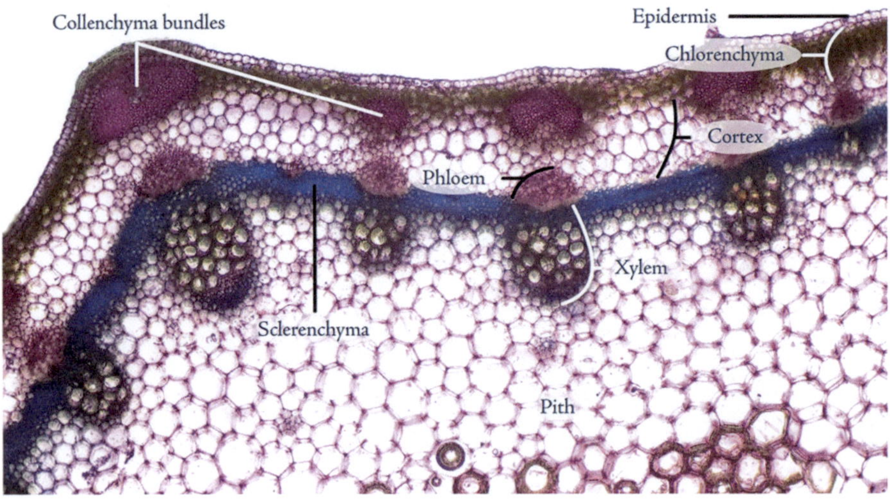

Fig. 11.155 Stem cross-section of a false sunflower (*Heliopsis helianthoides*) stained with toluidine blue

The vascular bundles are arranged in a circle, with a blue-coloured sclerenchyma, a bundle cap, on the outside. As a result of the staining, the phloem and the xylem are well separated. In the xylem, the rounded cross-sections of the vessels are prominent. There is a cambium between the phloem and the xylem. The inside of the stem is filled with pith parenchyma. The stem is supported by longitudinal bundles of collenchyma running under the epidermis and by bands of sclerenchyma connecting the adjacent vascular bundles (Figs. 11.155 and 11.156).

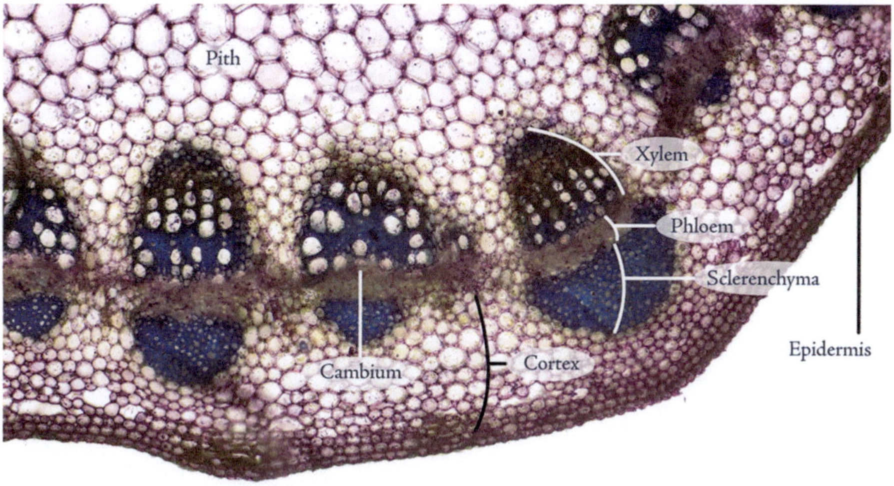

Fig. 11.156 Higher magnification image of a cross-section of the stem

11.2.2.6 White Campion (*Silene latifolia*)

The most common carnation in weed communities, cultivated areas, and roadsides is the white campion.

The monolayer of epidermis covering the young stem contains multicellular hairs (trichomes). Underneath, the layer of the chlorenchyma, containing many chloroplasts, is observed. Further inwards is the unicellular starch sheath, with dark blue-staining sclerenchyma attached to its inner surface. In the parenchyma filling the stem, the vascular bundles are arranged in a circle. The outward facing phloem, the inward facing xylem, and the cambium between them are clearly visible (Figs. 11.157 and 11.158).

In the cross-section of the older stem, there is a thick hypodermis beneath the epidermis, followed inwards by a cortex parenchyma. Inwards, the starch sheath is clearly recognisable, followed by a layer of sclerenchyma that develops from the bundle caps and becomes increasingly closed. The circular vascular bundles are connected by sclerenchymatised tissue, and large cells of the pith parenchyma can be observed in the middle (Figs. 11.159 and 11.160).

11.2.2.7 Claspingleaf Pondweed (*Potamogeton perfoliatus*)

Make a cross-section of the stem of a claspingleaf pondweed, then examine it with and without stained with toluidine blue with a light microscope.

The outside of the stem is covered with a single layer of epidermis with a thin, permeable cuticle. The main mass of the stem is the aerenchyma. In this, the large air chambers are bounded by parenchymal plates, creating extensive channels along the longitudinal axis of the stem. At the junction of the parenchymal plates, a single spherical cell is observed. The air chambers gradually decrease in size as they move

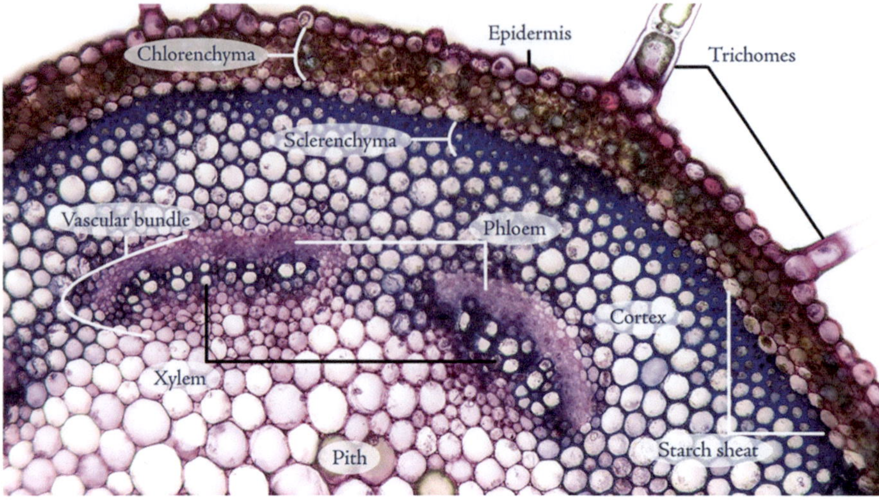

Fig. 11.157 Cross-section of a young stem of a white campion (*Silene latifolia*) stained with toluidine blue

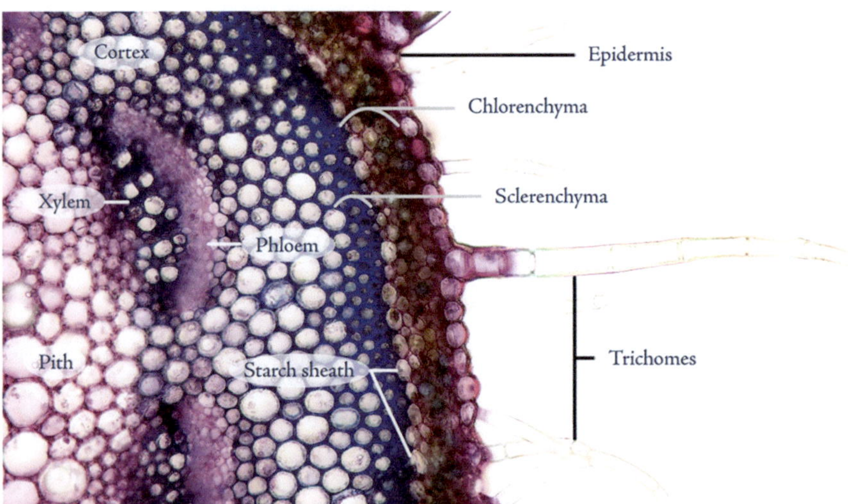

Fig. 11.158 Cross-section of a young stem of a white campion (*Silene latifolia*) stained with toluidine blue

away from the surface of the stem. Moving inwards, the parenchyma cells contain more and more starch granules.

The stele acts as a kind of solidifying cable that resists the movement of water. It is surrounded by the innermost cell row of the cortex, the dark-staining endodermis. The stele consists of parenchyma cells filled with starch granules and radially arranged vascular bundles. The latter are surrounded by a ring of blue-coloured

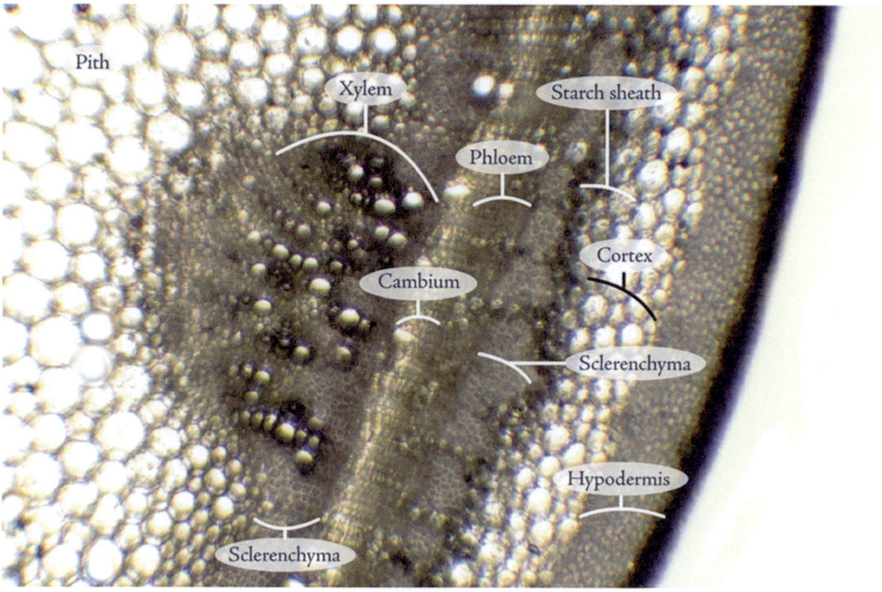

Fig. 11.159 Unstained cross-section of an old stem

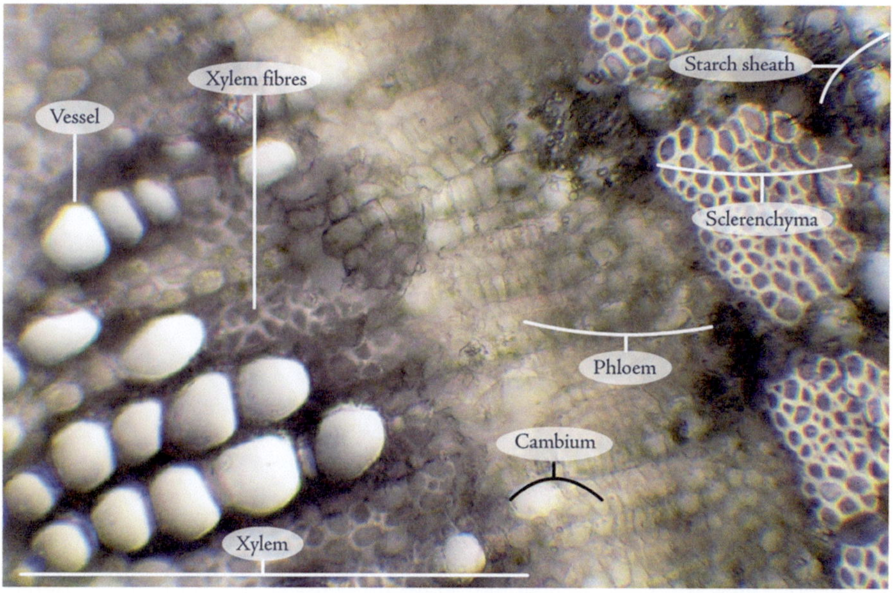

Fig. 11.160 High magnification image of the vascular bundle of a white campion (*Silene latifolia*)

sclerenchyma. In the vascular bundles, in addition to a significant amount of phloem, there are only large aquifer cavities and slits instead of true xylem (Figs. 11.161 and 11.162).

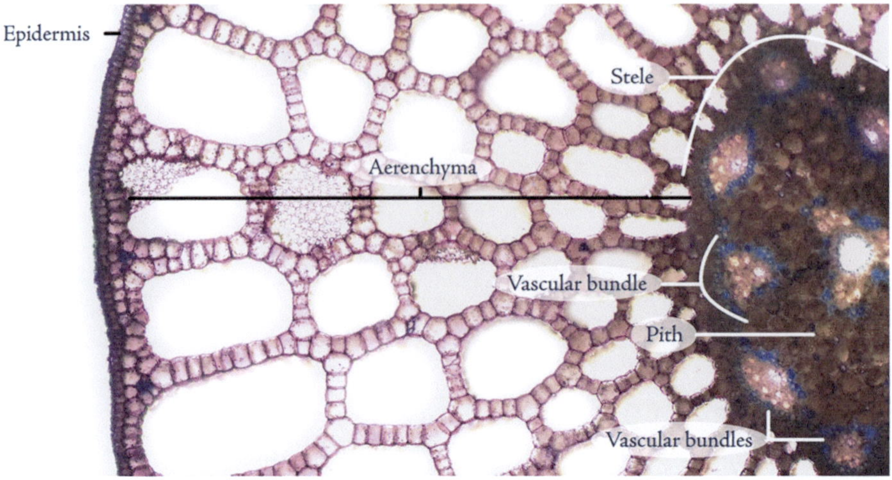

Fig. 11.161 Cross-section of a claspingleaf pondweed (*Potamogeton perfoliatus*) stem stained with toluidine blue

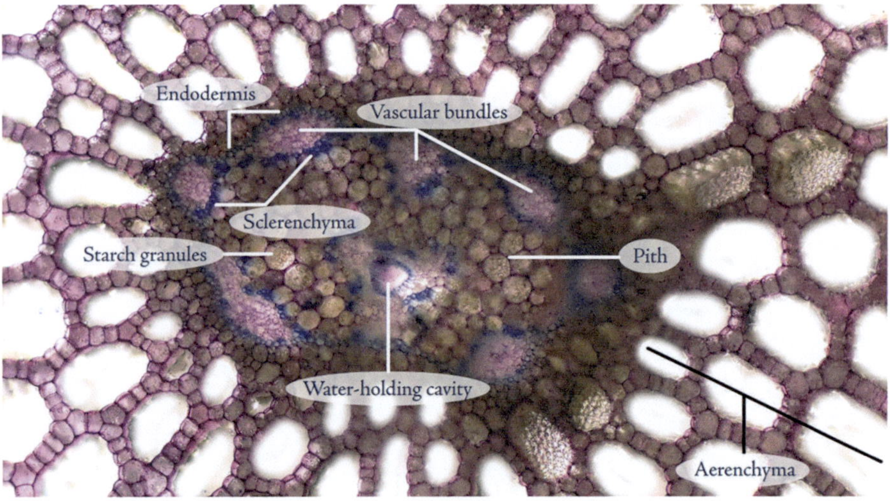

Fig. 11.162 Stele of a claspingleaf pondweed (*Potamogeton perfoliatus*) stem

11.2.2.8 Wild Pansy (*Viola tricolor*)

The wild pansy is a common herb in gardens and woodland clearings across Europe. The name comes from the fact that the petals of the flower are mostly tri-coloured, yellow, violet and white, but bicoloured, yellow-white and purple-white flowers are also common. The plant's soft stems are easy to make thin sections. The section is covered on the outside by a single layer of epidermis, under which some cellular layers of chlorenchyma are visible. Between and on the inner side of the circularly arranged vascular bundles, a blue-coloured sclerenchyma is observed. In the

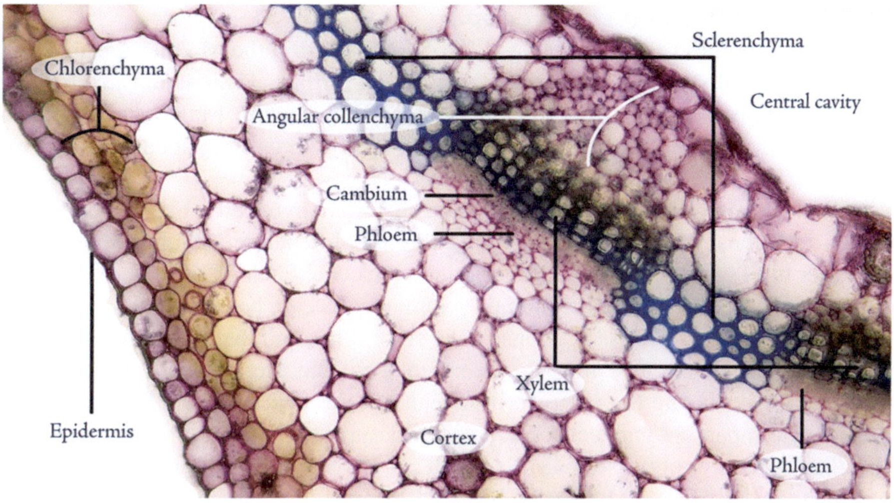

Fig. 11.163 Cross-section of a wild pansy (*Viola tricolor*) stem stained with toluidine blue

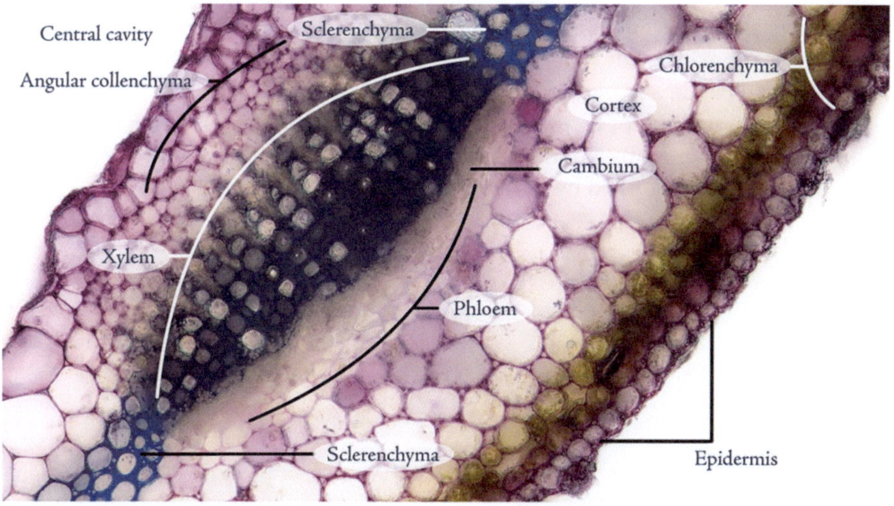

Fig. 11.164 High magnification image of a vascular bundle

vascular bundle, the light phloem faces towards the cortex and the dark xylem faces inwards. In the latter, the cross-sections of the wide cavity vessels and the narrower cavity tracheids are clearly visible. The darker part of the xylem facing inwards is the protoxylem, while the metaxylem faces outwards. Within the periphery of the vascular bundles is the pith, consisting of large rounded parenchyma cells, with intercellular spaces forming in the middle and then becoming a central cavity (Figs. 11.163 and 11.164).

11.2.2.9 Cornflower (*Centaurea* sp.)

Most of the cornflower species (Fig. 11.165) are found in the Mediterranean, from where they are distributed as far north as northern Europe. Their stems are stiff and harder to prune due to the strengthening elements they contain. Make a cross-section of the cornflower stem and examine it with a light microscope, stained with toluidine blue.

Below the monolayer epidermis, bundles of collenchyma are observed under the stem projections, while elsewhere some cellular chlorenchyma is observed. Below the latter is a layer of large rounded parenchyma cells. Each of the unicircular vascular bundles bears a light blue-stained sclerenchyma cap, below which follows the phloem part. Here the cross-sections of the sieve tubes and the companion cells are clearly visible. Below the cambium the xylem is visible. Here, the cross-sections of the dilated vessels and the narrower tracheids are arranged in rows. Between them are the xylem parenchyma cells. Each vessel is surrounded by a sclerenchymatised sheath. The interior of the stem is filled with a pith consisting of thin-walled rounded parenchyma cells (Figs. 11.166, 11.167, and 11.168).

11.2.2.10 Garlic Mustard (*Alliaria petiolata*)

Garlic mustard is native to Europe, western and central Asia, and northwest Africa, and is an invasive weed in North America (Fig. 11.169). It can be abundant in shrublands and forest edges, but is also common in shady and semi-shady areas in urban environments, parks, and gardens.

The stem is suitable for histological examination. Thin cross-sections and longitudinal sections can easily be made from the soft stem, which can be stained with toluidine blue and examined by light microscopy.

Fig. 11.165 Cornflower (*Centaurea* sp.)

Fig. 11.166 Cross-section of a cornflower (*Centaurea* sp.) stem stained with toluidine blue

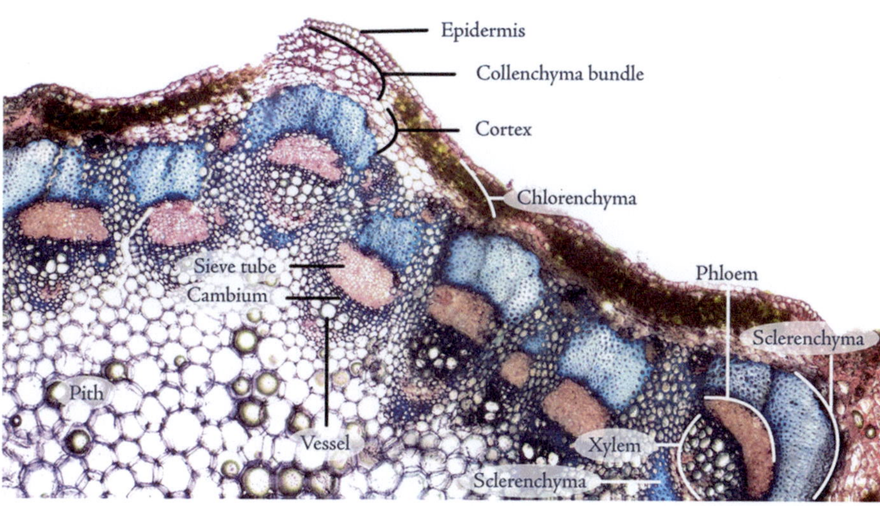

Fig. 11.167 Histology of the cornflower (*Centaurea* sp.) stem

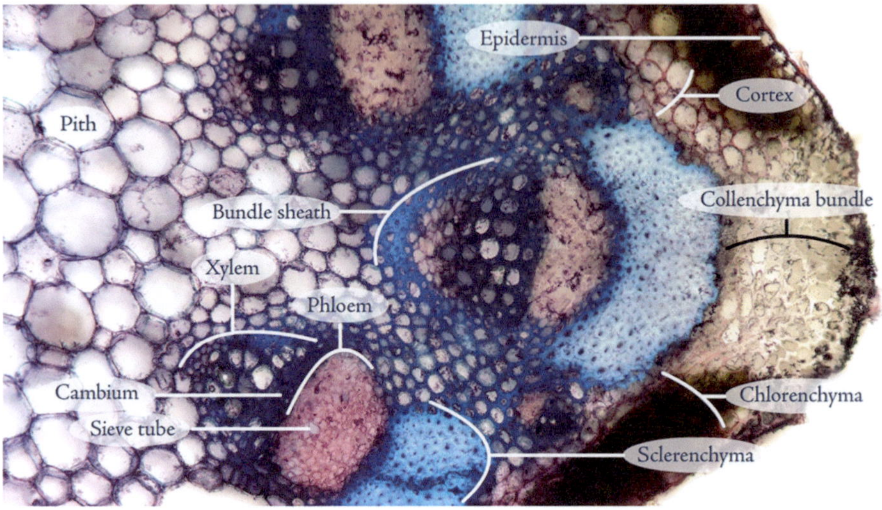

Fig. 11.168 High magnification image of the stem cross-section

Fig. 11.169 Garlic mustard (*Alliaria petiolata*)

The stem is covered on the outside by a single-celled epidermis, under which is the cortical parenchyma. The near-surface of the latter is transformed into a chlorenchyma rich in chloroplasts. The circular vascular bundles are connected by a ring of sclerenchyma. The outer part of the vascular bundles is covered by sclerenchyma, below which the phloem and xylem follow inwards. The xylem is clearly recognisable from the wide cavity of the vessels. The central part of the stem is filled with

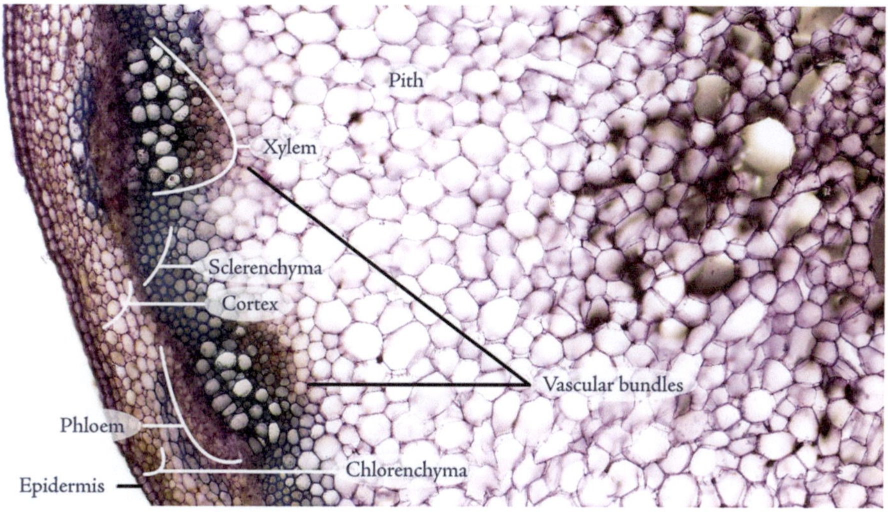

Fig. 11.170 Cross-section of the stem stained with toluidine blue

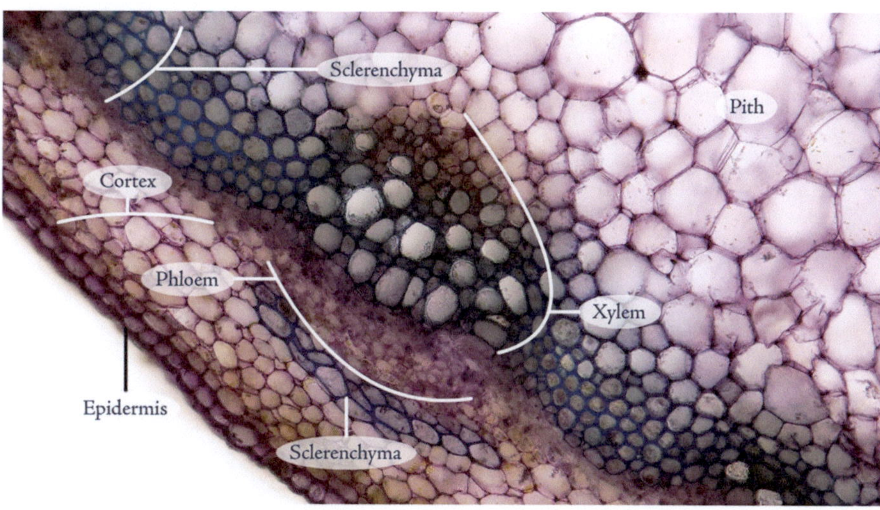

Fig. 11.171 Enlarged image of a vascular bundle

pith parenchyma, inside which intercellular spaces develop (Figs. 11.170, 11.171, and 11.172).

You can also easily make longitudinal sections from the stem. With this toluidine staining, the cell wall thickening of the vessels can be clearly observed in addition to the individual tissues (parenchyma, xylem, and phloem) (Fig. 11.173).

11.2.2.11 Common Purslane (*Portulaca oleracea*)

The fleshy, creeping stem of the common purslane (Fig. 11.174) are easy to make a section of. Examine it with a light microscope by staining it with toluidine blue! The

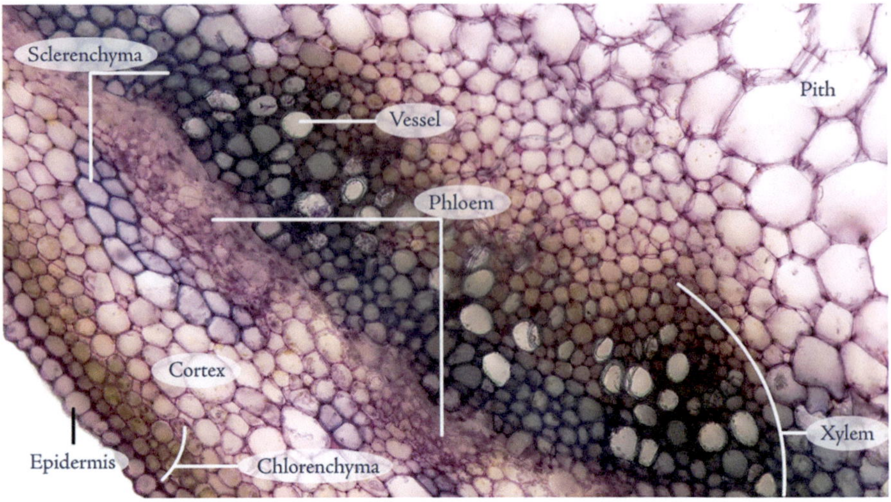

Fig. 11.172 Detail of a cross-section of the stem stained in toluidine blue

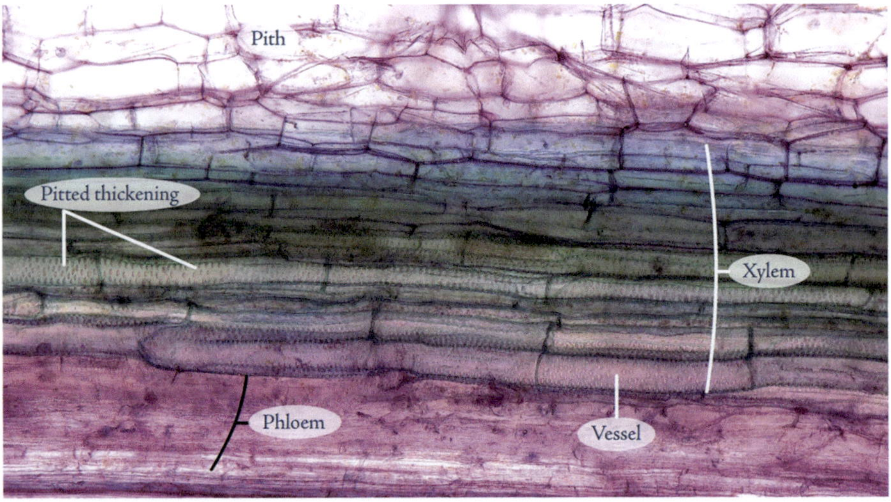

Fig. 11.173 Longitudinal section of the stem

stem shows large thin-walled cells of the cortical parenchyma beneath the monolayer epidermis, which often contain druse crystals. The innermost cell row of the cortex is the starch sheath, in whose cells many starch granules are visible. The open vascular bundles, containing the cambium, are arranged in a circle. Pith parenchyma fills the stem, and druse crystals are also often observed in its cells (Fig. 11.175).

11.2.2.12 Marsh-Marigold (*Caltha palustris*)

A member of the buttercup family, the marsh-marigold is a spectacular-looking perennial herbaceous plant that lives in wetlands, marshes, bogs, and streamsides

Fig. 11.174 Common purslane (*Portulaca oleracea*)

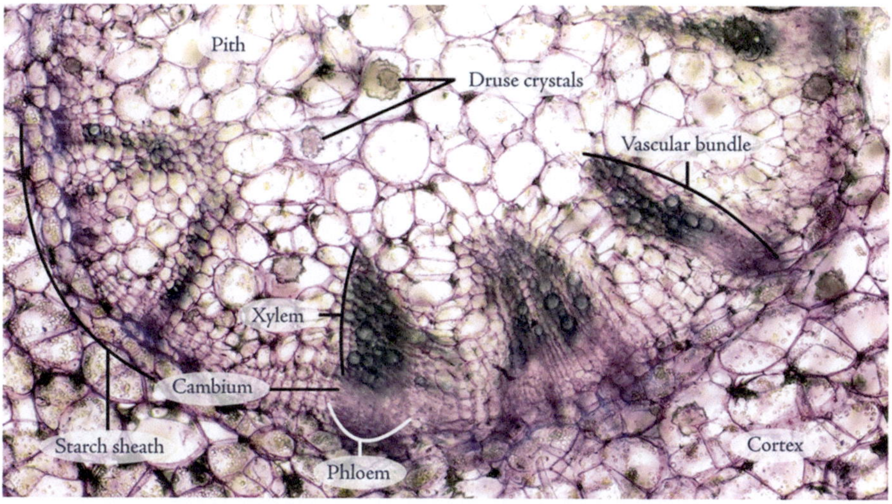

Fig. 11.175 Cross-section of the stem stained with toluidine blue

(Fig. 11.176). Make cross-sections of the stem, stain it with toluidine blue, and study it with a light microscope!

A chlorenchyma is observed under the monolayer of epidermis. Its cells contain a large number of chloroplasts. Next comes the aerenchyma, typical of aquatic and marsh plants, in which air-filled air chambers are separated by single-cell-thick tissue sheets.

Fig. 11.176 Marsh-marigold (*Caltha palustris*) on a stream bank

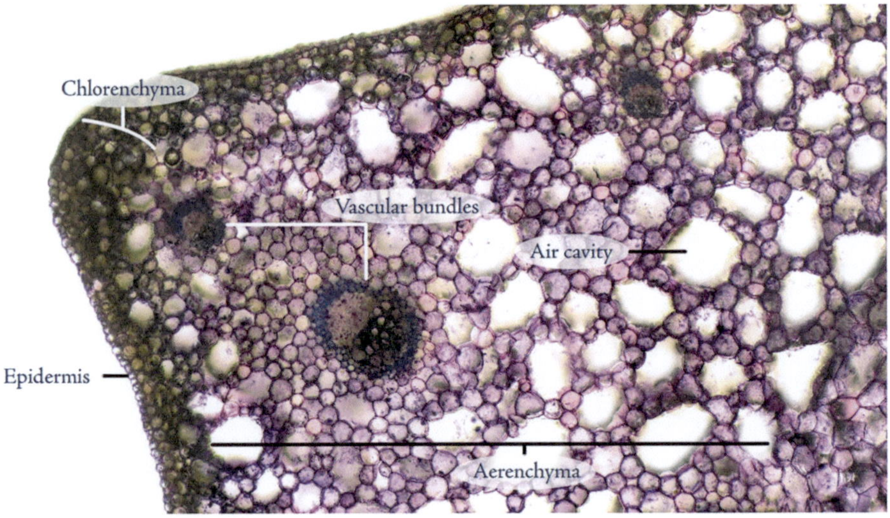

Fig. 11.177 Cross-section of the stem stained with toluidine blue

Smaller and larger vascular bundles are already visible near the stem surface. They are surrounded by a sheath of sclerenchyma and do not contain a cambium. Looking outwards, the pale purple-coloured phloem is visible, in which the cross-sections of the sieve tubes and the companion cells can be discerned under higher magnification. In the xylem, we can observe wide cavity vessels, narrower cavity tracheids, and xylem parenchyma. The cortical and pith parenchyma are not separated (Figs. 11.177, 11.178, and 11.179).

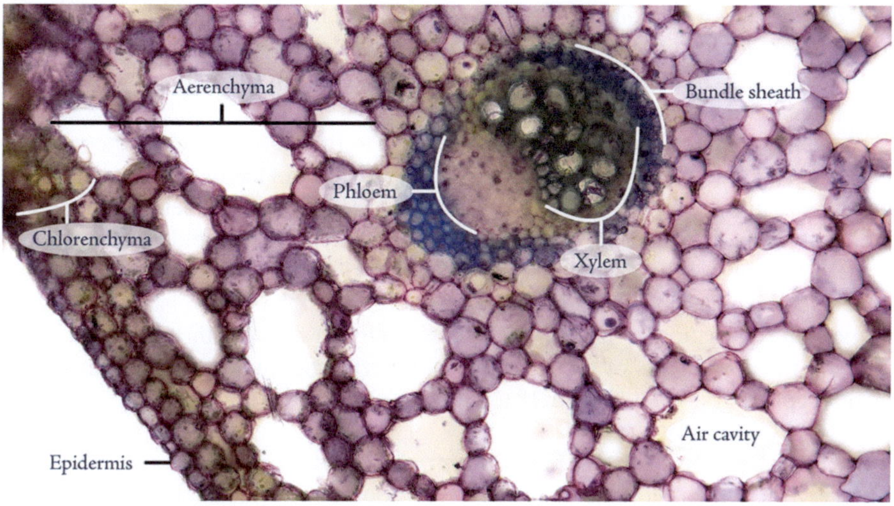

Fig. 11.178 Higher magnification image of a stem of a marsh-marigold (*Caltha palustris*)

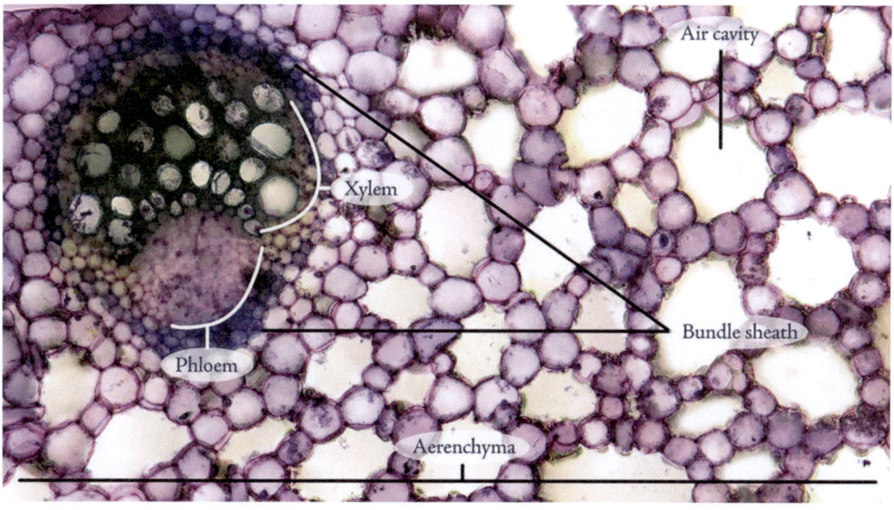

Fig. 11.179 High magnification image of a cross-section of a marsh-marigold (*Caltha palustris*) stem

11.2.2.13 Common Nettle (*Urtica dioica*)

Thin cross-sections of the angular stem of nettle can be easily made, and after staining with toluidine blue, they should be examined by light microscopy.

The stem is covered on the outside with a single layer of epidermis, under which a chlorenchyma layer may appear. In some places in the cortex, longitudinal bundles of collenchyma can be observed beneath the epidermis, forming protrusions on the surface of the stem.

A pinkish bundle of collenchyma is attached to the outer surface of the vascular bundles. The vascular bundles have light-coloured phloem facing towards the cortex and dark xylem facing inwards. A cambium is observed between them. In the xylem, the cross-sections of the dilated vessels, the narrow cavity tracheids, and the thick-walled, narrow cavity xylem fibres are clearly visible. The vascular bundles are connected by a dark blue ring of sclerenchyma. Within the ring of vascular bundles is the pith, consisting of large, rounded parenchyma cells, with intercellular passages in the upper parts of the central part. Later in the middle of the stem central cavity appears (Figs. 11.180, 11.181, and 11.182).

11.2.2.14 Red Deadnettle (*Lamium purpureum*)

The red deadnettle is common in glades, roadsides, dry woods, and acacia woods. It is a common weed in gardens and urban green areas (Fig. 11.183). Flowering from March to the end of September. Its mass appearance makes it one of the most important early bee-feeders. The anthocyanin content makes the plant reddish and the stems are rectangular.

Soft stems can be easily cross-sectioned for light microscopy. Stain the sections with toluidine blue!

Below the unilayered epidermis, a pink-staining cross-section of a bundle of collenchyma is observed under the corners of the rectangular stem, while elsewhere the chlorenchyma is visible. The cells of the latter are packed with chloroplasts. Beneath

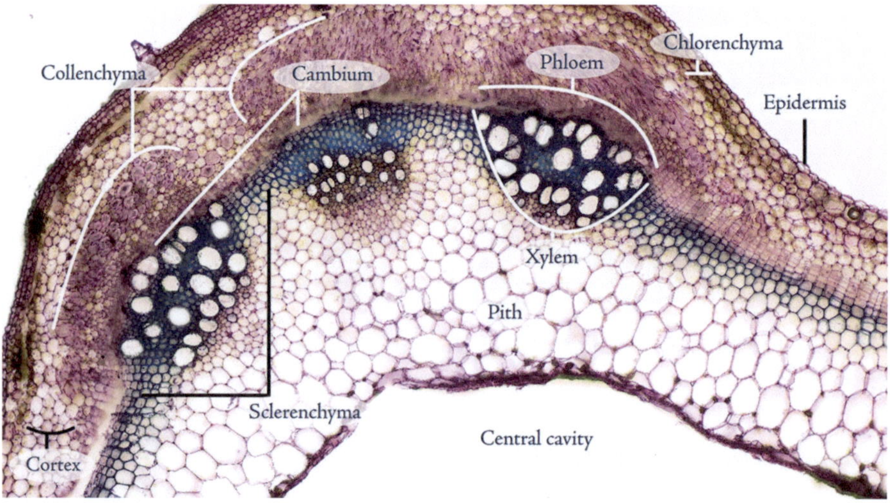

Fig. 11.180 Cross-section of the stem stained with toluidine blue

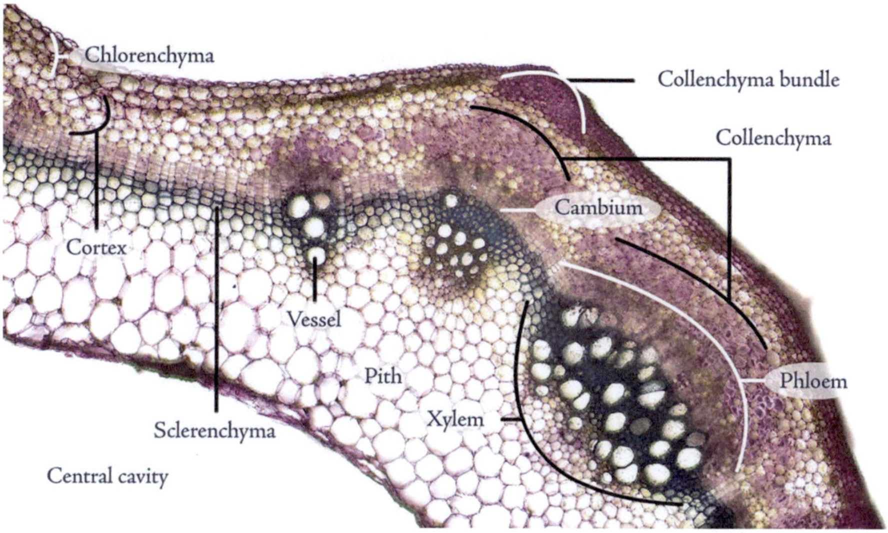

Fig. 11.181 Cross-section of the stem

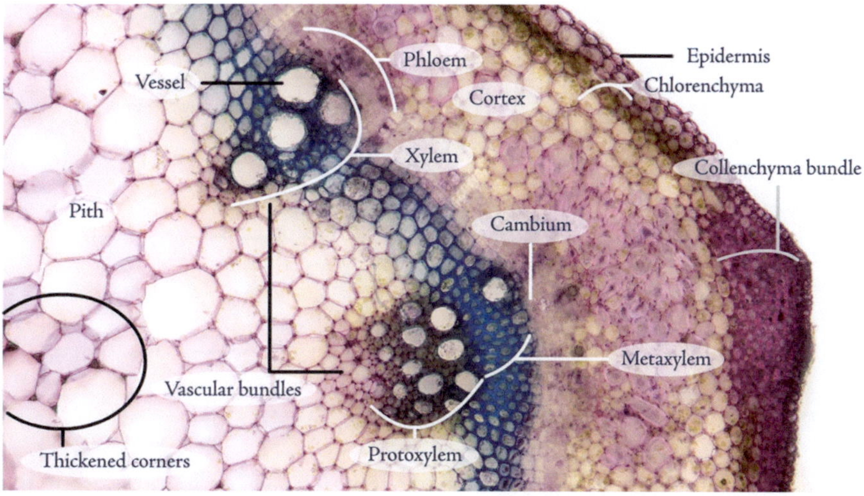

Fig. 11.182 Cross-section of the stem at higher magnification

the layer of large rounded cells of the cortical parenchyma are the unicircular vascular bundles. These are connected by a dark blue band of sclerenchyma. The darker-coloured protoxylem is visible in the inner part of the vascular bundles, which continues outwards in the metaxylem. In the xylem, the large cavity vessels and the narrower cavity tracheids are clearly visible, arranged in radial rows.

Fig. 11.183 Red deadnettle (*Lamium purpureum*)

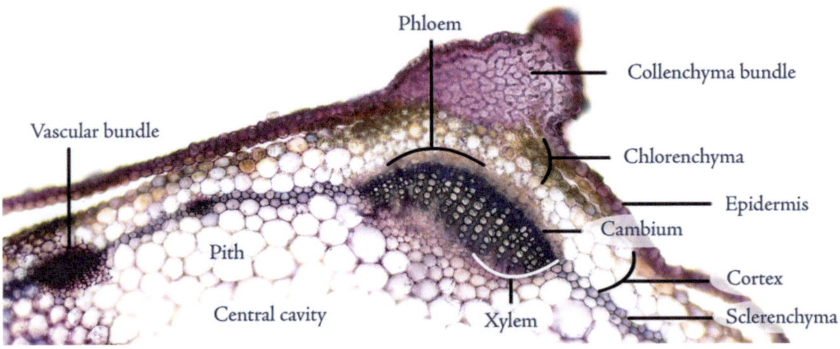

Fig. 11.184 Stem cross-section stained in toluidine blue

They are separated by xylem parenchyma cells. The xylem is followed by the cambium and then the lighter phloem. The inward facing part of the vascular bundles is connected to the angular collenchyma. The inside of the stem is formed by pith parenchyma, which is torn in the middle to form a central cavity (Figs. 11.184 and 11.185).

11.2.2.15 Common Peony (*Paeonia officinalis*)

The common peony or garden peony occurs in woodland clearings and is also common in gardens. It is a 40–80 cm tall perennial plant. The stem is covered on the outside with a single layer of epidermis. The outer surface of its cells is covered

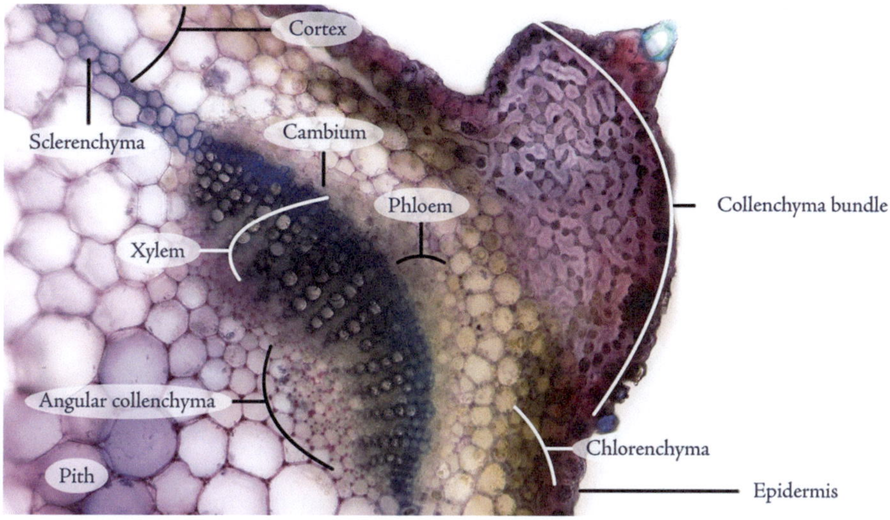

Fig. 11.185 Image of the vascular bundle of a red deadnettle (*Lamium purpureum*)

with a thick cuticle. Beneath the epidermis is a chlorenchyma, followed by a cortex with parenchymatous cells. At its inner border, a broad sclerenchyma band extends to support the stem. Smaller vascular bundles are attached to its outer side. In the pith parenchyma, which fills the inside of the stem, larger vascular bundles can be observed (Figs. 11.186 and 11.187).

11.2.2.16 Meadow Buttercup (*Ranunculus acris*)

Meadow buttercup is common in wet meadows and pastures (Fig. 11.188). The golden-yellow flowering plant is 30–100 cm tall and flowers from May to November. It is avoided by grazing animals because of its toxicity. It loses its toxicity when dried. Make a thin cross-section of the stem and examine it with a light microscope by staining it with toluidine blue.

In the cross-section, a small magnification shows that most of the stem is composed of parenchyma and the vascular tissue system has a bundle structure. Underneath the monolayer epidermis, the cortex is formed by chlorenchyma. Many chloroplasts are observed in its cells. The central cylinder is filled with increasingly larger round, thin-walled parenchyma cells. Intercellular passages with ruptures are observed at the contact corners of the cells. In the central part of the stem, the central cavity formed by tissue rupture is visible.

Vascular bundles are arranged in a circle. The xylem elements are coloured blue-green by the toluidine blue staining, while the phloem's are pale purple.

At higher magnification, it is also possible to observe that between the xylem and the phloem there is a few rows of dividing tissue, the cambium, which adds new elements to the open vascular bundle as the stem develops.

Some of these vascular bundles are surrounded by a sclerenchymatous sheath, which is thickest around the phloem, here it turns light blue. The bundle sheath

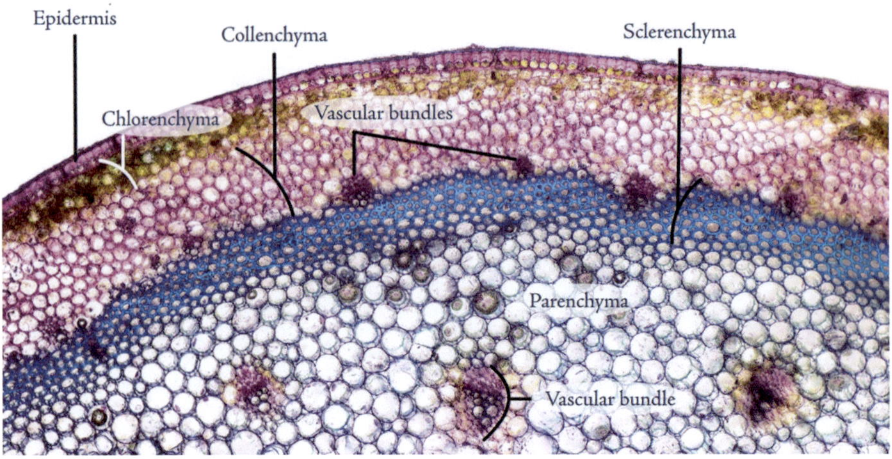

Fig. 11.186 Stem cross-section of a peony (*Paeonia officinalis*) stained in toluidine blue

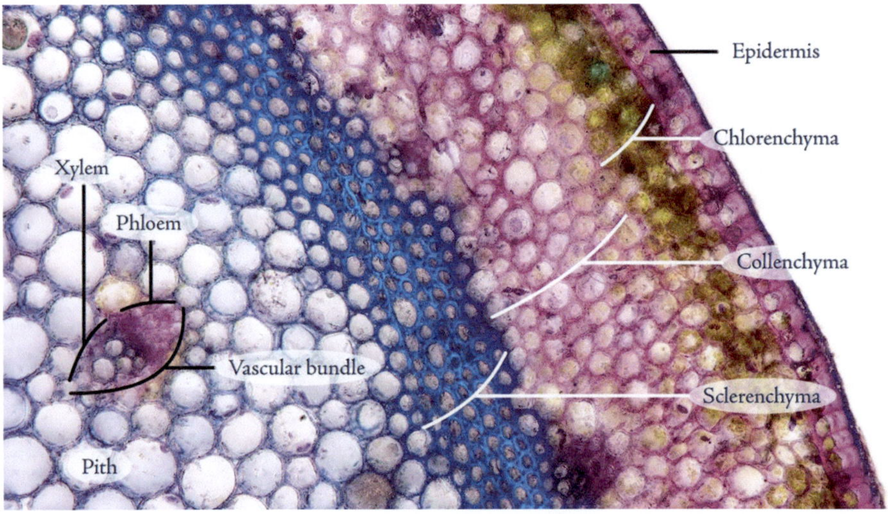

Fig. 11.187 Higher magnification image of the stem cross-section of the peony (*Paeonia officinalis*)

consists of cells with thinner walls in the line of the cambium and around the wood part. The latter is called the permeable zone. In the phloem, there are broader sieve tubes with angular cross-sections and adjacent, plasma-filled companion cells. The phloem parenchyma is absent, which is typical of vascular bundles in plants of the family Ranunculaceae.

In the xylem, the larger cavity vessels are sometimes arranged in two or three rows, roughly radially, and are adjacent to each other and to tracheids or xylem parenchyma. The compressed elements of the protoxylem are seen at the border of

Fig. 11.188 Meadow buttercup (*Ranunculus acris*)

Fig. 11.189 Stem section stained with toluidine blue

the xylem and the bundle sheath. Its elements are much smaller than the later developing metaxylem elements (Figs. 11.189, 11.190, and 11.191).

11.2.2.17 Crown Vetch (*Securigera varia*)

The crown vetch is a plant of the legume family, living in dry grasslands and weed communities (Fig. 11.192). The grooves on its stem are well reflected in its tissue structure as well.

At the longitudinal protrusions of the stem, under the single-layered epidermis longitudinal collenchyma bundles run. Below them, collateral open vascular bundles are arranged in a circle. At the indentations of the stem, under the epidermis,

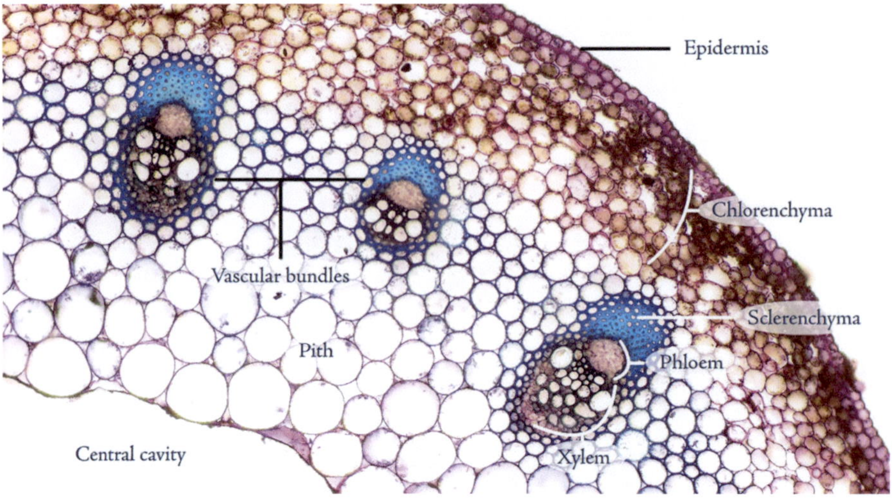

Fig. 11.190 Detail of stem cross-section of a buttercup (*Ranunculus acris*), stained with toluidine blue

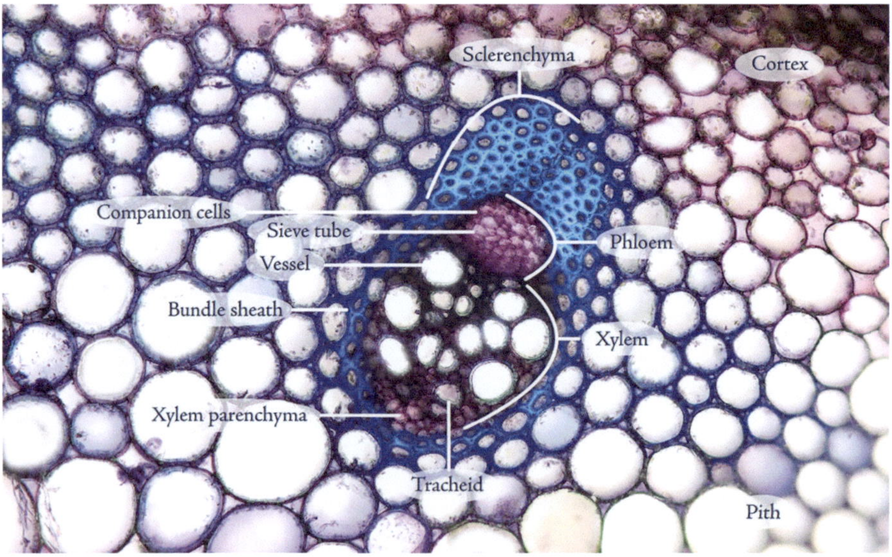

Fig. 11.191 Vascular bundle of meadow buttercup (*Ranunculus acris*)

there is green chlorenchyma followed by a one or two cell layer cortex. On the outer surface of the vascular bundles, a sclerenchyma cap can be observed. Further inside, there is the purplish phloem, the cambium, and then the dark blue xylem. In the latter, the wide-lumen vessels and the narrower tracheid's round cross-sections, and the dark groups of xylem fibres are clearly recognisable. A sclerenchyma strip connects the vascular bundles laterally. The interior of the stem is filled with a pith cavity and the remnants of the pith (Figs. 11.193, 11.194, and 11.195).

Fig. 11.192 Crown vetch (*Securigera varia*)

Fig. 11.193 Toluidine blue-stained cross-section of the crown vetch stem (*Securigera varia*)

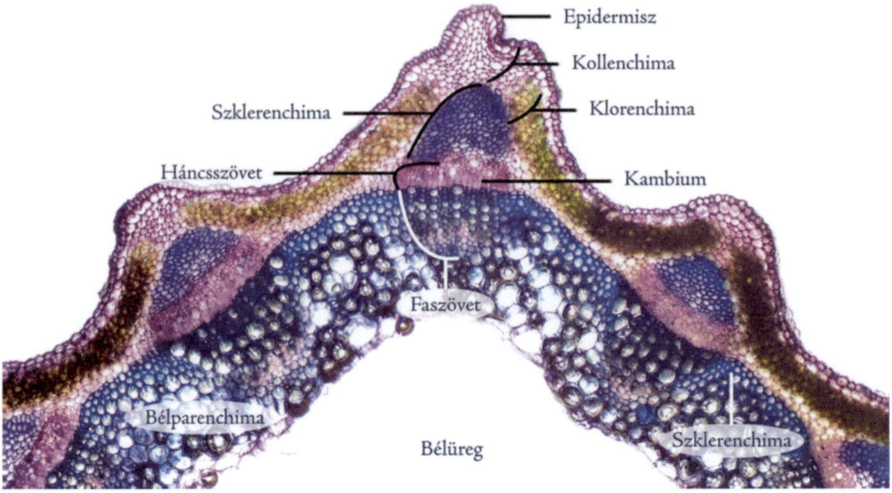

Fig. 11.194 Tissue structure of the crown vetch (*Securigera varia*) stem

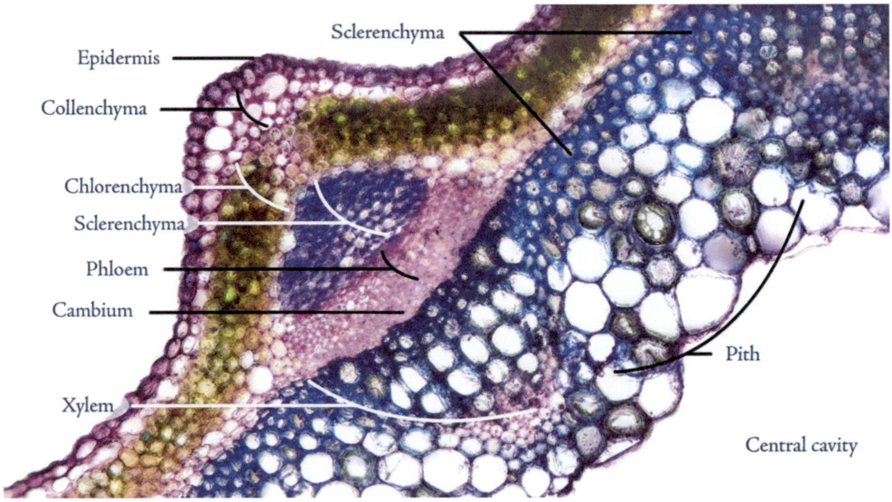

Fig. 11.195 Higher magnification image of the stem (*Securigera varia*) cross-section

11.2.2.18 Spring Ragwort (*Senecio leucanthemifolius* subsp. *vernalis*)

From the stem of the spring ragwort (Fig. 11.196), we can easily prepare thin cross-sections, which after staining with toluidine blue, it is advisable to examine with a light microscope. The section is covered from the outside by a single-layered epidermis, under which chlorenchyma forms most of the cortex. In the cortex, under the epidermis, longitudinal collenchyma bundles can be observed, which form protrusions on the surface of the stem. The circularly arranged vascular bundles are surrounded by a sclerenchyma bundle sheath. In the vascular bundle, looking

Fig. 11.196 Spring ragwort (*Senecio leucanthemifolius* subsp. *vernalis*)

towards the cortex, there is the light phloem, and inwardly, the dark xylem. In the latter, the wide-lumen vessels and the narrower tracheids' cross-sections can be clearly seen. Looking inward, there is the darker protoxylem and outward to the cortex, the metaxylem. Inside the circle of vascular bundles, there is the pith made up of large, rounded parenchyma cells, in the middle of which intercellular spaces form, and then with the appearance of a pith cavity, the stem becomes hollow (Figs. 11.197, 11.198, and 11.199).

11.2.2.19 Cucumber (*Cucumis sativus*)

The Solanaceae and the Cucurbitaceae have special, bicollateral vascular bundles. Thus, this type of vascular bundle can also be studied on a cross-section made from the stem of a cucumber, which belongs to the Cucurbitaceae. Let us make a cross-section from the stem and stain it with toluidine blue! In the stem, the cortex follows under the epidermis, which is separated from the pith parenchyma by a sclerenchyma cylinder sheath (Fig. 11.200).

The bicollateral bundles are located in the pith parenchyma. Their characteristic feature is that a phloem part is connected to both the outer and inner sides of their xylem. Cambium can only be found between the outer phloem and the xylem (Figs. 11.201 and 11.202).

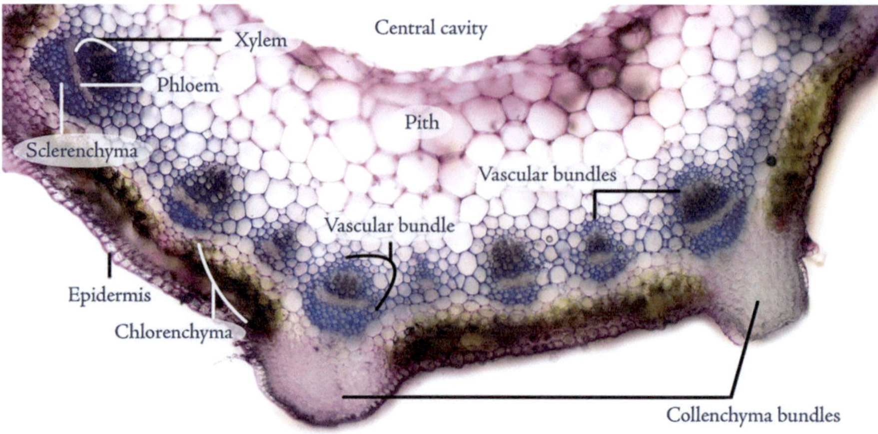

Fig. 11.197 Cross-section of the stem of spring ragwort (*Senecio leucanthemifolius* subsp. *vernalis*) stained with toluidine blue

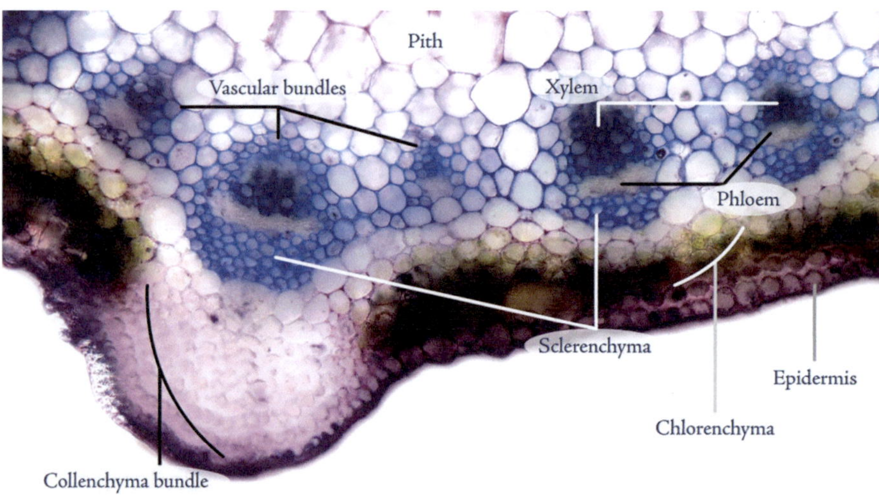

Fig. 11.198 Higher magnification image of the stem cross-section

11.2.2.20 Hoary Cress (*Lepidium draba*)

The hoary cress (Fig. 11.203), a weed widespread almost worldwide in temperate and Mediterranean zones, belongs to the cruciferous family. Let us make a cross-section of its stem and examine it with a light microscope after staining with toluidine blue!

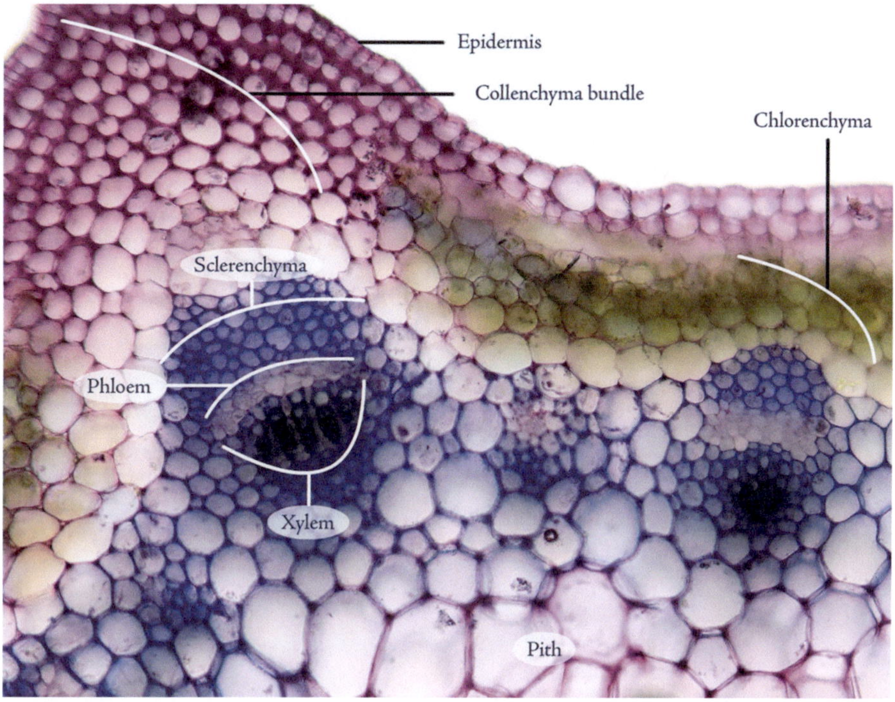

Fig. 11.199 High magnification image of the stem cross-section

Fig. 11.200 Cross-section of the cucumber (*Cucumis sativus*) stem

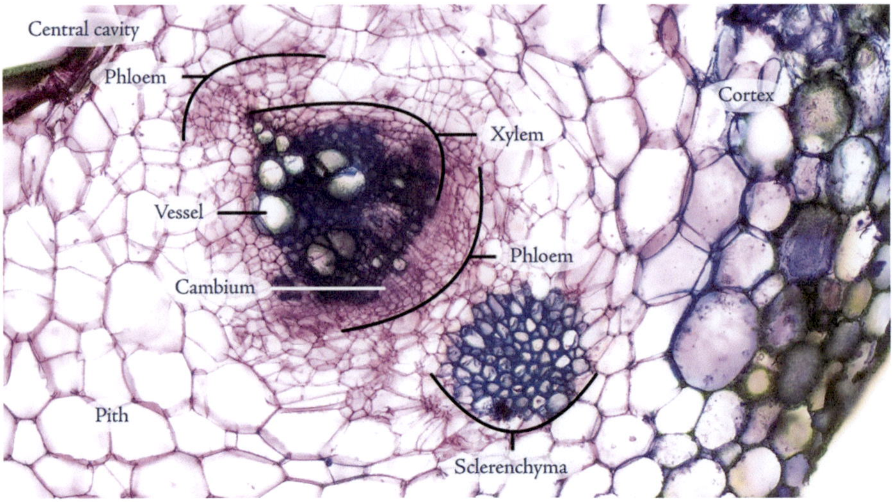

Fig. 11.201 Structure of a bicollateral vascular bundle

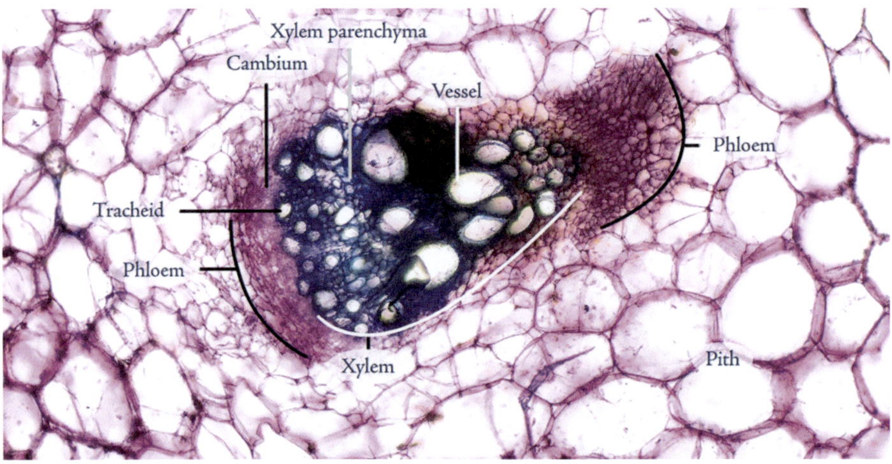

Fig. 11.202 Structure of a bicollateral vascular bundle

The stem is covered externally by a single-layered epidermis, under which collenchyma bundles run beneath the ribs. Between these, under the epidermis, there is chlorenchyma and a few rows of cortex cells.

On the outward-facing part of the vascular bundles arranged in a circle, a flat, medium-blue staining sclerenchyma cap can be observed. Below it is the purplish phloem, followed by the xylem containing many vessels. The interior of the stem is filled with pith. In its central part, air-filled intercellular spaces can be observed. Between the vascular bundles, light blue-stained sclerenchyma parts can be seen, which contribute to the stiffening of the stem (Figs. 11.204, 11.205, and 11.206).

Fig. 11.203 Hoary cress (*Lepidium draba*)

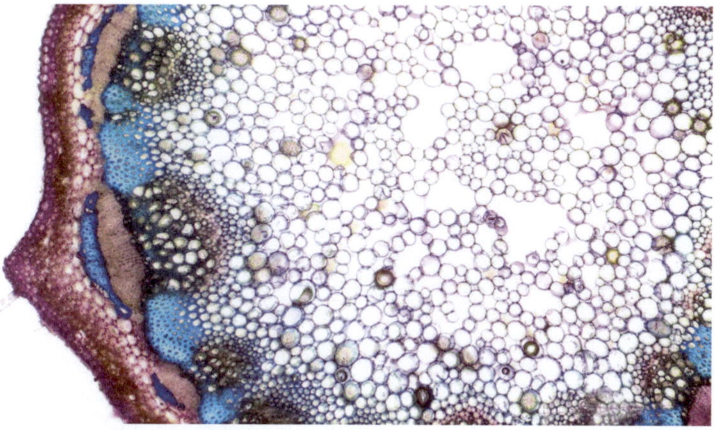

Fig. 11.204 Cross-section of hoary cress (*Lepidium draba*) stem stained with toluidine blue

11.2.3 Secondary Growth in Dicot Stem

11.2.3.1 Liana Type Stem Thickening

In dicots, there can be several forms of secondary stem thickening. The individual types are named after plants on which the given thickening is characteristic. During the development of secondary thickening of the *Aristolochia* or liana type, the cambium ring is formed by the parenchyma cells between the bundles and by the vascular cambium. The meristem parts between the bundles only produce parenchyma

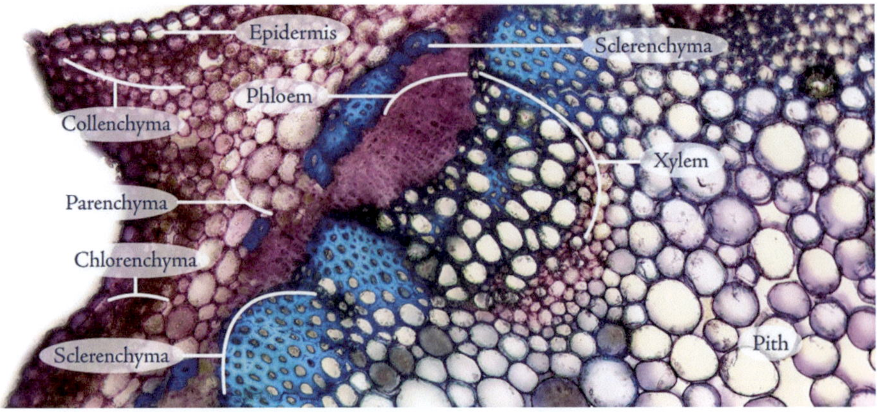

Fig. 11.205 Histological structure of the hoary cress (*Lepidium draba*) stem

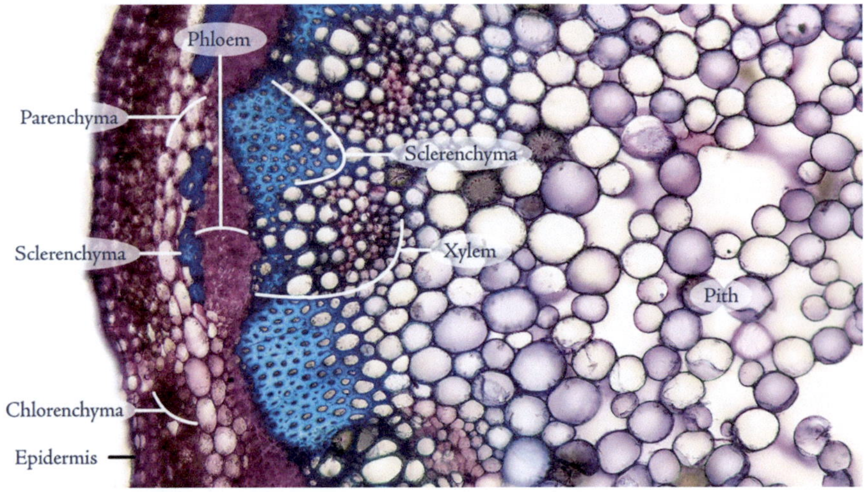

Fig. 11.206 Histological structure of the hoary cress (*Lepidium draba*) stem

cells. The cambium ring remains functional for years, so during the light microscopic examination of the cross-section of the perennial stem, the annual ring structure of the xylem can be clearly observed within the continuously growing, increasingly larger vascular bundles.

11.2.3.1.1 Birthwort (*Aristolochia clematitis*)

The birthwort is a perennial herbaceous plant that can grow up to half a meter or even a metre (Fig. 11.207). This common plant in floodplains, shrubs, and edge vegetation blooms from May to September. It can often be encountered in cities as well.

Fig. 11.207 Birthwort (*Aristolochia clematitis*) with fruiting shoot

Let us make a thin cross-section from the stem of a birthwort, then examine it with a light microscope after staining it with toluidine blue and covering it!

On the section, a few cell rows of hypodermis can be observed under the single-layered epidermis. Then come the parenchymal cells of the primary cortex, which contain many chloroplasts. The innermost cell layer of the cortex forms the starch sheath. We can see the sclerenchyma under the cortex dyed blue. Cross-sections of the open vascular bundles standing in a circle, surrounded by parenchymal cells, can be observed inside the stem. Cambium can also be found between the inward-looking xylem part and the outward-looking phloem part. The cross-sections of the wide-cavity vessels in the xylem part are clearly recognisable (Figs. 11.208 and 11.209).

In the case of an older stem cross-section, the initiation of the interfascicular cambium can already be observed between the bundles. We can recognise this from the fact that tangential division planes appear in the middle of the round parenchymal cells here, indicating the cells dividing in two (Fig. 11.210).

11.2.3.1.2 Grape (*Vitis vinifera*)

The grape is an important agricultural plant cultivated since ancient times, which originates from the Mediterranean vine (*Vitis silvestris*) (Fig. 11.211). From the very young shoot of the climbing shrub with tendrils, we can easily make a thin cross-section, which we examine under a light microscope after staining with toluidine blue.

According to the ribbing of the young grape shoot, the cross-section of the stem also has a wavy outline, so protruding (rib) and recessed (groove) details alternate.

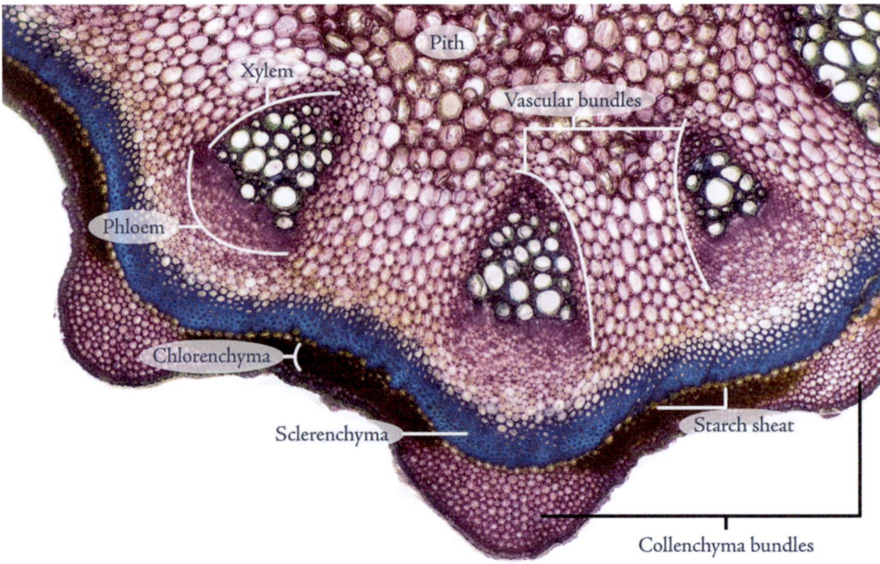

Fig. 11.208 Cross-section of young birthwort (*Aristolochia clematitis*) stem stained with toluidine blue

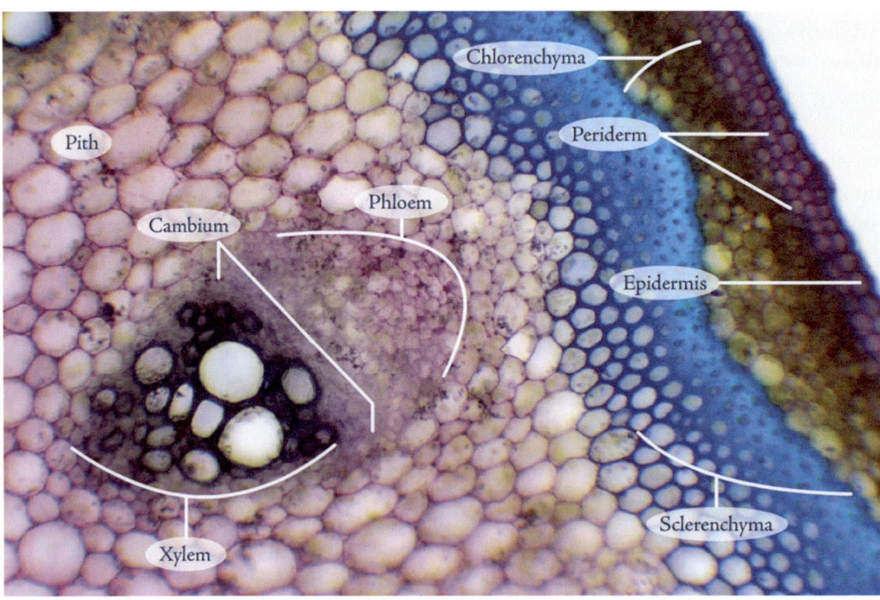

Fig. 11.209 High magnification image of young birthwort (*Aristolochia clematitis*) stem

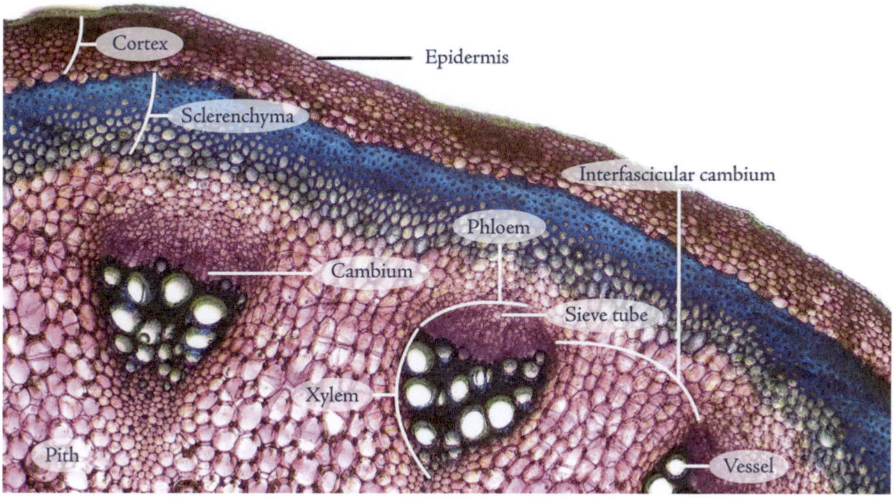

Fig. 11.210 Image of older birthwort (*Aristolochia clematitis*) stem

Fig. 11.211 Grape (*Vitis vinifera*)

Under the single-layered epidermis, there is angular collenchyma compressed in the direction of the ribs; chlorenchyma containing chloroplasts in the direction of the grooves, and a few rows of parenchyma underneath. Within the mentioned tissue areas, we can observe bundles of sclerenchyma (the so-called phloem crown) formed from thicker-walled cells belonging to the transport tissue system arranged in a circle (Figs. 11.212, 11.213, and 11.214). The elements of these bundles in the very young stem still store the prepared nutrients, participated in the transport, but later abandoning this function, partly transformed into thick-walled solidifying elements, fibers.

Fig. 11.212 Detail of a young grape (*Vitis vinifera*) stem cross-section

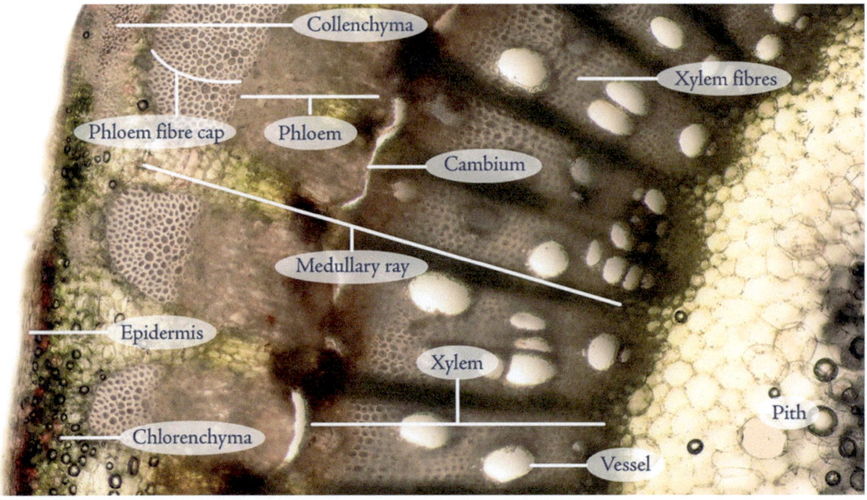

Fig. 11.213 Structure of a young grape (*Vitis vinifera*) stem cross-section

On the first-year mature shoots of the grape, the formation of the periderm does not start in the epidermis or the layer under the epidermis, but in a deeper tissue. In the 1-year-old shoot, the inward part of the phloem crown becomes capable of division and laterally supplemented by dividing cells, it closes into a continuous periderm-forming tissue, cork cambium (Fig. 11.215).

Therefore, in the stem of the grapevine, the cork cambium forms on the outer part of the vascular system and produces outward cork. Under the sclerenchyma bundles, we can see the few cell-layered cork, the single-cell-row cork cambium, then within it the living phloem elements of the vascular system, cambium, wide-cavity elements containing xylem part and medullary ray fragments (Fig. 11.215).

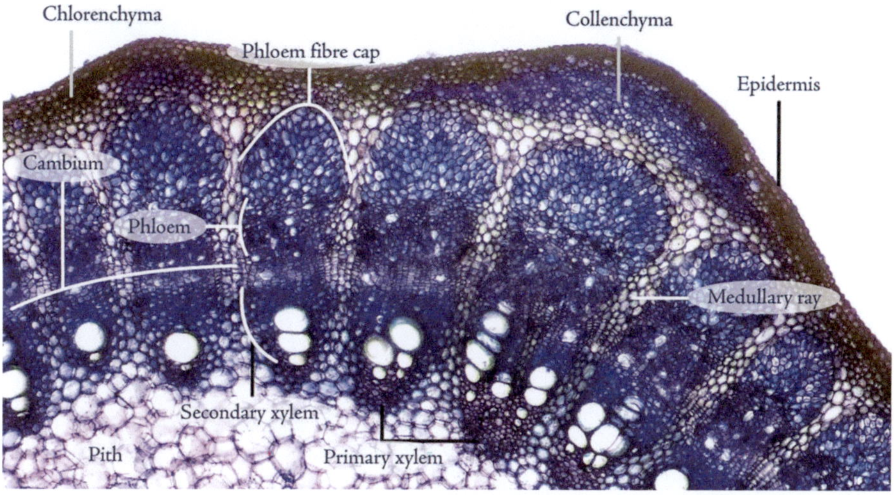

Fig. 11.214 Cross-section of a young grape (*Vitis vinifera*) stem stained with toluidine blue

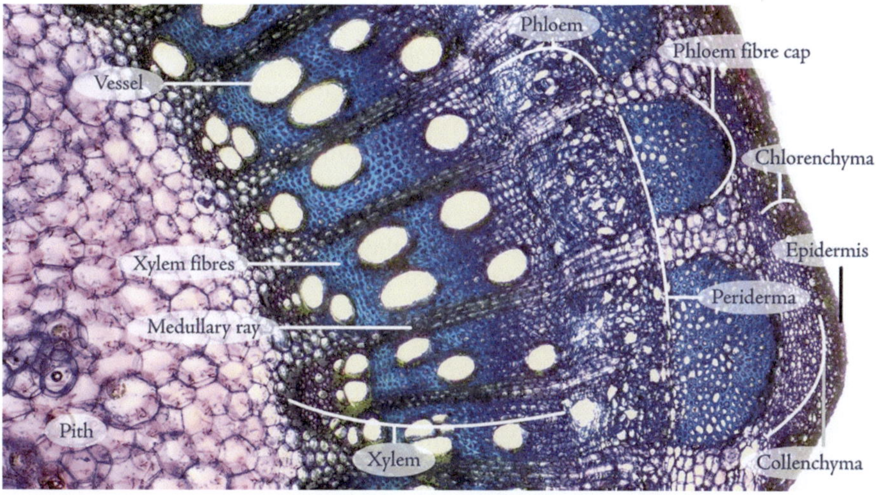

Fig. 11.215 Cross-section of a grape (*Vitis vinifera*) stem with developing periderm

The material exchange between the periderm of the grape and the central tissues ceases, so the tissues located outside gradually peel off, usually in the form of longitudinal strips. The vascular system is bundle-structured.

Although the closed cambium ring forms, its inter-bundle parts produce parenchyma cells, which participate in the formation of inter-bundle medullary rays. The bundle structure ensures the appropriate flexibility of the stem.

Fig. 11.216 Old man's beard (*Clematis vitalba*)

11.2.3.1.3 Old Man's Beard (*Clematis vitalba*)

The old man's beard is a common, liana-like plant in groves, woods, and gardens (Fig. 11.216).

Its upper stem has six prominent ribs and six furrows. The epidermis, which is associated with lamellar collenchyma, is found on the outermost part. Beneath the purplish-coloured collenchyma is a layer of light blue sclerenchyma, which forms a continuous mantle across the cross-section of the stem, although not of equal thickness. Beneath this, the pith parenchyma can be observed, which consists of thicker-walled parenchyma cells.

There are 12 vascular bundles, 6 of which are larger under the ribs and 6 smaller ones opposite the grooves. Both types of bundles contain cambium, these are collateral open vascular bundles. The protoxylem on the inner side of the vascular bundle can be recognised by its narrower elements.

In the metaxylem, which is located outwards, wide cavity vessels can be observed. The xylem is separated from the phloem, which is less stainable and therefore light-coloured, by violet-blue cambium. The cambium is present not only in the vascular bundles, but also in the medullary rays between the bundles, thus forming a continuous mantle. However, its function is very slow and is mainly confined to the vascular bundles.

Little cortical parenchyma can be observed on the outer side of the cortical tissue, especially in the large limbs. This is followed on the outer side by the already mentioned layer of sclerenchyma. The cortical parenchyma within the collenchyma of the grooves contains greenish bodies (Figs. 11.217, 11.218, and 11.219).

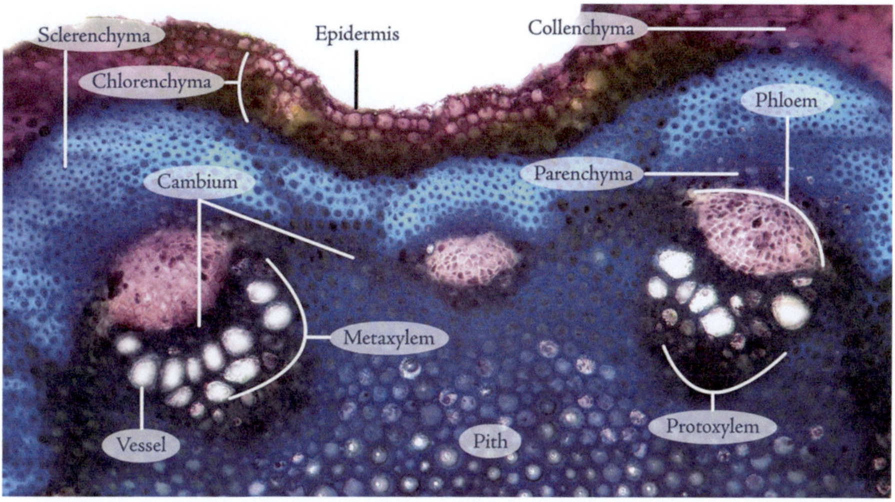

Fig. 11.217 Cross-section of the old man's beard (*Clematis vitalba*) stem stained with toluidine blue

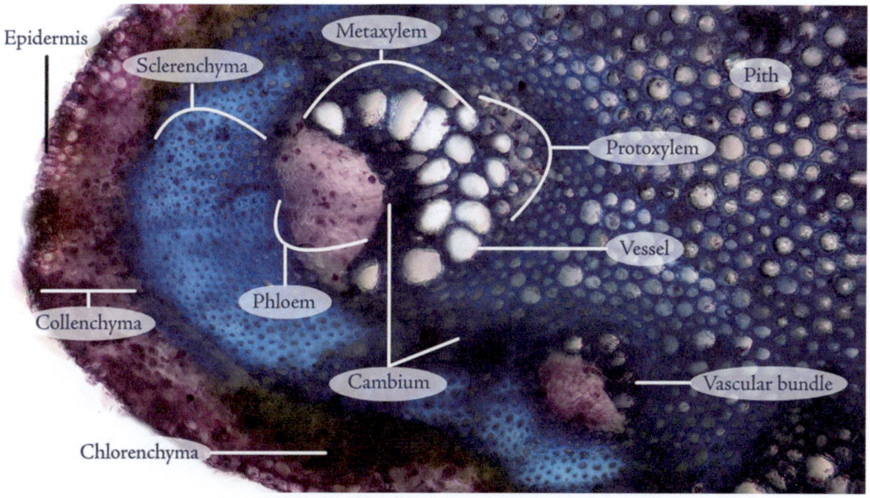

Fig. 11.218 Cross-section of the old man's beard (*Clematis vitalba*) stem stained with toluidine blue

11.2.3.2 Sunflower Type Stem Thickening

In the case of secondary thickening of the sunflower type, as in the case of the *Aristolochia* type, the cambium ring is formed by the cambiums of vascular bundles and the parenchyma cells between bundles that have recovered their ability to divide. The closed cambium ring creates new (smaller) secondary vascular bundles between the initially existing bundles. As the stem thickens, the original bundles also continue to grow, so that several vascular bundles of different sizes can be seen in the stem.

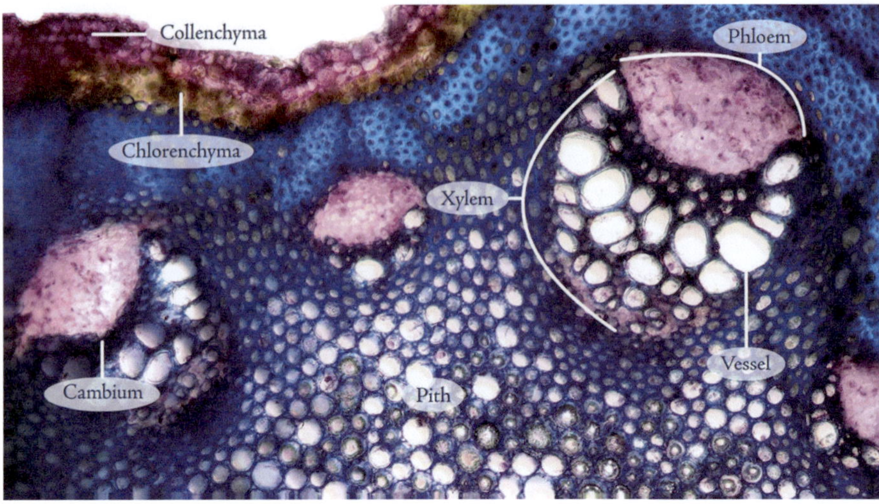

Fig. 11.219 Cross-section detail of the old man's beard (*Clematis vitalba*) stem stained with toluidine blue

Fig. 11.220 Cross-sections of sunflower's (*Helianthus annuus*) different-sized vessel bundles

11.2.3.2.1 Common Sunflower (*Helianthus annuus*)

Let us examine the section made from the sunflower peduncle without staining and stained with toluidine blue! On the section, the closed cambium ring and the different-sized vascular bundles cross-sections "wearing" the sclerenchyma cap can be clearly observed (Figs. 11.220, 11.221, and 11.222).

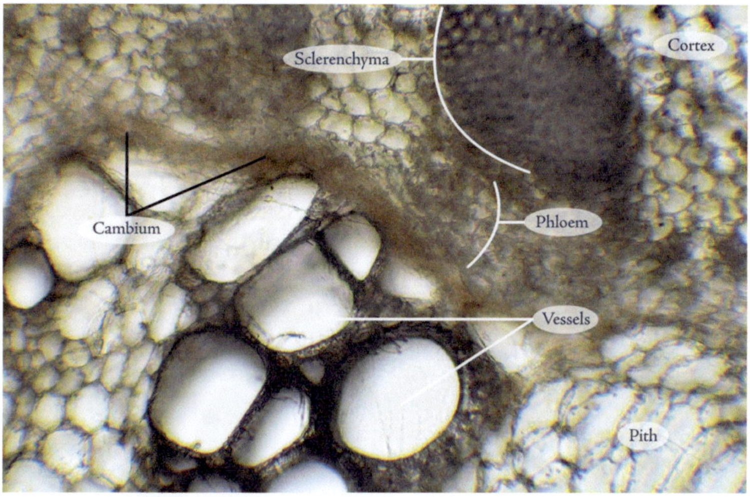

Fig. 11.221 An enlarged image of a vessel bundle

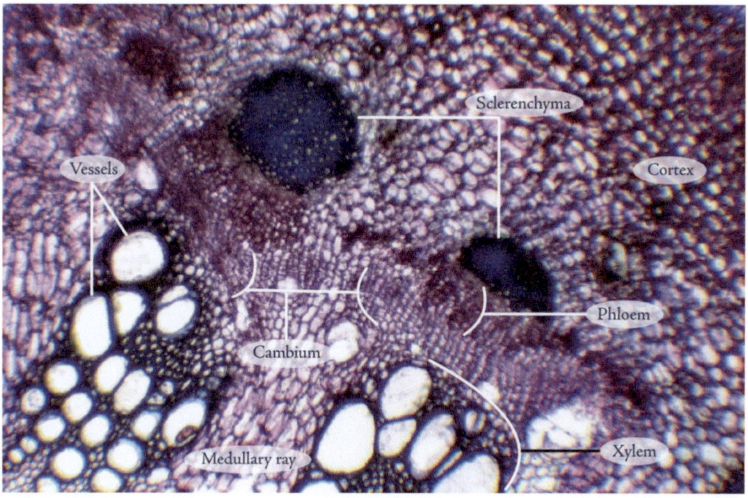

Fig. 11.222 Detail of a sunflower (*Helianthus annuus*) stem cross-section stained with toluidine blue

11.2.3.3 Ricinus Type Stem Thickening

During the secondary stem thickening of the Ricinus type, the cambiums of the vascular bundles located in the stem and their continuation form a closed cambium ring, which produces a continuous wood part inward and a continuous bark part outward. The primary xylem remnants form a pith crown towards the pith, while the phloem remnants form a cortex crown towards the cortex. The latter are more difficult to recognise in the light microscopic section because they are compressed as the stem thickens.

11.2.3.3.1 Banner Milkvetch (*Astragalus onobrychis*)

The stem of the banner milkvetch (Fig. 11.223) is bordered by a single-layered epidermis from the outside. Below it, at the protrusions of the stem, there are collenchyma bundles, and below them, vascular bundles run in a circle. At the indentations of the stem, chlorenchyma can be seen under the epidermis. The collateral open vascular bundles standing in one circle are connected laterally by a sclerenchyma band, which strengthens the stem. We can also observe that the cambiums of the vascular bundles form a closed cambium ring with the parenchyma cells between them, according to the type of stem thickening of the Ricinus. The cambium ring forms a uniform phloem outward and a uniform xylem inward. These tissue cylinders are very thin in the banner milkvetch, so the type of stem thickening can only be recognised here with a more thorough examination (Figs. 11.224, 11.225, 11.226, and 11.227).

11.2.3.3.2 Japanese Knotweed (*Fallopia japonica*)

The Japanese knotweed is a large, soft-stemmed, perennial plant species. It is native to East Asia, but has also been introduced to Europe, where it is considered an invasive species (Fig. 11.228). The dioecious plant, which can reach around 2 m in size, spreads mainly through its laterally spreading rhizome. Its overwintering buds are located on this, from which upright shoots of up to two metres develop by summer. It spreads in human-modified, unfavourable habitats, on railway embankments, spoil heaps, neglected gardens, and roadsides. It degrades natural plant communities in near-natural aquatic habitats. Let us make cross-sections from very young and older stem parts and examine these stained with toluidine blue with a light microscope!

Fig. 11.223 Banner milkvetch (*Astragalus onobrychis*)

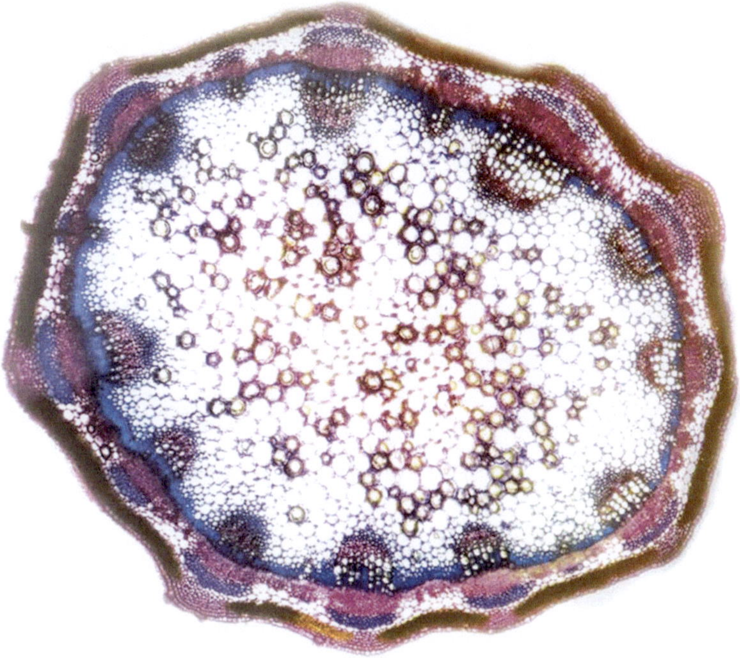

Fig. 11.224 Cross-section of the stem of the banner milkvetch (*Astragalus onobrychis*) stained with toluidine blue

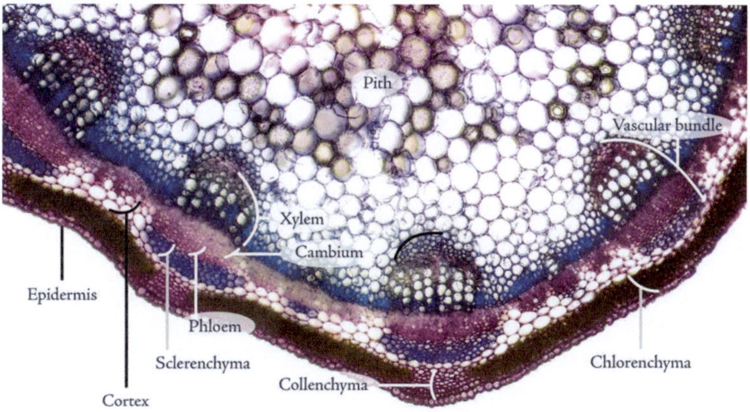

Fig. 11.225 Histological structure of the stem of the banner milkvetch (*Astragalus onobrychis*)

The outer surface of the young stem is still covered by a single-layered epidermis, under which chlorenchyma and cortex appear. In the case of the vascular tissue, the bundle structure is still recognisable because the vascular bundles are separated from each other by medullary rays (Fig. 11.229).

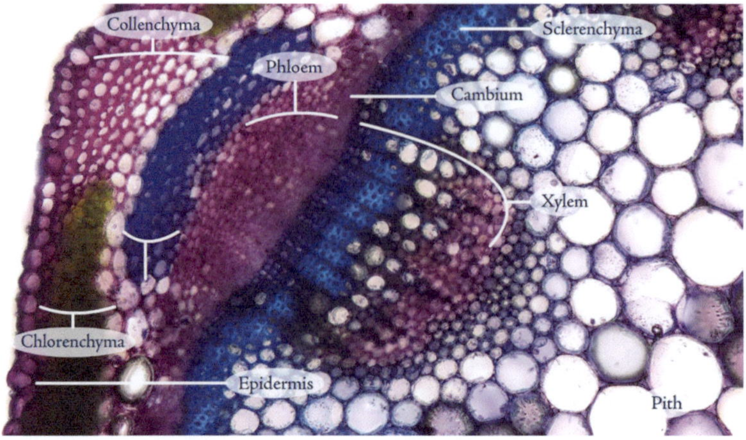

Fig. 11.226 High magnification image of a vessel bundle and its surroundings

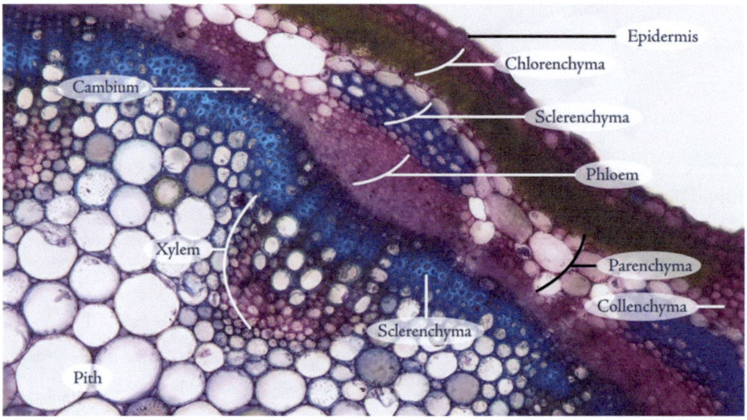

Fig. 11.227 Detail of the cross-section of the stem of the banner milkvetch (*Astragalus onobrychis*)

Between the phloem and the xylem, there is a cambium. In the phloem, the sieve tubes and the darker companion cells can be recognised under stronger magnification. A sclerenchyma cap that stains light blue can be observed on its outer surface. In the xylem, there are vessels, tracheids, and a large amount of xylem fibre. The inside of the stem is filled with large parenchyma cells called pith, which later ruptures as the stem becomes hollow. In the older stem, a uniform xylem and phloem form, and the epidermis is replaced by secondary dermal tissue, the periderm (Figs. 11.230, 11.231, and 11.232).

Fig. 11.228 Japanese knotweed (*Fallopia japonica*)

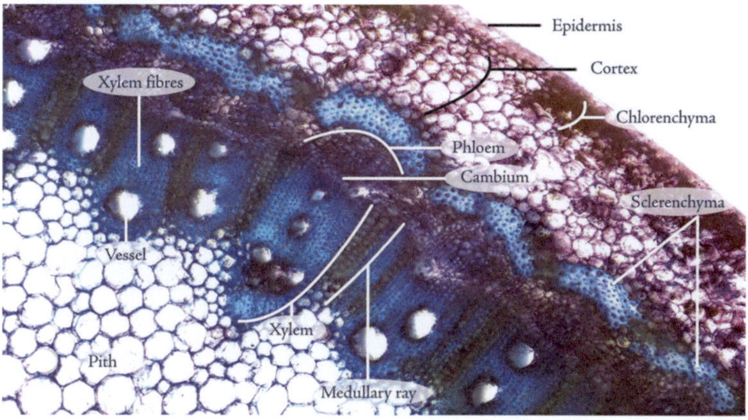

Fig. 11.229 Cross-section of young Japanese knotweed (*Fallopia japonica*) stem stained with toluidine blue

11.2.3.3.3 Black Elder (*Sambucus nigra*)

The black elder is a sprawling shrub that grows 3–10 m high, rarely a tree. The black elder has excellently adapted to secondary habitats affected by human interventions (Fig. 11.233). Its 1-year-old branches are green, the older ones are covered with grey secondary dermal tissue, on which conspicuous lenticels can be observed.

Later, the secondary dermal tissue, the periderm, appears between the torn epidermis (Fig. 11.234). When the branches are broken, their soft white inside, the pith, can be observed with the naked eye. Let us make a cross-section from the young, still green branch and examine it with a light microscope, stained with toluidine blue!

The stem is bordered by a single-layered epidermis. Underneath this is the chlorenchyma that gives the green stem its color. Opposite to the primary xylem, the dark

Fig. 11.230 Cross-section of older Japanese knotweed (*Fallopia japonica*) stem

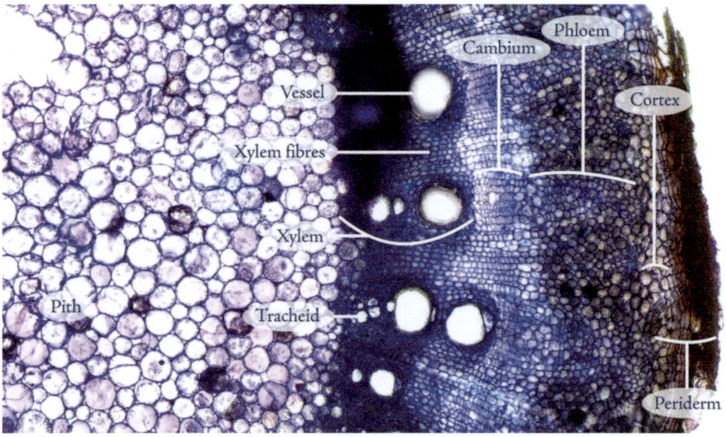

Fig. 11.231 Cross-section of an older Japanese knotweed (*Fallopia japonica*) stem stained with toluidine blue

blue-staining, sclerenchymatised primary phloem residue can be observed, while between them the closed cambium ring develops outward secondary phloem and inwardly forms secondary xylem. During the development of the branch, these become uniform phloem and xylem. The radial darker medullary rays can be clearly seen in the secondary xylem (Figs. 11.235 and 11.236).

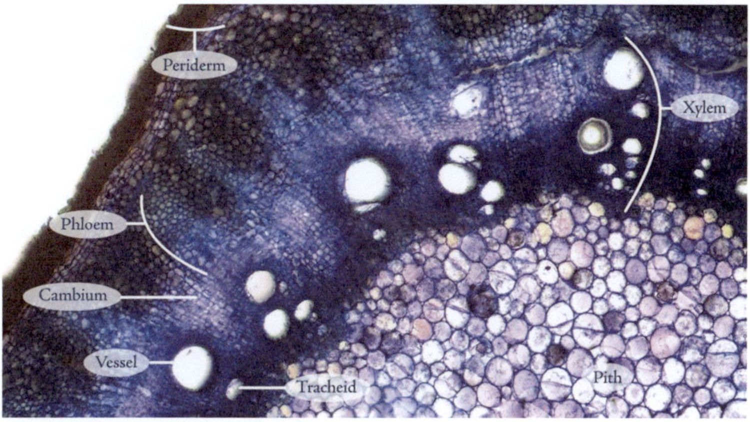

Fig. 11.232 Cross-section of an older Japanese knotweed (*Fallopia japonica*) stem stained with toluidine blue

Fig. 11.233 Black elder (*Sambucus nigra*) bush among acacia trees

It is not always easy to get thin section from the harder, woody branches. In such cases, the use of upper lighting technique can be useful. In this case, we do not need to prepare a thin section from the branch, even a 1 or 2 mm thick preparation is suitable to study the tissue structure of the branch. Cut the branch perpendicularly to the surface of the branch with a razor blade, then make another section one millimetre from the cut surface! In this way, we get to the 1 mm thick cross-section of the branch. Put the slice in a solution of toluidine blue for 10–15 s, then place it on a slide and examine it without cover with a light microscope with upper lighting! The

Fig. 11.234 The periderm appears between the torn epidermis on the older branch

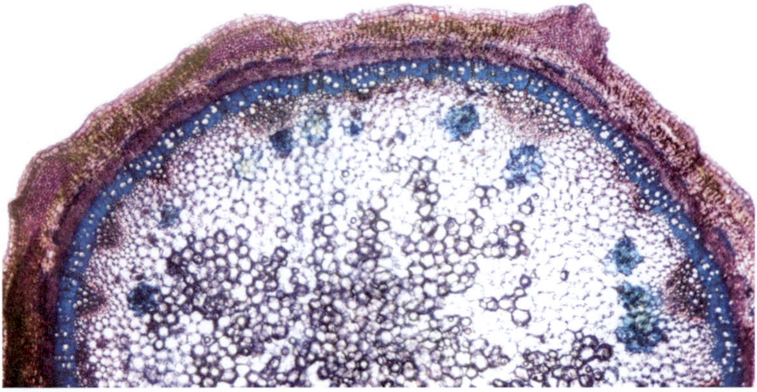

Fig. 11.235 Cross-section of a young black elder (*Sambucus nigra*) stem stained with toluidine blue

central part of the elder branch cross-section is filled with the light blue-stained pith, which consists of large, rounded parenchyma cells. The xylem, which contains the cross-sections of the vessels and tracheids appearing as dark holes, is also light blue. The radial medullary rays can be clearly seen in the xylem, which can continue in the greenish phloem. The cambium ring is located between the xylem and the phloem. Outside the secondary phloem, dark blue-stained sclerenchyma islands can be seen, which are remnants of the primary phloem. Moving outward, layers of purple cortex cells follow. Outside the cortex cells, we find the three-layer secondary dermal tissue, the periderm. Its middle layer is the blue-stained cork cambium, which inwardly produces the flattened cell phelloderm (brownish cells), and outwardly the cork (greenish-brown cells). The primary dermal tissue, the epidermis, can still be observed on the surface of the stem above the periderm (Fig. 11.237).

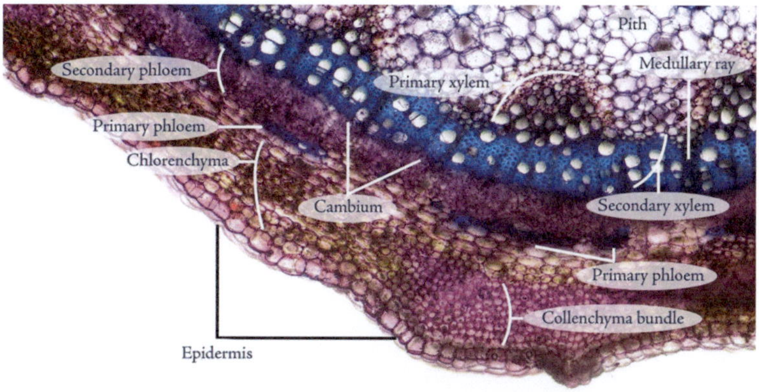

Fig. 11.236 Tissue structure of a young black elder (*Sambucus nigra*) stem

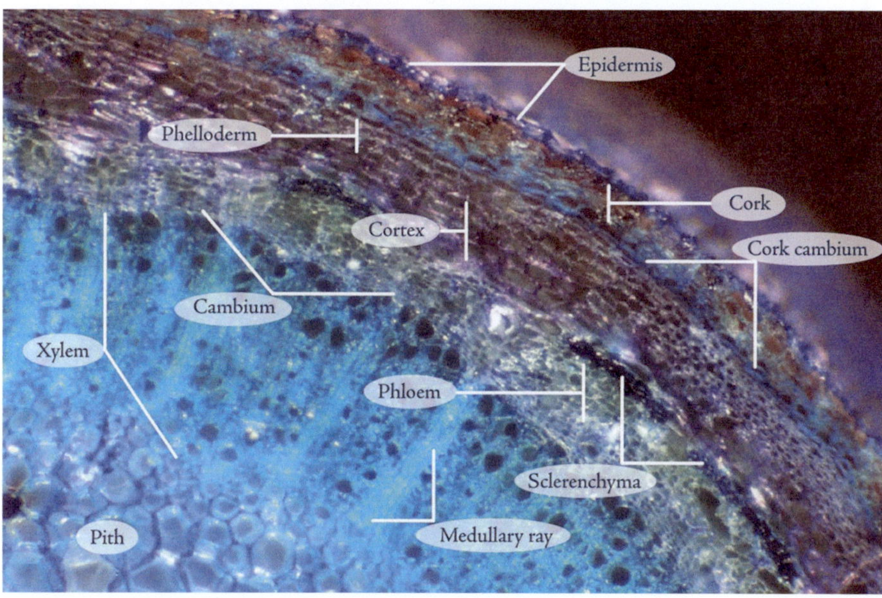

Fig. 11.237 Tissue structure of an old black elder (*Sambucus nigra*) stem

11.2.3.3.4 Black Nightshade (*Solanum nigrum*)

The black nightshade is a weed belonging to the Solanaceae family (Fig. 11.238). The annual plant is common in fallows, ploughlands, and cities. Prepare a cross-section and a longitudinal section of the stem of the black nightshade, stain with toluidine blue, and examine these with a light microscope!

Under the single-layered epidermis is the few cell-row chlorenchyma, followed by the layer of large, rounded, thin-walled cells of the cortex. Beneath small groups of dark blue coloured sclerenchyma cells is a pale purple band of the phloem, produced by the closed cambium ring. Inwards from the cambium ring, in the

Fig. 11.238 Black nightshade (*Solanum nigrum*)

Fig. 11.239 Toluidine blue-stained black nightshade (*Solanum nigrum*) stem cross-section

blue-stained xylem, the wide-lumen vessels, the narrower tracheids, the cross-sections of xylem fibres and xylem parenchyma cells are clearly recognisable. On the inner side of the xylem, the darker stained primary xylem tissue can be seen, with pale purple stained phloem tissue further in. Like the gourd family, the nightshade family is also originally characterised by bicollateral vascular bundles, so a phloem can be observed on both the outer and inner sides of the xylem part. The inner phloem parts can also be recognised later, with the formation of the cambium ring, on the cross-section. The inside of the stem is filled with pith, which torns in the middle (Figs. 11.239, 11.240, and 11.241).

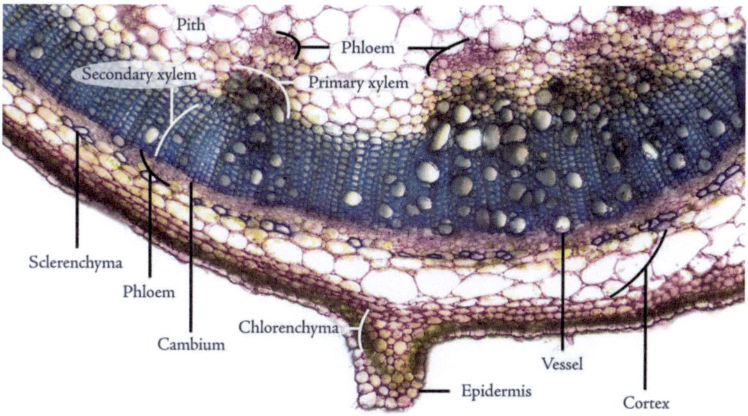

Fig. 11.240 Tissue structure of black nightshade (*Solanum nigrum*) stem

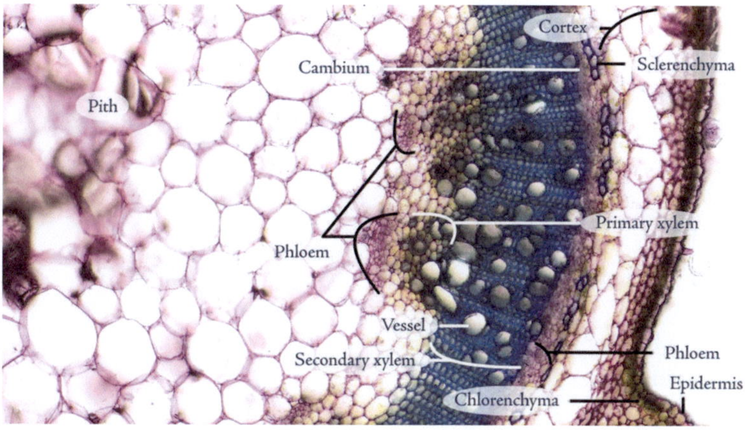

Fig. 11.241 Higher magnification image of the black nightshade (*Solanum nigrum*) stem cross-section

The radial longitudinal section clearly shows the characteristic tissues, cell types, cell wall thickenings, cell boundaries, and the transverse walls of the individual tube members (Figs. 11.242 and 11.243).

11.2.3.3.5 Geranium (*Pelargonium zonale*)

The surface of the young, not yet secondarily thickened stem is covered by a single-layered epidermis, under which follows a 5–6 cell row of chlorenchyma cells. The inner boundary of the cortex is strengthened by a sclerenchyma ring. The open vascular bundles are located on its inner side. The phloem faces outward, the xylem inward, and between the two, a 4–5 cell layer thick cambium can be seen (Fig. 11.244).

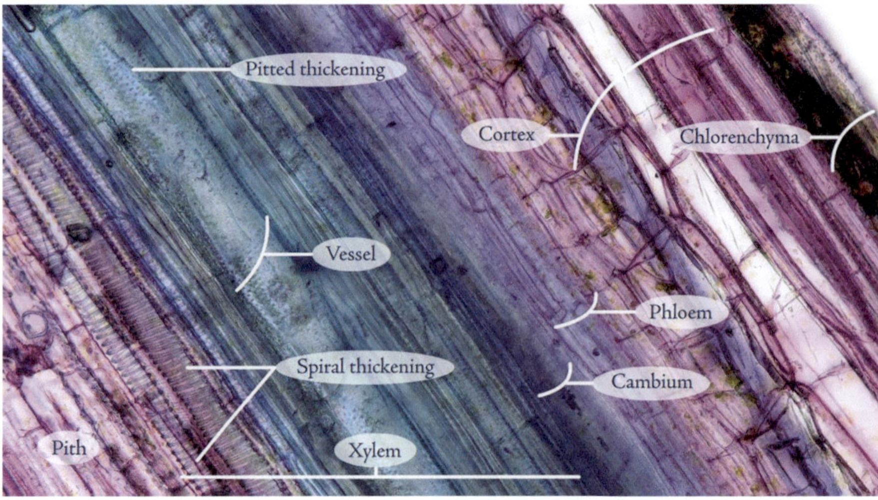

Fig. 11.242 Radial longitudinal section from the stem of black nightshade (*Solanum nigrum*)

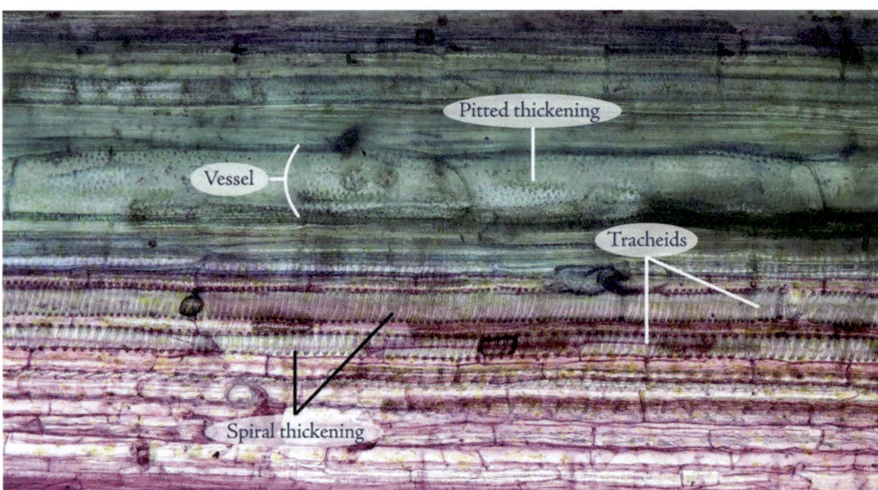

Fig. 11.243 Radial longitudinal section from the stem of black nightshade (*Solanum nigrum*)

Later, directly under the epidermis, a single-cell row cork cambium develops, which will create the secondary dermal tissue, the periderm (Fig. 11.245). On the cross-section of the older thickened stem, we can already observe the change in the cortex. The place of the epidermis breaking off during thickening is taken over by the periderm. The cork cambium develops cork outward. In the case of geranium, the cork cambium does not produce phelloderm, so we can see the parenchymal

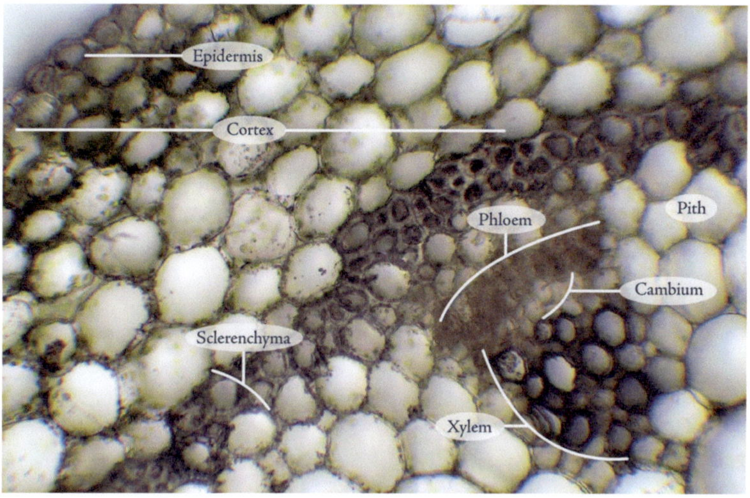

Fig. 11.244 Cross-section of a young geranium (*Pelargonium zonale*) stem

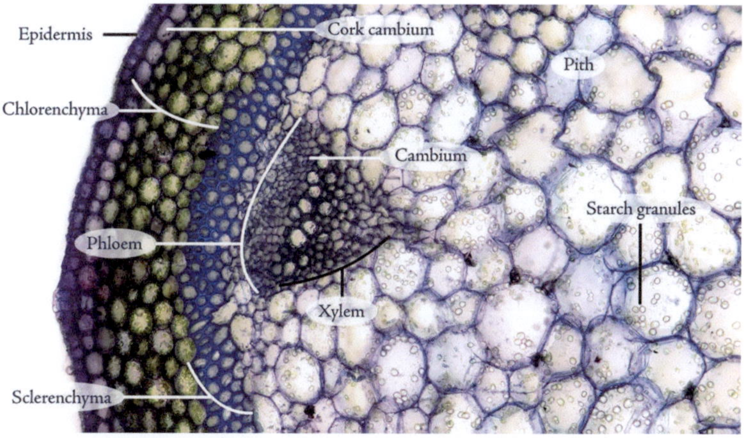

Fig. 11.245 Cross-section of a young geranium (*Pelargonium zonale*) stem with developed cork cambium

cells of the cortex underneath it. As we have already experienced in the case of the young stem, a sclerenchyma ring can be found at the inner boundary of the bark. Inside this, the vascular tissues standing in a circle can be observed. The cambiums, along with the parenchyma between them, closed into a cambium ring.

Later, the cambium produces a uniform secondary phloem outward and a continuous secondary xylem inward (secondary stem thickening of the Ricinus type). The secondary phloem and xylem produced by the cambium ring are well separated from the vascular bundles' primary phloem and xylem (Figs. 11.246 and 11.247).

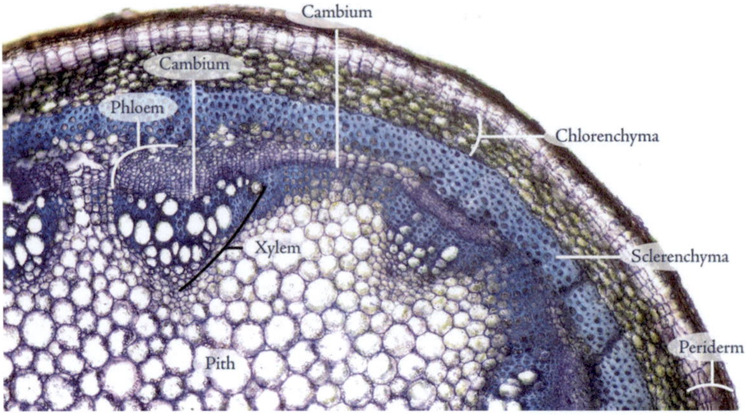

Fig. 11.246 Cross-section of a secondarily thickened old geranium (*Pelargonium zonale*) stem

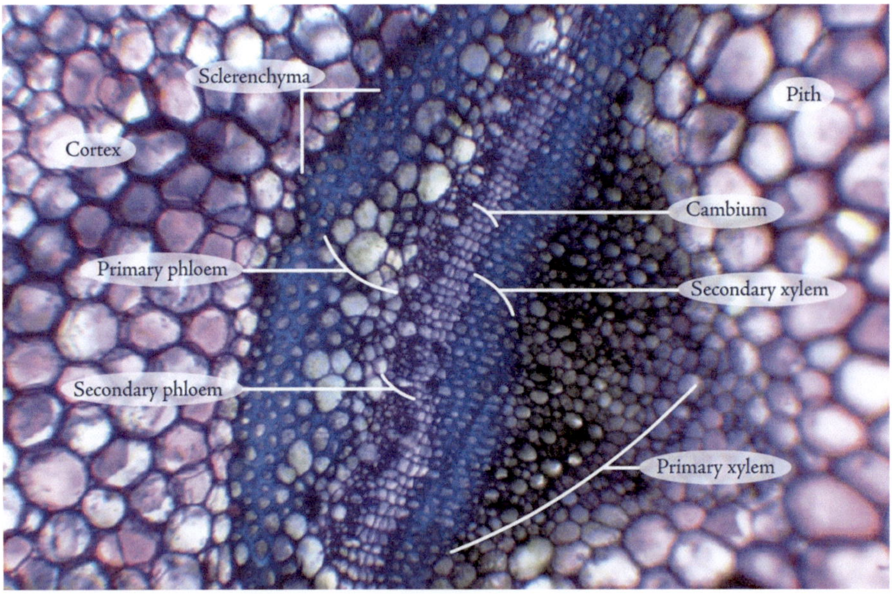

Fig. 11.247 High magnification image of the cross-section of a secondarily thickened old geranium (*Pelargonium zonale*) stem

11.2.3.3.6 Pepper (*Capsicum annuum*)

Fresh peppers can be purchased almost all year round in stores, so obtaining the peduncle needed for the examination is a simple task. From the peduncle of the pepper (Fig. 11.248), make a section, then stain it with toluidine blue, and examine it with a light microscope!

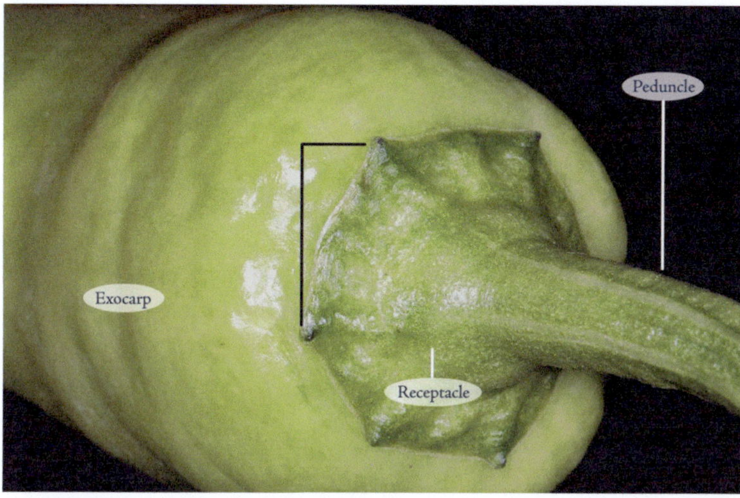

Fig. 11.248 The pepper (*Capsicum annuum*) peduncle is a good subject for sectioning

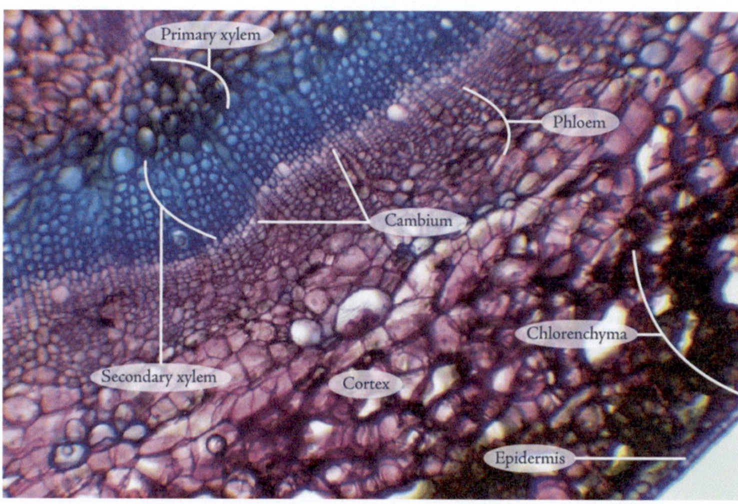

Fig. 11.249 Cross-section of a pepper (*Capsicum annuum*) peduncle stained with toluidine blue

When studying the microscopic section, a colourful picture unfolds before our eyes. The plant part is covered from the outside by a single-layered epidermis, underneath which the chlorenchyma and the cortex can be seen. The pepper stem is also characterised by secondary thickening of the Ricinus type, so a uniform secondary phloem (pink layer) forms on the outer side of the cambium ring, and a uniform secondary xylem (blue layer) forms on the inner side. The primary xylem bundles can be clearly identified in the section, while the remnants of the primary phloem bundles are less identifiable due to compression (Fig. 11.249).

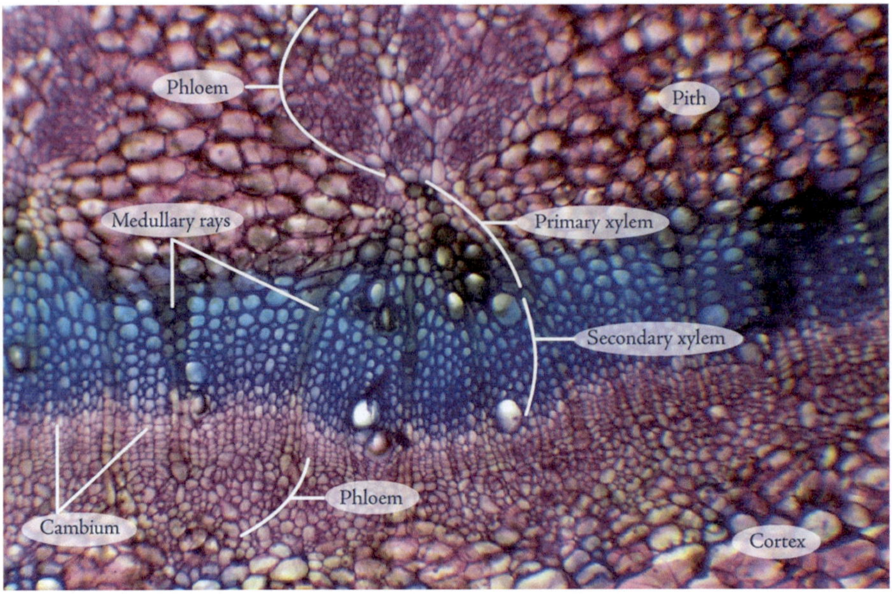

Fig. 11.250 Cross-section of a pepper (*Capsicum annuum*) peduncle stained with toluidine blue

Pepper belongs to the nightshade family, so there were originally bicollateral bundles containing two phloem parts in the young stem. The inner phloem parts of the vascular bundles did not compress, so these can be clearly identified in the section (Fig. 11.250).

11.2.3.3.7 Tomato (*Solanum lycopersicum*)

The peduncle and stem of the tomato are also suitable for histological examinations. Make a cross-section from the stem and examine it with a light microscope after staining with toluidine blue!

The stem is covered from the outside by a single-cell layer of epidermis, on which multicellular glandular hairs can be observed. Underneath the epidermis is a thin 1–2 cell layer of chlorenchyma, followed by a cortex made up of large round cells. Like the pepper, the tomato also belongs to the nightshade family, so here also originally bicollateral vascular bundles form in the young stem. Later, a closed cambium ring forms from the cambium of the vascular bundles and the parenchyma between the bundles, which produces a uniform secondary phloem outward and a uniform secondary xylem inward.

The primary phloem compresses, but the darker staining parts of the primary xylem can be recognised on the inner side of the secondary xylem. Inwards we can observe the internal phloem of original bicollateral bundles. The inside of the stem is filled with pith (Figs. 11.251, 11.252, and 11.253).

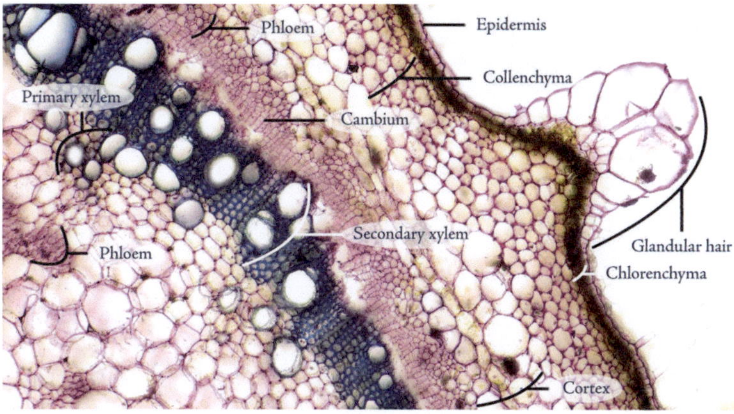

Fig. 11.251 Cross-section of a young tomato (*Solanum lycopersicum*) stem stained with toluidine blue

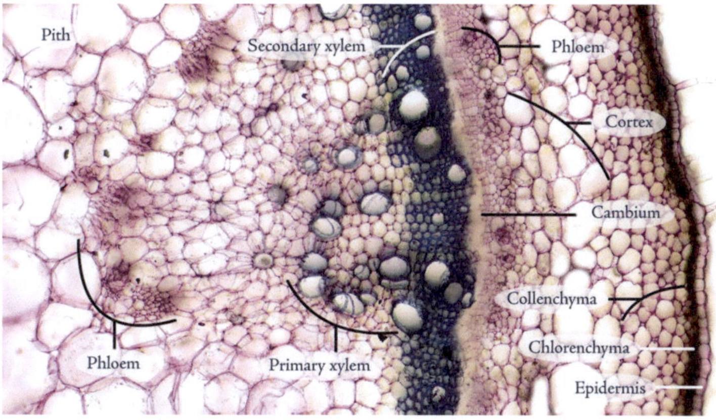

Fig. 11.252 Cross-section of a young tomato (*Solanum lycopersicum*) stem

11.2.3.3.8 Petunia (*Petunia* sp.)

The petunia is a plant originating from South America, which is grown as an ornamental plant all over the world (Fig. 11.254). The plant genus belongs to the nightshade family, so bicollateral bundles appear first in its developing stem. Let us make a cross-section of the young stem and examine its tissue structure with a light microscope!

The stem is covered by a single-layered epidermis covered with protective hairs. Below it is the chlorenchyma and the cortex consisting of large, rounded cells. The stem has a Ricinus type thickening, thus it has a closed cambium ring, which forms a continuous phloem outward and a continuous xylem inward. These are secondary tissues because there were only bicollateral bundles in the stem originally. Their primary phloem and xylem tissues, however, are no longer clearly distinguishable.

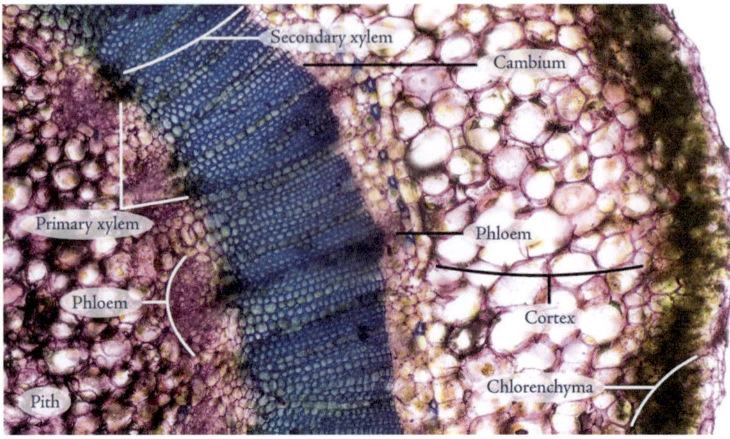

Fig. 11.253 Cross-section of an older tomato (*Solanum lycopersicum*) stem

Fig. 11.254 Petunia (*Petunia* sp.)

The only exceptions to this are the phloem parts of the bicollateral bundles facing towards the inside of the stem, which are clearly visible on the blue-stained inner border of the xylem. The inside of the stem is filled with large parenchymal cells, the pith (Figs. 11.255 and 11.256).

In the older stem, the secondary xylem becomes thicker and a layer of dark blue-staining sclerenchymatised cells appears in the inner layer of the cortex (Figs. 11.257 and 11.258).

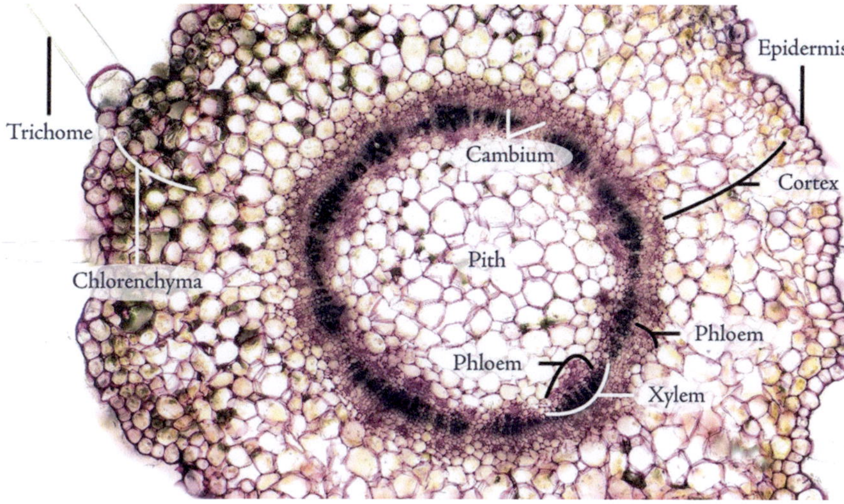

Fig. 11.255 Cross-section of a young petunia (*Petunia* sp.) stem stained with toluidine blue

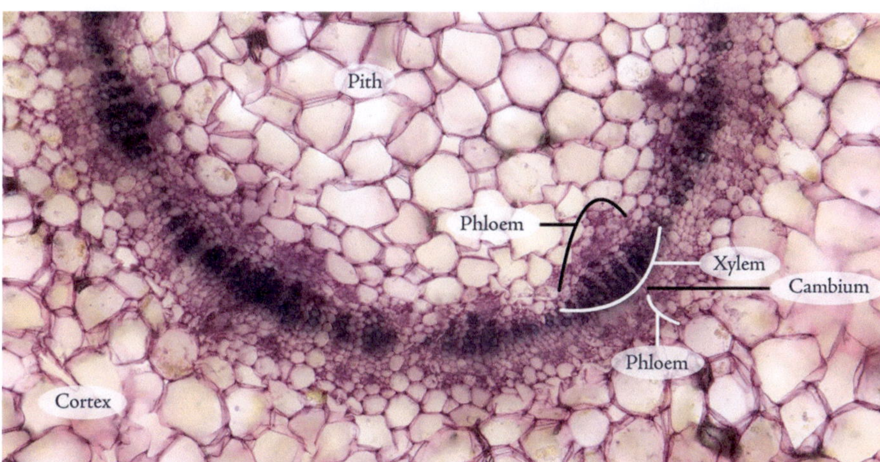

Fig. 11.256 Central part of a young petunia (*Petunia* sp.) stem

11.2.3.3.9 Mistletoe (*Viscum album*)

The mistletoe is an evergreen plant widespread throughout Europe, belonging to the Santalaceae family (Fig. 11.259). Its spherical, 60–100 cm diameter bushes can most often be observed in the crowns of poplars, willows, and acacia trees. The plant is partial parasitic (hemiparasite), which means that it only absorbs mineral substances dissolved in water from the xylem of its host plant, not organic substances. This also implies that it is green in color and capable of independent photosynthesis. Its stem is special in the sense that during its thickening, the

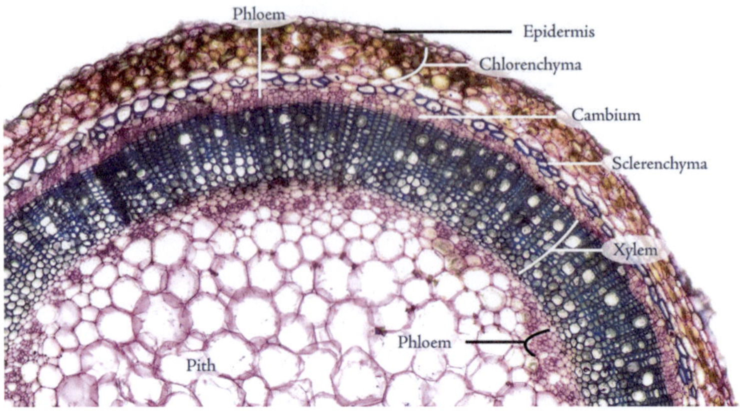

Fig. 11.257 Cross-section of an older petunia (*Petunia* sp.) stem stained with toluidine blue

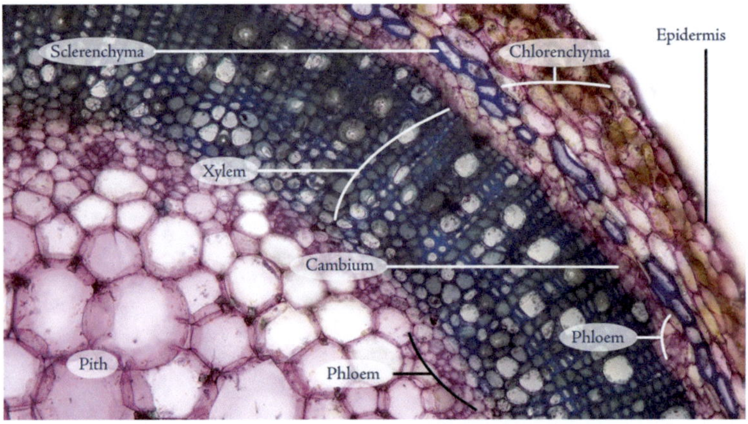

Fig. 11.258 High magnification image of the cross-section of the old stem

single-layered epidermis retains its ability to divide, so it does not tear off and is not replaced by secondary dermal tissue, the periderm.

Let us make a cross-section of the young and older stem of the mistletoe and examine it stained with toluidine blue with a light microscope!

The outer surface of the epidermal cells is covered by a thick cuticle. Below them is the thick cortex, followed by the open vascular bundles in the case of a young stem. In these, the phloem and the xylem can be clearly identified due to their different staining. Sclerenchyma islands that stain light blue are attached to both the outer and inner parts of the vascular bundles.

In the cross-section of the older stem, it can be observed that the closed cambium ring has already formed, which produces phloem outward and the continuous xylem inward, thus forming the Ricinus type stem thickening. The older stems are also covered by a single-layered epidermis (Figs. 11.260, 11.261, and 11.262).

Fig. 11.259 Mistletoe (*Viscum album*)

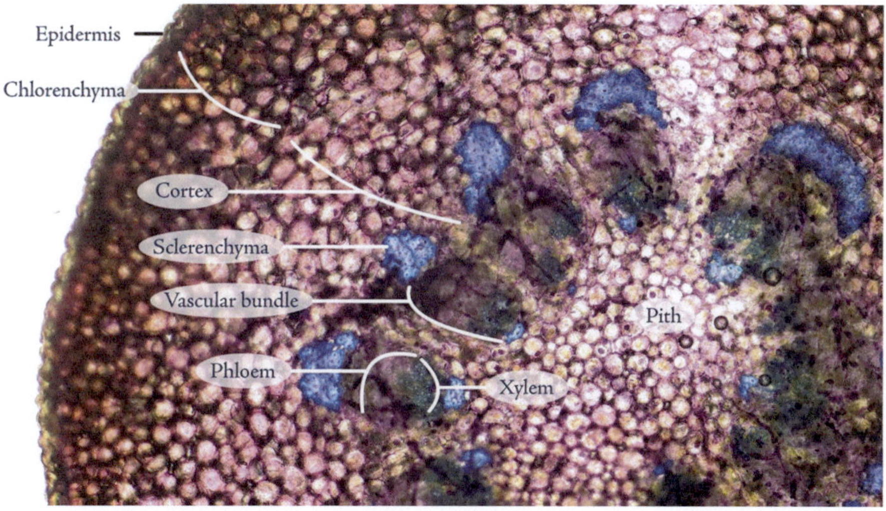

Fig. 11.260 Cross-section of a young mistletoe (*Viscum album*) stem stained with toluidine blue

11.2.3.3.10 Trumpet Creeper (*Campsis radicans*)

The trumpet creeper is a climbing shrub belonging to the Bignoniaceae family, originating from the southeastern part of the USA. It was brought to Europe in the seventeenth century. The plant is capable of climbing up to 10–12 m high by clinging with its aerial roots. Its large flowers are trumpet-shaped, rich in nectar, and bloom in July–August (Fig. 11.263).

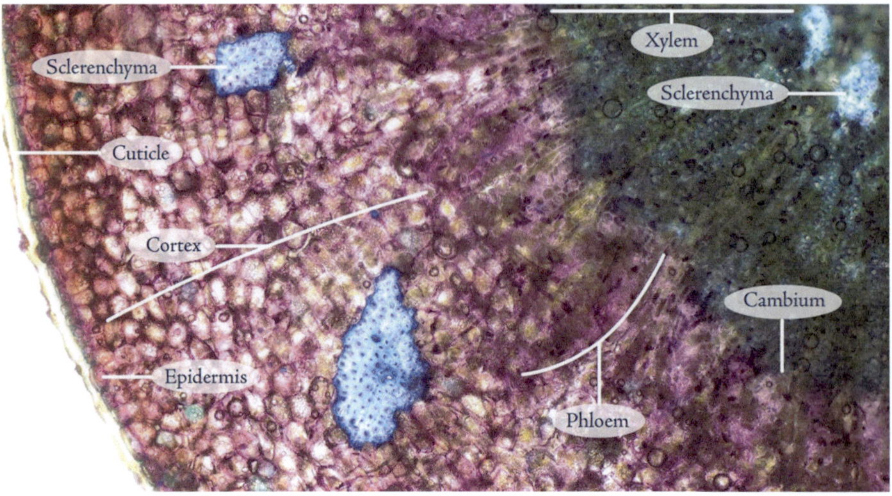

Fig. 11.261 Cross-section of an old mistletoe (*Viscum album*) stem

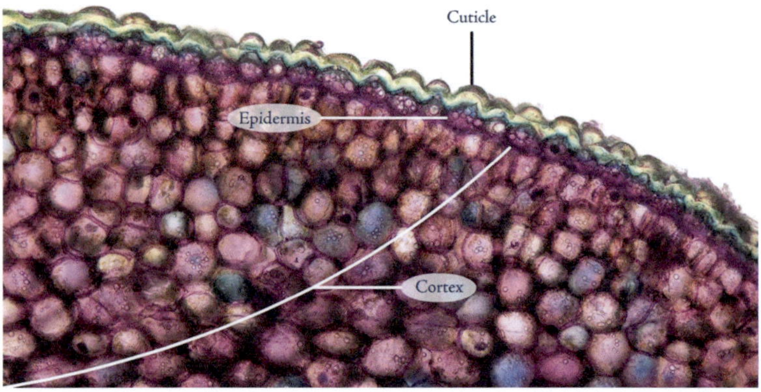

Fig. 11.262 Cross-section of an older mistletoe (*Viscum album*) stem with the epidermis and the cortex parenchyma

Let us make a section from the young, green stem of the trumpet creeper, then stain it with toluidine blue and examine it with a light microscope!

On the cross-section of the stem, moving from the outside in, we first see the single-layered epidermis, then the hypodermis located underneath it, which consists of lamellar collenchyma. We can identify the chlorenchyma based on its green colour.

Sclerenchyma groups are located in an island-like manner at the inner boundary of the cortex. The closed cambium ring forms the uniform phloem outward and the continuous xylem inward. The large, rounded parenchyma cells of the pith are located in the central part of the stem.

Fig. 11.263 Trumpet creeper (*Campsis radicans*) and its stem

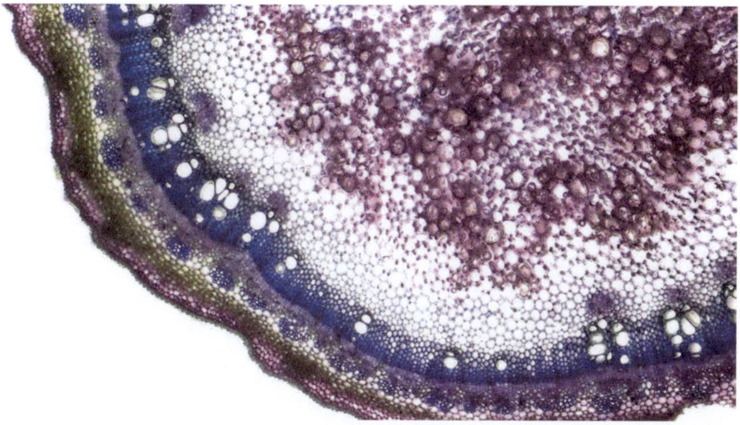

Fig. 11.264 Cross-section of a young trumpet creeper (*Campsis radicans*) stained with toluidine blue

The trumpet creeper is characterised by the Ricinus type of secondary thickening form. Thus, in the case of very young shoots, the separate vascular bundles can still be found, while later the cambium forms a continuous phloem and xylem. Especially in thicker stems, the primary xylem extending into the pith can be clearly identified.

We can identify parts of the phloem inward from the primary xylem, which clearly indicates that the first vascular bundles to appear in the stem were bicollateral. The inner phloem parts of these form the pale purple-coloured phloem visible at the border of the pith (Figs. 11.264, 11.265, 11.266, 11.267, and 11.268). During the thickening of the stem, the cortex also transforms in a characteristic way, with secondary dermal tissue, or periderm, appearing in place of the epidermis that tears due to stem thickening.

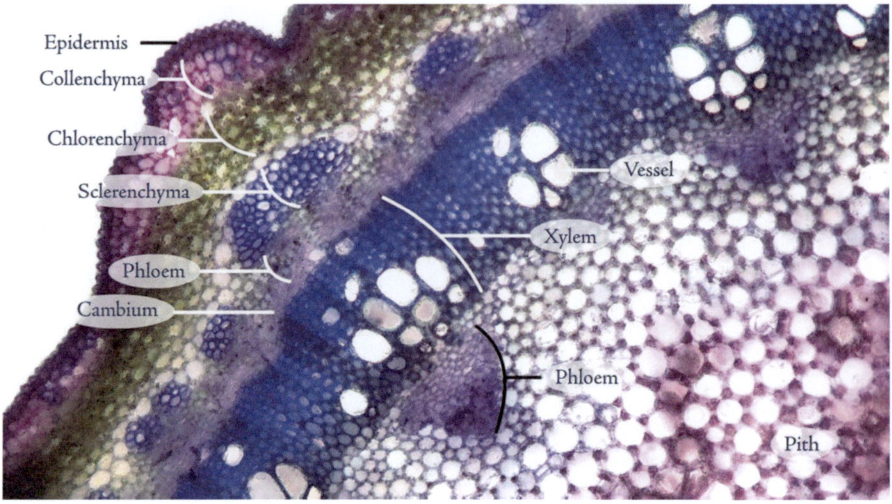

Fig. 11.265 Tissue structure of a young trumpet creeper (*Campsis radicans*) stem

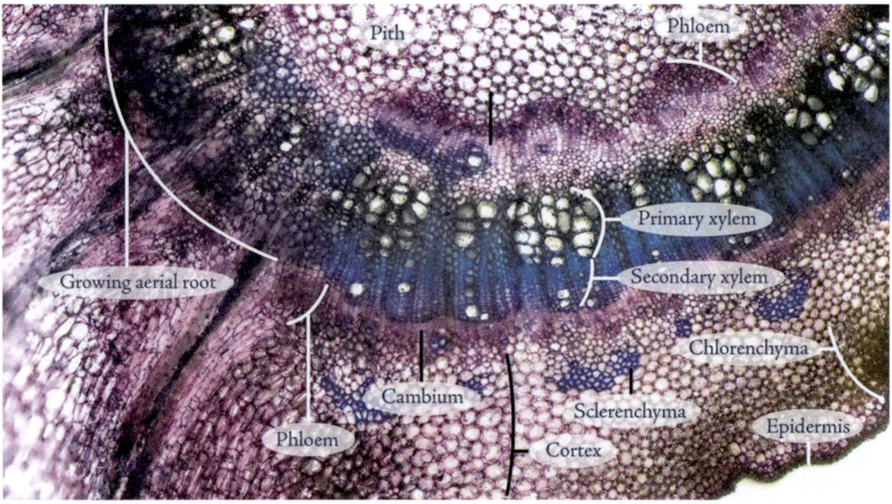

Fig. 11.266 Tissue structure of an older trumpet creeper (*Campsis radicans*) stem at the site of aerial root formation

11.2.3.3.11 Horse Chestnut (*Aesculus hippocastanum*)

The Ricinus-type secondary stem thickening is most common among herbaceous plants, but there are exceptions, such as in the case of the horse chestnut. The horse chestnut is a large, lushly foliaged tree of parks, which produces its showy inflorescences in spring. This species, belonging to the soapberry family, spread from the mountainous regions of the Balkan Peninsula and Western Asia throughout Europe. Let us make a cross-section from its very young and older stem and examine it with a light microscope after staining with toluidine blue! The vascular bundle structure

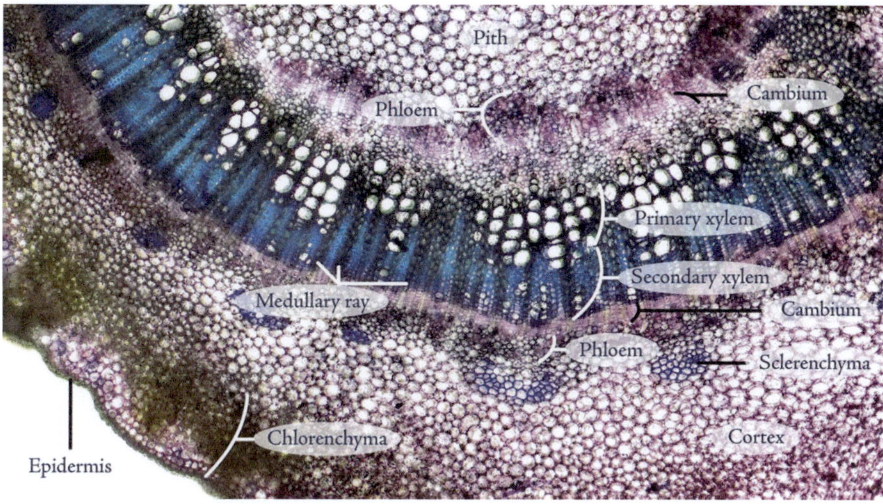

Fig. 11.267 Tissue structure of an older trumpet creeper (*Campsis radicans*) stem

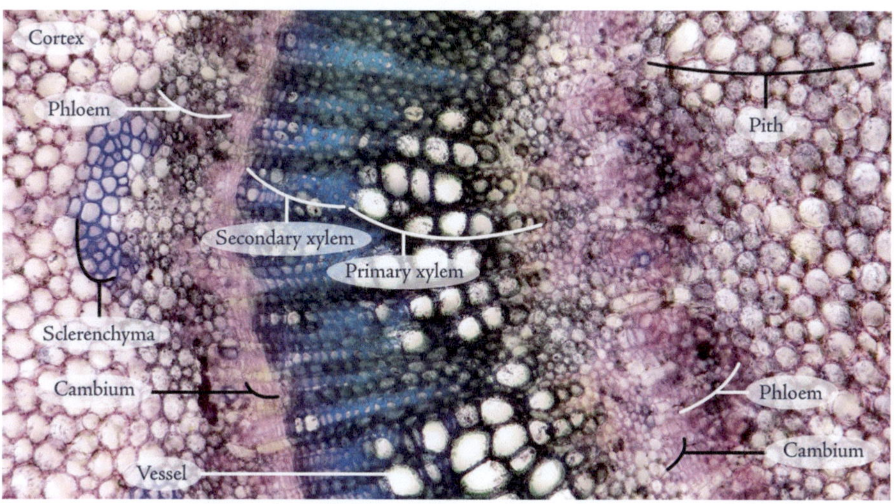

Fig. 11.268 High magnification image of the older trumpet creeper (Campsis radicans) stem cross-section

is easily recognisable at first glance on the cross-section of the young stem. The stem is still covered externally by the single-layered epidermis, underneath which follows the layer of chlorenchyma composed of multiple cells. Further in, the cortex composed of larger rounded parenchyma cells can be seen. The open vascular bundles stand in a circle, while two opposing vascular bundles embedded in the pith parenchyma can also be seen (Figs. 11.269, 11.270, and 11.271). In the two vascular bundles, the xylem with the wide-cavity vessels, and the outward-facing phloem are clearly distinguishable at high magnification.

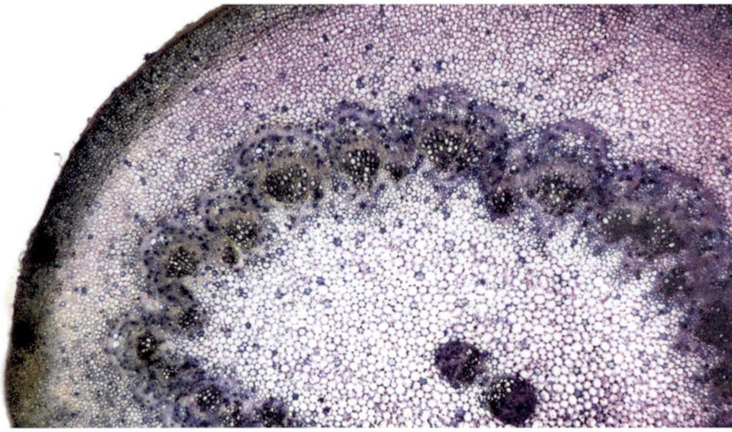

Fig. 11.269 Cross-section of a young horse chestnut (*Aesculus hippocastanum*) stem stained with toluidine blue

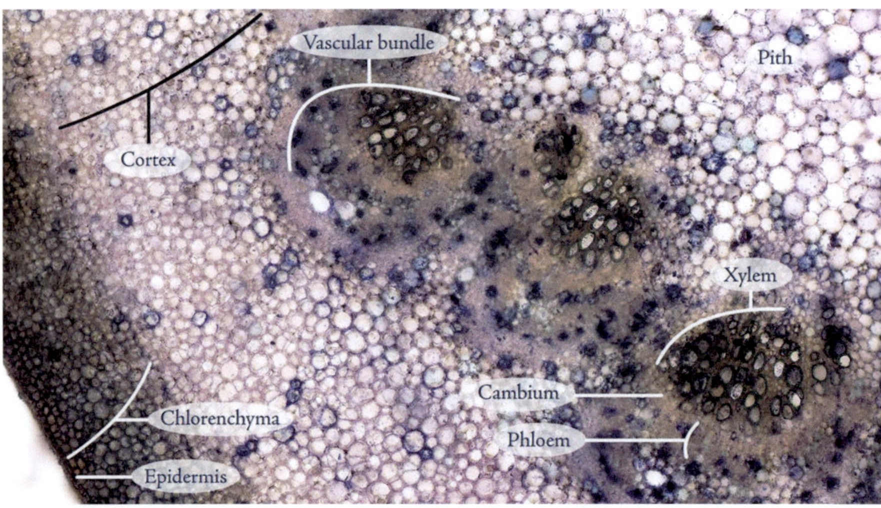

Fig. 11.270 High magnification image of the vascular bundles

The older branch is covered externally by the multi-layered cork, under which the cork cambium and the phelloderm are located. Underneath the parenchymatic cells of the cortex, a ring of the blue-stained sclerenchyma can be seen. Moving further in, the phloem, as a continuous phloem body, the cambium, and the xylem, as a unified xylem body, follow (Fig. 11.272).

11.2.3.4 Linden-Type Stem Thickening

The linden-type secondary thickening form is mostly found in trees. These plants have a procambium that creates the primary transport elements, which does not

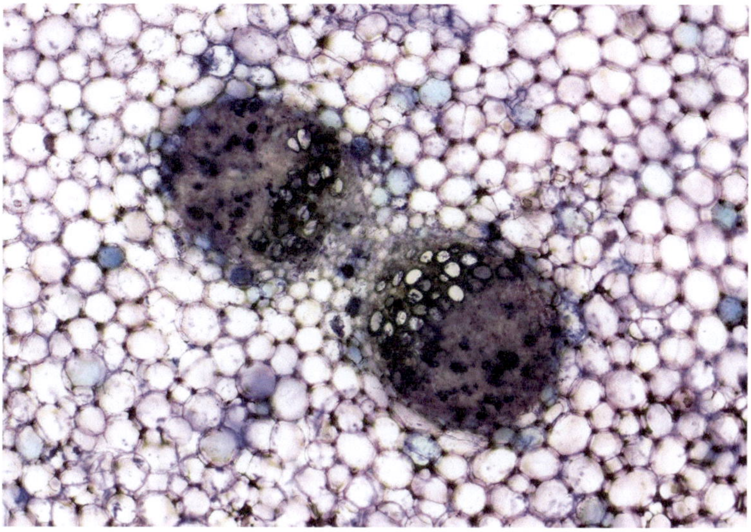

Fig. 11.271 Cross-section of two vascular bundles in the pith parenchyma

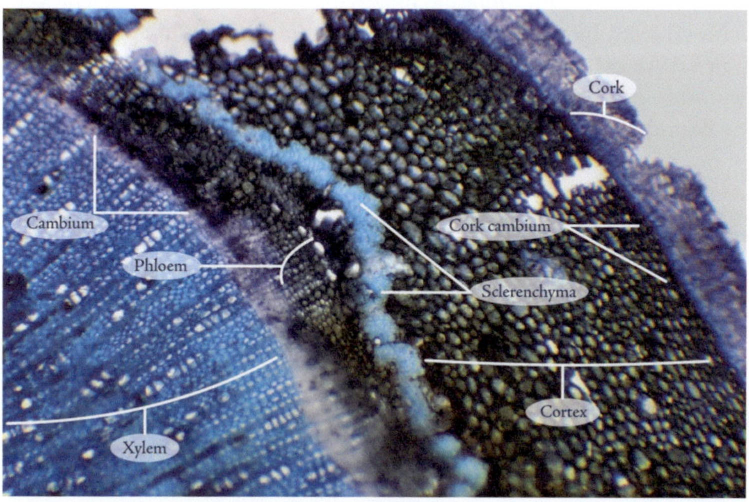

Fig. 11.272 Cross-section of an older horse chestnut branch

segment into bundles, but forms a continuous cylinder, and from the very beginning, it develops a uniform phloem ring outward and a uniform xylem ring inward. Outside the phloem part, in the woody stem, there is the primary cortex made up of parenchyma cells, while the outermost layer is the primary dermal tissue, the epidermis. As the stem thickens over time, the outer layer of the epidermis breaks away, and a secondary dermal tissue, periderm, forms in its place.

11.2.3.4.1 Forsythia (*Forsythia* sp.)

Forsythia is a plant belonging to the olive family (Fig. 11.273). The flowers of the forsythia bloom before or at the same time as leafing, along the twigs. Make cross-sections from very young shoots and paint them with toluidine blue!

The tissue structure of the stem can be well studied with a light microscope. Under the single-layered epidermis follows the cortex, in whose outer cell rows the layer of cells containing many chloroplasts forms chlorenchyma. Moving inward, dark blue staining sclerenchyma islands can be observed, under which the uniformly arranged phloem can be seen. The cambium ring under the phloem produces the phloem outward and the also uniform xylem inward. A sclerenchyma layer is attached to the inner side of the xylem, followed by the pith filling the inner part of the stem. The innermost part is the central cavity formed by the breaking away of the pith. Vascular bundles cannot be observed in the vascular tissue system (Figs. 11.274 and 11.275). This indicates that during the development of the shoot, the meristem (procambium) formed a continuous cylinder sheath, primary cambium, and this produced the phloem and the xylem.

11.2.3.4.2 Cherry (*Prunus avium*)

Make a section under the bud of the young green branch of the cherry (Fig. 11.276) and examine it stained with toluidine blue with a light microscope! The very young stem is still covered by a single-layered epidermis, under which chlorenchyma can be seen, in whose cells many chloroplasts can be observed.

In the cortex parenchyma, club-shaped calcium oxalate crystals often appear. At the inner boundary of the bark, sclerenchyma islands coloured light blue can be

Fig. 11.273 Forsythia (*Forsythia* sp.)

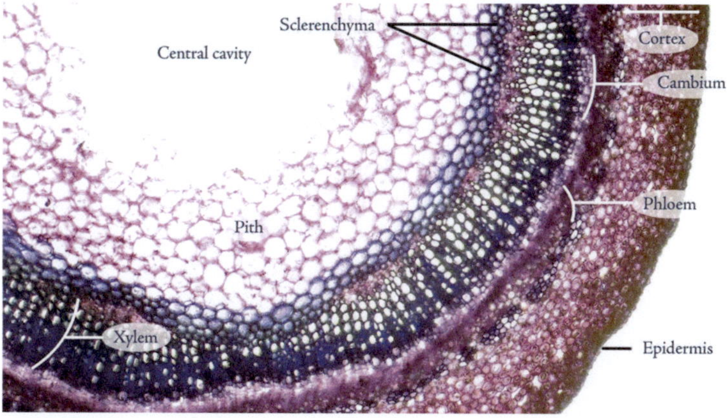

Fig. 11.274 Cross-section of a forsythia (*Forsythia* sp.) stem stained with toluidine blue

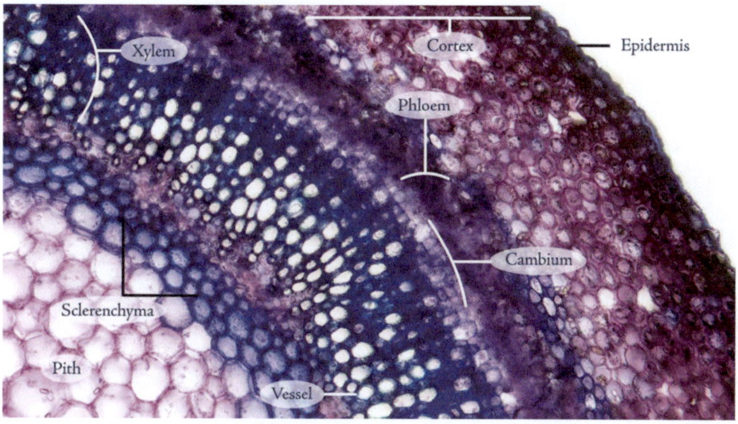

Fig. 11.275 Tissue structure of a forsythia (*Forsythia* sp.) stem

seen. Further inside, the continuous phloem and xylem appear, with the cambium ring clearly recognisable between them. This primary cambium originates from the dividing tissue of the shoot apex, the procambium. The primary cambium does not completely transform into vascular tissues, but dividing cells also remain in it, which here form a continuous ring between the primary xylem and phloem. In the inner part of the xylem body, the first appearing xylem elements (protoxylem), which tear and compress during stem growth, appear darker. Their place is taken by the later developing xylem elements (metaxylem). Inside, the rounded parenchyma cells of the pith fill the stem (Figs. 11.277, 11.278, and 11.279).

Fig. 11.276 The object of sectioning is the green branch part under the cherry (*Prunus avium*) bud

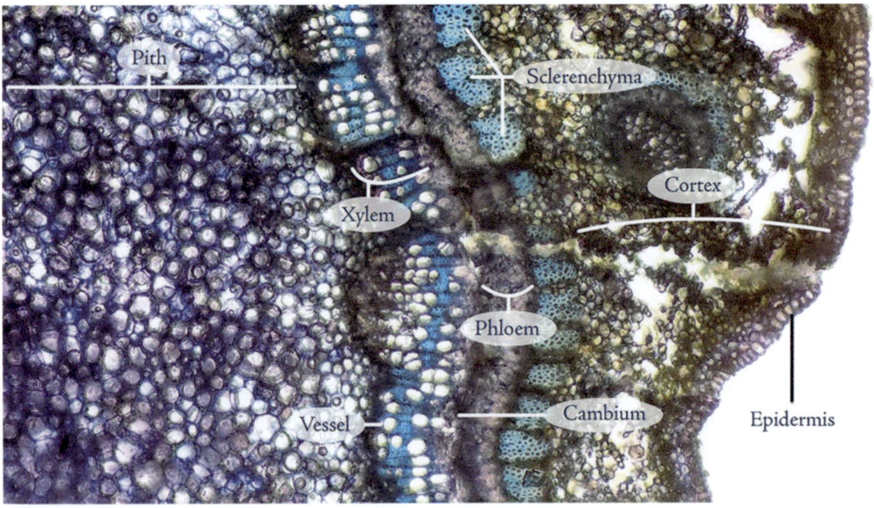

Fig. 11.277 Cross-section of a young cherry (*Prunus avium*) branch stained with toluidine blue

11.2.3.4.3 Common Mallow (*Malva sylvestris*)

The common mallow is widespread throughout Europe, which is a soft-stemmed, perennial medicinal plant species. It is common in weed associations (Fig. 11.280).

On the cross-section made from the young stem, the single-layered epidermis, which contains many hairs, can also be observed (Fig. 11.281).

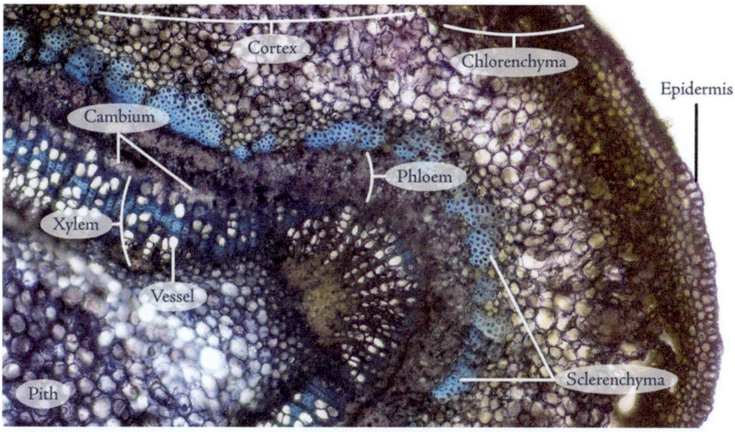

Fig. 11.278 Higher magnification image of the cross-section of a cherry (*Prunus avium*) branch

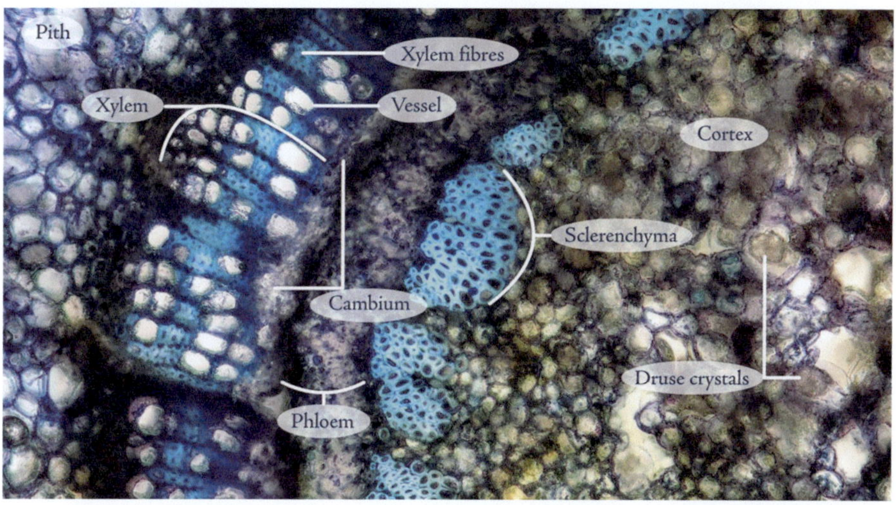

Fig. 11.279 High magnification image of a cross-section of a cherry (*Prunus avium*) branch

Underneath it, a chlorophyll-containing parenchyma cells, the chlorenchyma, develops. Later, a secondary dermal tissue, periderm, develops under the epidermis. With the tearing of the epidermis, this will protect the stem from the outside (Fig. 11.282).

Inside the chlorenchyma, a light blue colouring sclerenchyma ring can be observed. Inside this, the continuous layer of phloem, the cambium ring, and the also continuous xylem can be seen.

In the xylem, we can observe wide-cavity vessels and tracheids arranged in radial rows. Among them, thick-walled xylem fibres can be seen. Lighter coloured, 1–2 cell-thick medullary rays also appear here, which continue in the radial direction in the phloem. The central part of the stem is filled with pith rich in intercellular spaces (Figs. 11.281 and 11.282).

Fig. 11.280 Common mallow (*Malva sylvestris*)

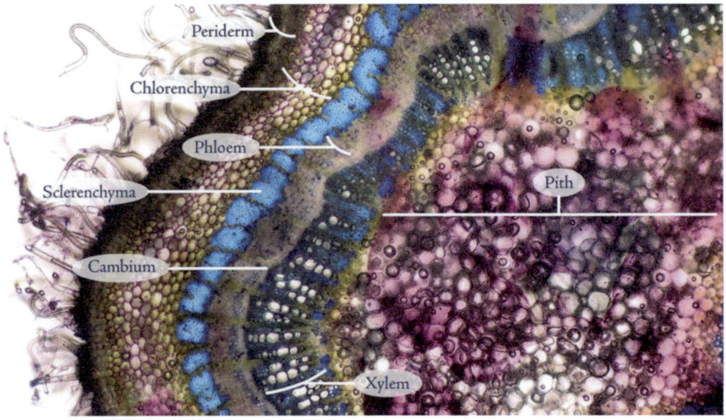

Fig. 11.281 Cross-section of a young common mallow (*Malva sylvestris*) stem stained with toluidine blue

11.2.3.4.4 False Acacia (*Robinia pseudoacacia*)

The false acacia is a leguminous tree native to the southeastern regions of the United States, but it has also been introduced to Europe, where it is mainly used as firewood. Let us examine a cross-section of its trunk! Here, the outer layer of rhytidome and the inner xylem of the trunk (Fig. 11.283) are clearly distinguishable. The outer part of the rhytidome is made up of cork layers, while the inner part is made up of the phloem and the cambium that forms it. The organic materials are transported in

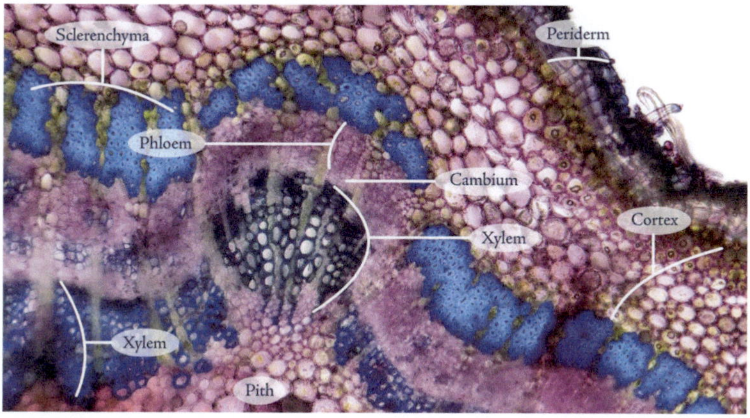

Fig. 11.282 High magnification image of a cross-section of a common mallow (*Malva sylvestris*) stem

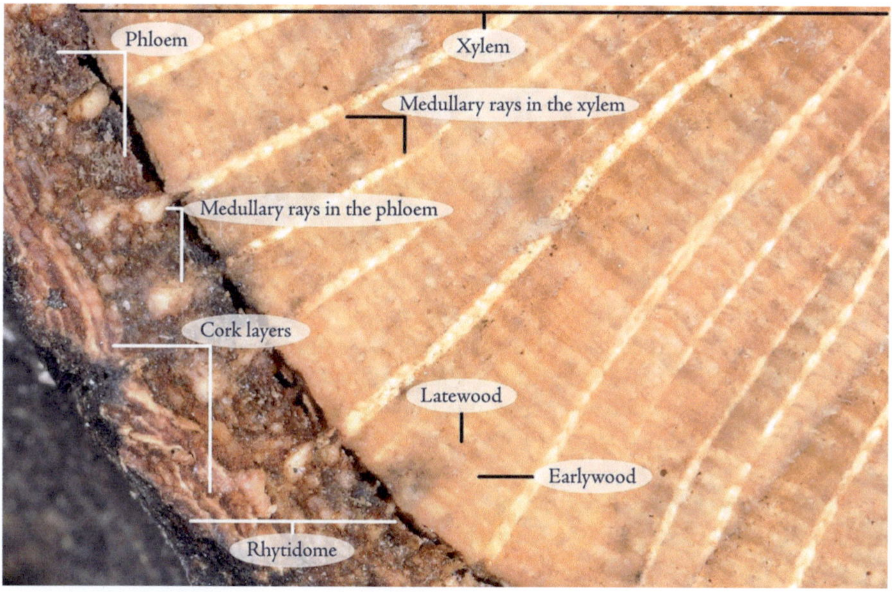

Fig. 11.283 Cross-section of a false acacia (*Robinia pseudoacacia*) trunk

the thin, cylinder-like phloem of the trunk, so if we remove even a 1 cm ring of a tree's bark, it inevitably causes the plant to die. Below the phloem, there is also a thin layer of meristem (cambium) arranged in a cylinder-like manner, which develops the phloem outward and the xylem inward. Only the thin layer of phloem next to the cambium participates in material transport, the previously formed layers are pushed outward and become part of the bark (rhytidome) by tearing. In contrast, the layers of xylem formed year after year remain and thicken the trunk. The solution of mineral salts taken up from the soil is transported in the xylem to the shoot. Unlike the phloem, not only the parts next to the cambium but also the inner layers of the trunk participate in material transport, but the central part of the trunk is already given by dead xylem. Radial medullary rays run in the trunk, which participate in the nutrition and metabolism of the transport elements. The formation of lighter and darker concentric circles, the annual rings in the trunk, can be traced back to the change of seasons. In the spring, following bud burst, the plant's water demand significantly increases, so at this time the dividing tissue forms large-cavity transport cells, the lighter colored layer of which is called earlywood. At the end of summer, after flowering and fruit ripening, the tree's water demand decreases, then narrow-cavity transport cells are formed, which give the darker coloured latewood. Therefore, a light and a dark coloured ring is formed in the trunk in 1 year. Let us examine the cross-section of the trunk with a magnifier! Due to the magnification, we can clearly observe the medullary rays running radially in the trunk, which continue in the phloem. Between their light spots, the dark coloured phloem can be seen. The cork layers can be observed in the outer part of the bark. The earlywood's light and the latewood's dark rings are clearly distinguishable in the trunk (Fig. 11.283).

Let us make a section from a young false acacia branch, then we examine it with a light microscope after staining with toluidine blue! As a result of the staining, the individual parts of the cross-section are clearly separated. From the outside, the branch is covered by the periderm, in which the purple-stained cork cambium and the greenish-brown cork it produces can be clearly distinguished.

Under the periderm, the cortex can be seen, followed inward by the sclerenchyma, which is the innermost layer of the bark.

Under the pink phloem, the cambium and the xylem can be found. In the xylem, the cross-sections of the large vessels and the darker coloured, radially running medullary rays, which continue less conspicuously in the bark, can be clearly recognised. The section was made from a young branch, so the xylem is not too thick yet, and the pith, which consists of parenchyma cells, can still be recognised in the central part of the branch (Figs. 11.284 and 11.285).

11.2.3.4.5 Black Pine (*Pinus nigra*)

Let us make a cross-section of a young branch of black pine (Fig. 11.286), then stain the sections with toluidine blue! We examine these with a light microscope, with transmitted and top lighting!

Moving inward from the outside on the cross-section of the pine branch, we can observe the periderm, the cortex parenchyma, the phloem, the cambium, and the xylem. Cross-sections of longitudinal resin ducts can be seen in the cortex, the phloem, and the xylem. The xylem appears homogeneous, there are no vessels in it. The uniform image is only interspersed with the cross-sections of the resin ducts. The xylem is built up of square, five- or hexagonal tracheids, which form radial rows, with a few thin medullary rays in between.

The xylem resin ducts are lined with thin-walled epithelial cells, which secrete the resin dissolved in the essential oil. There are no longitudinal parenchyma cells

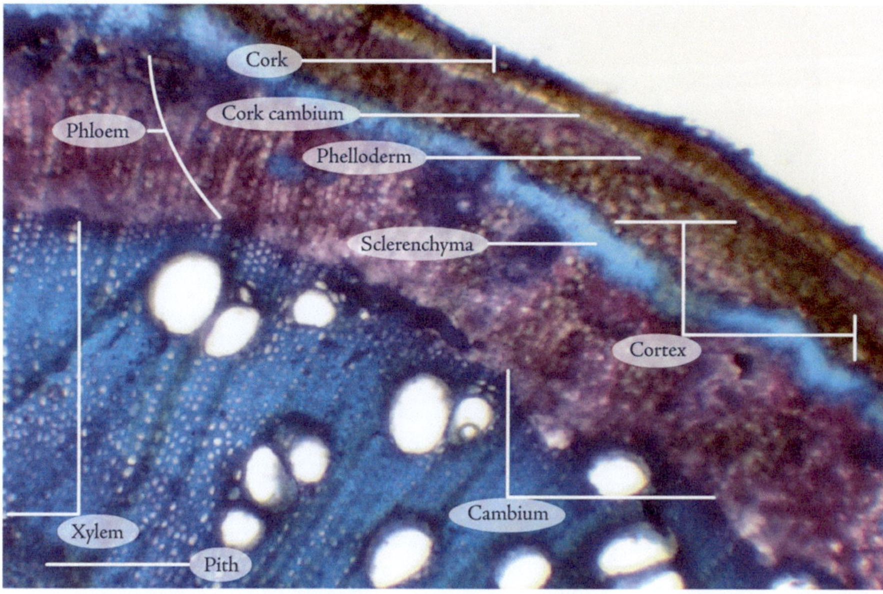

Fig. 11.284 Higher magnified image of the false acacia (*Robinia pseudoacacia*) trunk cross-section

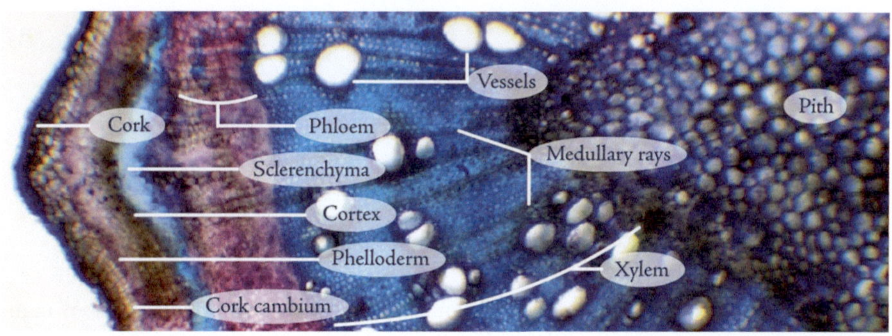

Fig. 11.285 Light microscopic image of a cross-section of a false acacia (*Robinia pseudoacacia*) branch stained with toluidine blue

in the xylem. On one side of the annual ring boundary, this year's earlywood has a wide cavity tracheids, on the other side, last year's latewood has narrow cavity tracheids.

The boundary of the xylem is clearly visible with the multi-cellular cambium, and outside of it is the phloem, which is made up of sieve tubes and larger phloem parenchyma cells (Figs. 11.287, 11.288, and 11.289).

Let us make a radial longitudinal section of a young branch of black pine, then examine the sections stained with toluidine blue with a light microscope! As we move down from the top on the longitudinal section image of the pine branch, we

Fig. 11.286 Black pine (*Pinus nigra*)

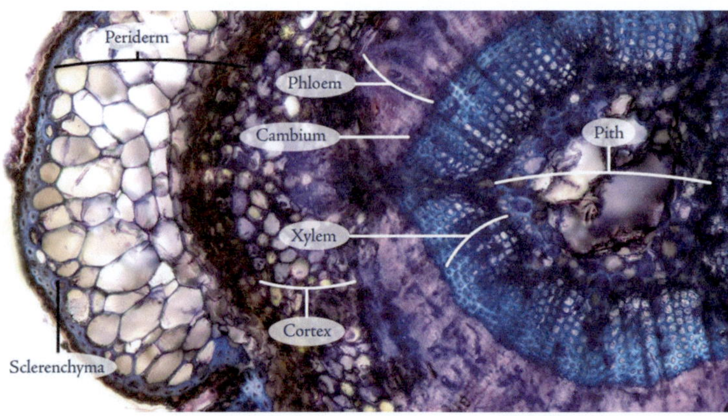

Fig. 11.287 Cross-section of a black pine (*Pinus nigra*) branch stained with toluidine blue

can clearly see the greenish cortex, the purplish phloem, the lighter layer of cambium, and the light blue coloured xylem can be recognised (Fig. 11.290). Gymnosperms do not have vessels capable of transporting large amounts of water, so unlike angiosperms, not only the annual ring next to the cambium participates in water transport, but also the ones located further inside, which together are capable of transporting a significant amount of water.

11.2.3.4.6 Linden (*Tilia* sp.)

Let us make cross-sections of a few millimeters from a young linden branch, stain it with toluidine blue, and then examine it with a light microscope with top lighting! Moving from the outside to the inside of the section, the triple-layered periderm (cork, cork cambium, and phelloderm) and the cortex are clearly separated. The cork cambium at the site of lenticel formation forms complementary cells. Under

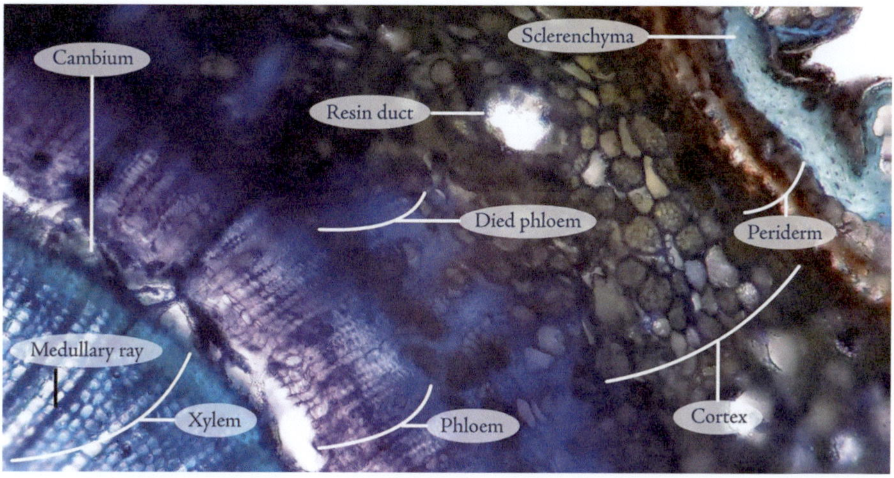

Fig. 11.288 High magnification image of the cross-section of a black pine (*Pinus nigra*) branch

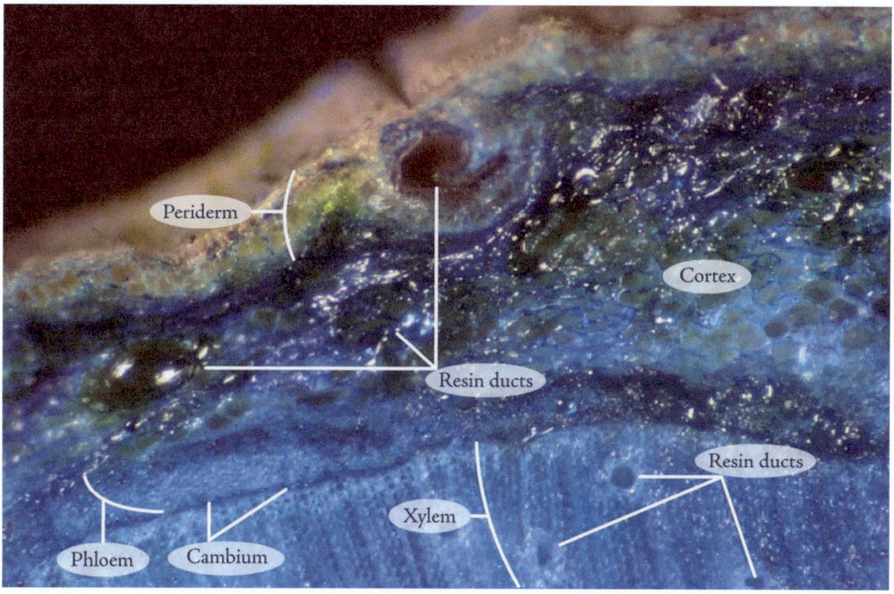

Fig. 11.289 Top-lit image of the cross-section of the pine branch

the cortex, the mass of phloem can be observed, only the part next to the cambium forms a cylinder sheath, the parts located further out have split. The gaps between the phloem islands are filled by shiny medullary funnels developing from the medullary rays. Under the cambium, the blue-stained xylem can be seen (Figs. 11.291 and 11.292).

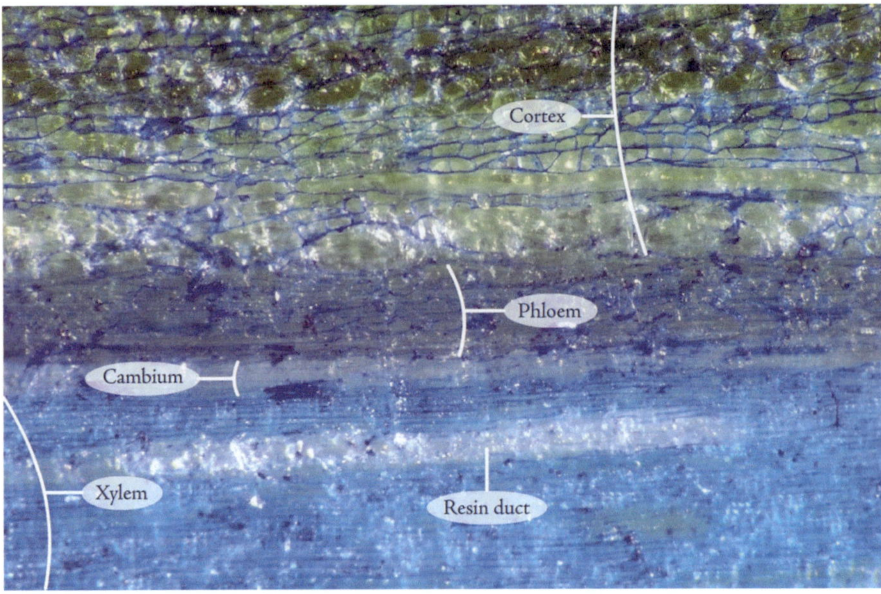

Fig. 11.290 Radial longitudinal section of a pine branch

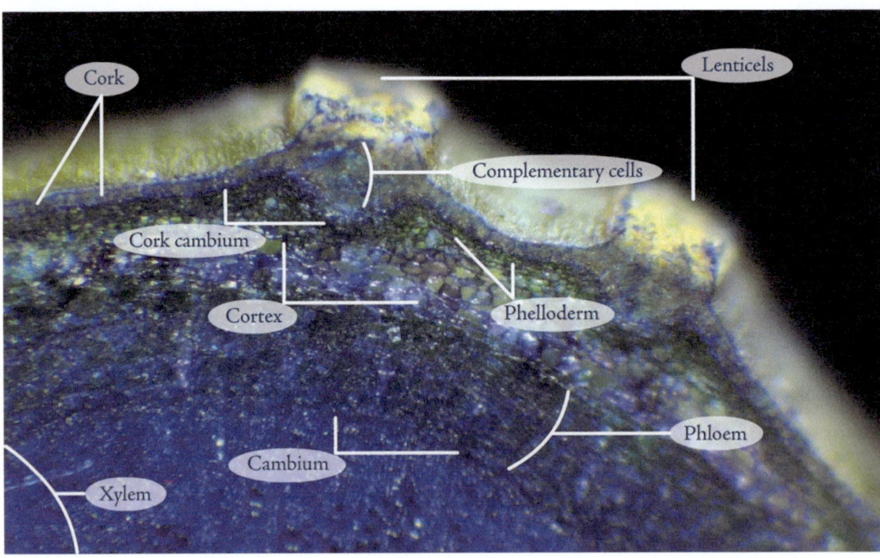

Fig. 11.291 Top-lit image of a linden (*Tilia* sp.) branch stained with toluidine blue

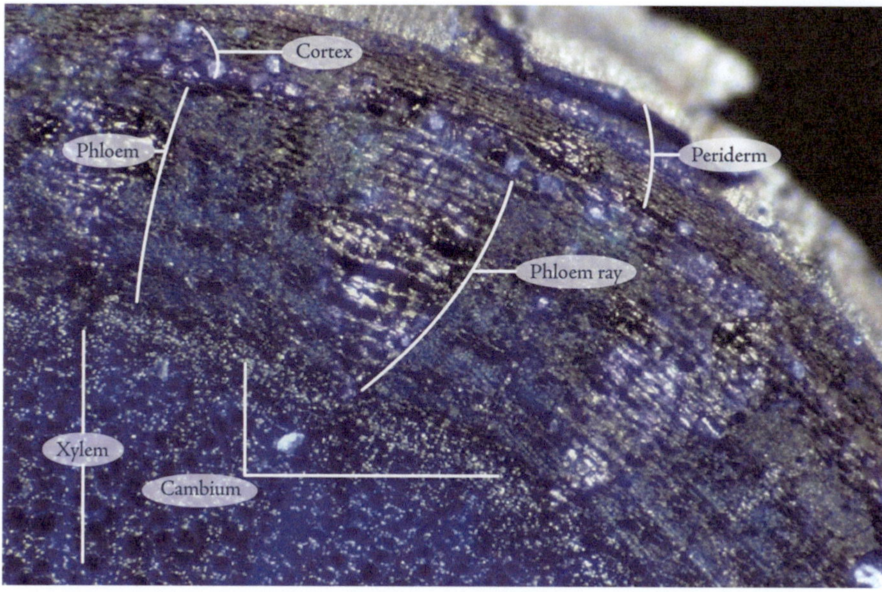

Fig. 11.292 Histological structure of a linden (*Tilia* sp.) branch

Fig. 11.293 Common ash (*Fraxinus excelsior*) branch

11.2.3.4.7 Common Ash (*Fraxinus excelsior*)

Let us make a cross-section of the common ash branch (Fig. 11.293) and examine it with a light microscope after staining with toluidine blue!

In the case of common ash (Fig. 11.294), the procambium that creates the primary transport elements does not form bundles, but forms a continuous cylinder,

Fig. 11.294 Cross-section of a common ash (*Fraxinus excelsior*) branch stained with toluidine blue

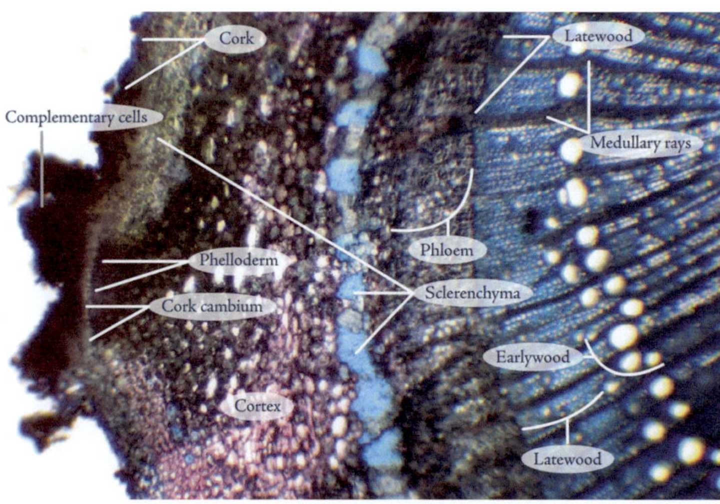

Fig. 11.295 Tissue structure of a common ash (*Fraxinus excelsior*) branch

and from the beginning, it develops a uniform phloem outward and a uniform xylem inward. Outside the phloem tissue is the primary cortex made up of parenchyma cells, while the outermost layer is the epidermis.

As the stem thickens, the outer epidermis tears and is replaced by cork, then by rhytidome (Figs. 11.295 and 11.296). The majority of the cross-section of the branch is made up of the centrally located xylem, only those closer to the circumference of the branch participate in the transport of nutrients. The phloem is located outside the cambium and still forms a uniform phloem body here. The phloem is

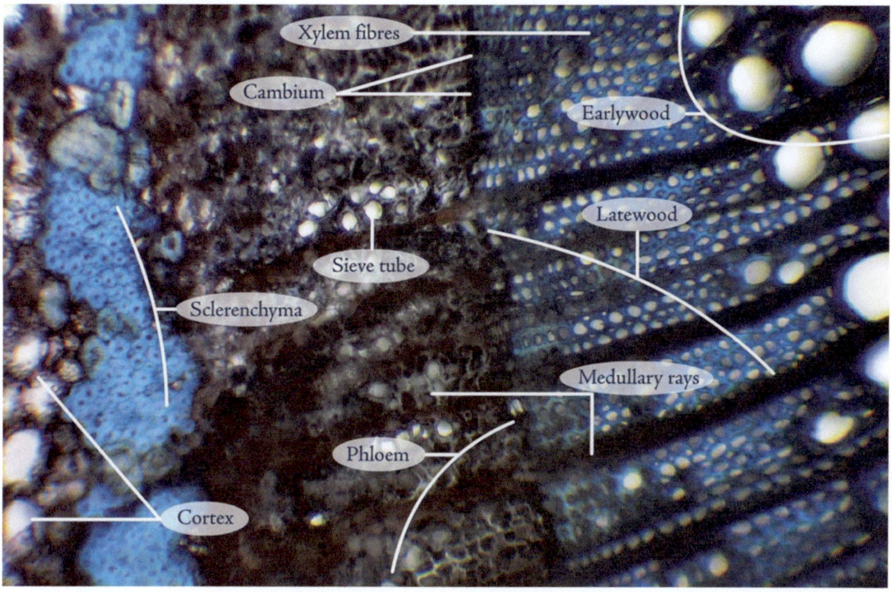

Fig. 11.296 High magnification image of the vascular tissue of the common ash (*Fraxinus excelsior*)

made up of sieve tubes, companion cells, and phloem parenchyma cells. Darker medullary rays cut through it radially. Later, due to the circumferential expansion of the trunk, it tears and only the newly formed phloem elements close to the cambium participate in material transport. The dead, torn older parts are important in the formation of the rhytidome.

The basic structure of the xylem is provided by the xylem fibres, which are important for support, among which xylem parenchyma cells can also be observed. In the cross-section of the xylem, the transport elements, the narrower tracheids, and the wide cavity vessels can be clearly recognised with a light microscope. These are arranged in bands according to the annual-ring structure of the tree trunk. Between them run radial medullary rays, of which the longer ones (primary rays) run from the pith to the bark, while the shorter ones (secondary rays) only bridge a few annual rings. The rays are one or only a few cell rows thick (Fig. 11.296).

The common ash tree belongs to the ring-porous trees because the wide cavity vessels only occur in the early wood of the annual ring. As a result, the early- and latewood are clearly separated from each other. This characteristic allows for the clear identification of the annual ring boundaries.

11.2.3.4.8 Oleander (*Nerium oleander*)

In the very young stem of the oleander, we can already observe a uniform phloem and xylem body, which is created by the primary cambium formed from the procambium. Due to the continuous operation of the cambium, the stem thickens secondarily. Initially, the stem is covered by the epidermis, under which the cortex can be seen. Its outer layer transforms into collenchyma. Below it, there is the

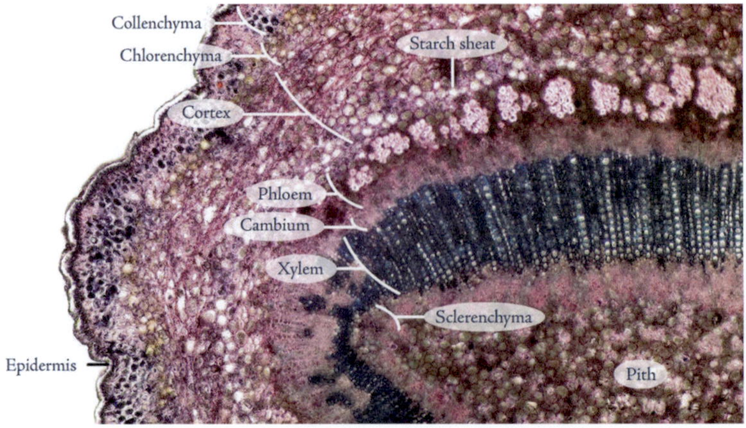

Fig. 11.297 Cross-section of a young oleander (*Nerium oleander*) stem stained with toluidine blue

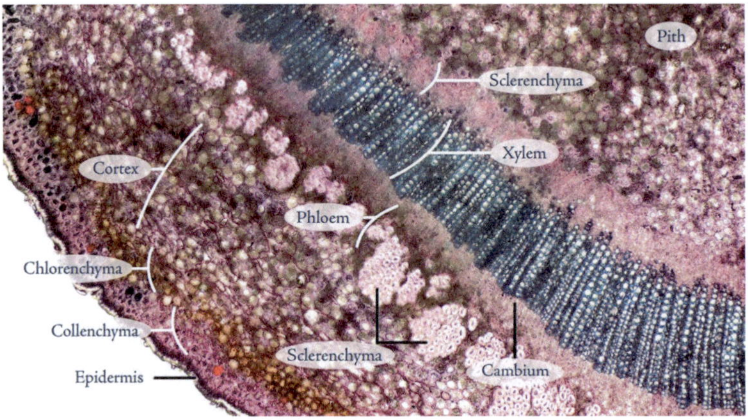

Fig. 11.298 Tissue structure of a young oleander (*Nerium oleander*) stem

chlorenchyma. The innermost cell row of the cortex is the starch sheath, consisting of cells with a high starch content. Below the cortex, there is the phloem segmented with sclerenchyma islands, followed by the cambium and the xylem. A sclerenchyma ring is connected to the inner side of the xylem, followed by the pith, consisting of many starch-rich parenchyma cells (Figs. 11.297 and 11.298).

In the older stem, a cork cambium develops from the epidermis, which creates the secondary dermal tissue, the periderm, directly under the epidermis (Fig. 11.299). Later, the epidermis peels off, and its place is taken by the periderm (Fig. 11.300).

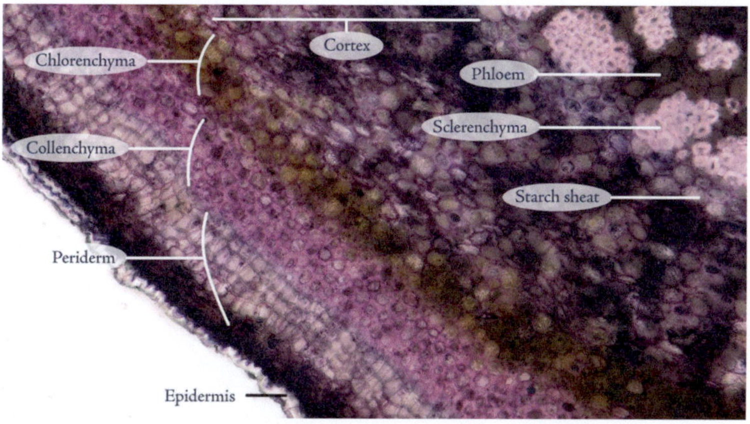

Fig. 11.299 On the cross-section of the older oleander (*Nerium oleander*) stem, the periderm developing under the epidermis can already be observed

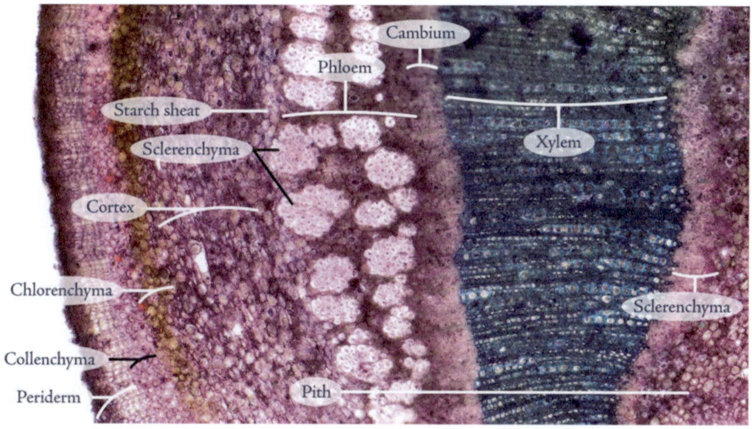

Fig. 11.300 Tissue structure of an older oleander (*Nerium oleander*) stem

11.2.3.4.9 Thuja (*Thuja* sp.)

On the cross-section of the thuja stem, we can study the structure of the secondarily thickening gymnosperm stem. Make cross-section from the branch, stain with toluidine blue, and examine it with a light microscope!

Moving from the outside in on the cross-section of a thuja branch, we can observe the periderm, the cortex parenchyma, the phloem, the cambium, and the xylem. The xylem contains only narrower tracheids. In the xylem, darker medullary rays run radially, while the centre of the branch is filled with pith parenchyma (Figs. 11.301 and 11.302).

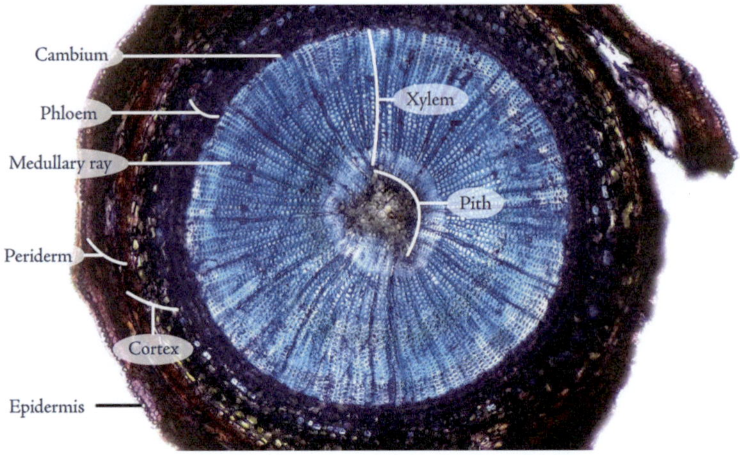

Fig. 11.301 Cross-section of a thuja (*Thuja* sp.) branch stained with toluidine blue

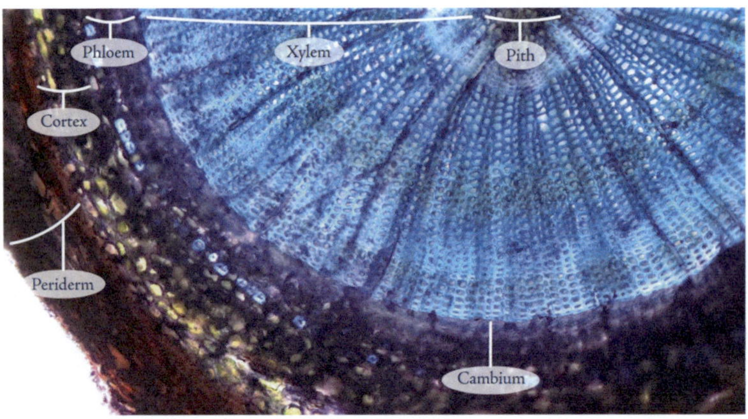

Fig. 11.302 Higher magnification image of the cross-section of a thuja (Thuja sp.) branch

11.2.3.4.10 Box Elder (*Acer negundo*)

Make a section from a young box elder branch (Fig. 11.303), then examine it with a light microscope after staining with toluidine blue! The young box elder branch is covered from the outside by the epidermis, followed inward by the periderm. Further inside, the cortex‘s parenchymatic and sclerenchymatised layers follow. Although the box elder is a diffuse-porous tree, the annual ring boundary is clearly recognisable on the section because most of the wide-cavity vessels are located in the earlywood. On the cross-section of the box elder branch, the latewood containing narrower cavity transport elements can be seen first from the cambium towards the inside of the branch (the section shown in the picture was made at the end of June),

Fig. 11.303 Young box elder (*Acer negundo*) branch

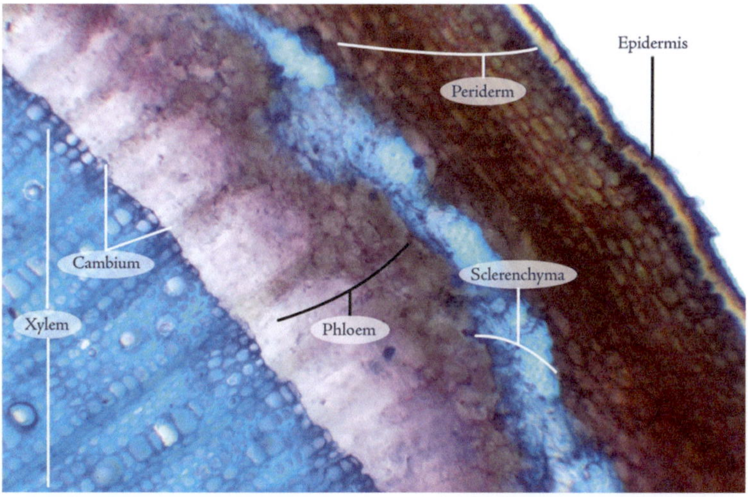

Fig. 11.304 Cross-section of a box elder (*Acer negundo*) branch stained with toluidine blue

followed by the early wood containing wider elements. The annual ring boundary can be observed next to the earlywood. Inside this, the latewood from the previous year is located (Figs. 11.304 and 11.305). Medullary rays run radially on the section, and the centre of the branch is filled with pith parenchyma.

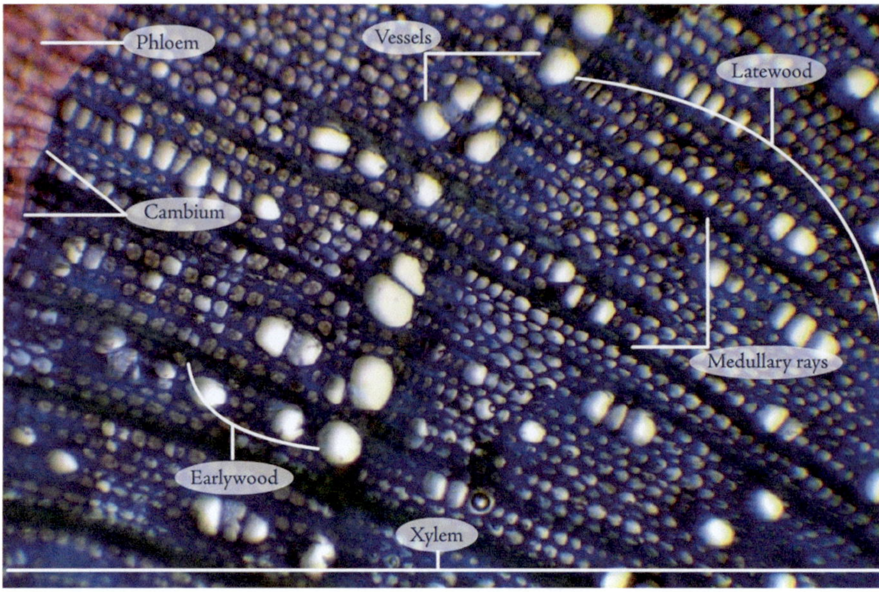

Fig. 11.305 Cross-section of a box elder (*Acer negundo*) branch at the annual ring boundary

12 Examination of Plant Cells and Cell Organelles

12.1 Epidermis of the Purple Onion Leaf

Separate the fleshy leaves of the purple onion from each other, then prepare an epidermis peel from one of the outer surfaces! To do this, scratch a square into the epidermis, then try to peel off the piece of epidermis by grabbing the corner of the square with tweezers! Place the peel on a slide, and examine it in a drop of water with a magnifying glass and a light microscope (Figs. 12.1, 12.2, and 12.3)!

On the scrape, the tightly closing epidermal cells can be clearly seen, occasionally the cell nucleus can also be observed (Fig. 12.4). In some places, we can also see stomata, which have no significant role in the underground organ.

The blue, purple, and red colours of plants are mostly caused by **anthocyanin**s, which have acid-base indicator properties and accumulate in the cell sap of epidermis cells. The purple onion also owes its colour to this. Anthocyanins are water-soluble coloured compounds in the cell sap. Their colour changes depending on the pH of the medium. From a strongly alkaline pH to a strongly acidic pH, they can turn yellow, green, blue, purple, and red. In nature, plant parts containing anthocyanins can be red, purple, and blue. Anthocyanins cause, for example, the red colour of geraniums, the blue of common chicory, the purplish colour of common bugloss and violets. Anthocyanins are glycosides consisting of coloured anthocyanidin and one or two colourless sugar parts.

To examine the anthocyanin content of purple onion, prepare an epidermis peel from the purple side of the leaf and examine it under a light microscope covered with a drop of water! Look for purple cells containing anthocyanin in the field of view, then drop a drop of sodium hydroxide solution next to the covered epidermis! To prepare the alkali solution, dissolve a sodium hydroxide tablet in 1 cm^3 of water! As a result of the alkali seeping under the cover slip, the colour change in the cells occurs within a few seconds, then with the destruction of the cell membrane, the anthocyanin, which has turned greenish, flows out of the cells (Fig. 12.5).

G. Kriska, *Teaching Biology*, https://doi.org/10.1007/978-3-031-84085-2_12

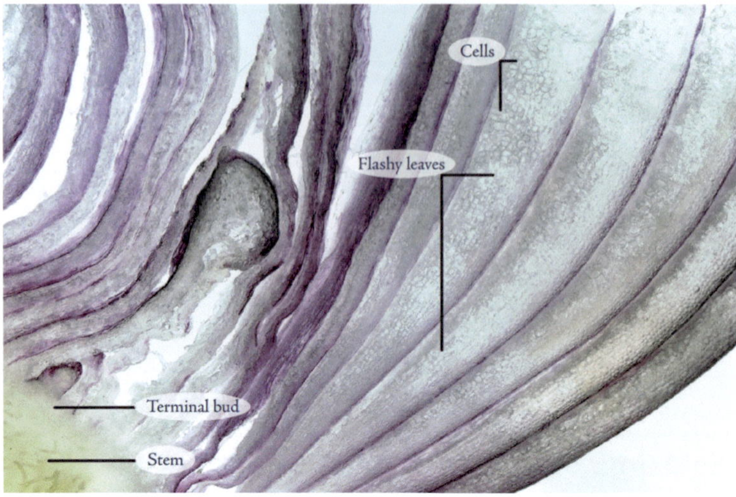

Fig. 12.1 Cross-section of the purple onion under a magnifying glass

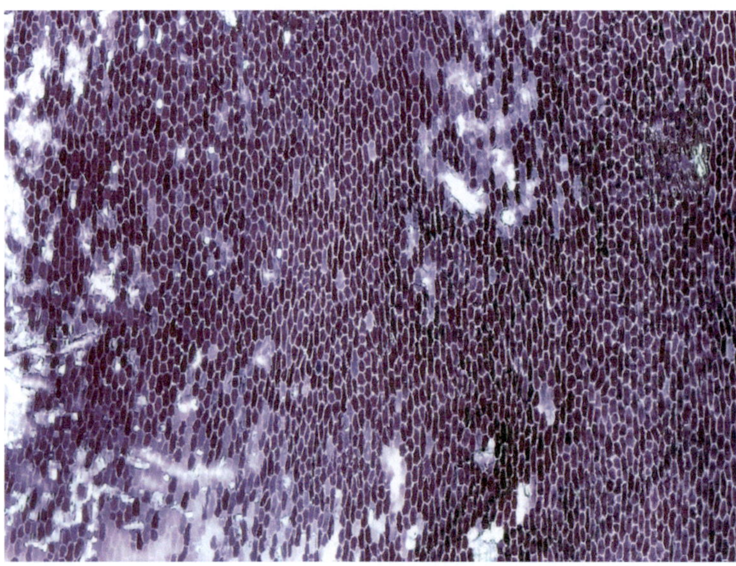

Fig. 12.2 Purple onion epidermal cells under a magnifying glass

12.2 Anthocyanin Content of Red Cabbage

Prepare a rough, multi-cellular section with a razor blade from the purple part of the cabbage, then drop a sodium hydroxide (NaOH) solution next to the section placed on the slide and covered with a drop of water. To prepare the alkali solution, dissolve a sodium hydroxide tablet in 1 cm^3 of water. The alkali gradually seeps under

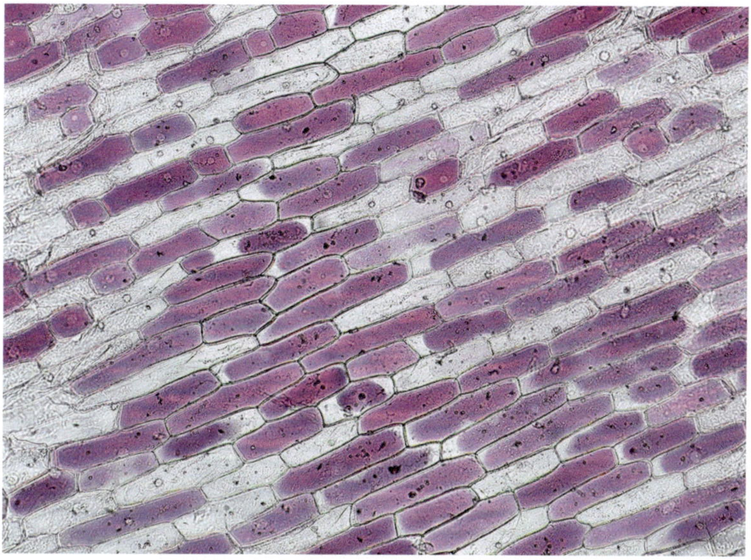

Fig. 12.3 Purple onion epidermal cells under light microscope

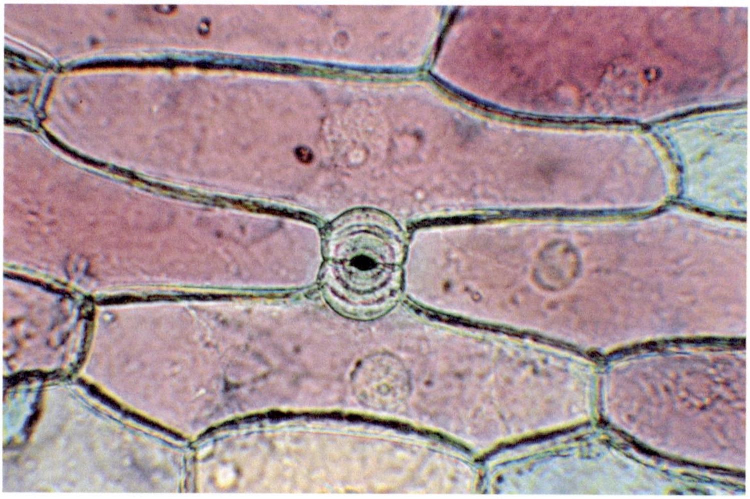

Fig. 12.4 Stomata in the epidermis of the purple onion

the cover slip, so the purple, blue, green, and finally yellow colour change appears simultaneously in the cells (Fig. 12.6). The experiment can also be performed with 20% acetic acid. Then the purplish colour of the cabbage section turns bright red.

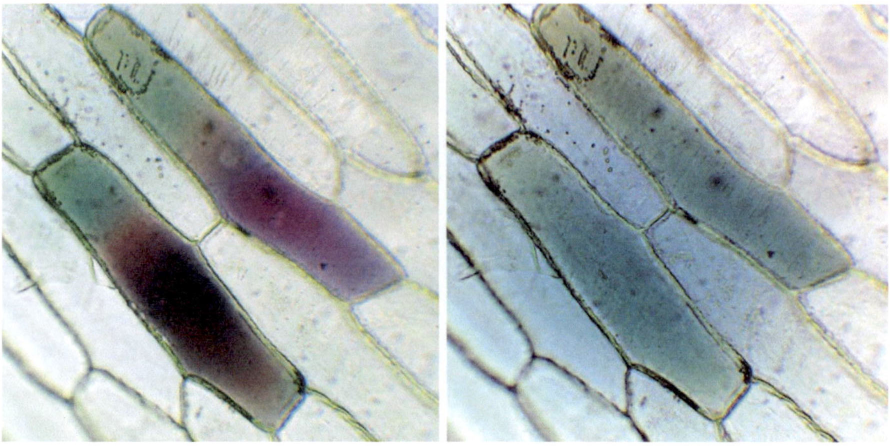

Fig. 12.5 The colour change in the cells due to the effect of alkali solution

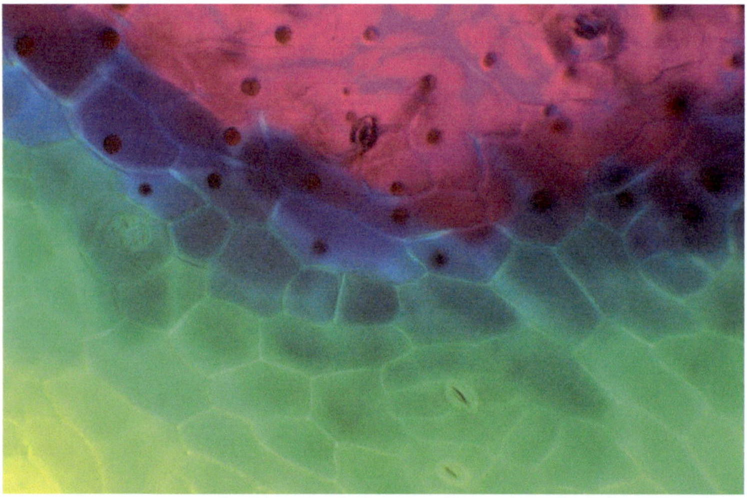

Fig. 12.6 Colour change of red cabbage section under the effect of sodium hydroxide solution

12.3 Central Vacuole Filled with Cell Sap

Prepare an epidermis peel from the fleshy lower leaf of purple onion, then place it in a 1% aqueous neutral red solution for 1–2 min! Rinse the pieces of epidermis with water, spread the peel with a dissecting needle on a slide, then cover it in a drop of water! In the field of view of the microscope, look for the **central vacuole** filled with red-dyed cell sap and the light cytoplasm!

The acidic solution of the central vacuole is dyed pink with the neutral red solution, thus under a light microscope, the cytoplasm and the central vacuole filled with

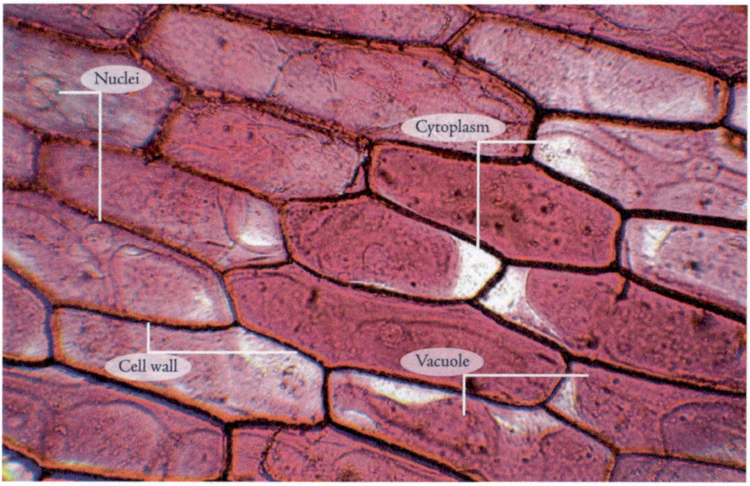

Fig. 12.7 Epidermis of purple onion dyed with neutral red solution

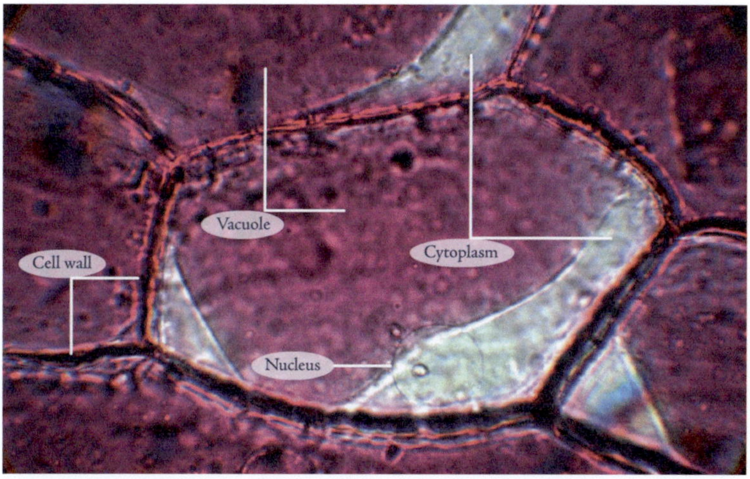

Fig. 12.8 High magnification light microscopic image of a single cell

cell sap become distinguishable from each other in the cells (Figs. 12.7 and 12.8). In some cells, the central vacuole does not cover the nucleus, thus we can also conclude that the nucleus is located in the plasma.

We have already seen above that the cell cavity filled with cell sap in epidermal cells often becomes so extensive in size that the cytoplasm forms only a cytoplasmic tube under the cell wall. We can also be convinced of this by examining the peel of the narcissus epidermis (Figs. 12.9 and 12.10).

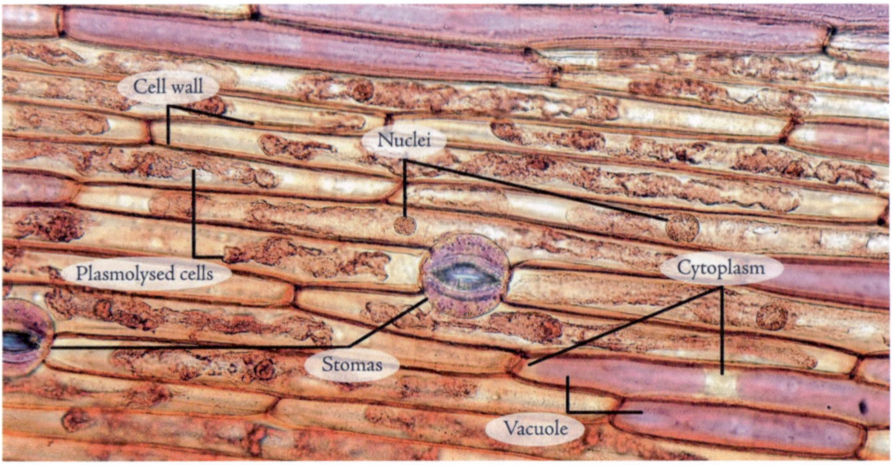

Fig. 12.9 Light microscopic images of daffodil (*Narcissus* sp.) epidermis peel dyed with neutral red solution. Damaged cells can be seen on the orange parts, which do not give the expected reaction

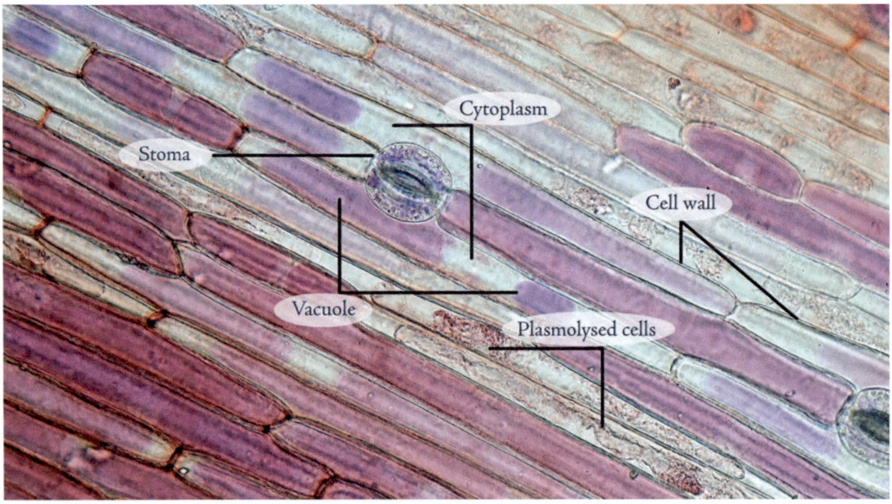

Fig. 12.10 Light microscopic image of daffodil (*Narcissus* sp.) epidermis peel dyed with neutral red solution

Let us make a peel from a narcissus leaf, then paint it with neutral red solution and examine it with a light microscope! In a significant part of the peel, damaged cells can be observed, from whose cell cavity the liquid has departed, so we can observe the shrunken cytoplasm within the cell. Therefore, for the observation of the painting with neutral red, we need to find a part where intact epidermal cells can be seen.

12.4 Plasmolysis

If the plant gets little water, then with the decrease of the water content of the central vacuole filled with cell sap, the tension state of the cells also decreases, the plant begins to wilt. In the case of significant water loss of the cells, the cytoplasm separates from the rigid cell wall and it shrinks. This process is called **plasmolysis**. The plasmolysed cells, if they get water, are able to absorb it, as a result of which the central vacuole fills with water and when the central vacuole filled with cell sap pushes the cytoplasm back to the cell wall, the cell stretches out. This process is called deplasmolysis. The form of plasmolysis depends on the state of the cytoplasm, primarily its viscosity. The less viscous cytoplasm separates from the cell wall with a convex surface, curling up, which results in convex plasmolysis (Figs. 12.11 and 12.12).

The more viscous cytoplasm remains adhered to the cell wall in several places during plasmolysis, resulting in concave plasmolysis (Figs. 12.13 and 12.14).

In both cases of plasmolysis, the cytoplasm can be connected to the cell wall by thin plasma threads (Hechtian strands). We perform plasmolysis with a 10% solution of potassium chloride and calcium chloride on an onion leaf epidermal peel. The process of plasmolysis takes place very quickly. Initially, concave plasmolysis develops under the influence of both salt solutions, with the plasma separating sharply from the cell wall in most cells. However, later the cytoplasm in the cells plasmolysed with the potassium chloride solution curls up, the image of convex plasmolysis is demonstrating. This phenomenon is explained by the fact that the cytoplasm not only releases water, but also absorbs ions, potassium (K^+) and calcium (Ca^{2+}) ions, and since the hydration shell of potassium ions is larger than that of calcium ions, the former return more water to the cytoplasm. As a result, the

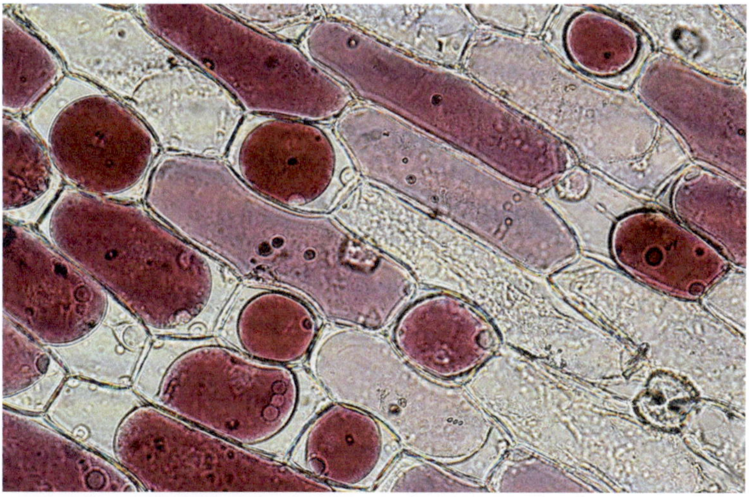

Fig. 12.11 Convex plasmolysis of purple onion epidermis peel

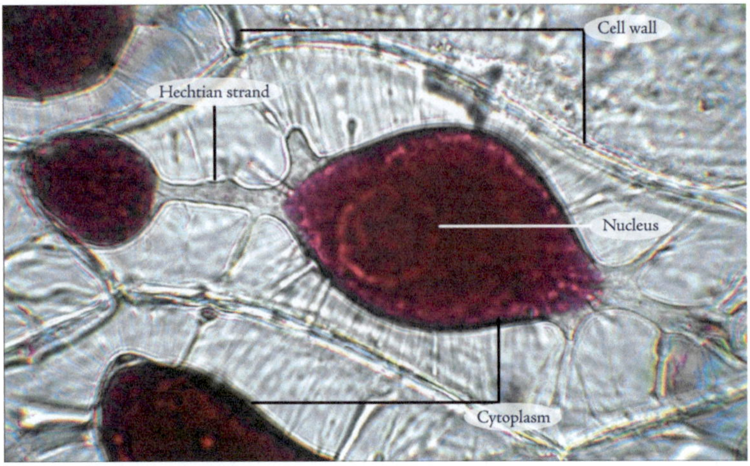

Fig. 12.12 Convex plasmolysis

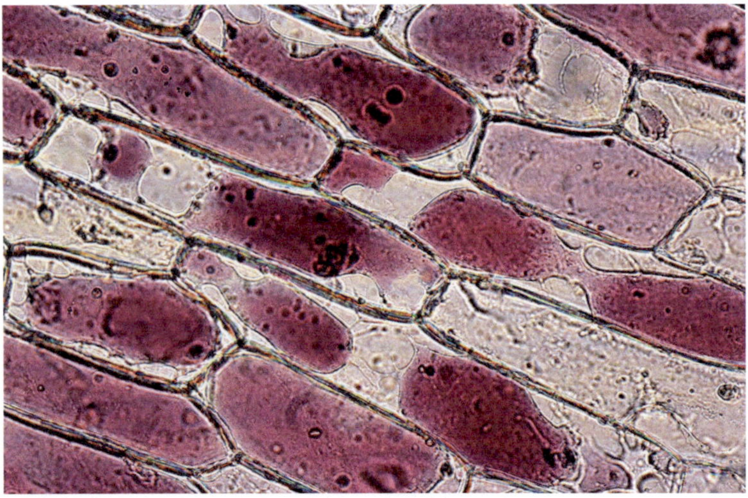

Fig. 12.13 Concave plasmolysis of purple onion epidermis peel

viscosity of the cytoplasm decreases and it becomes rounded. Plasmolysis requires a living cell membrane, so if the peel is placed in concentrated ethanol or hot water for two minutes, it loses its semi-permeable nature and plasmolysis can no longer be induced.

Prepare an epidermis peel from the fleshy lower leaf of a purple onion or purple onion!

When preparing the peel, first scratch a square a few millimetres in side length into the surface of the leaf with a scalpel, then reach under the epidermis at its corner and pull off the peel with tweezers! We can also easily obtain the epidermis

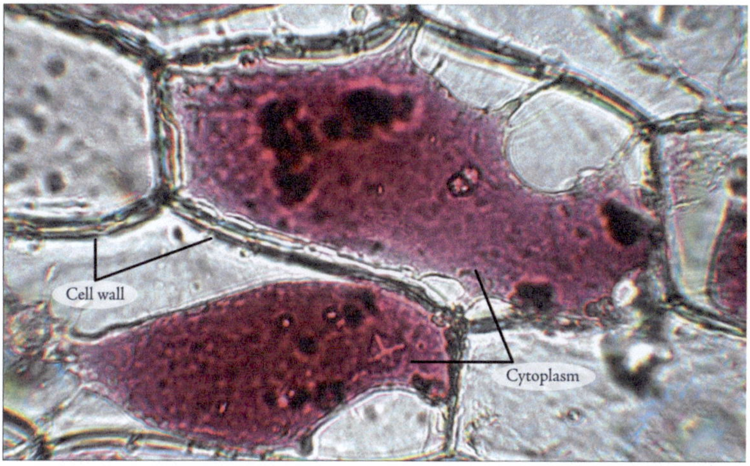

Fig. 12.14 Concave plasmolysis

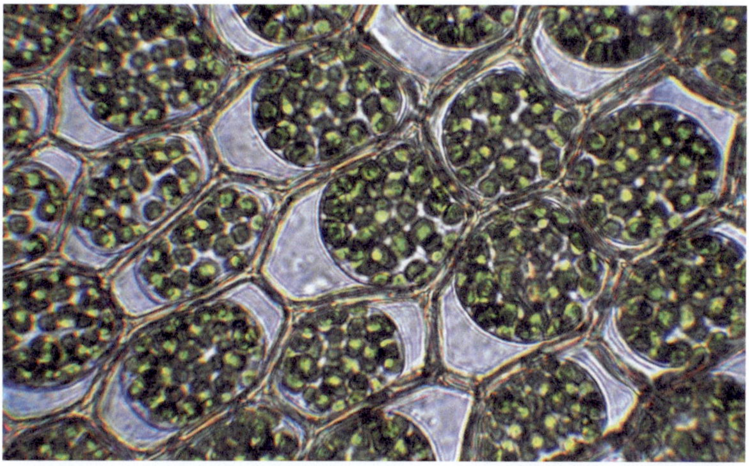

Fig. 12.15 Plasmolysis of moss leaf cells

needed for the experiment if we separate the fleshy lower leaves of the purple or purple onion and use the epidermis membrane that comes off the inner surface of the lower leaves during the experiment. The plasma of the cells found here does not contain purple anthocyanins, so the result of the experiment will be less spectacular in their case. Plasmolysis can also be well examined on the single-layered moss leaf (Fig. 12.15).

Film

Concave plasmolysis of the epidermis of the lower leaf of the purple onion: https://youtu.be/s5tr3eNU4HI

Concave, then convex plasmolysis of the epidermis of the lower leaf of the purple onion: https://youtu.be/DEQwKfz3Bn8

12.5 Cell Nucleus

Prepare an epidermis peel from the fleshy lower leaf of the onion, then place it in a methyl green solution for 1–2 min! Rinse the peel with clean water, place it on a slide, and examine it with a light microscope!

The cells of the skin tissue contain large cell nuclei, but these are not always easy to recognise without staining. Therefore, it is advisable to use methyl green, methyl blue, or iodine solution staining for the microscopic examination of cell nuclei. Methyl green is a microscopic dye that binds to DNA, turning the cell nucleus bluish-green, where the majority of the cell's DNA content is concentrated (Figs. 12.16 and 12.17).

12.6 Plasma Movement

The American waterweed is a popular aquarium ornamental plant, available strand by strand from ornamental fish sellers. The plant also lives freely, mainly in the streams of the plains where it can multiply en masse. With its overgrowth the aquatic plant life becomes uniform, which also leads to the impoverishment of the animal world. The thin leaves stand in triple whorls on its shoot. Let us pick a leaf from the plant and examine it with a light microscope! The mesophyll is reduced, and the majority of the leaf surface is only given by the lower and upper epidermis. The

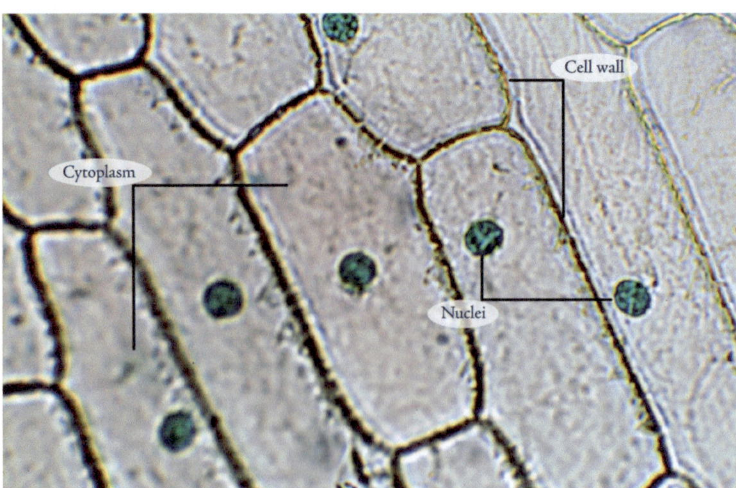

Fig. 12.16 Cell nuclei stained with methyl green in onion epidermis peel

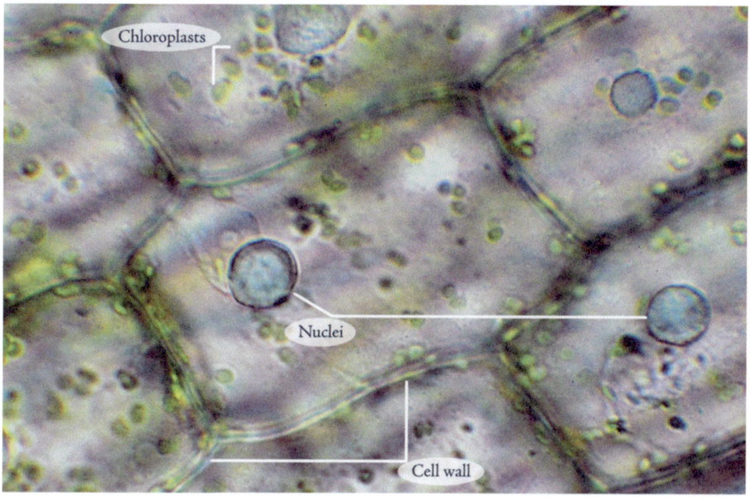

Fig. 12.17 Cell nuclei stained with methyl blue in the leaf of the American waterweed (*Elodea canadensis*)

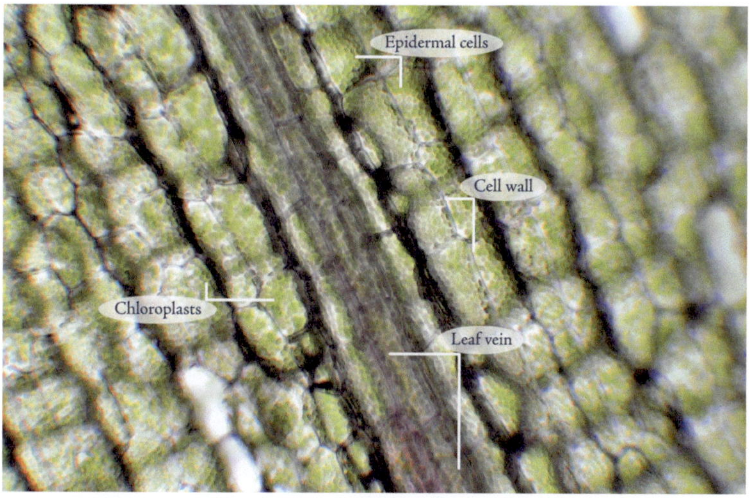

Fig. 12.18 Light microscopic image of a leaf fragment of American waterweed (*Elodea canadensis*)

mesophyll only appears near the larger leaf veins in a thickness of one or two cell rows. When examined with a light microscope, the elongated transport elements of the leaf veins can be clearly observed. The epidermis does not contain stomata, but interestingly, many chloroplasts can be observed in the epidermal cells (Figs. 12.18, 12.19, and 12.20). These perform constant movement in the cells due to the circular flow of the cytoplasm.

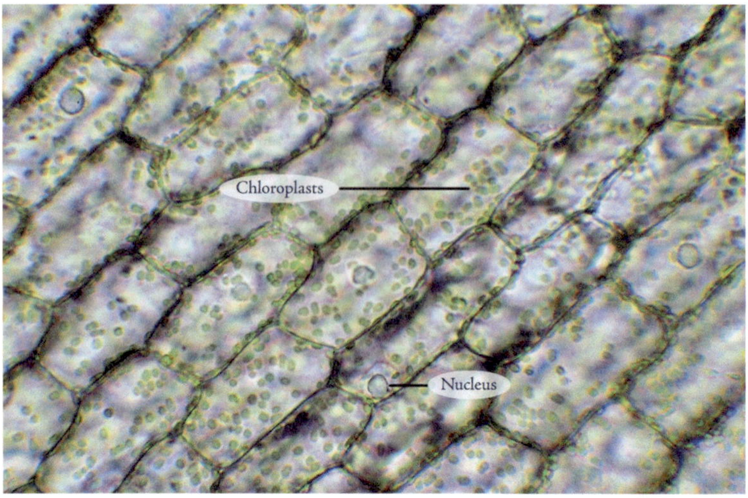

Fig. 12.19 Detail of an American waterweed (*Elodea canadensis*) leaf with cell nuclei

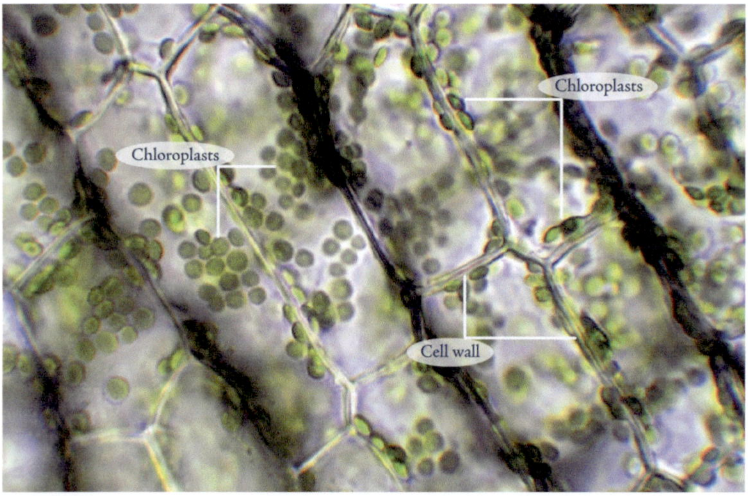

Fig. 12.20 High magnification light microscopic image of a leaf fragment of American waterweed (*Elodea canadensis*)

Film

Cytoplasmic movement in Canadian waterweed leaf: https://youtu.be/IucFrRFesqI

12.7 Starch Grains: Bean (*Phaseolus vulgaris*)

During the germination of the bean, the material and energy needs of growth are provided by the reserve nutrients accumulated in the cotyledons. In the bean cotyledon, starch accumulates in the largest quantity among the reserve nutrients in the colourless plastids (leucoplasts). Starch is produced by the effect of light during photosynthesis in the chloroplasts. The starch produced during the day breaks down into sugar (sucrose) at night under the effect of digestive enzymes and reaches the plant's storage organs, the roots, rhizomes, seeds, through the vascular tissues. Starch is formed again at these places, which accumulates in the colourless plastids formed from the plastid primordia.

The shape, size, and structure of the resulting starch grains are characteristic of the individual plant species.

If the crystallisation focus of the starch is in the middle, we speak of concentric, while if it is peripheral, we speak of eccentric starch grains. The starch grains of the bean are concentric, while colourless plastids with eccentric structure can be found in the potato tuber.

Soak bean seeds in water for twenty minutes! Then cut the seed in half, make a scrape from the cut surface of the cotyledon, and spread this in water dropped onto a slide! After covering, study the structure of the starch grains under a microscope (Fig. 12.21)!

Apply Lugol's solution to one edge of the cover slip and use a strip of filter paper to draw it across the preparation from the opposite edge of the cover slip! As a result of the Lugol's solution, the starch grains turn blue-violet.

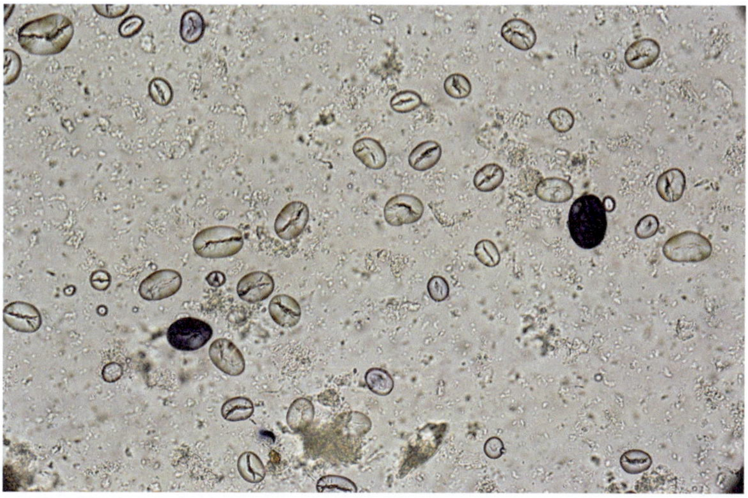

Fig. 12.21 Light microscopic image of the concentric starch grains of bean (*Phaseolus vulgaris*)

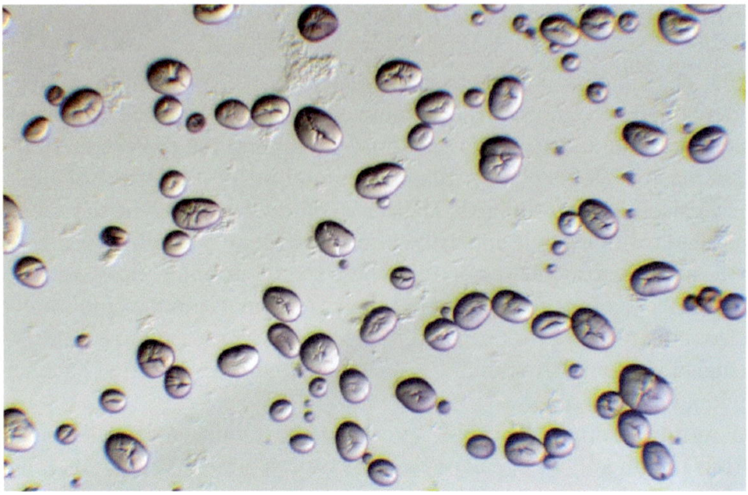

Fig. 12.22 The blue-purple stained starch grains of a bean (*Phaseolus vulgaris*) in a light microscope

During the reaction, the triiodide ions (I^{3-}) found in Lugol's solution interact with the helical structure of the starch amylose molecules, resulting in a blue-violet molecular complex (Fig. 12.22).

During staining, we should not use too concentrated Lugol's solution, because over-staining will turn the starch grains completely black, and thus their structure can no longer be studied.

During germination, the starch is converted into maltose by the action of the amylase enzyme. This process progresses from the outside of the starch grain inward, so during germination the starch grains become smaller and smaller, and the so-called corroded starch grains appear. The transformation of the starch grains is associated with another process, the transformation of colourless plastids into chloroplasts. The starch grains of the bean, which are actually formed in the colourless plastids, are capable of transforming into chloroplasts under the influence of light, so until the appearance of the leaves, the cotyledons can supply the developing plant not only with the nutrients stored inside them, but also with the organic materials produced during photosynthesis.

12.8 Transformation of Colourless Plastids in the Bean Cotyledon

Make a cut with a razor blade from the green cotyledon of the bean and examine it with a light microscope! The chloroplasts developed from the colourless plastids can be clearly seen in the ground tissue cells under the single-layered epidermis (Fig. 12.23). These contribute to the nutrition of the developing plant through their photosynthesis. The reduced size corroded starch grains can be observed on the

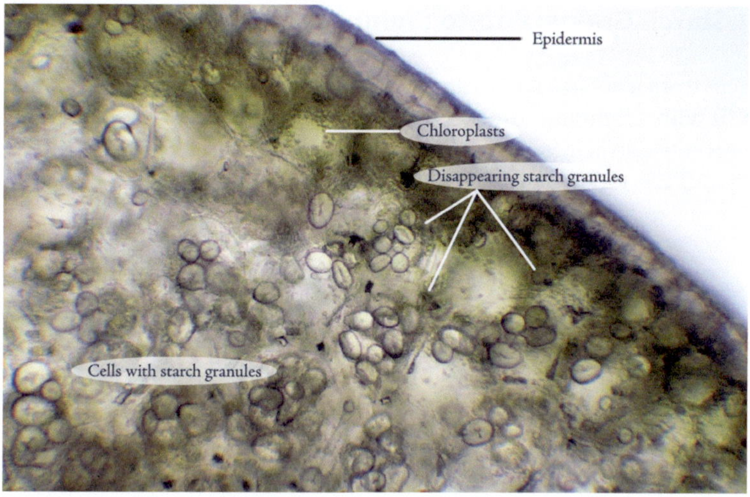

Fig. 12.23 Light microscopic section from the shrinking cotyledon of the bean (*Phaseolus vulgaris*)

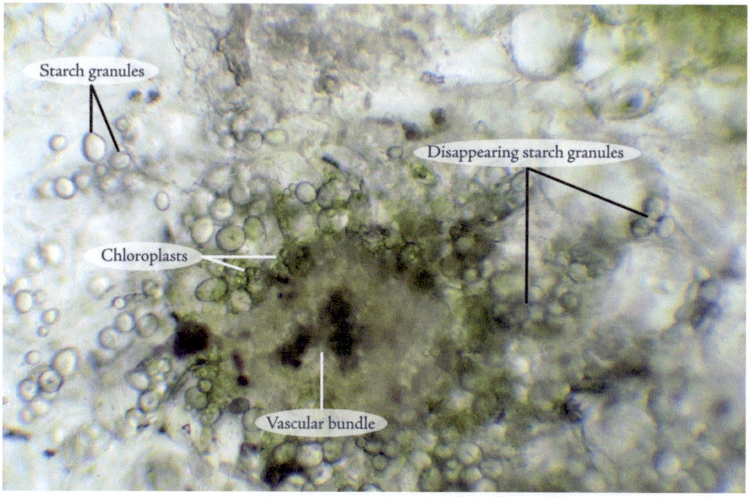

Fig. 12.24 Light microscopic section from the inside of the shrinking cotyledon of the bean (*Phaseolus vulgaris*)

section. The corrosion of the starch grains and the formation of chloroplasts can be best observed in the cells closer to the surface of the cotyledon, as well as around the vascular bundles running in the cotyledon. The sugar formed from the starch is transported by the vascular bundles to the other parts of the plant, which is why the intensive transformation can be observed in their vicinity (Fig. 12.24).

12.9 Starch Grains: Potato (*Solanum tuberosum*)

Make a cut from a potato tuber with a razor blade. Stain it with Lugol's solution and examine it with a light microscope with a cover slip! The storage parenchyma cells and the eccentric structured starch granules can be clearly observed on the preparation (Figs. 12.25, 12.26, and 12.27).

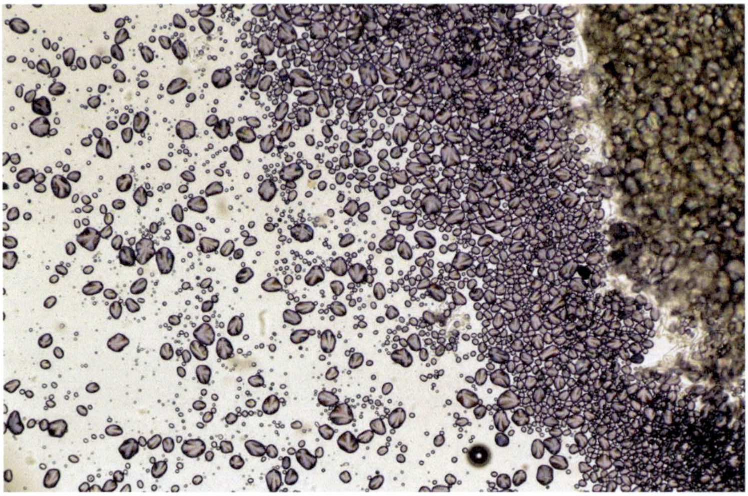

Fig. 12.25 Starch grains of the potato (*Solanum tuberosum*) stained with Lugol's solution in a low magnification image

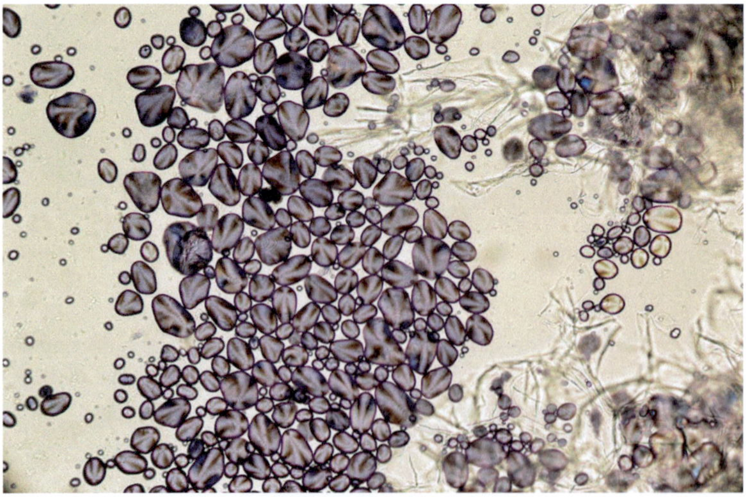

Fig. 12.26 Potato (*Solanum tuberosum*) starch granules stained with Lugol's solution

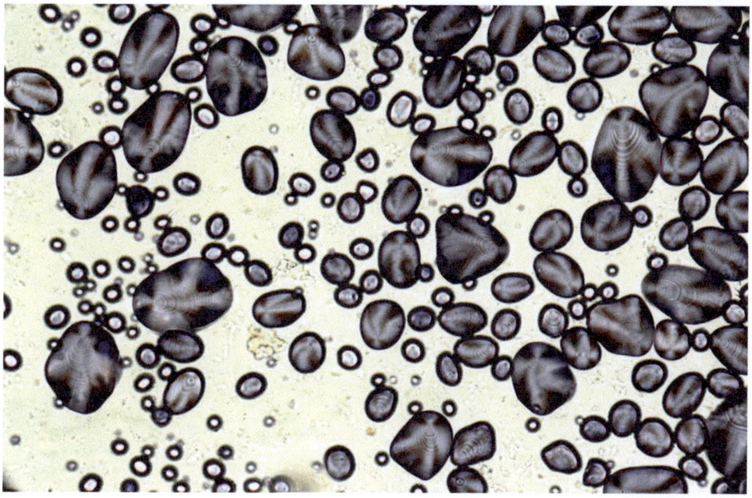

Fig. 12.27 Potato (*Solanum tuberosum*) starch granules stained with Lugol's solution in a high magnification image

Drop Lugol's solution to one edge of the cover slip and suck the solution across the preparation with a strip of filter paper from the opposite edge of the cover slip! Under the effect of Lugol's solution, the starch granules turn bluish-purple, or black in case of over-staining. During the reaction, the triiodide ions (I^{3-}) in the Lugol's solution interact with the helical structure of the starch amylose molecules, resulting in a bluish-purple molecular complex.

The stored starch in the cells of the potato tuber is found in colourless plastids, which are capable of transforming into chloroplasts under light. If the potato tuber is exposed to light, this transformation begins after a few days, resulting in the greening of the cell layers beneath the potato's periderm (Fig. 12.28).

If we make a section from the greened part, we can follow the transformation of the colourless plastids into chloroplasts and the reduction in the size of the starch granules as they approach the surface of the potato tuber with a light microscope (Figs. 12.29 and 12.30).

During staining, we should not use too concentrated Lugol's solution, because over-staining will turn the starch grains completely black, and thus their structure can no longer be studied.

During germination, the starch is converted into maltose by the action of the amylase enzyme. This process progresses from the outside of the starch grain inward, so during germination the starch grains become smaller and smaller, and the so-called corroded starch grains appear. The transformation of the starch grains is associated with another process, the transformation of colourless plastids into green plastids. The starch grains of the potato, which are actually formed in the colourless plastids, are capable of transforming into green plastids under the influence of light.

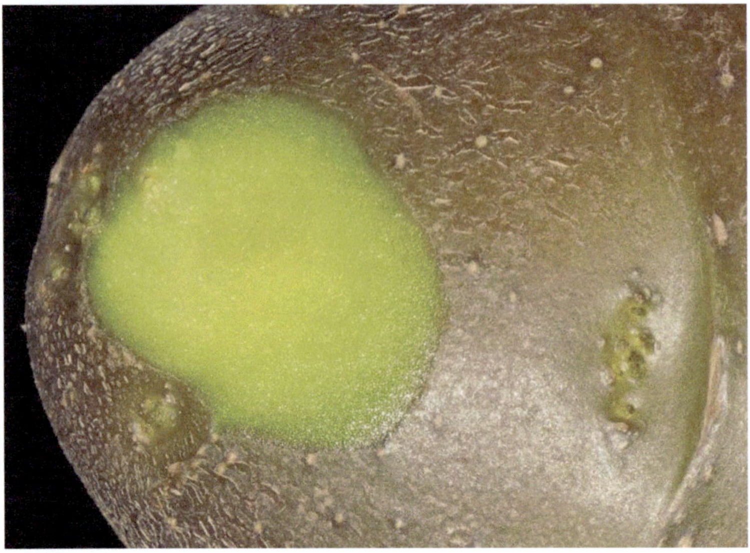

Fig. 12.28 Greening of a potato (*Solanum tuberosum*) tuber under light

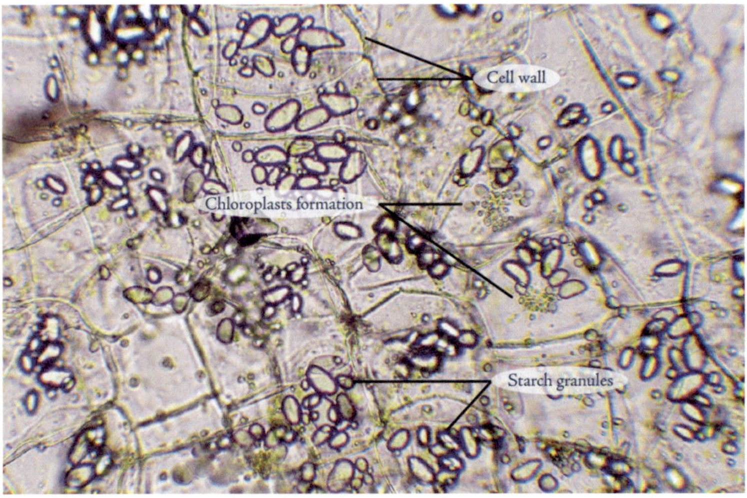

Fig. 12.29 Under the effect of light, potato (*Solanum tuberosum*) starch granules transform into chloroplasts

12.10 Starch Grains: Maize (*Zea mays*)

Prepare a scrape from the nutritive tissue of a pre-soaked maize grain, then examine the sample, spread on a slide in one drop of water with a light microscope! It can be clearly seen that the starch granules of maize are rounded and have a concentric structure similar to that of bean starch granules (Fig. 12.31). These also turn purple under the effect of Lugol's solution.

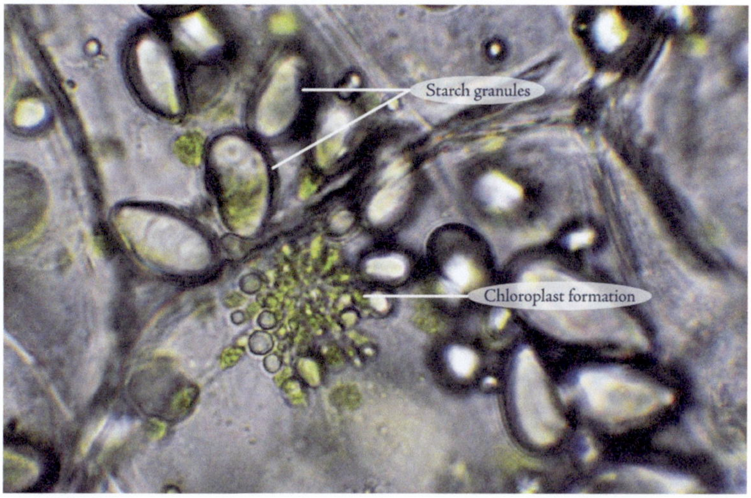

Fig. 12.30 High magnification light microscopic image of a centre of chloroplast formation

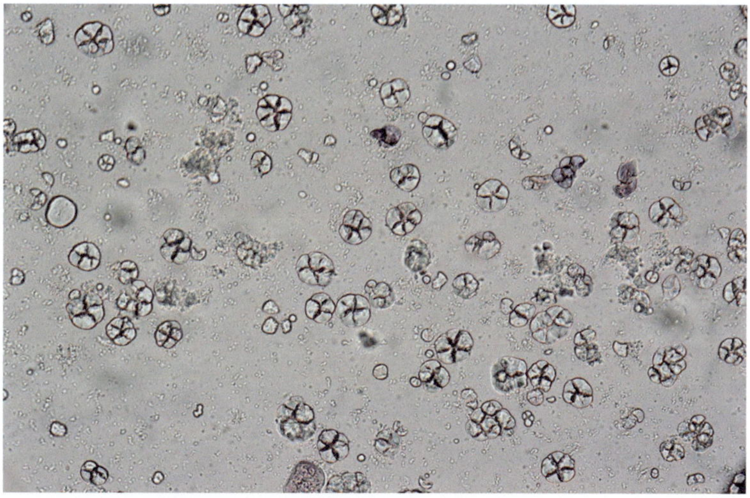

Fig. 12.31 Concentric structured starch granules of maize (*Zea mays*)

12.11 Starch Grains: Cypress Spurge (*Euphorbia cyparissias*)

The body of the cypress spurge (Fig. 12.32) is networked with milk ducts, which contain a white milky fluid. In addition to dissolved and emulsified droplets, the liquid also contains solid components, such as microscopic-sized starch granules (Figs. 12.33 and 12.34).

Fig. 12.32 Cypress spurge (*Euphorbia cyparissias*)

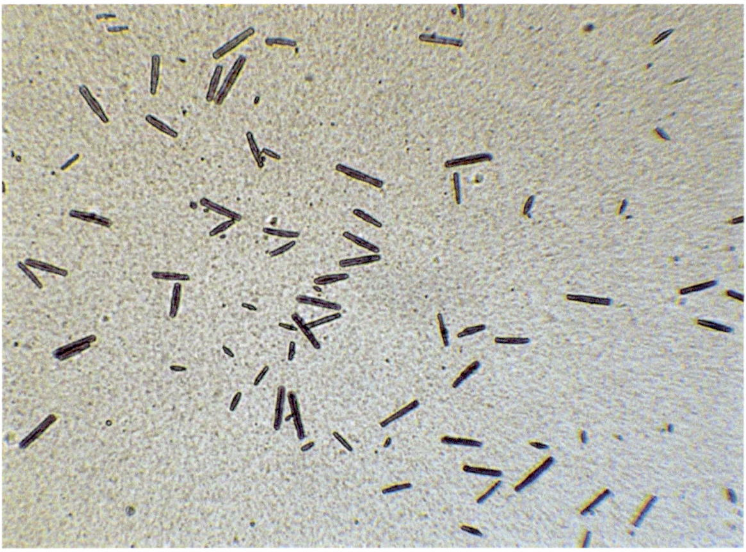

Fig. 12.33 Starch granules stained with Lugol's solution in the milky sap of the cypress spurge (*Euphorbia cyparissias*)

Cut off a shoot of a plant, then prepare a smear on a slide from the white milky sap that oozes out of the cut surface! Examine the smear with a light microscope! Add a drop of Lugol's solution to the sap, then examine the preparation under cover!

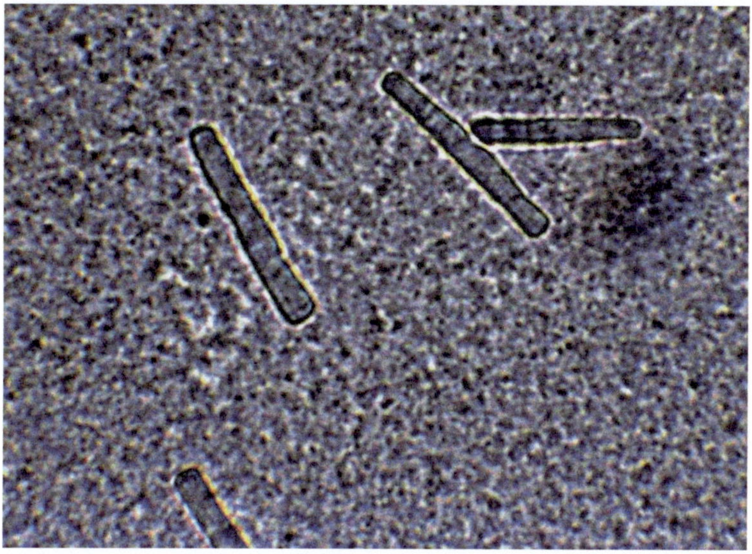

Fig. 12.34 High magnification image of the starch grains

12.12 Examination of Druses

12.12.1 Calcium Oxalate Crystals in Onion Leaf

The cells of plants can contain crystals of various materials and shapes in the cell cavity. The most common are crystals of calcium oxalate and calcium carbonate. Calcium oxalate crystals are most often needle or prism-shaped.

There are prism-shaped calcium oxalate crystals in the tunic leaf of an onion. During light microscopic examination of the dry leaf, prism-shaped crystals appear even at low magnification (Figs. 12.35 and 12.36).

The thinner, transparent parts of the leaf are most suitable for observation, where the dried cell layers interfere less with the recognition of the crystals. The dead cells are filled with air, which should be removed, that makes the crystals even more visible. For this, it is advisable to soak the dry leaf in 70% alcohol an hour before the examination. If we forget about the preliminary soaking, we can also achieve very good results with alcohol dropped onto the sample.

Place a dry onion leaf on a slide in a drop of water, then cover it with a cover slip. Drop concentrated sulfuric acid onto the edge of the cover slip, which slowly seeps under the cover slip and soaks the sample. Under the influence of concentrated sulfuric acid, oxalic acid is displaced from calcium oxalate, as a weak acid (oxalic acid) salt, by the strong acid (sulfuric acid), so oxalic acid and calcium sulfate (gypsum) are formed in the reaction. Calcium sulfate does not dissolve in water, but

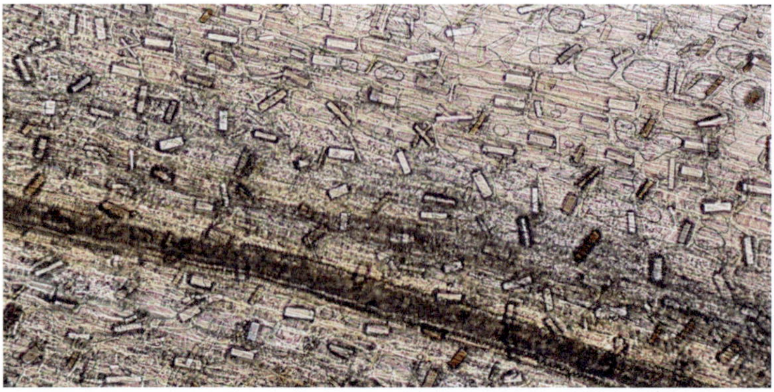

Fig. 12.35 Calcium oxalate crystals in the epidermis of an onion (*Allium cepa*)

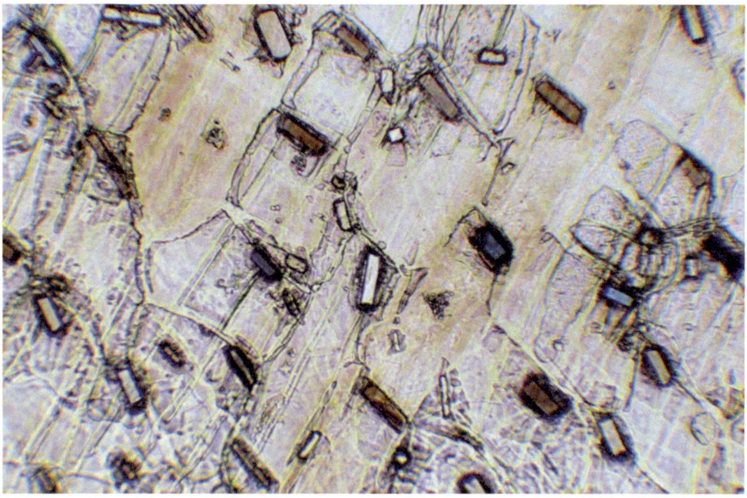

Fig. 12.36 High magnification image of calcium oxalate crystals

forms needle-shaped crystals, which cause the appearance of tiny "stars" in place of the prism-shaped oxalate crystals.

The process can be best observed in those prism crystals that have been washed out of the leaf (Figs. 12.37, 12.38, 12.39, and 12.40). The process of crystal transformation takes place gradually. First, it we can see that the surface of the columnar calcium oxalate crystals becomes "spiky" (Fig. 12.39).

Film

Transformation of calcium oxalate crystal into calcium sulphate crystal under the influence of sulphuric acid in onion leaf: https://youtu.be/yy-HegLM3T8

Fig. 12.37 Under the influence of concentrated sulfuric acid, calcium oxalate crystals transform into calcium sulfate needle crystals

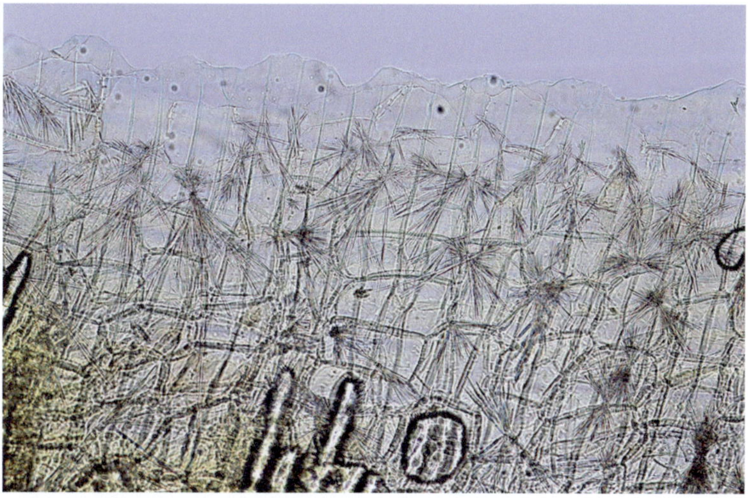

Fig. 12.38 Higher magnification image of the needle crystals

12.12.2 Calcium Carbonate Crystals

When examining the surface of a rubber tree (*Ficus elastica*) leaf under a stereomicroscope, clusters of crystals appear as white spheres in the leaf's epidermis (Fig. 12.41).

Let us make a section from a rubber tree leaf perpendicular to the side veins, and examine the preparation with a light microscope! In the two-layered skin tissue of

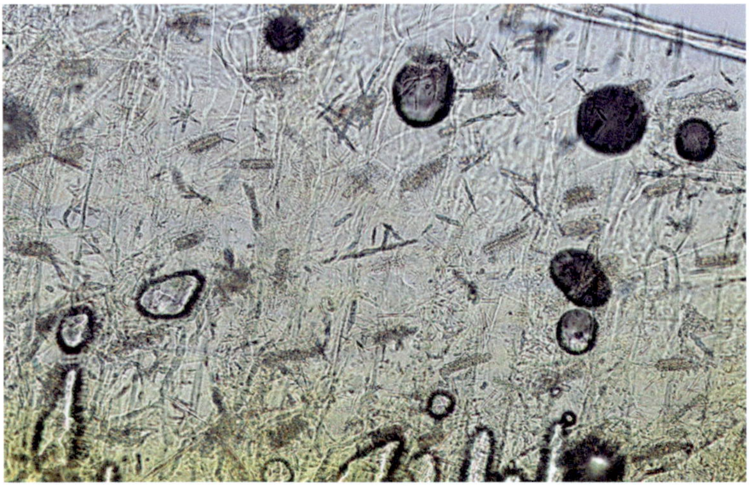

Fig. 12.39 The "spiky" calcium oxalate prism crystals already indicate the crystal transformation

Fig. 12.40 Calcium sulfate needle crystals in the onion (*Allium cepa*) leaf

the rubber tree, there are large crystal-containing cells, in which a calcium carbonate cystolith is attached to a stalk (Figs. 12.42 and 12.43). When 25% hydrochloric acid is drawn through the section, the crystal inclusion dissolves with effervescence.

Film

Reaction of rubber tree calcium carbonate cystolith with hydrochloric acid: https://youtu.be/Jv51u64ABOY

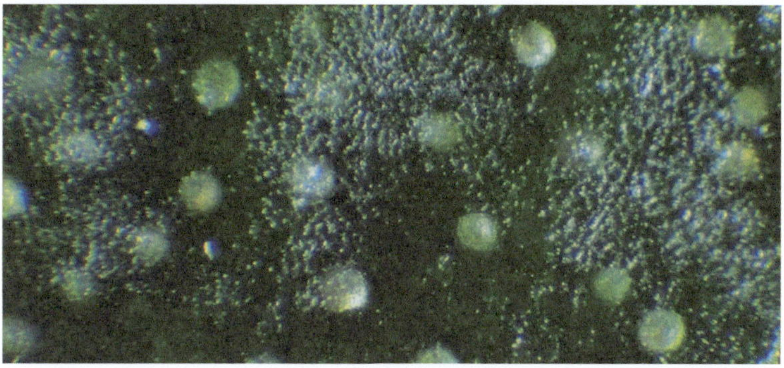

Fig. 12.41 Stereomicroscopic image of the surface of a rubber tree (*Ficus elastica*) leaf

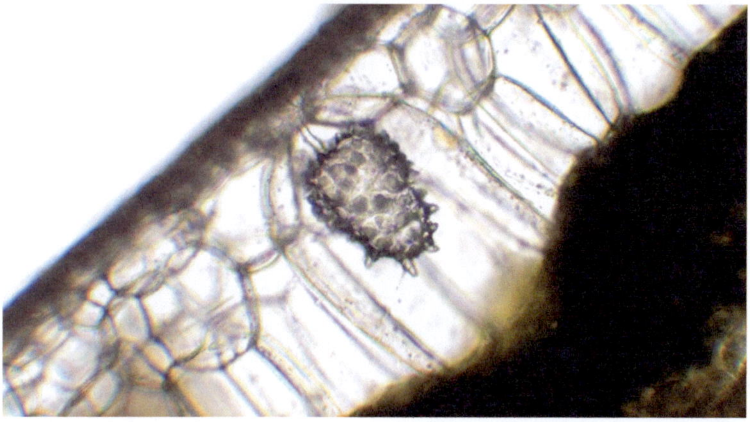

Fig. 12.42 Cystolith in the rubber tree (*Ficus elastica*) epidermis

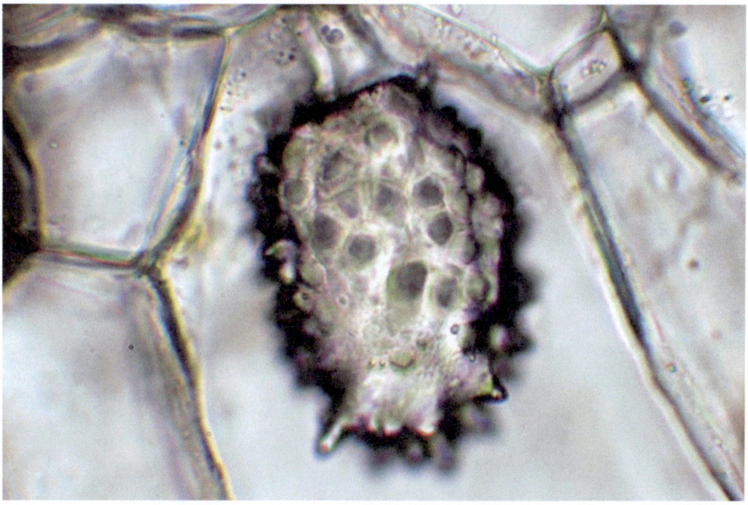

Fig. 12.43 High magnification image of the cystolith

12.13 Epidermis Examinations

12.13.1 Stoma of the Great Stonecrop (*Hylotelephium maximum*)

The *H. maximum* is characterised by three subsidiary cells surrounding the guard cells in the epidermis (anisocytic stomata). The *H. maximum* is a very good subject for examination, as we can easily make a peel from its leaf. With a light microscope, the three subsidiary cells located around the guard cells can be well observed, their identification is helped by the numbers on the figure (Figs. 12.44 and 12.45).

The *Hylotelephium maximum* tolerates drought well, and its mucous parenchyma cells store a lot of water in its leaf. Their drought tolerance is also aided by the CAM (Crassulacean Acid Metabolism), during which they are able to bind atmospheric carbon dioxide taken up through the stomata at night in the form of malic acid.

During the day, with closed stomata, carbon dioxide is released from the malic acid and is used in photosynthesis.

This metabolic pathway is useful for plants living in precipitation-poor conditions, as they can keep their stomata closed during the day, reducing their evaporative water loss. The CAM in desert succulent plants is quite common, especially in cacti and euphorbias. The metabolic process was originally studied in plants belonging to the Crassulaceae family.

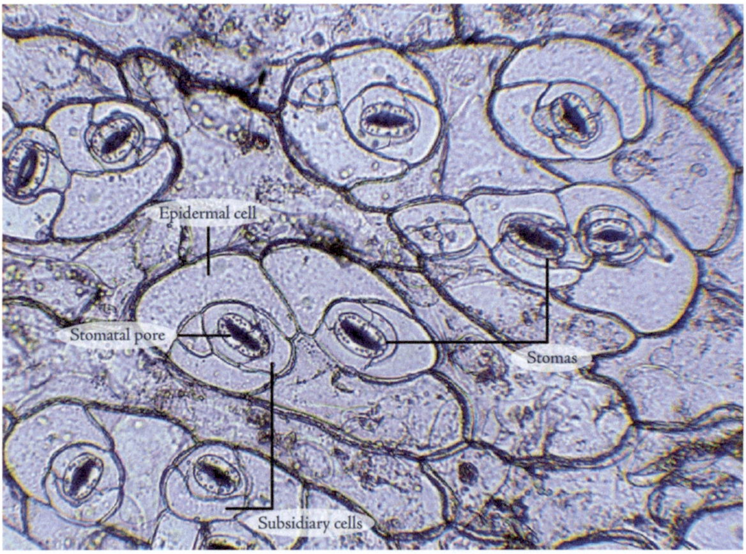

Fig. 12.44 Great stonecrop (*Hylotelephium maximum*) epidermis peel

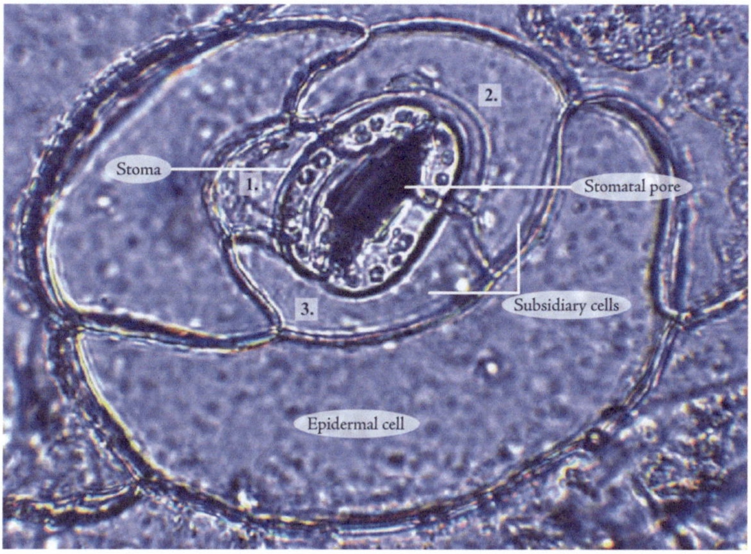

Fig. 12.45 High magnification image of the great stonecrop (*Hylotelephium maximum*) gas exchange opening

12.13.2 Epidermal and Glandular Hairs: Geranium (*Pelargonium zonale*)

Make a cut with a razor blade from the soft stem or petiole of a geranium! After covering, study the epidermal and glandular hairs with a light microscope! On the section, three different epidermal formations of the geranium can be seen: the pointed, multicellular epidermal hair, the long-stalked glandular hair of roughly the same length, and the much smaller short-stalked glandular hair (Figs. 12.46 and 12.47).

The essential oil secretion of the glandular hairs accumulates under the cuticle of the gland's unicellular head. The essential oil released upon touch gives the geranium a strong scent.

12.13.3 Epidermal Hairs: Orange Mullein (*Verbascum phlomoides*)

The orange mullein is covered with branched hairs, which give the plant a silvery appearance (Fig. 12.48).

Scrape the surface of the leaf and examine it with a light microscope, either dry or dispersed in water. The branched hairs develop from a single cell, which forms the multi-level spirally branching formation through serial divisions (Fig. 12.49). After its formation, the epidermal hair dies and becomes silvery due to the air inside it.

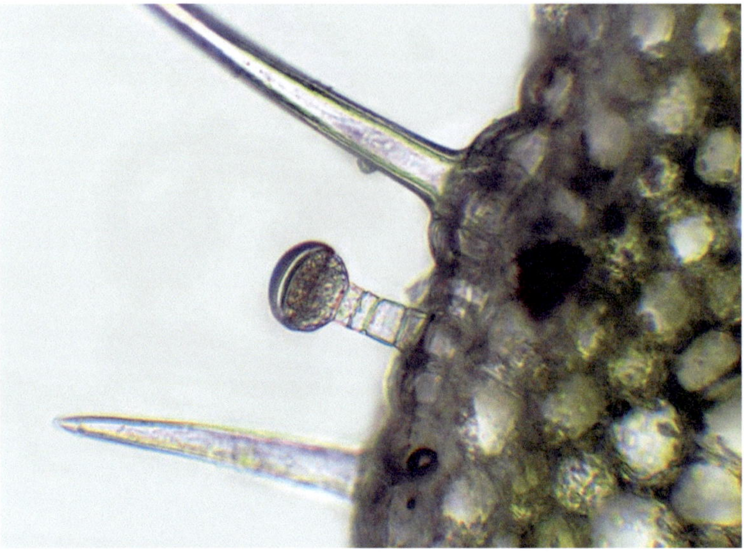

Fig. 12.46 Geranium (*Pelargonium zonale*) short-stalked glandular hair and two epidermal hairs

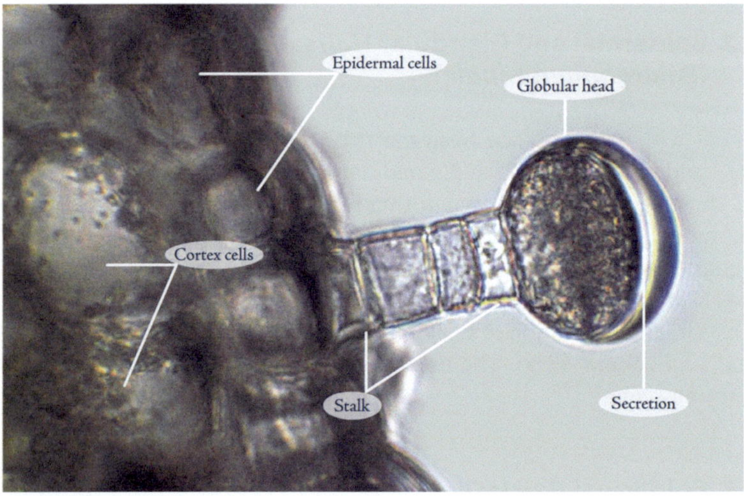

Fig. 12.47 Geranium (*Pelargonium zonale*) short-stalked glandular hair

Thanks to the hairs, the strongly reflective hair layer protects the leaf from overheating and reduces transpiratory water loss. Similar branched hairs can be found on the veins of plane tree leaves.

Fig. 12.48 Orange mullein (*Verbascum phlomoides*) inflorescence covered with branched hairs

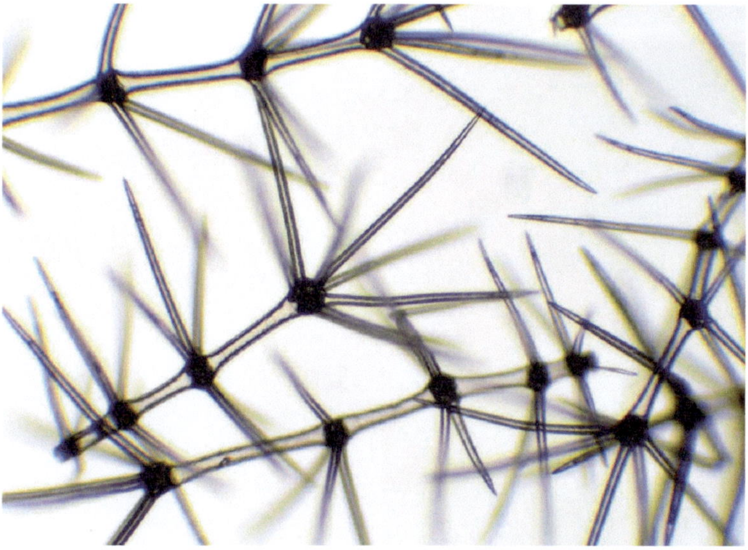

Fig. 12.49 Branched hairs of the orange mullein (*Verbascum phlomoides*)

12.13.4 Epidermal Hairs: Silver Berry (*Elaeagnus angustifolia*)

The shoots of the silver berry (Fig. 12.50) are covered with special epidermal hairs, the star and scale hairs, which give the plant a silvery appearance when they die (Figs. 12.51 and 12.52).

Fig. 12.50 Flowering silver berry (*Elaeagnus angustifolia*)

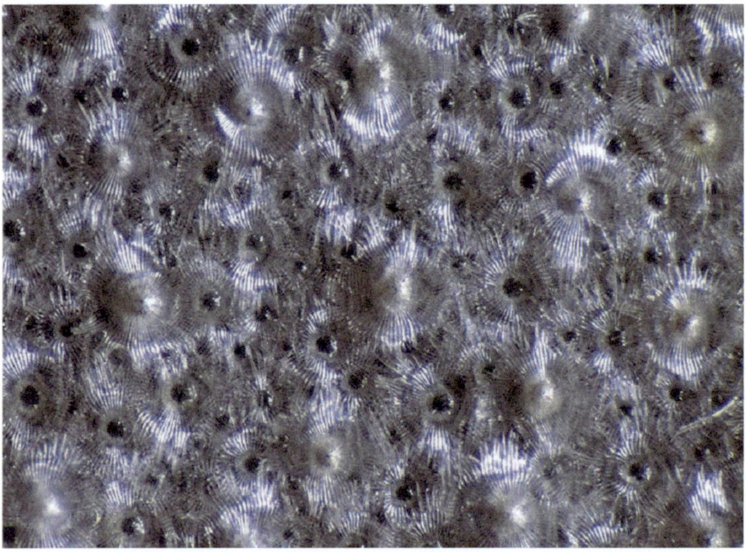

Fig. 12.51 The underside of the silver berry (*Elaeagnus angustifolia*) leaf is covered in multiple layers of scales

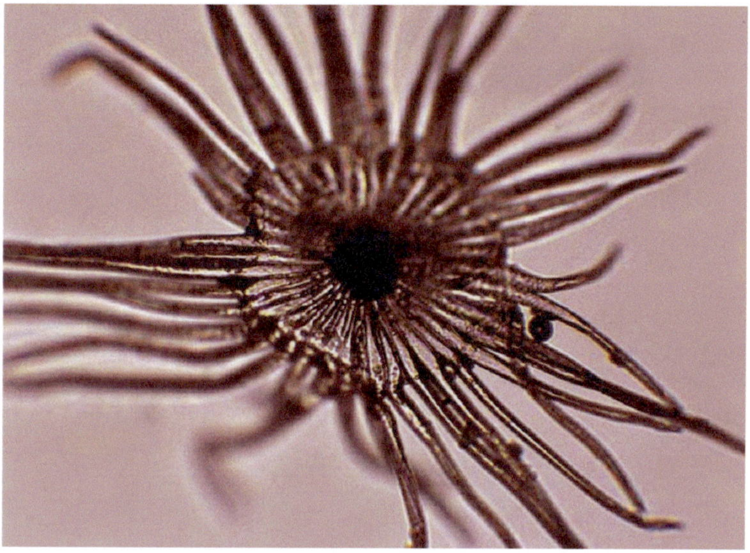

Fig. 12.52 Silver berry (*Elaeagnus angustifolia*) scale

The star hair consists of one or more basal cells and the radially connected, horizontally positioned ray cells. In scales, many ray cells are connected to form a shield-like structure.

The leaf blade of the silver tree is covered by scale in several layers, which can be studied with a light microscope under top lighting. For microscopic examination, a scrape can also be made from the leaf surface.

12.13.5 Common Nettle (*Urtica dioica*) Hair

The common nettle is a widespread weed and medicinal plant throughout Europe (Fig. 12.53). It is common in fresh, damp forests, weed communities, marshes, and along stream banks.

The body of this perennial plant, which can reach a height of one and a half meters, is covered with hairs and nettle hairs (Figs. 12.54, 12.55, and 12.56). The latter cause a painful, burning, itching sensation and the appearance of small rashes upon contact.

The single-celled nettle hair extends from an epidermal cushion. Its narrowing end breaks like glass due to its silicon dioxide content. Upon contact, the broken-tipped nettle hair injects formic acid and histamine into the skin, causing irritation.

Fig. 12.53 Common nettle (*Urtica dioica*) population

Fig. 12.54 Nettle hairs and smaller hairs on the underside of the common nettle (*Urtica dioica*) leaf

12.14 Germinating of Daffodil (*Narcissus* sp.) Pollen

During the pollination of the daffodil, the pollen grain reaches the stigma. The adhesion of the pollen grain is facilitated by microscopic-sized spikes and other protrusions (papillae) that form on its surface, as well as the sticky surface of the stigma

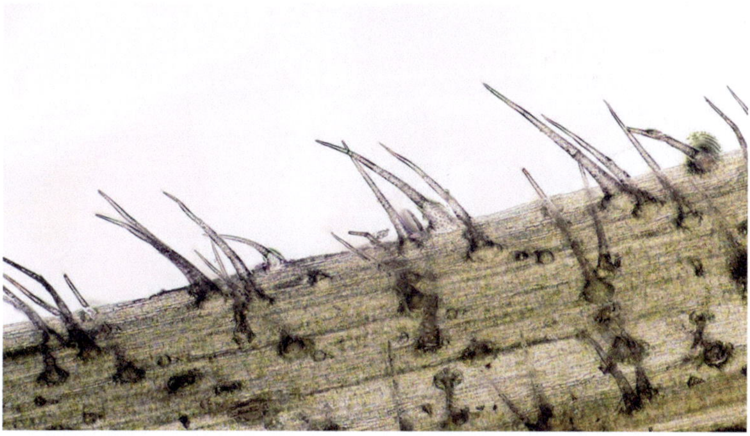

Fig. 12.55 The common nettle's (*Urtica dioica*) hairs on the stem

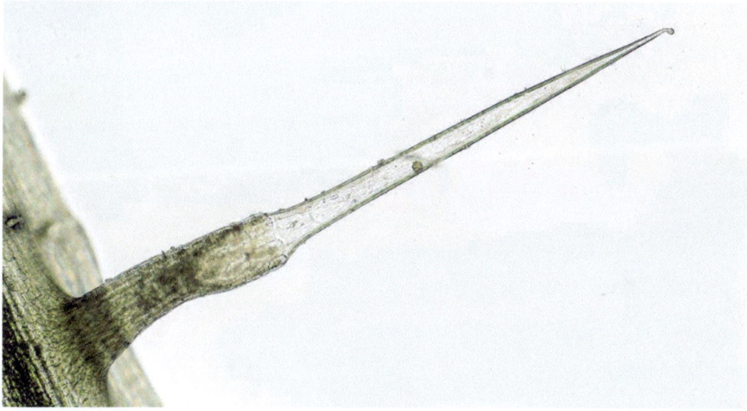

Fig. 12.56 Nettle hair of common nettle (*Urtica dioica*)

(Figs. 12.57 and 12.58). Cut off the flower's stigma and gently crush it on a slide with a cover slip, then examine it under a light microscope!

The first step in fertilisation is the division of the cell found in the pollen into a vegetative and a generative cell. The vegetative cell drives a pollen tube towards the developing ovule in the ovary, where fertilisation takes place with the participation of sperms that form in the pollen tube. These latter are created by the division of the generative cell.

We can also artificially induce the development of the pollen tube. Official recipes for pollen tube germinating specify a possible sucrose concentration from 1 to 20 weight percent and generally recommend the use of dilution series to determine the effective sugar concentration. In addition, they provide the ideal mineral salt composition of the forcing solution. The literature recommends the following for setting this up: This solution should be prepared fresh, just before use.

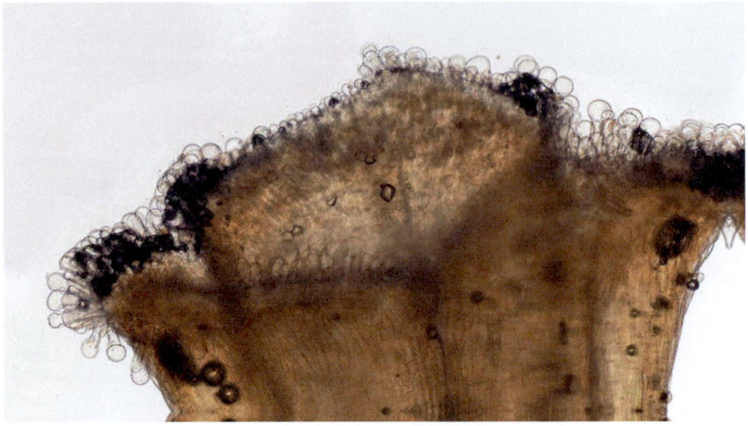

Fig. 12.57 Daffodil (*Narcissus* sp.) stigma under a light microscope. The surface-enhancing papillae that help the pollen adhere can be clearly observed on the surface of the stigma

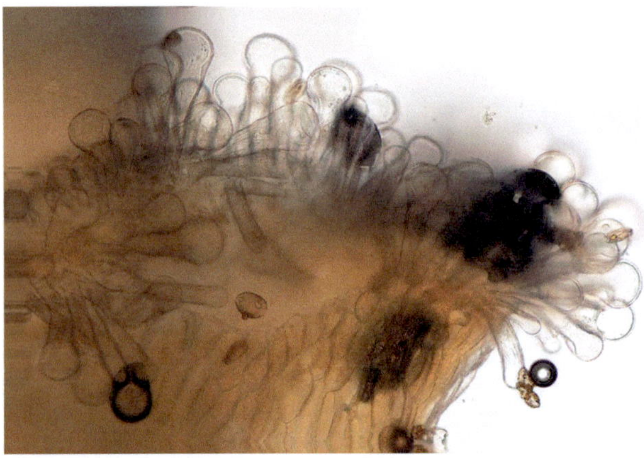

Fig. 12.58 A high magnification image of the daffodil (*Narcissus* sp.) stigma

To prepare 1 L, dissolve 0.417 g of calcium nitrate, 0.200 g of boric acid, 0.101 g of potassium nitrate, 0.217 g of magnesium sulphate, and 3.5 cm^3 of 1.0 M ammonium hydroxide solution in water. When mixed with the sucrose solution, the pH of the mineral salt solution should be 8.8.

Instead of the above complicated method, we propose a simpler one that requires only two substances.

Mix a level teaspoon of sucrose (sugar) and a heaped spoonful of gelatine in 1 dl of tap water.

The gelatine will not completely dissolve in the solution, but our solution will still be sufficiently viscous for a droplet test.

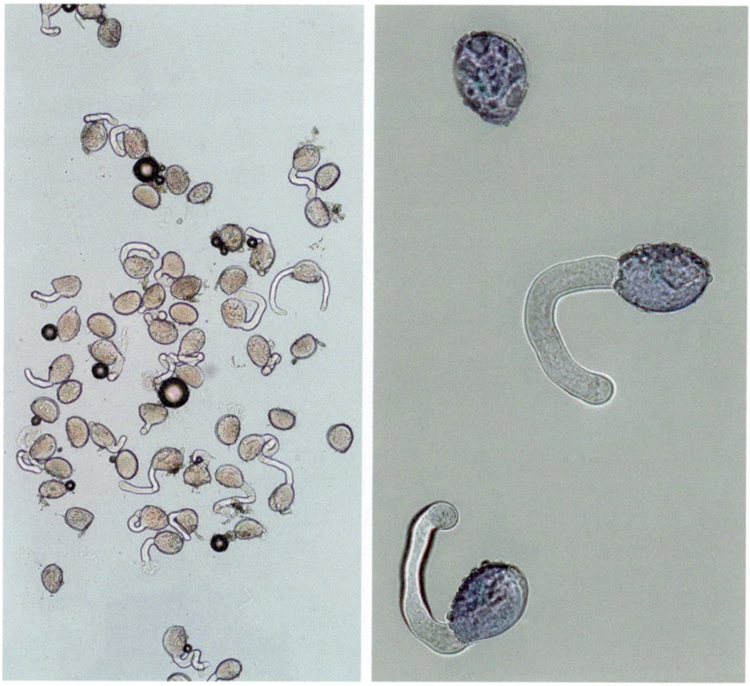

Fig. 12.59 Examination of daffodil (*Narcissus* sp.) pollen tube formation under a light microscope

After stirring for 1–2 min, place a drop of the sugar gelatine solution on a cover slip, then sprinkle it with pollen. Then place the inverted cover slip on a slide on a frame made of wet cotton wool or paper handkerchief strips, which prevents the hanging drop from touching the slide. It is even simpler if we make the support for the cover slip from strips cut out of a plastic sponge. It is sufficient to support the cover slip on two sides with the wet sponge, as this can also provide the humid environment necessary for the experiment, thanks to which our hanging drop will not dry out. Under a microscope, we can continuously observe the tube growth (Figure 12.59). Once several pollen tubes have developed, we can further examine the preparation covered with a cover slip dyed with methyl green.

It is worth making a microscopic film recording of the formation of the pollen tube. If we speed this up tenfold, this peculiar phenomenon becomes more perceptible.

Film

Formation of narcissus pollen tubes at ten times speed, light microscopic video recording: https://youtu.be/iw9cVT9XoHg

Biochemical Tests 13

13.1 Carbohydrates

The majority of the organic matter in the biosphere is made up of **carbohydrates**. These include the primary energy source of cells, the **glucose**, the stored nutrient **starch**, the **glycogen,** and the structural material of plant cells and organisms, the **cellulose**. Carbohydrate molecules can participate in the formation of important macromolecules in life processes by linking with other compounds.

The name carbohydrates originates from their general formula [$Cn(H_2O)m$]. Based on their functional groups, they were previously thought to be compounds of carbon and water. According to their functional groups, they can be **aldoses** (containing the functional group of aldehydes, formyl group) and **ketoses** (containing keto group). Based on their size, we can talk about **simple carbohydrates** (which cannot be further broken down by acidic hydrolysis, **monosaccharides**) and **complex carbohydrates** (which can be broken down into monosaccharides by acidic hydrolysis, e.g. starch and cellulose). Among the complex carbohydrates, disaccharides consist of two monosaccharides (e.g. sucrose), a few of the **oligosaccharides** and many hundreds or thousands of simple carbohydrates make up the **polysaccharides**. The simplest monosaccharides are the 3-carbon **trioses** (e.g. glyceraldehyde and dihydroxyacetone), which are important metabolic intermediates in living cells. Among the 5-carbon monosaccharides (pentoses), the **ribose** and **deoxyribose** are biologically very important, which are components of the genetic material, the **RNA** and **DNA**. Their molecules can be open-chain and cyclic.

The most common and the monosaccharide found in the largest quantities in living organisms is the 6-carbon (hexose) glucose. In its cyclic state, it has two configurational isomers (α-D-glucose and β-D-glucose), which can transform into each other through the open form, establishing equilibrium. In this way, in the aqueous solution of D-glucose, three isomers in dynamic equilibrium with each other can be found along with the open-chain form. α-D-glucose and β-D-glucose differ in the position of the glucosidic hydroxyl group formed from the formyl group (and

G. Kriska, *Teaching Biology*, https://doi.org/10.1007/978-3-031-84085-2_13

capable of reverting to it): in α-D-glucose it is axial, and in β-D-glucose it is equatorial (Fig. 13.1).

The sweetest monosaccharide, also 6-carbon **fructose**, which is mainly found in fruits and honey, contains not an aldehyde group but a keto group in its open-chain form (Fig. 13.2). Fructose is capable of isomerising to glucose in an alkaline medium, a process that takes place in living cells with the help of enzymes.

Disaccharides are formed by the condensation of two monosaccharides (with the elimination of water) are formed by one molecule's glycosidic –OH group's H-atom being replaced by another monosaccharide. The ether bond formed between the two monosaccharides by water elimination is called a **glycosidic bond**.

If both monosaccharides participate in the formation of the bond with the glycosidic –OH, their rings cannot open in aqueous solution, thus their reducing formyl group does not appear (non-reducing disaccharides: e.g. sucrose). In the case of reducing disaccharides (e.g. cellobiose, maltose, and lactose), the molecules contain glycosidic –OH, the ring can open, resulting in the appearance of an aldehyde group. The **cellobiose** is composed of two β-D-glucose molecules in such a way that the –OH- (glycosidic –OH) group attached to the 1st C-atom of one ring is connected to the –OH group attached to the 4th carbon atom of the other ring during water elimination. Cellobiose can be considered a building unit of cellulose. The **maltose** or malt sugar is composed of two α-D-glucose molecules with a 1 → 4 bond. Maltose is a building unit of starch (Fig. 13.3).

The well-known **sucrose** or beet sugar or cane sugar is formed from one α-D-glucose and one β-D-fructose with a 1–2 bond. Lactose or milk sugar is formed from beta-D-galactose and alpha-D-glucose with a 1–4 bond (Fig. 13.4).

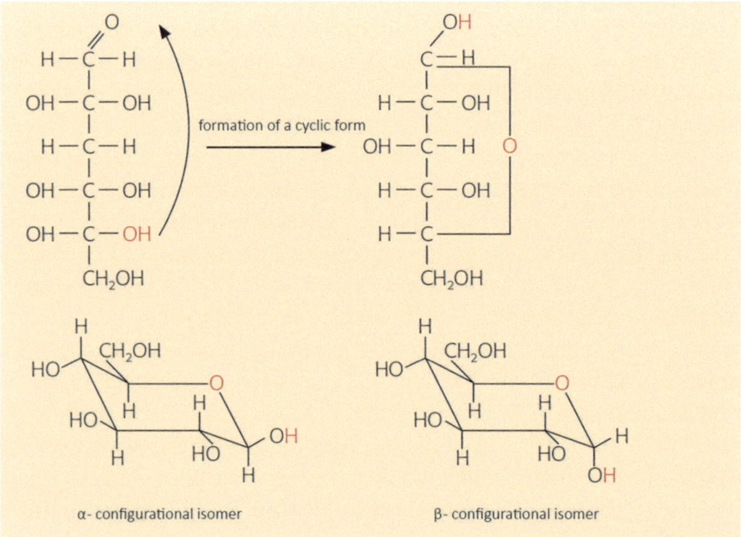

Fig. 13.1 D-glucose has two configurational isomers in the cyclic state (α-D-glucose and β-D-glucose), which can transform into each other through the open form into each other

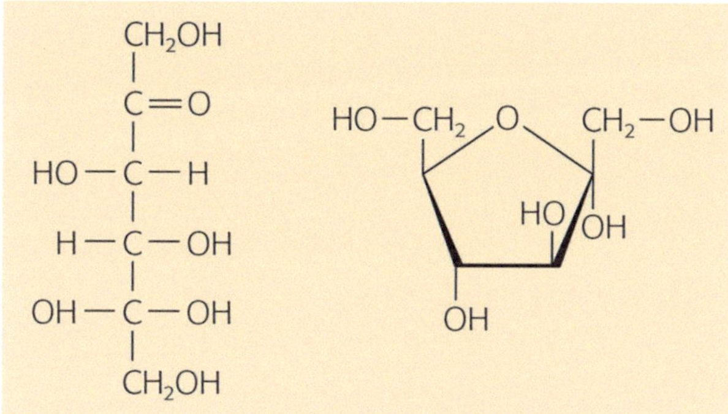

Fig. 13.2 The structural formula of D-fructose and α-D-fructose

Fig. 13.3 The structural formula of maltose and cellobiose

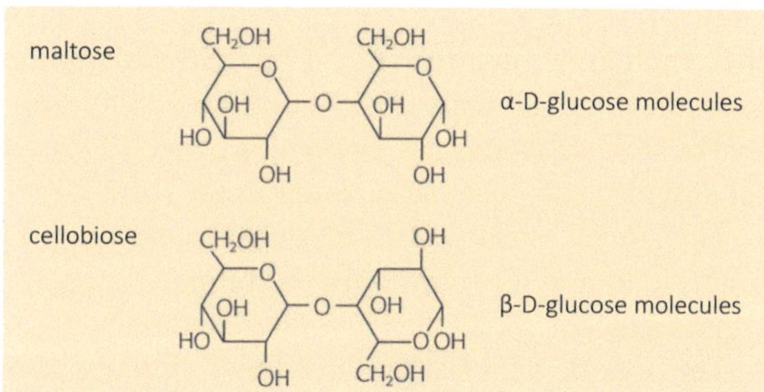

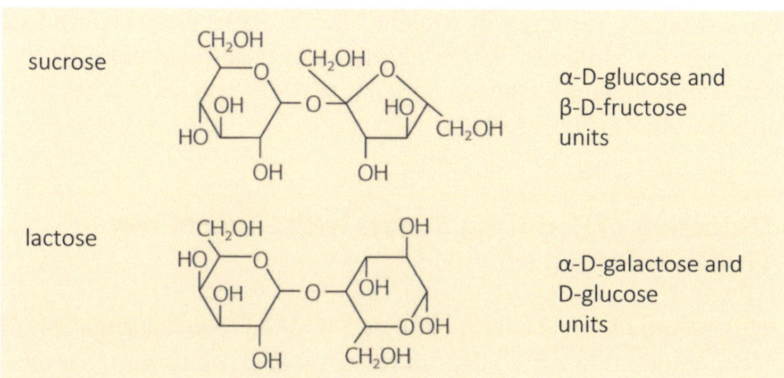

Fig. 13.4 The structural formula of sucrose and lactose

The most common carbohydrates are the **polysaccharides**, in which several hundred or several thousand monosaccharides are connected by glycosidic ether bonds.

In starch molecules, several hundred alpha-D-glucose are connected by 1–4 bonds. Since the hydroxyl group in the α-D-glucose units is axial (along the axis), the starch chain has a helical structure. This "coiling" ensures that the starch can fit in a relatively small space, making it an ideal, space-saving reserve nutrient in terms

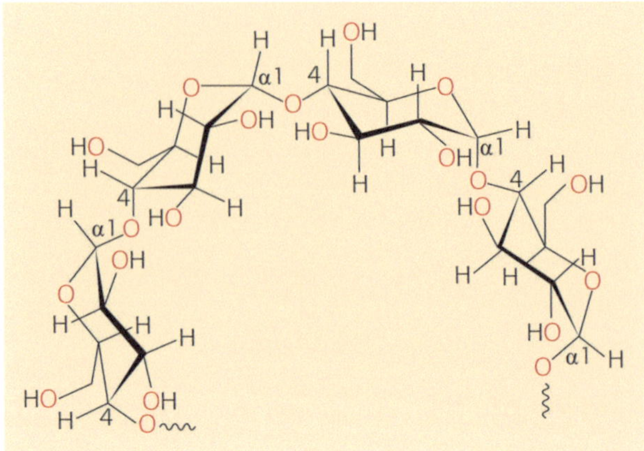

Fig. 13.5 Structural formula of a starch molecule fragment

of its biological function (Fig. 13.5). Starch granules consist of two types of starch molecules, the helically coiled, non-branching **amylose** and the branching **amylopectin** with an average of 1–6 bonds every 12 glucose units. Amylopectin does not, but amylose can dissolve in water in the form of colloid particles.

The stored carbohydrate of animals and fungi is glycogen, which has a similar structure to amylopectin, only the branches appear more frequently, on average in every 6 glucose molecules in its molecules.

The organic material produced by living organisms, biomass, is about half made up of cellulose, whose molecules are built up from many thousands of beta-D-glucose units. In cellulose, the glucose units are connected to each other at a 180° angle with 1–4 bonds, resulting in straight, fibrous molecules. The molecules are linked by numerous H-bonds, which gives a very stable structure to the water-insoluble material, making cellulose an ideal scaffold polysaccharide (long fibres can be formed from it) (Fig. 13.6).

13.1.1 Detection of Reducing Sugars with Silver Mirror (Tollen's) Test

In a clean, degreased test tube, add enough 2 mol/dm^3 concentration of ammonia solution (NH_3 solution) to a 5% silver nitrate ($AgNO_3$) solution so that the initially precipitated yellowish-brown silver oxide (Ag_2O) just dissolves as a silver-diamine complex ($[Ag(NH_3)_2]OH$)!

$$AgNO_3 + 2NH_4^+ + 2OH- = Ag_2O + 2NH_4NO_3 + H_2OAg_2O + 2NH_4^+ + 2OH^- = 2\left[Ag(NH_3)_2\right]OH + H_2O$$

To the prepared reagent, add a pinch of glucose! Heat the test tube in a water bath at 80–90 °C! A shiny silver coating forms on the wall of the test tube (Fig. 13.7).

Explanation: In the process, the formyl group of glucose

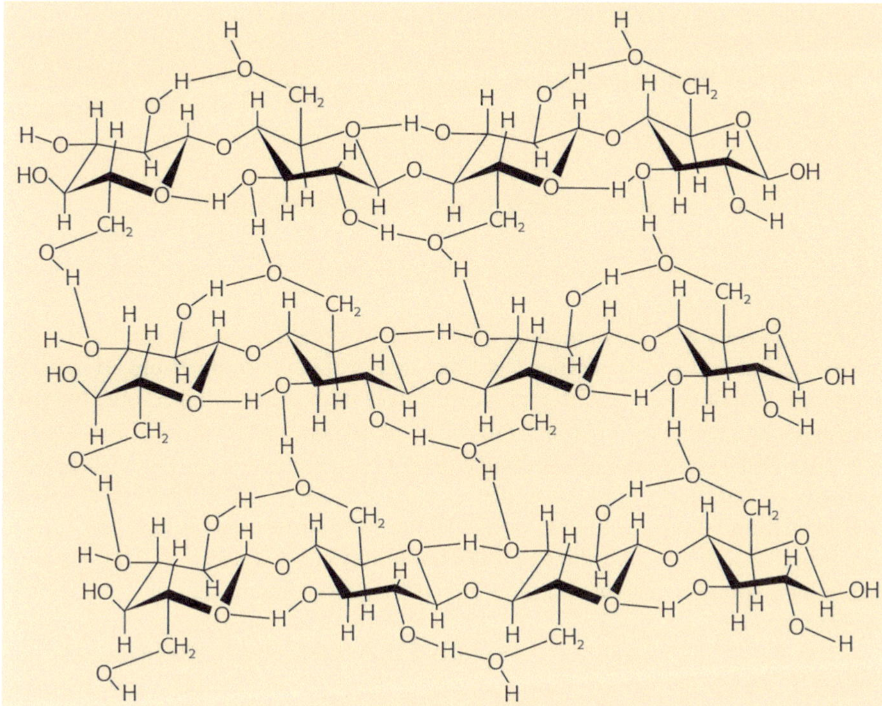

Fig. 13.6 Structural formula of cellulose fragments connected by H-bonds

Fig. 13.7 Silver mirror test in test tubes placed in a water bath

is oxidised to a carboxyl group, while reducing the silver ions to metallic silver.

$$R—CHO + 2\left[Ag(NH_3)_2\right]OH \rightarrow 2\underset{\blacklozenge}{Ag} + R—COOH + 4NH_3 + H_2O$$

13.1.2 Detection of Reducing Sugars with Fehling's Test

Fill a test tube one-third full with finely chopped apple, carrot, or onion, then add 3–4 cm^3 of water (Fig. 13.8)!

Then carefully boil the contents of the test tube on a Bunsen burner! Pour 1 cm^3 of Fehling-I solution into another test tube, and drip in enough Fehling-II solution so that the initially excellent precipitate dissolves with a deep blue colour! Into this thoroughly shaken solution pour 1 cm^3 of plant extract, then carefully heat the test tube on a Bunsen burner! The initially deep blue solution turns greenish-yellow, then yellow, and finally a brick-red precipitate forms (Figs. 13.9, 13.10, and 13.11).

Explanation: The light blue Fehling-I solution is an aqueous solution of $CuSO_4$ [copper (II)-sulphate], and the Fehling-II solution is an aqueous $NaOH^-$ (sodium hydroxide) solution also containing the potassium-sodium salt of tartaric acid (K–Na-tartrate).

When the Fehling solutions are mixed, a light blue copper (II)-hydroxide precipitate [$Cu^{2+} + 2OH^- \rightarrow Cu(OH)_2$] forms first, then the salt of tartaric acid dissolves the precipitate by forming a complex with the Cu^{2+}-ions, resulting in a deep blue solution.

The aldehyde group of glucose reduces the Cu^{2+}-ions to Cu^+-ions, while it itself is oxidised to a carboxyl group:

Fig. 13.8 Finely chopped carrot

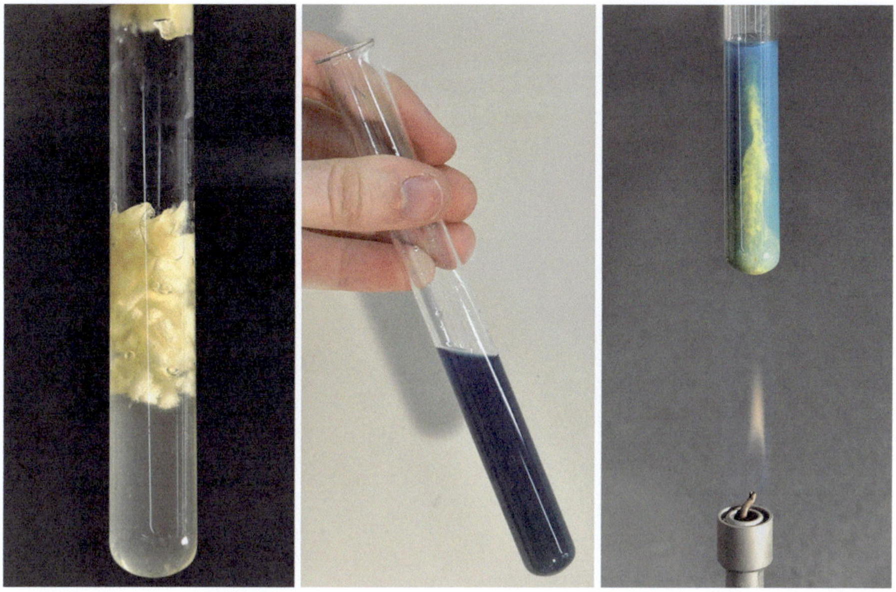

Fig. 13.9 The first stage of the Fehling test of the solution obtained by boiling apple pieces

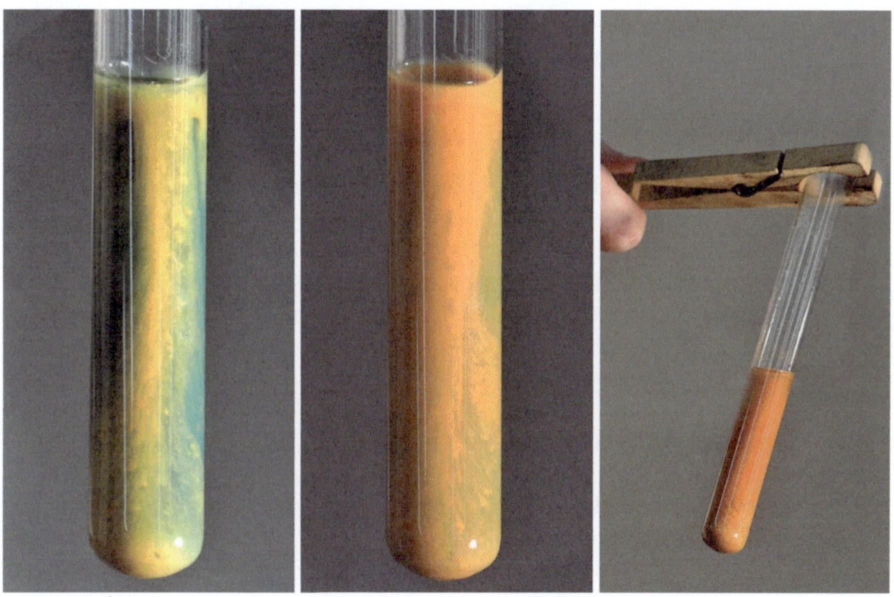

Fig. 13.10 The second stage of the Fehling test

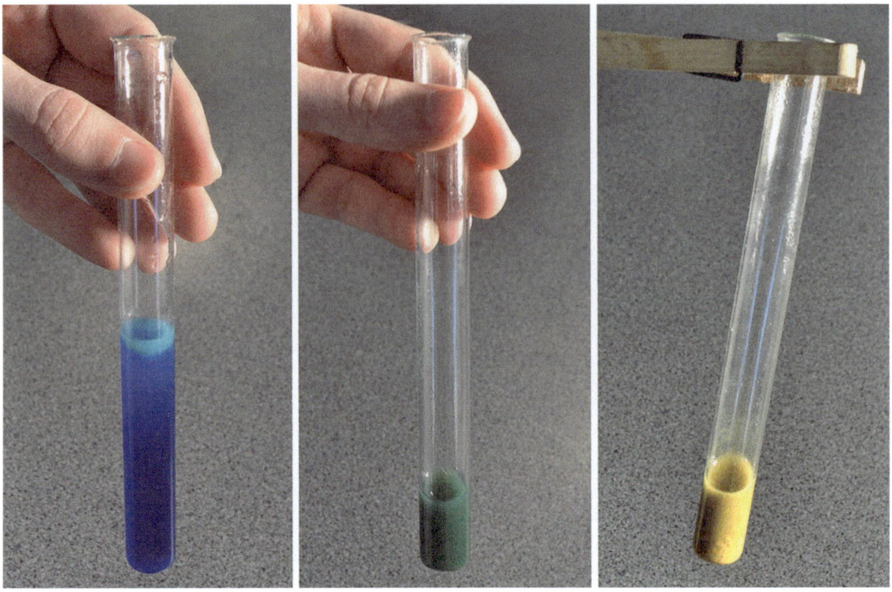

Fig. 13.11 The transitional colour changes of the Fehling test

$$2Cu^{2+} + 4OH^{-} + R—CHO \rightarrow 2CuOH + R—COOH + H_2O$$

As a result of the reaction, a yellow copper(I)-hydroxide (CuOH) precipitate forms, which turns into brick-red copper(I)-oxide, cuprous oxide (Cu_2O) upon heating:

$$2CuOH \rightarrow Cu_2O + H_2O$$

When heating the test tube, the deep blue solution first turns greenish due to the mixing of the original blue solution and the forming yellow precipitate, then the yellow becomes dominant, and finally the brick-red colour.

If we perform the experiment with pure glucose, the Cu^{+}-ions often get further reduced, and metallic copper (Cu) precipitates on the wall of the test tube (in the form of a copper mirror).

13.1.3 Detection of Fructose from Pear

Pears contain a particularly large amount of fructose, therefore the decoction obtained from the fruit pulp is suitable for the sulphuric acid detection of fructose. Prepare a pear decoction in the manner described in the previous experiment, then pour 2–3 cm^3 of it into a test tube! Then slowly pour 1–2 cm^3 of concentrated sulphuric acid under the plant extract! The density of the sulphuric acid is greater than that of the extract, so with careful pouring, the sulphuric acid collects under the fruit juice. At the contact surface of the two liquid phases, in the presence of fructose, a

Fig. 13.12 Appearance of a brown ring at the phase boundary of pear juice and sulphuric acid

brown ring forms due to the carbonising effect of the sulphuric acid (Fig. 13.12). Glucose and generally aldoses do not, or only weakly produce the brown ring. Protective gloves and safety glasses must be worn when performing the experiment!

13.1.4 Detection of Starch with Iodine Test

Prepare a thin starch solution by heating water-soluble starch powder! First, put a pinch of starch into a test tube of water, then heat the solution to boiling on a Bunsen burner! Cool the test tube under a tap, and pour 2–3 cm^3 of the clear solution into another test tube! Dilute the water-soluble part of the starch, the amylose-containing solution, by 2–3 times! Drop 1–2 drops of potassium iodide iodine solution (Lugol's solution), which turns blue as a result (Fig. 13.13).

Shake the solution, then continue to dilute it with water until its dark blue colour changes to light blue! Then drop Lugol's solution into the obtained solution! If the dark blue discolouration appears again, then the solution needs to be diluted further. We have a properly stained starch solution for later examination if the dark blue discolouration does not appear after the addition of another drop of Lugol's solution. Heat the obtained solution slightly on a Bunsen burner until the blue colour disappears! Then cool the solution under a tap until the blue colour reappears (Fig. 13.14)! The test can be performed several times in succession.

Explanation: The non-polar iodine molecules dissolve only to a very small extent in water, so Lugol's solution contains potassium iodide (KI) in addition to iodine

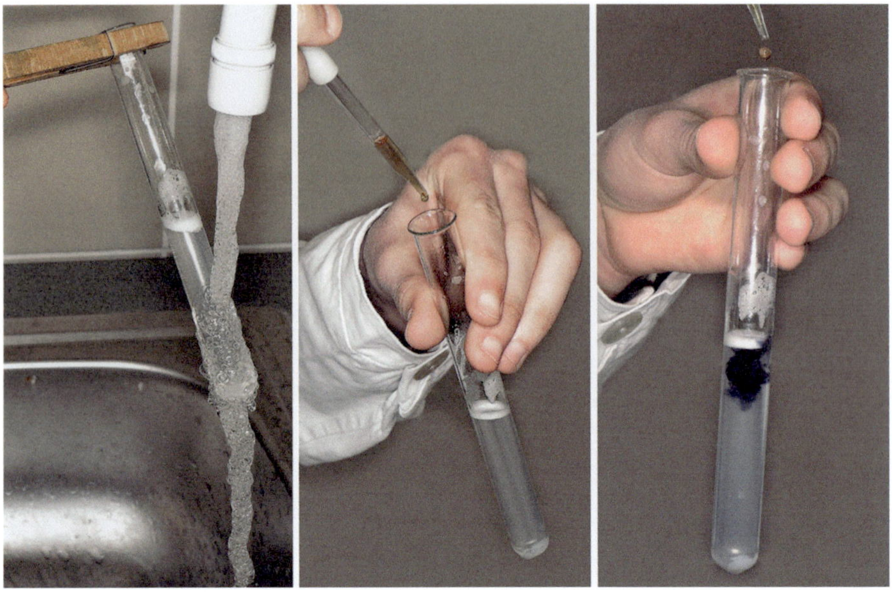

Fig. 13.13 We drop potassium iodide solution into the cooled starch solution

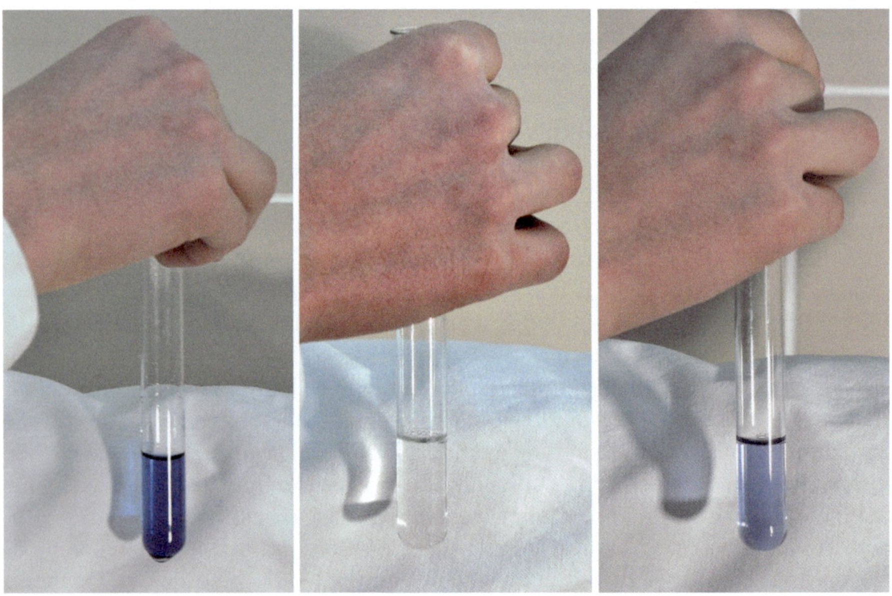

Fig. 13.14 The colour change of the diluted starch solution stained with Lugol's solution due to heating and cooling

and water, which facilitates the dissolution of iodine. The iodide ions (I^-) formed during the dissolution of potassium iodide attach to the iodine (I_2) molecules to form triiodide ions (I^{3-}), which dissolve well in water. Lugol's solution is brown in colour due to the iodine dissolved in it. The non-polar iodine molecules (I_2) just fit into the, "amylose tubes" wound up in a helix are fixed in its apolar cavity with weak secondary bonds.

As a result of the connection, the electron structure of the iodine molecules changes, they absorb light of a different wavelength, and can be excited with light of a different wavelength, so their colour becomes blue.

As a result of heating, the iodine molecules leave the amylose tubes due to thermal motion, so the solution becomes colourless. With the cooling of the solution a blue colour change appears again, because the connection between iodine and the amylose helices is re-established. The test can only be repeated several times in a row with the appropriate amylose-iodine ratio. If the amount of iodine is low in the solution compared to amylose, then despite cooling, the blue colour change does not return. If we then add more Lugol's solution to the solution, then a blue colour change indicates that the failure was due to the low iodine content of the solution (Fig. 13.15).

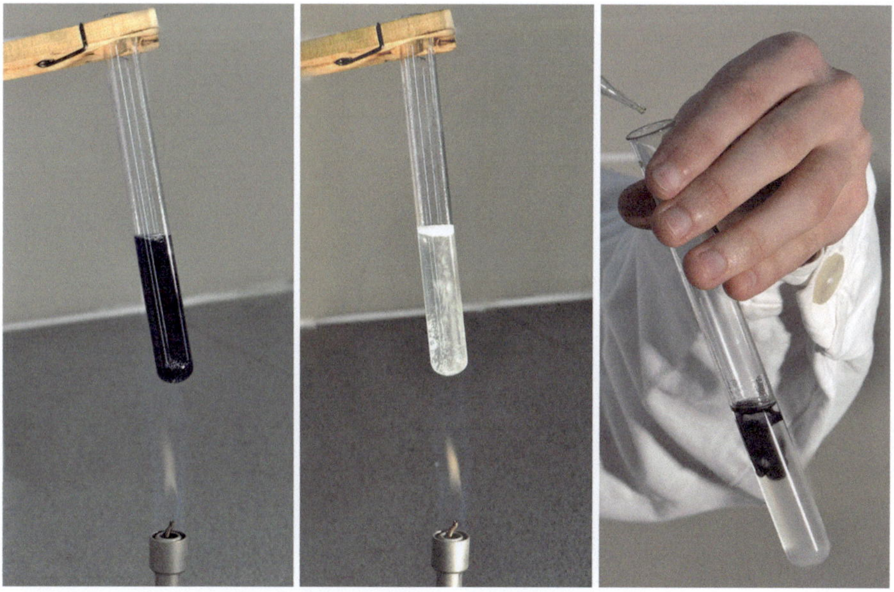

Fig. 13.15 The concentrated starch solution boiled with Lugol's solution becomes colourless, but when cooled it does not turn blue again, only if we add more Lugol's solution to it after cooling the solution

13.1.5 Examination of Potato Tuber and Banana with Lugol's Solution

We can perform the detection of starch with an iodine test on the potato tuber decoction as well, which should be prepared in the same way as we already described it in the Fehling reaction for the apple, the onion, and the carrot. Starch can also be directly detected from plant parts with high starch content. Cut a potato tuber in half, then drop Lugol's solution onto the cut surface! After 1–2 min, wash off the Lugol's solution! During the washing, not only the Lugol's solution, but also the starch grains released from the cells during the cut are removed, while the starch grain groups coloured blue by the diffused iodine become visible to the naked eye in the intact cells (Fig. 13.16).

There is also an opportunity for light microscopic examination of the potato tuber and the starch-containing seeds, which we wrote about in detail in the cell biology chapter (Fig. 13.17). In addition to the potato tuber, we can also detect the starch grains in the banana pulp with an iodine test (Fig. 13.18).

13.1.6 Examination of Flour with Lugol's Solution

Place a small pile of wheat flour on a watch glass, then create a small indentation in the centre! Drop some Lugol's solution into this! The blue-black discolouration that

Fig. 13.16 Dropping Lugol's solution onto the cut surface of the potato tuber

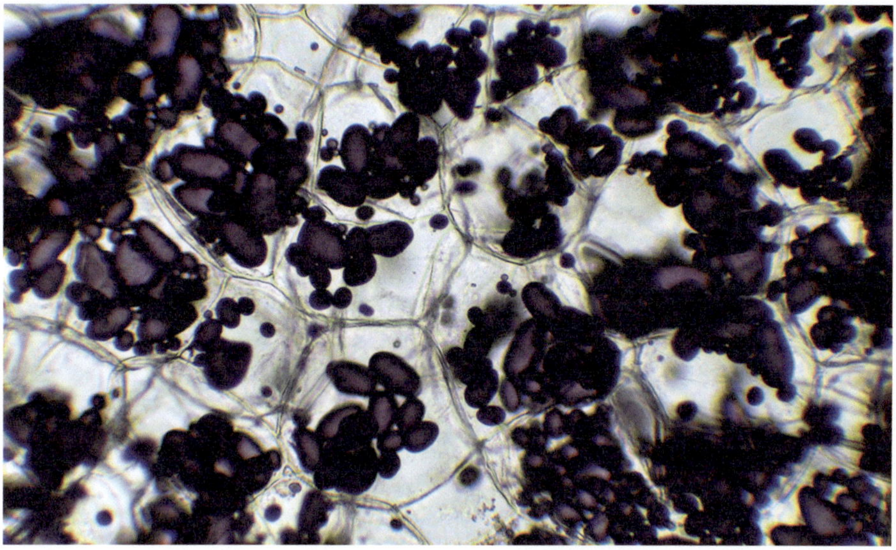

Fig. 13.17 Starch grains stained with Lugol's solution in the cells of the potato tuber

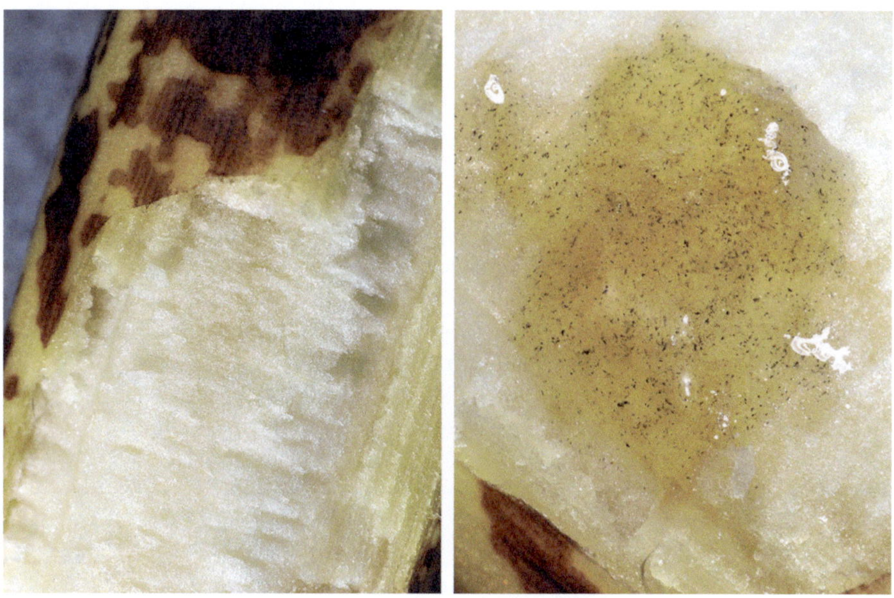

Fig. 13.18 The starch grains in the banana pulp are outlined with Lugol's solution dyed blue

Fig. 13.19 Examination of flour with Lugol's solution

appears as a result of the droplet indicates the high starch content of the flour (Fig. 13.19).

13.1.7 Examination of Cellulose

For the examination of cellulose fibres, we can use cotton labelled 100% cotton, or scrapings made from filter paper. By scraping the surface of the filter paper with a scalpel or knife, we can quickly obtain a pinch of cellulose fibres. Soak the fibres in Lugol's solution for two minutes, then wring them out, and spread the fibres out on a slide with a scalpel! Drop w = 75% (weight percent) sulfuric acid onto some of the fibres, causing the fibres to swell and turn blue (Fig. 13.20).

Explanation: The sulfuric acid causes some of the cellulose chains to be broken up and helically twisted, which give a blue colour reaction with iodine.

13.2 Proteins

13.2.1 Preparation of Egg White Solution

Separate an egg yolk and white, then pour the raw egg white into a beaker and dilute it to 200 cm^3! Stir the solution, then filter it through cotton! The protein solution filtered through the loose cotton ball will be almost clear, making the obtained protein solution suitable for further examinations (Fig. 13.21).

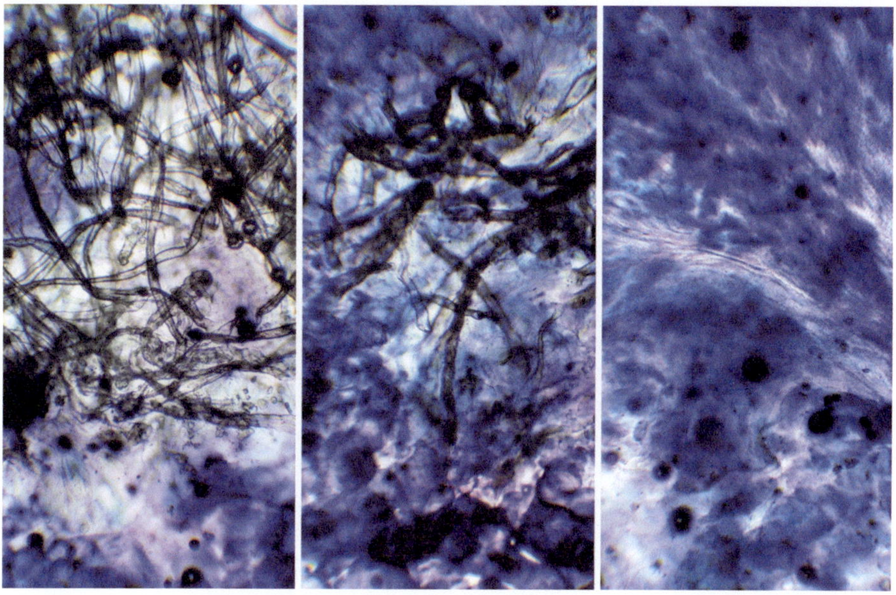

Fig. 13.20 The initially well-recognisable fibres disappear due to the effect of sulfuric acid

Fig. 13.21 Diluted egg white solution and its filtration through cotton

13.2.2 Examination of Egg White Solution with Biuret Test

The biuret test is a chemical reaction suitable for detecting peptide bonds. To perform the test, pour 2–3 cm^3 of egg white solution into a test tube, then add 1 cm^3 w = 20% sodium hydroxide (NaOH) solution! After shaking the solution, drip w = 1% copper (II) sulphate ($CuSO_4$) solution into the test tube until the appearance of a violet colour (Fig. 13.22)! Shake the contents of the test tube after each drop by tapping the test tube against our palm!

This can prevent the deep blue copper (II) complex formed from too much copper (II) sulphate in the solution from suppressing the original colour change. Dipeptides consisting of two amino acids and amino acids, with the exception of histidine, serine, and threonine, do not give the biuret reaction.

However, we can detect peptide chains consisting of at least three amino acids to show with a biuret test. During the reaction, the non-bonding electron pairs in the nitrogen-containing groups of proteins form a coordinate bond with the empty d-orbitals of copper (II) ions, resulting in the formation of a purple complex (Fig. 13.23).

The biuret test can also be used to determine the concentration of proteins, as the peptide bond in proteins usually occurs with the same frequency per unit mass of substance. The intensity of the purple colour, and consequently the light absorption at 540 nm, is directly proportional to the protein concentration. The simplest compound is biuret, which gives the reaction. Biuret can be produced by heating urea,

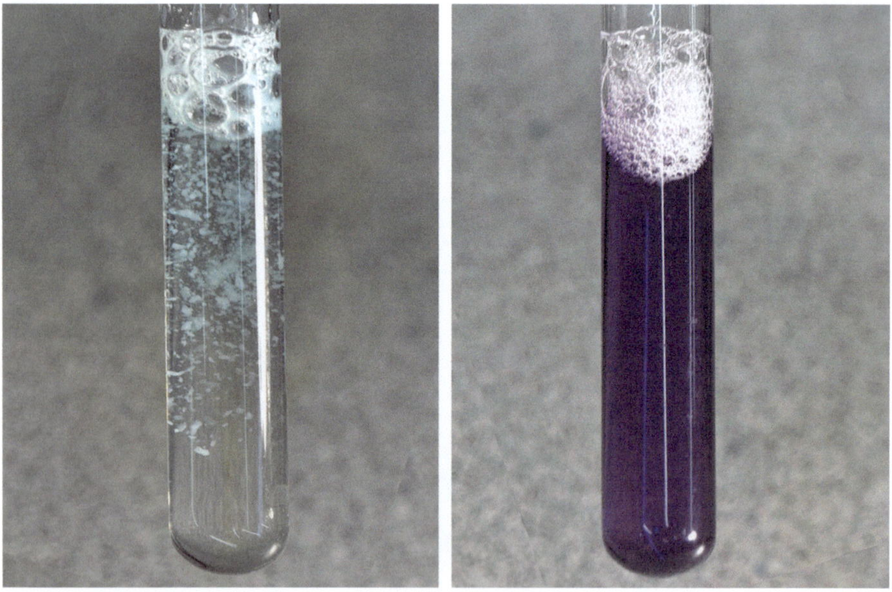

Fig. 13.22 The steps of the biuret reaction in two images

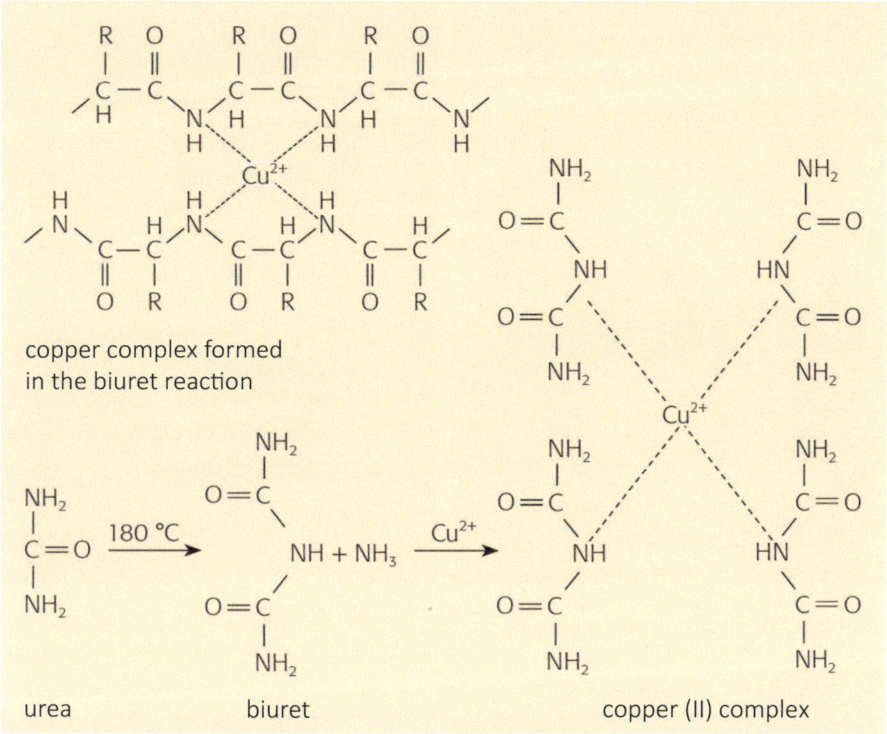

Fig. 13.23 Copper complexes formed in the biuret reaction

with ammonia (NH_3) being formed. In an alkaline medium, biuret forms a pink complex compound with copper (II) sulphate.

13.2.3 Detection of Protein From Plant and Animal Sources

Place a small amount of curd, sour cream, cheese, finely chopped potato or bean seeds on a watch glass and drip w = 20% sodium hydroxide solution onto it! Drip w = 1% copper (II) sulphate solution onto the alkalinised samples! Observe the appearance of the purple discolouration on the surface of the test materials, which indicates a positive biuret reaction (Figs. 13.24, 13.25, and 13.26)!

Pour 2–3 cm^3 of milk into a test tube and perform the biuret test in the same way as for the egg white solution! Upon adding to the copper (II) sulphate solution, the characteristic purple discolouration of the reaction appears (Fig. 13.27).

If too much copper (II) sulphate solution is dripped into the test tube during the dripping process, the blue colour of the copper complex will dominate instead of the purple colour change (Fig. 13.28).

Fig. 13.24 Dripping copper sulphate onto the alkalinised cheese surface

Fig. 13.25 Biuret reaction of cheese slice and curd

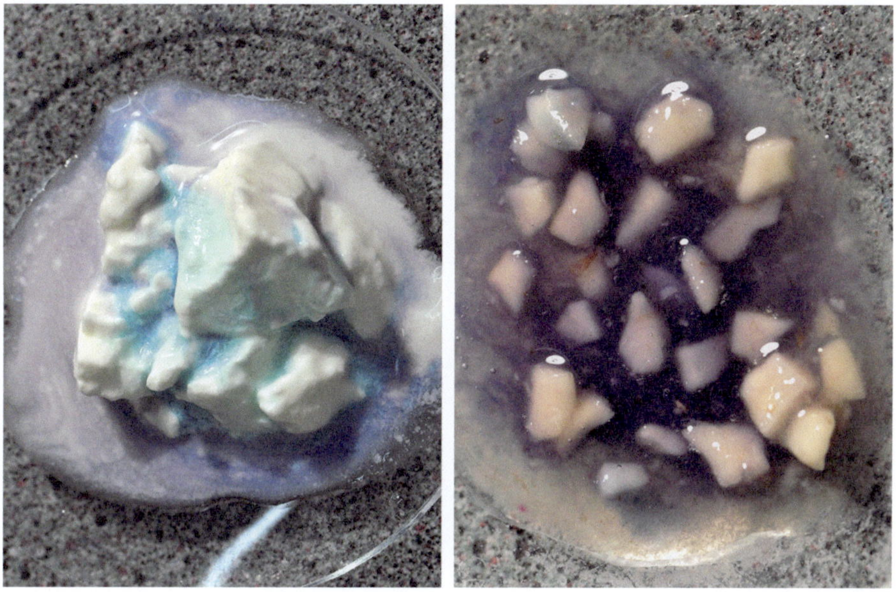

Fig. 13.26 Biuret reaction of sour cream and potato pieces

Fig. 13.27 Biuret test performed with milk

Fig. 13.28 The purple colour turns blue due to the excess copper sulphate solution added to the milk

13.2.4 Detection of Nitrogen Content in Proteins

To a small amount of chopped bean seeds, boiled egg white, a pinch of curd, or minced meat placed in a test tube, add 1–2 cm^3 w = 40% sodium hydroxide solution! Then gently heat the test tube with a Bunsen burner, while placing a moistened universal indicator paper at its mouth (Fig. 13.29)!

During the alkaline hydrolysis of proteins, amino acids are formed, and from the amino acids, ammonia (NH_3) develops. During the hydrolysis of ammonia in aqueous solution, hydroxide ions are formed alongside ammonium ions, causing an alkaline reaction that results in a blue-green colour change on the indicator surface.

13.2.5 Detection of Protein Sulphur Content

Sulphur-containing amino acids (cysteine and methionine) often participate in the construction of protein molecules. From these, hydrogen sulphide (H_2S) is released by boiling in an alkaline medium, which gives a black lead sulphide (PbS) precipitate with lead-containing salts. Pour 1 cm^3 w = 0.5% lead acetate ($Pb(CH_3COO)_2$) or lead nitrate [$Pb(NO_3)_2$] solution into a test tube, then drip w = 20% sodium hydroxide (NaOH) solution into it until the initially formed white lead hydroxide [$Pb(OH)_2$] dissolves!

Step 1: $Pb(NO_3)_2 + 2\ NaOH = Pb(OH)_2 + 2\ NaNO_3$

Fig. 13.29 Detection of protein nitrogen content

Step 2: $Pb(OH)_2 + 2\ NaOH = Na_2[Pb(OH)_4]$

Add 1–2 cm^3 of egg white solution to the solution, then gently heat the test tube on a Bunsen burner! As a result of the heating, a black lead sulphide (PbS) precipitate forms (Fig. 13.30).

13.2.6 Detection of Protein by Xanthoproteic Reaction

Add a few cm^3 of egg white solution to a test tube, add a little (approx. 1 cm^3) concentrated nitric acid, then gently heat it on a Bunsen burner! The use of protective gloves and goggles is mandatory for this experiment!

Proteins precipitate under the influence of concentrated nitric acid (HNO_3). They turn strongly yellow due to the effect of heating.

The test tube experiment can be performed with finely chopped bean seeds and onions. Under the influence of nitric acid, a white precipitate forms from the egg white solution, which shows a slight yellowing over time. Upon heating, a strong yellow colour appears (Fig. 13.31). The reaction is only given by proteins that include aromatic side-chain amino acids (phenylalanine, tyrosine, and tryptophan) in their structure (Fig. 13.32).

Under the influence of concentrated nitric acid, the aromatic ring is nitrated and turns into a yellow nitro compound. The type of reaction is electrophilic substitution, in which the nitronium cation reacts with the aromatic ring.

Fig. 13.30 Detection of sulphur from milk and egg white solution. The initially white or transparent solution turns brown and then black due to the heating

Fig. 13.31 Steps of the xanthoproteic reaction performed with egg white solution

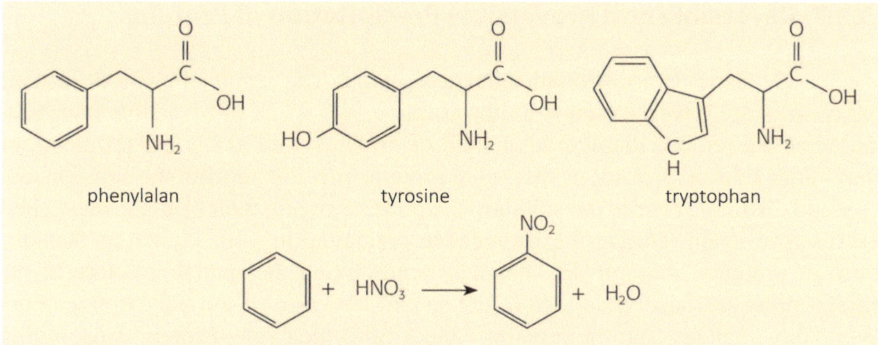

Fig. 13.32 Amino acids giving the xanthoproteic test and the reaction equation

Fig. 13.33 The result of the xanthoproteic reaction performed with flour, cottage cheese, and sour cream

Place a little flour, cottage cheese, and sour cream on a watch glass and drip concentrated nitric acid on them! As a result of the tests, yellow discolouration appears to a lesser or greater extent, indicating a positive xanthoprotein reaction (Fig. 13.33).

13.2.7 Reversible and Irreversible Precipitation of Proteins

Proteins are sensitive to external influences, which can result in changes in their spatial structure, precipitation from the solution. We speak of reversible precipitation when we remove the hydration shell of proteins with alcohol, light metal, or ammonium salts, as a result of which the proteins precipitate from the solution, but if we add distilled water to the solution, the proteins regain their original water shell and dissolve again. During the irreversible precipitation (also known as denaturation) of proteins, some of the chemical bonds fixing the spatial structure of the protein molecules also break down, the spatial structure of the polypeptide chain irreversibly changes, and the proteins cannot fulfil their role. Protein denaturation can be triggered by heat, heavy metal salts, and acids.

13.2.7.1 Precipitation of Protein with Light Metal Salts

Pour 2–2 cm^3 of protein solution into two test tubes! In the first test tube, slowly add saturated sodium sulphate (Na_2SO_4) solution until the solution becomes cloudy due to the precipitation of the protein (Fig. 13.34)!

Dissolve the precipitated protein by adding distilled water! Repeat the experiment with a saturated magnesium sulphate ($MgSO_4$) solution in the case of the other test tube! The test can be performed with saturated sodium chloride (NaCl) and ammonium sulphate [$(NH_4)_2SO_4$] solution, but in this case, the change is less noticeable.

Fig. 13.34 Reversible precipitation of egg white solution with saturated sodium sulphate solution and alcohol

13.2.7.2 Precipitation of Protein with Alcohol

The original protein solution can be seen in the first test tube, and the precipitated protein in the second.

Pour 1 cm^3 of w = 96% ethyl alcohol into a test tube containing 2 cm^3 of protein solution! The protein precipitates under the influence of alcohol (Fig. 13.34). Pour 1 cm^3 of the solution into another test tube and dilute it with distilled water! The dilution causes the precipitate to dissolve, indicating that the addition of alcohol caused a reversible precipitation. The test can also be performed with denatured alcohol.

13.2.7.3 Protein Precipitation by Heating

Pour 5 cm^3 of protein solution into a test tube and heat it in the flame of a Bunsen burner! The heating causes the protein to precipitate from the solution as a white precipitate. Pour the precipitated protein into a beaker and dilute it with distilled water to 100 cm^3! Observe whether the precipitated protein dissolves!

13.2.7.4 Protein Precipitation with Sulphuric Acid

Pipette 1 cm^3 of concentrated sulphuric acid into a test tube, then add 5 cm^3 of protein solution! Gently shake the contents of the test tube! The use of protective gloves and goggles is mandatory for this experiment! Concentrated mineral acids, with the exception of H_3PO_4, irreversibly precipitate proteins from their solutions (Fig. 13.35). The precipitate formed during precipitation dissolves in excess sulphuric acid, so

Fig. 13.35 Irreversible precipitation of egg white solution with heat and concentrated sulphuric acid

the sulphuric acid accumulating at the bottom of the test tube clears due to the dissolution of the protein.

13.2.7.5 Protein Precipitation with Heavy Metal Salts

Pour 5 cm^3 of protein solution into three test tubes! Add dropwise w = 0.5% lead nitrate [$Pb(NO_3)_2$] to the first test tube, w = 1% copper (II) sulphate ($CuSO_4$) solution to the second, and w = 3% silver nitrate ($AgNO_3$) solution to the third! Precipitate forms in all three test tubes. Add distilled water to the precipitated solution!

Proteins are precipitated by heavy metal salts even at low concentrations (Fig. 13.36). The process is irreversible (irreversible), the protein cannot be dissolved by adding distilled water.

13.3 Lipids

Sudan III (Fig. 13.37), a non-polar microscopic dye, dissolves in fats with a reddish colour, making it suitable for detection.

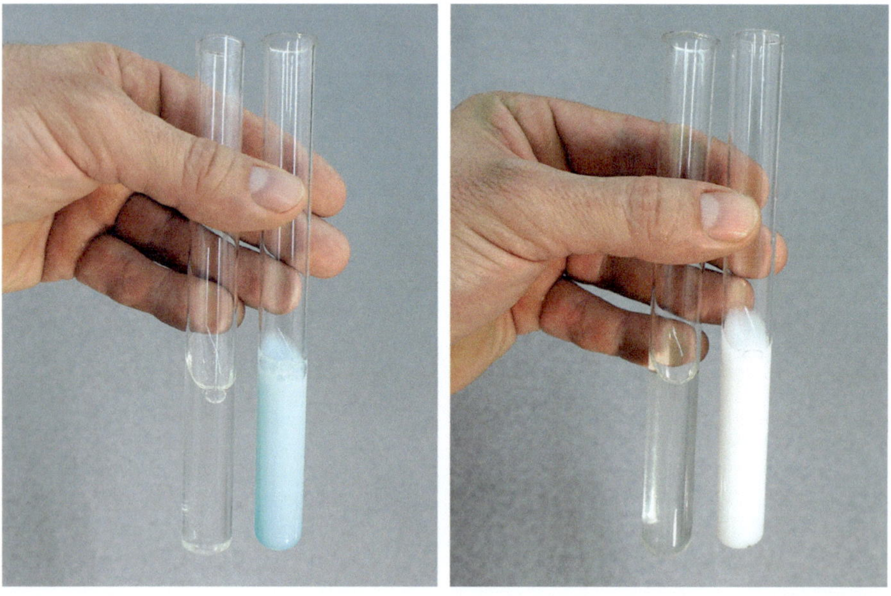

Fig. 13.36 Precipitation of egg white solution under the influence of copper (II) sulphate and lead nitrate

Fig. 13.37 Structural formula of the Sudan III molecule

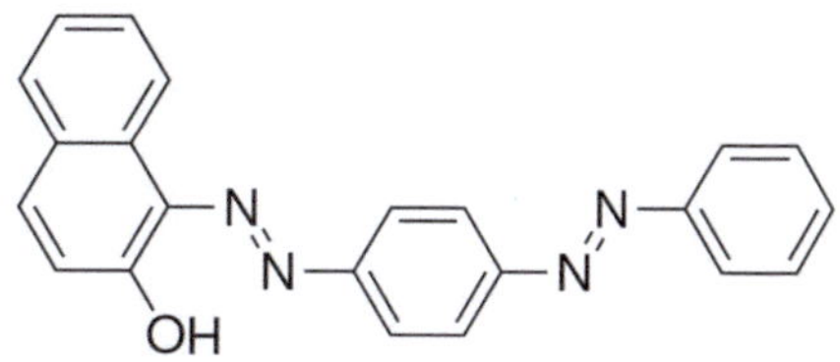

Fig. 13.38 After mixing cooking oil and water, the two liquid phases separate again

13.3.1 Detection of Fats and Oils with Sudan III Dye

Pour 6–8 cm^3 of water into a test tube, then 1–2 cm^3 of cooking oil! At the bottom of the test tube, the harder water can be observed, while the cooking oil is clearly separated from the water at the top. By shaking the contents of the test tube well, the water and oil mix, but the two types of liquid quickly separate into two liquid phases after shaking (Fig. 13.38).

Drop Sudan III dye solution into the test tube, then shake its contents! It can be clearly observed that the dye colours the cooking oil red and the water-cooking oil mixture formed at the boundary of the two liquid phases (Fig. 13.39).

An emulsion is a system of a liquid forming a continuous medium and another liquid/liquids finely dispersed (dispersed) in it in the form of tiny droplets (which do not mix with the former). The components of the emulsion can separate into separate liquid phases when stationary.

The separation can be prevented with various surfactants. In the case of milk, protein molecules are the surfactants that stabilise the fat droplets, preventing them

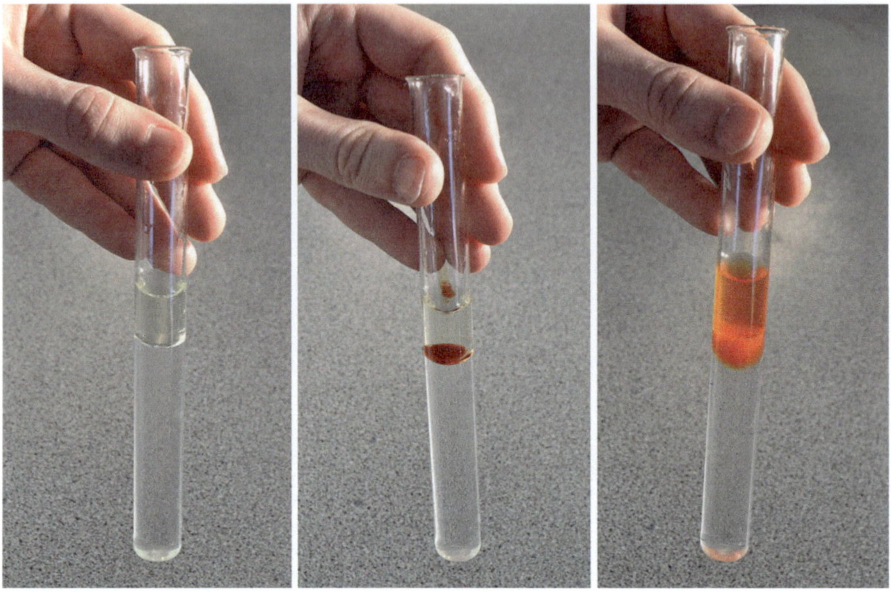

Fig. 13.39 First picture: Water and cooking oil in the test tube. Second picture: Dropping Sudan III dye solution into the test tube. Third picture: Distribution of the dye after shaking the test tube

from merging into larger droplets. Bile acids are also surfactants, which stabilise the fat droplets formed by the movements of the digestive system (emulsification), thus, the fat-digesting enzyme (lipase) can efficiently perform the digestion of fats on a large surface. Most detergents are also surfactants, so if we add detergent to the cooking oil-water mixture, the liquid will not separate into two phases even after shaking (Fig. 13.40).

13.4 Nucleic Acids

13.4.1 Detection of DNA from a Banana

In the experiment, we extract the DNA from the cells of the banana fruit. First, we mash half a banana, then in a cooking cup, we mix a little water, dish soap, and a teaspoon of table salt (NaCl). The soap solution is added to the banana puree and we thoroughly mix the mixture (Fig. 13.41).

The dish soap breaks down the structure of the cell membrane and the nuclear membrane, thus the DNA molecule is released from the cells. The effect of sodium chloride increases the solubility of DNA molecules, while proteins precipitate from the solution. This is called salting out, as a result of which the DNA remains in the supernatant part of the mixture (Fig. 13.42). We filter the banana mixture into a test tube and over the filtrate we layer concentrated (96%) ethyl alcohol. DNA does not dissolve in ethyl alcohol, so it precipitates at the phase boundary of the supernatant

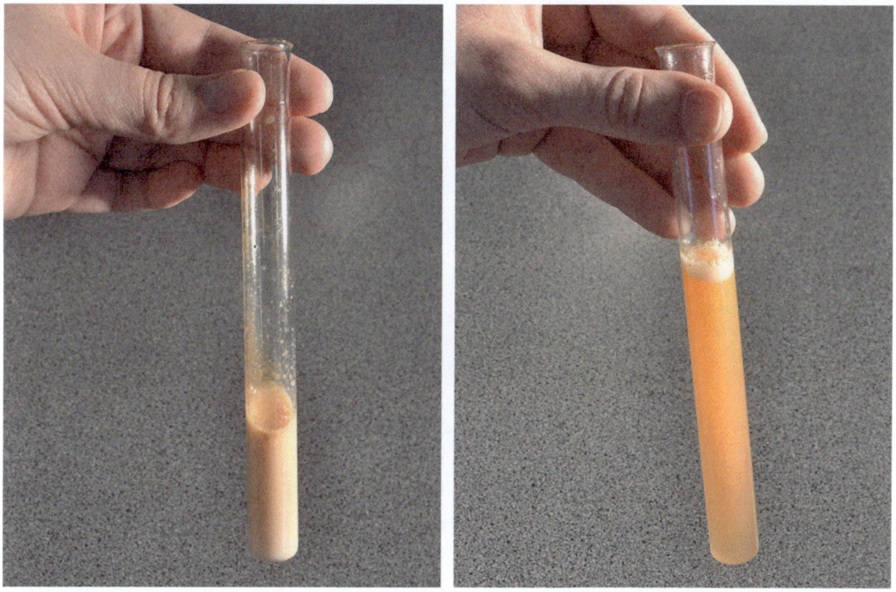

Fig. 13.40 First picture: Milk stained with Sudan III dye. Second picture: Cooking oil, water, and detergent created by shaking emulsion

Fig. 13.41 Table salt and detergent, and the mixed banana pulp mixture

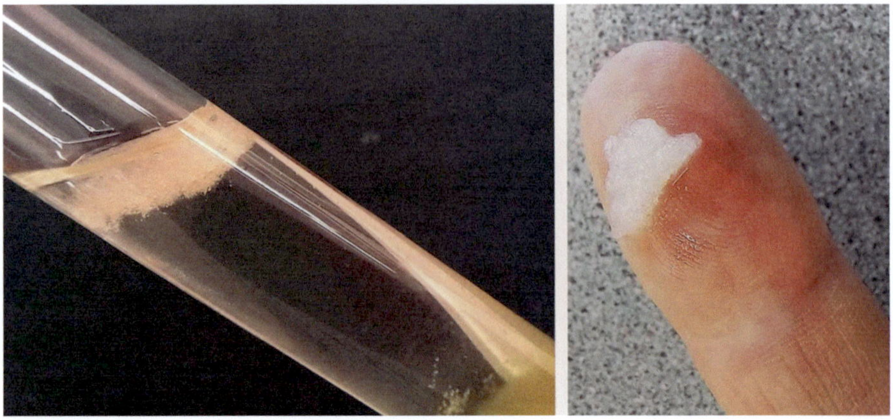

Fig. 13.42 DNA extracted from banana in the supernatant and on the fingertip

alcohol and the lower banana extract. The white, fibrous DNA mass can be lifted out of the test tube with a glass rod or tweezers.

13.5 Plant Dyes

13.5.1 Separation of Photosynthetic Pigments by Paper Chromatography

Chromatography is a chemical procedure for separating gas or liquid mixtures. During its application, the components to be separated, in our case the photosynthetic pigments, are distributed between two phases. One phase is stationary (filter paper), the other is mobile (chlorophyll solution), which flows over the stationary phase. The stationary phase is a solid material with adsorption properties, so we are talking about adsorption chromatography. The individual components are separated and migrate through the stationary phase according to how much they are adsorbed onto the stationary phase and what their solubility is in the mobile phase.

Those components that are more strongly adsorbed onto the solid phase, travel at a slower speed in the direction of liquid flow than those that are less or not at all adsorbed.

The individual components thus separate from each other in the direction of liquid movement according to their adsorptivity and concentrate into ring or band-shaped moving zones.

Chop up green leaves, grass blades are excellent subjects, then with the addition of a little quartz sand and 96% ethanol, grind it in a mortar! The resulting green solution is an alcoholic solution of photosynthetic dyes. The solution can be used as a running mixture in the future without filtering (Fig. 13.43). Pour the solution into a small cooking glass, then insert a strip of filter paper (Fig. 13.44).

Fig. 13.43 Chlorophyll solution extracted from grass blades

Fig. 13.44 Running chlorophyll solution on a strip of filter paper

In the experiment, the filter paper acts as a capillary adsorbent structure, in which the dye mixture is moved by capillary forces. The components of the solution, depending on their physical and chemical properties, bind to the surface of the filter paper to varying degrees. The more they bind, the shorter the distance they travel relative to the solution interface. Thus, the components of the mixture are separated (Fig. 13.45).

The experiment can also be performed by dropping the chlorophyll solution onto the centre of a piece of filter paper, thus the solution spreads out radially (Figs. 13.46

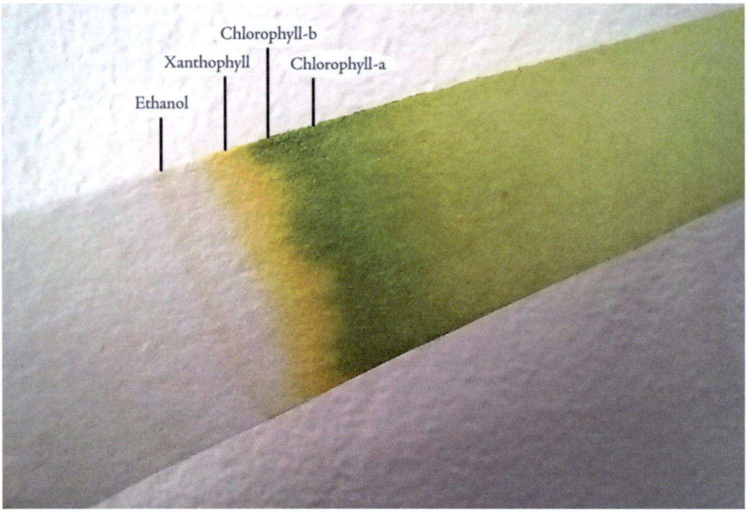

Fig. 13.45 The result of separating the dyes

Fig. 13.46 The beginning and final result of chromatography

and 13.47). The topmost or outermost yellow spot contains xanthophyll, the middle, yellow-green spot contains chlorophyll-b, the bottommost or innermost blue-green spot contains chlorophyll-a. Chromatographic separation should be allowed at least half an hour.

With the remaining chlorophyll solution, we can still perform a quick test tube experiment. The bright green solution turns brown when acid (acetic acid, hydrochloric acid, or sulphuric acid) is added (Fig. 13.48). The change is due to the same process that causes green leaves to turn brown in autumn. At this time, an enzyme extracts the magnesium ions (Mg^{2+}) bound by complex binding from the

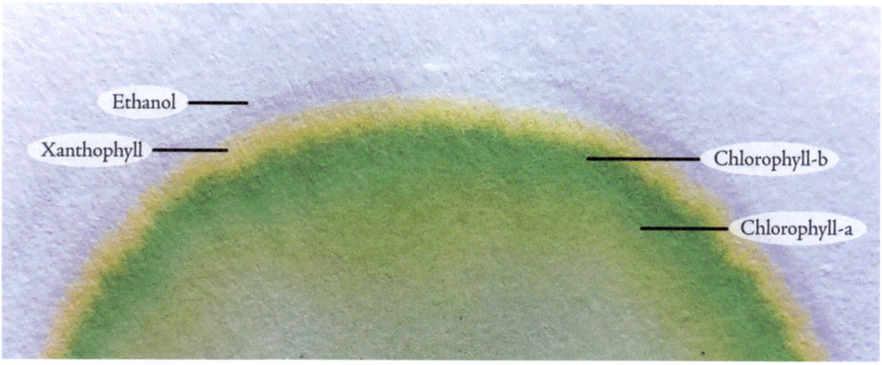

Fig. 13.47 A higher magnification image from the filter paper

Fig. 13.48 The green chlorophyll solution turns brown under the influence of acid

chlorophyll-a molecules and in their place, two H+ enter the molecule, resulting in the formation of greenish-brown pheophytin-a. Pheophytin-a can be found in small amounts under natural conditions in plant organisms and plays a significant role in the process of photosynthesis.

In an acidic environment, chlorophyll-a easily gives up its magnesium ion, while pheophytin-a is formed from it.

This process is attempted to be avoided during the processing of green food plants by shifting the pH towards alkaline.

13.5.2 Anthocyanin Tests

Anthocyanins found in cell sap are water-soluble coloured compounds that have indicator properties (Fig. 13.49). Their colour changes depending on the pH of the medium. From strongly alkaline pH to strongly acidic pH, they can turn yellow, green, blue, purple, and red. In nature, plant parts containing anthocyanins can be red, purple, and blue.

Anthocyanins are glycosides consisting of coloured anthocyanid and one or two colourless sugar parts. The sugar part promotes the solubility of the dye in water and stabilises the anthocyanid, which easily decomposes under the influence of light and heat.

In anthocyanins, one or two sugar parts can be attached to the anthocyanid, which can be a mono-, di-, and trisaccharide. The basic structure of the anthocyanidin is provided by the flavylium ion.

Different anthocyanins vary in the number of hydroxy and methoxy groups attached to the flavylium ion (Fig. 13.50).

There are several explanations for the biological role of anthocyanin accumulation. Flowers coloured by anthocyanins are more easily found by nectar-collecting

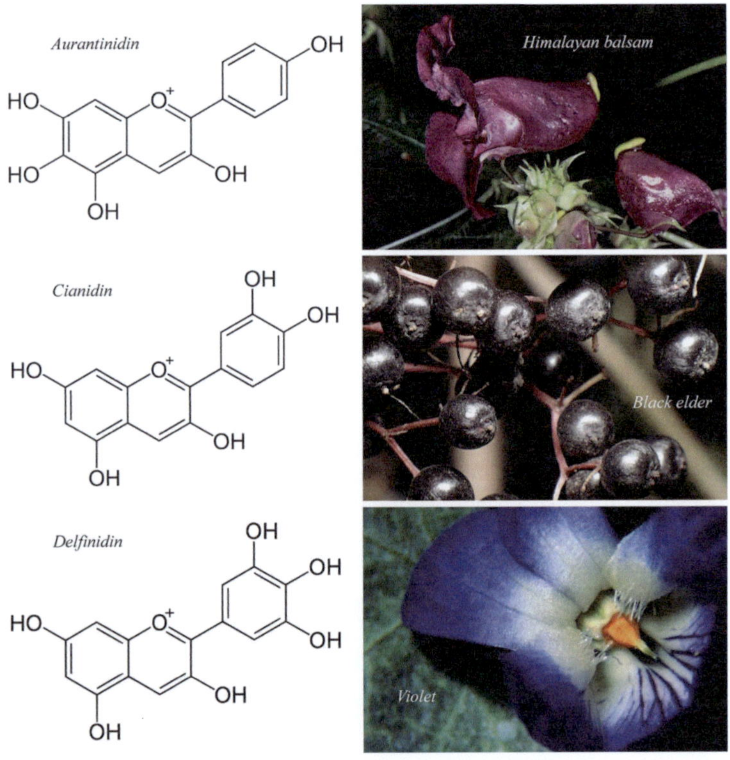

Fig. 13.49 Coloured anthocyanids and the plants "painted" by them

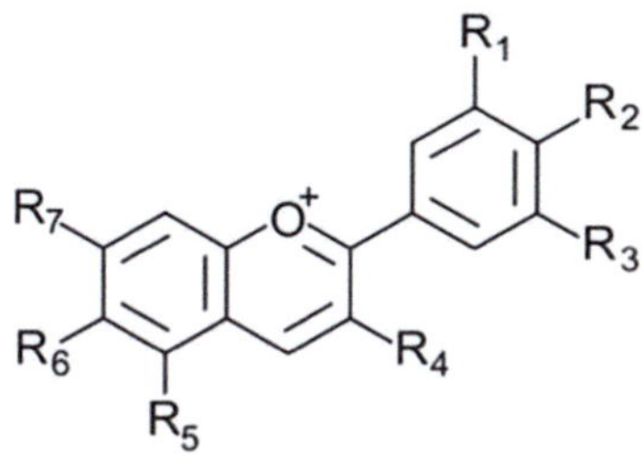

Anthocyanidins	R1	R2	R3	R4	R5	R6	R7
Aurantinidin (*Himalayan balsam*)	-H	-OH	-H	-OH	-OH	-OH	-OH
Cianidin (cherry, plum, grape, black elder)	-OH	-OH	-H	-OH	-OH	-H	-OH
Delfinidin (violet)	-OH	-OH	-OH	-OH	-OH	-H	-OH
Europinidin (leadwort, cape leadwort)	$-OCH_3$	-OH	-OH	-OH	$-OCH_3$	-H	-OH
Luteolinidin (broomcorn)	-OH	-OH	-H	-H	-OH	-H	-OH
Malvidin (grape)	$-OCH_3$	-OH	$-OCH_3$	-OH	-OH	-H	-OH
Pelargonidin (raspberry, geranium)	-H	-OH	-OH	-OH	-OH	-H	-OH
Peonidin (peony, morning glory)	$-OCH_3$	-OH	-H	-OH	-OH	-H	-OH
Petunidin (petunia)	-OH	-OH	$-OCH_3$	-OH	-OH	-H	-OH
Rosinidin	$-OCH_3$	-OH	-H	-OH	-OH	-H	$-OCH_3$

Fig. 13.50 The colourful components of different anthocyanins are derivatives of the flavylium ion, which differ in the molecular parts marked with the letter "R"

Fig. 13.51 The flowers of the common bugloss (*Anchusa officinalis*)

insects, such as bees, wasps, and flies, thus increasing the likelihood of plant pollination. During the flowering of borage plants, the colour of the flower can change; the bud and young flower of the spreading adder's tongue, the lungwort (*Pulmonaria officinalis*), and the common bugloss (*Anchusa officinalis*) (Fig. 13.51) are still reddish, while the older flowers are bluish-purple. The colour change is not due to pH change, but to biochemical processes that change the number of sugar components in the anthocyanin.

According to some opinions, the biological function of the colour change is that the appearance of multiple colours can contribute to attracting more insects involved in pollination, as different colours lure different insects.

Taking advantage of the indicator properties of anthocyanins, we can even change the dark blue flower colour of the common bugloss. If we place its flowers in the strongest, 20% vinegar available in grocery stores, we can observe after 2–3 min that the originally blue flowers turn burgundy, then become bright red after about ten minutes of soaking (Figs. 13.52, 13.53, and 13.54).

The vinegar diffuses into the cells of the petal leaves, where it shifts the pH of the cell sap in an acidic direction. As a result, the colour of the anthocyanins in the cell sap changes from blue to red. We can also achieve the colour change without dipping in vinegar by placing the cut plant stems in vinegar. In this case, the bugloss reaches the flower petal cells through its vascular bundles. Dissolve three sodium hydroxide tablets in 10 cm^3 of water, then soak an inflorescence of the bugloss in the resulting strongly alkaline solution. Under the influence of the strong alkali, the blue-hearted flowers turn green, then yellow (Fig. 13.54).

Fig. 13.52 Flowers of the common bugloss (*Anchusa officinalis*) soaked in vinegar

Fig. 13.53 Natural coloured and slowly reddening flowers of the common bugloss (*Anchusa officinalis*) soaked in 20% vinegar

Fig. 13.54 The flowers of the common bugloss (*Anchusa officinalis*) turn red from vinegar, while they turn yellow under the influence of a sodium hydroxide solution

Fig. 13.55 Wood violet (*Viola odorata*)

We can achieve even more beautiful colour changes with the grape hyacinth and the wood violet (*Viola odorata*). The violet is native to Europe and Asia, but it is also a widespread perennial plant in North America. This species, native to deciduous forests, is common in urban environments, in parks and gardens (Fig. 13.55).

Fig. 13.56 Experiments with the wood violet (*Viola odorata*) flower

The white colour of the violet flower is caused by the total reflection of light in the air-filled intercellular spaces. The bluish-purple colour is given by anthocyanins, which have acid-base indicator properties and accumulate in the cell sap of the epidermis cells. By soaking violet flowers in ammonia water and hydrochloric acid, we can create a colourful "bouquet of violets" (Figs. 13.56 and 13.57). The purple flower soaked in ammonia water turns bluish-green after 5–10 s, then turns completely yellow after 10–15 min.

The grape hyacinth, classified among the asparagus family, blooms in March and April (Fig. 13.58). This bulbous plant, originally living in grasslands, rocky places, and oak forests, is planted in gardens as an ornamental plant. Having escaped from there, it is now widespread and abundant in sunny urban green areas.

The inflorescence of the grape hyacinth (Fig. 13.58) is particularly suitable for demonstrating the pH-dependent colour change of anthocyanins. Let us first look at the effect of a not too strong alkali, ammonia water, on the flower colour. Pour a little of this chemical, which can also be purchased in paint shops, into a glass vial, then dip the inflorescence of the grape hyacinth into the liquid. The alkali seeping into the petals causes a greenish colour change after a few minutes (Fig. 13.59).

Continuing the experiment, we can repaint the purple inflorescences blue, green, and yellow. Under the influence of acid, the purple anthocyanin turns red. For this transformation, in the case of the grape hyacinth, we need to use a strong acid, 20% hydrochloric acid. The hydrochloric acid damages the plant tissues, so the red-painted inflorescence is less durable than those discoloured with ammonia (Fig. 13.60).

Fig. 13.57 The results of anthocyanin tests conducted with wood violet (*Viola odorata*) flowers

Fig. 13.58 Grape hyacinth (*Muscari neglectum*)

A characteristic of anthocyanin is that it discolours when oxidised. We can also illustrate this process with the inflorescence of the grape hyacinth.

Dissolve potassium permanganate in water, then acidify the resulting purple-red solution with a little sulphuric acid, or hydrochloric acid. If we soak the inflorescence of the grape hyacinth in the solution obtained for a longer period, 1–2 h, then it turns a whitish-brown colour (Fig. 13.61).

If the chopped red cabbage is boiled in water, the water is coloured dark purple by the released anthocyanins. These do not decompose under heat, so we get a solution with which we can continue experimenting.

Fig. 13.59 Original and ammonia-soaked grape hyacinth (*Muscari neglectum*) flower

Fig. 13.60 Results of anthocyanin tests conducted with grape hyacinth (*Muscari neglectum*) inflorescences

Fig. 13.61 Grape hyacinth (*Muscari neglectum*) inflorescence discoloured with potassium permanganate next to the original inflorescences

Fig. 13.62 The colour change of anthocyanin extracted from red cabbage in solutions with increasing pH

Pour 2–2 cm^3 of the obtained solution into five test tubes. To the first, add a few drops of 20% acetic acid, which turns the solution red.

In the second test tube, do not add anything to the boiled neutral pH purple solution. In the third test tube, by adding 3–4 drops of sodium hydroxide solution, the liquid turns blue-green.

In the 4th test tube, keep adding the alkali until it turns light green, and to the last one until we observe a yellow colour change. If we add the alkali to the solution more precisely than above, we can produce even more shades of colour (Figs. 13.62 and 13.63).

Fig. 13.63 The colours created by adding alkali and acid

13.6 Colloid Systems

We call colloids those materials in which particles ranging in size from about 1 nm to 500 nm are dissolved, dispersed in some dispersing medium. The dispersing medium in the case of aqueous colloids is water, but oils, fats, solid, and gaseous substances can also be the media of the dispersed particles (Table 13.1).

Depending on the type of dispersed particles, there are dispersion, macromolecular, and association colloids. In the continuous medium of dispersion colloids, there are particles bounded by gas, liquid, or solid surface.

The particles of the macromolecular system are macromolecules, such as polymer or protein molecules, which size falls into the colloid size range. In association colloids, there are surface-active molecules, which group into micelles to create colloid-sized particles. If in a colloid system the particles are less tightly bound to each other and the properties of the solvent, such as its fluidity, determine the properties, then we can talk about an incoherent (non-cohesive) system (e.g. sol state). In the coherent (cohesive) system, a continuous framework is formed as a result of tighter interconnection of the particles (e.g. gel state).

Adsorption refers to the binding, accumulation of materials at the interface (phase boundary).

13.6.1 Examination of the Adsorption of a Colloid System

Colloid particles have a large surface area, so they are generally considered good adsorbents. To 100 cm^3 of water poured into a beaker, add 6–8 drops of carmine or eosin microscopic dye! To the mixed red solution, add a teaspoon of activated charcoal powder, then filter the whole thing through filter paper! The filtered solution no longer shows the red colour, because the activated charcoal can be considered a

Table 13.1 Examples of different colloids

	Dispersed particles		
Dispersing medium	Solid	Liquid	Gas
Solid	Coloured glass	Butter	Foam rubber
Liquid	AgCl aqueous solution	Milk	Whipped cream
Gas	Colloidal smoke	Colloidal fog	–

Fig. 13.64 Adding activated charcoal to water stained with microscopic dye, then filtering it all through filter paper

Fig. 13.65 Red wine, activated charcoal red wine, and the filtered mixture

solid-gas (carbon-air) colloid system, which binds the microscopic dye on its huge surface (Fig. 13.64). Performing the experiment with red wine, the wine loses its red colour, resembling white wine in appearance (Fig. 13.65).

Physiological Examinations 14

14.1 Conditions for Germination and Germination Experiments

For germination, place wet cotton wool at the bottom of a Petri dish, then place the seeds to be germinated on it and replace the lid of the Petri dish (Fig. 14.1)!

The seeds set up in the Petri dish are constantly in wet conditions, and their oxygen supply is also ensured, so usually after 3 or 4 days, we can observe the developing seedlings (Figs. 14.2 and 14.3).

With these set-up germination experiments, the environmental factors necessary for germination can also be easily examined. Germination does not take place at low temperatures, which can be confirmed by germination experiments set up in the refrigerator at 7 °C and at room temperature. Oxygen is always needed for germination, so if we pour water on the seeds so that it covers them, germination will not take place despite the appropriate temperature and water supply, because the oxygen dissolved in the water to a small extent is not able to provide the significant amount of oxygen needed for germination. The effect can be further enhanced by layering cooking oil on the water surface, which completely prevents the atmospheric oxygen from dissolving in the water. The seeds do not germinate inside the fleshy fruits, even if the necessary amount of water, oxygen, and the appropriate temperature are provided by opening them. The reason for this is that there are germination inhibitors in the fruit wall. Therefore, if thin apple slices are placed on the seeds in a germination experiment set-up in a Petri dish, germination will not take place. In the case of above-ground germination types, the embryonic root part starts to develop first during germination. The root developing from this breaks through the seed coat near the umbilicus and anchors the germinating plant in the soil (Fig. 14.2). This is followed by the stem part below the cotyledon (hypocotyl) starting vigorous growth in beans and lentils, as a result of which the cotyledons and the shoot located above them emerge from the soil (Figs. 14.4 and 14.5).

The material and energy needs of growth are still provided by the reserve nutrients accumulated in the cotyledons. In the bean cotyledon, starch accumulates in the

G. Kriska, *Teaching Biology*, https://doi.org/10.1007/978-3-031-84085-2_14

Fig. 14.1 Setting up pea (*Pisum sativum*) germination in a Petri dish

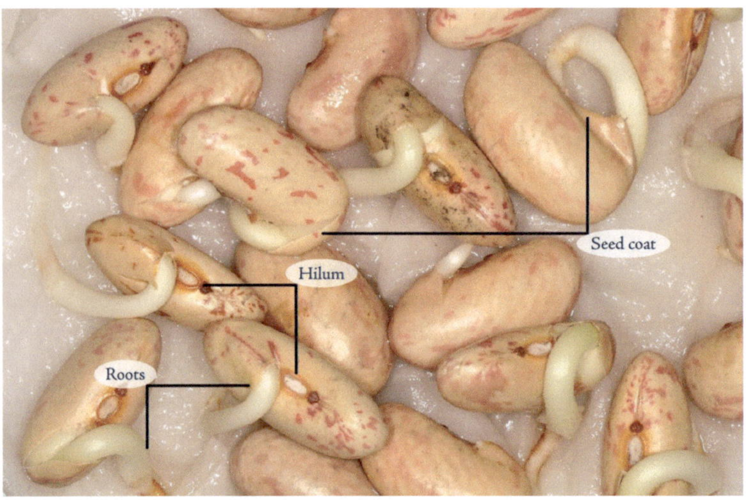

Fig. 14.2 Bean (*Phaseolus vulgaris*) seeds that have started to germinate

largest quantities among the reserve nutrients in the colourless plastids (leucoplasts). Starch is produced in the chloroplasts during photosynthesis under the influence of light. The starch produced during the day is broken down into sugar (sucrose) at night by digestive enzymes and reaches the plant's storage organs, the roots, rhizomes, seeds, through the transport tissues. Starch is formed again at these places, which accumulates in the colourless plastids developing from the plastid primordia.

The bean seed is covered by a seed coat on the outside, and inside it contains the embryo, which consists of the two cotyledons and the germ hidden between them.

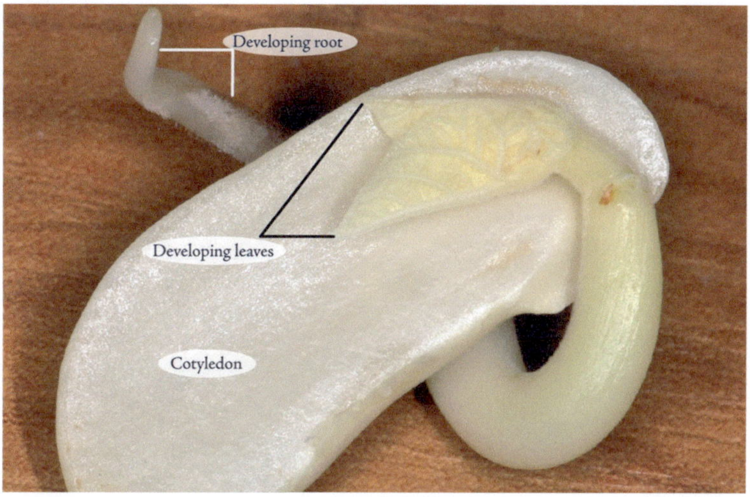

Fig. 14.3 Structure of a bean (*Phaseolus vulgaris*) sprout

Fig. 14.4 Appearance of the shoot in the second stage of germination

The two main parts of the germ are the plumule, from which the shoot (leafy stem) will be, and the radicle, from which the root will be.

The root developing from the radicle comes out of the seed near the umbilicus. The **umbilicus** is an indentation on the seed coat, which indicates the former connection point of the funiculus and the ovule. Lateral side roots appear on the root, which anchor the plant. Later, the stem part below the cotyledons starts vigorous growth, and this stem part lifts the cotyledons and the leaves visible between them out of the seed.

Fig. 14.5 Prominent bean (*Phaseolus vulgaris*) cotyledons and leaf primordia

Fig. 14.6 Degrading bean (*Phaseolus vulgaris*) cotyledons under the first leaves

The bean cotyledons turn green on the surface and photosynthesise until the first leaves develop. Later, they are degraded and fall off the plant (Fig. 14.6).

Fig. 14.7 The large heart-shaped cotyledons of the radish (*Raphanus sativus*)

In other plants, such as radishes, the cotyledons that rise to the surface are large and photosynthesise intensively (Fig. 14.7).

The germination of peas is similar to beans, with the difference that the cotyledons do not rise together with the leaves, but remain in the soil to nourish the developing plant (Figs. 14.8 and 14.9).

In the formation of the corn grain, the seed and the fruit wall also participate. Therefore, this is actually not a seed, but a fruit. Unlike beans and peas, the nutrients stored in the corn kernel are not in the cotyledon, but in the seed's nutritive tissue. The cotyledon is a shield-like formation that separates the germ from the yellow nutritive tissue of the corn. During germination, the rupturing pericarp first reveals the root covered with a root sheath (Fig. 14.10).

Later, the coleoptile starts to develop at the other pole of the germ, also emerges from the grain. Unlike beans and peas, lateral roots grow from the epicotyl in corn. The root hairs are clearly visible in the maturation zone of the roots, and the slimy root cap can be seen at the root tip (Fig. 14.11). The germination of the wheat grain is similar to that of corn (Fig. 14.12).

14.2 Demonstrating the Root Elongation Zone

Draw dots or small lines with an alcohol marker on the root of a three–four-day-old bean sprout, starting from the tip at one millimetre intervals!

Then put the germinating plant back into a Petri dish lined with wet filter paper! After half a day, examine the markings drawn on the root!

Fig. 14.8 Pea (*Pisum sativum*) germ shoot and root

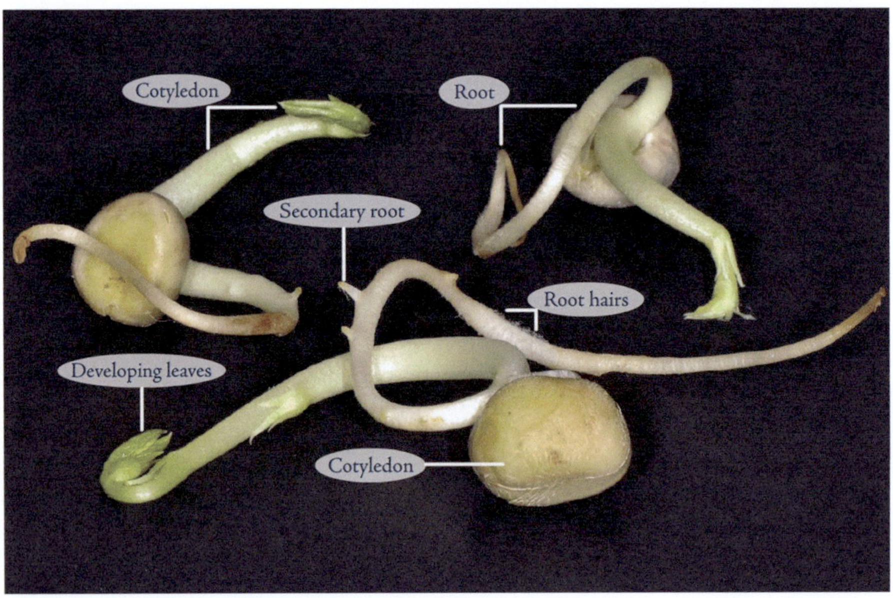

Fig. 14.9 Germinating pea (*Pisum sativum*) seeds

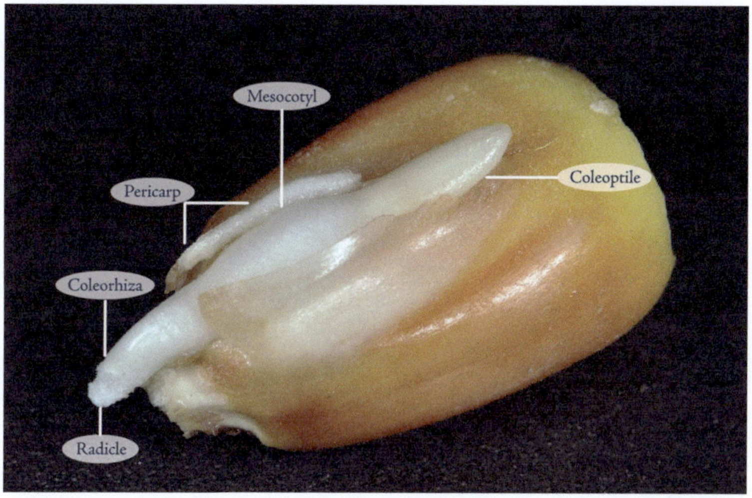

Fig. 14.10 The initial stage of corn (*Zea mays*) kernel germination

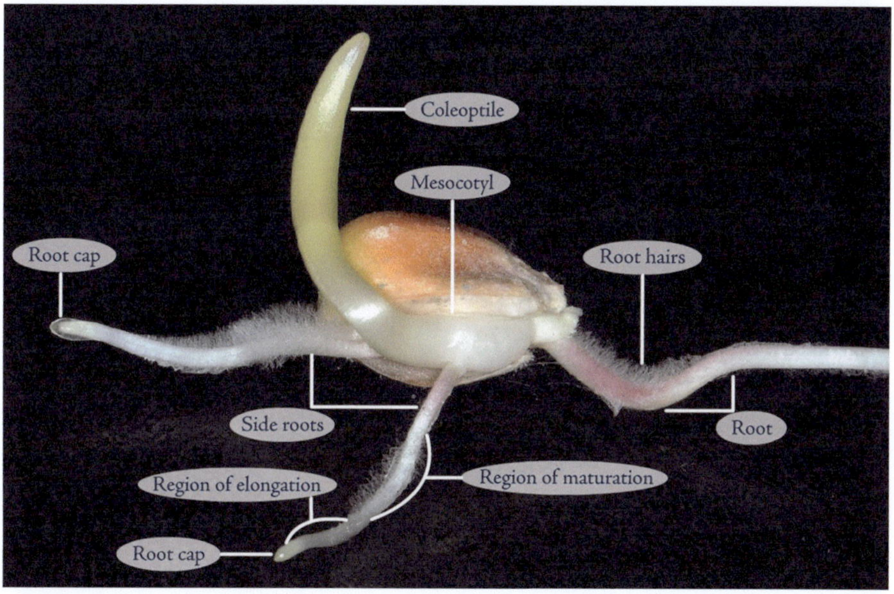

Fig. 14.11 The later stage of corn (*Zea mays*) kernel germination

During growth, the greatest elongation of the root is above the root tip, so after half a day, we can observe significant length growth in the so-called elongation zone. On the rest of the root, the markings have not visibly moved apart. Within half a day, the lateral roots have started to develop vigorously in the branching zone (Fig. 14.13).

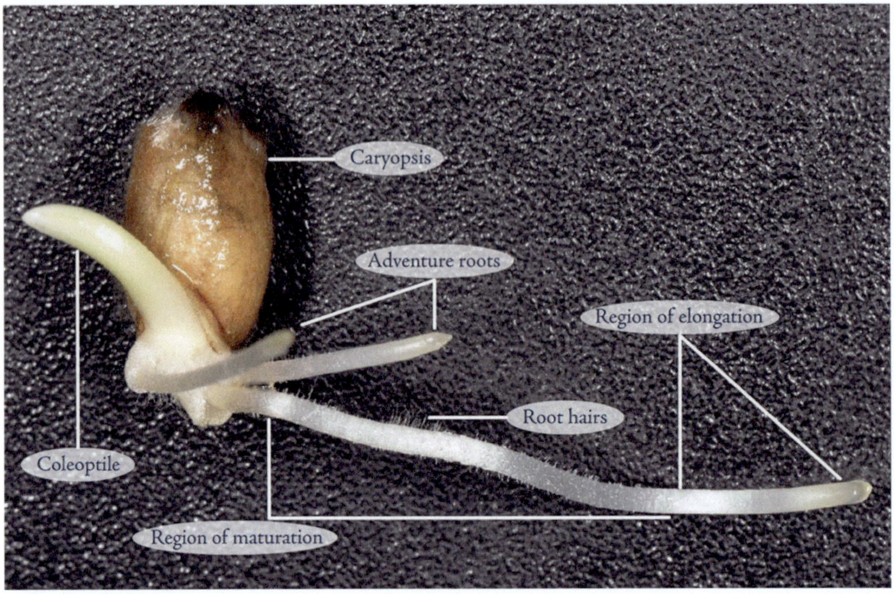

Fig. 14.12 Germinating wheat (*Triticum* sp.) grain

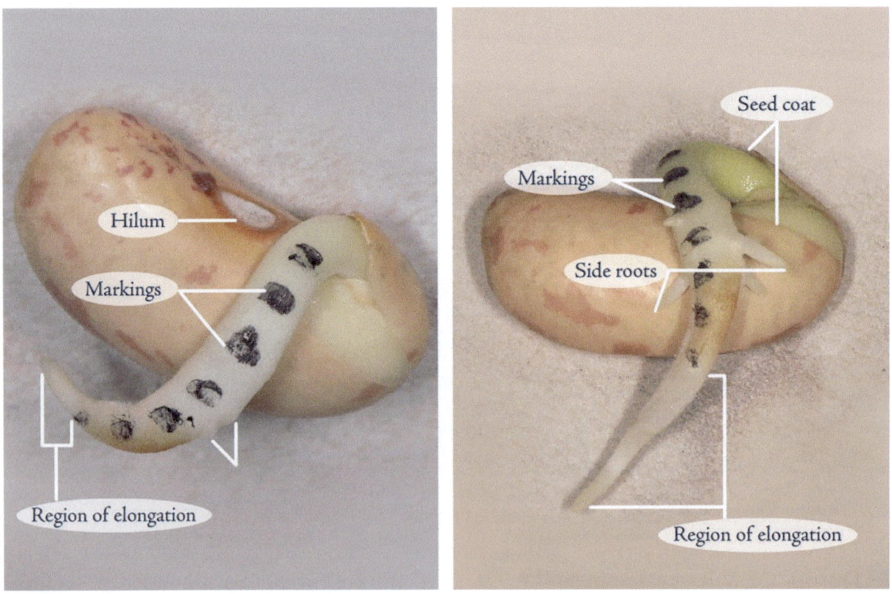

Fig. 14.13 Half a day passed between the two bean (*Phaseolus vulgaris*) sprout pictures. The significant length growth in the elongation zone can be clearly observed

14.3 Detection of Respiratory Carbon Dioxide with Lime Water

For the experiment, the lime water (calcium hydroxide solution) should always be freshly prepared. Mix 1 g of calcium hydroxide [$Ca(OH)_2$], or calcium oxide (CaO) with a little distilled water, then dilute it to 100 cm^3 (Fig. 14.14)!

Calcium hydroxide and calcium oxide easily carbonate with the carbon dioxide found in the air:

$$Ca(OH)_2 + CO_2 = CaCO_3 + H_2O$$

$$CaO + CO_2 = CaCO_3,$$

therefore, we can only get a suitable concentration of calcium hydroxide solution using not too old chemicals.

We can also produce lime water by throwing pieces of metal calcium into water:

$$Ca + 2H_2O = Ca(OH)_2 + H_2$$

After the whitish solution obtained in this way has settled, filter it through filter paper! The resulting clear solution is lime water (calcium hydroxide solution). Pour 5 cm^3 of lime water into a test tube, then blow air into the solution with a straw! The

Fig. 14.14 Preparation of lime water

carbon dioxide in the exhaled air causes the lime water to become cloudy due to the forming and precipitating calcium carbonate ($CaCO_3$).

$$Ca(OH)_2 + CO_2 = CaCO_3 + H_2O$$

14.4 Examination of Plant Respiration in a Test Tube

As the first step of the germination process, the seeds absorb water and swell. At this point, their degrading metabolism becomes intense, and they produce carbon dioxide (CO_2).

The majority of the produced carbon dioxide comes from fermentation processes before the seed coat cracks, and from biological oxidation (cellular respiration) afterwards.

To detect the carbon dioxide emitted by the seeds, set up the following experiment! Soak two thin strips of filter paper in lime water, then drop phenolphthalein indicator solution on them! The phenolphthalein turns red due to the alkaline pH of the solution. Place the reddened filter paper strips in two test tubes, then put dry bean seeds next to one and bean seeds soaked in water for a day next to the other! Then seal both test tubes with a rubber stopper! The swollen bean seeds use up the oxygen content of the air in the test tube during their respiration, while producing carbon dioxide.

The latter turns the calcium hydroxide of the lime water into calcium carbonate, while the pH of the filter paper strips changes from alkaline to neutral. As a result, the originally red phenolphthalein gradually fades (Fig. 14.15). The test can also be performed with germinating seeds (e.g. mustard seeds) and small cut plant parts (e.g. potato tuber, onion, carrot, and apple).

14.5 Respiration and Photosynthesis of Green Plant Parts

During their respiration, plants produce carbon dioxide, while during their photosynthesis, they produce oxygen and bind carbon dioxide. In the following, we will examine with the help of a test tube experiment whether the rate of photosynthetic carbon dioxide binding or respiratory carbon dioxide production is greater in the case of illumination.

Place a filter paper strip soaked in limewater and dyed red with phenolphthalein into three test tubes, then put finely chopped potato into one, and green leaf pieces into another!

Nothing goes into the third test tube, this will be the control. Place these three test tubes in sunlight, or illuminate them with a bright lamp! Place a fourth test tube filled with a dyed filter paper strip and green leaf pieces in a dark place! After 10–15 min, examine the contents of the test tubes!

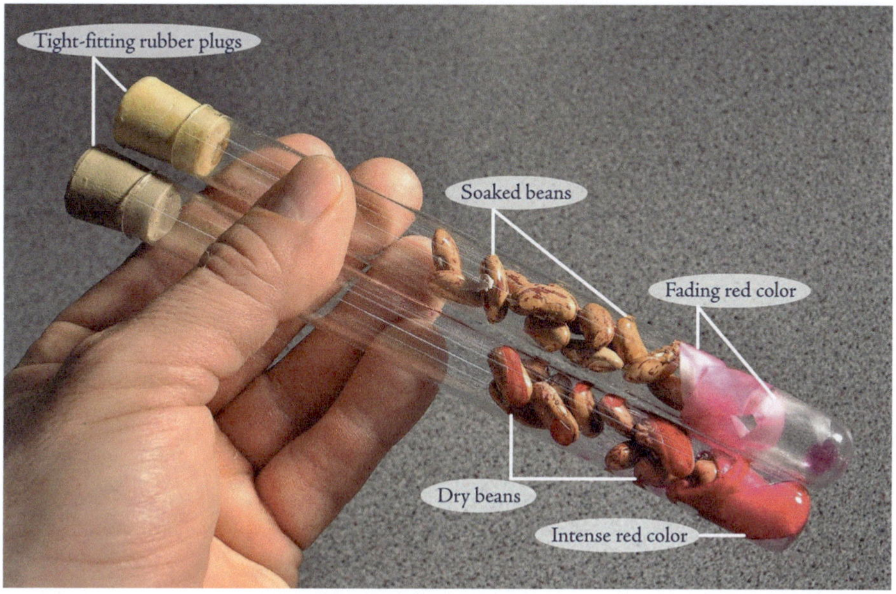

Fig. 14.15 Examination of plant respiration in a test tube

Of the three illuminated test tubes, only in the first one, containing the potato pieces, did the originally red filter paper strip fade in colour. This suggests that the green plant parts bound the carbon dioxide released as a result of respiration during their photosynthesis, thereby the carbon dioxide concentration did not increase in the second test tube (Fig. 14.16).

In the test tube stuffed with leaf pieces placed in the dark, photosynthesis does not take place due to the lack of light, therefore the carbon dioxide produced during plant respiration gradually fades the originally red filter paper strip (Fig. 14.17).

The process takes a longer time (10–20 min), because there is no direct need for light for photosynthetic carbon dioxide binding, so this process continues for a while even after the cessation of illumination.

14.6 Examination of Plant Material Transport

It is also worth examining the tulip petal itself with a light microscope. At lower magnification, light stripes appear, inside which we can identify vascular bundles at higher magnification (Figs. 14.18 and 14.19).

The vascular bundles are also suitable for colouring these tepals by supplying dyed water to the tepals. After a few hours, a colourless pattern emerges on the petals of a white tulip placed in a solution of toluidine blue or red eosin (Fig. 14.20).

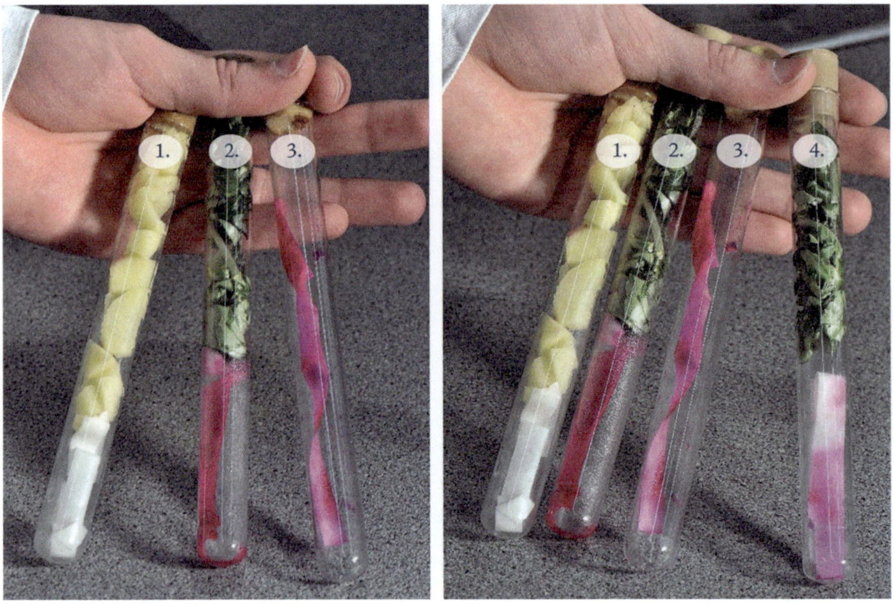

Fig. 14.16 Image 1: In the test tube containing leaf pieces, the filter paper strip soaked in limewater and dyed red with phenolphthalein does not fade in colour, similar to the control. Image 2: In the test tube containing green plant parts placed in the dark (fourth test tube), the filter paper strip slowly fades in colour

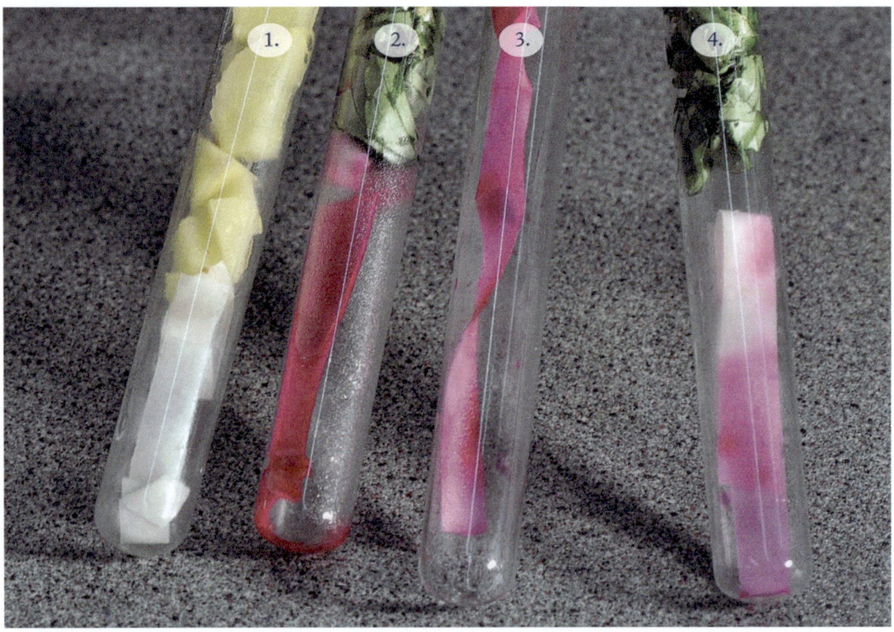

Fig. 14.17 The colour change of the filter paper strips can be clearly observed in the magnified image

Fig. 14.18 Light microscopic image of a tulip (*Tulipa gesneriana*) tepal

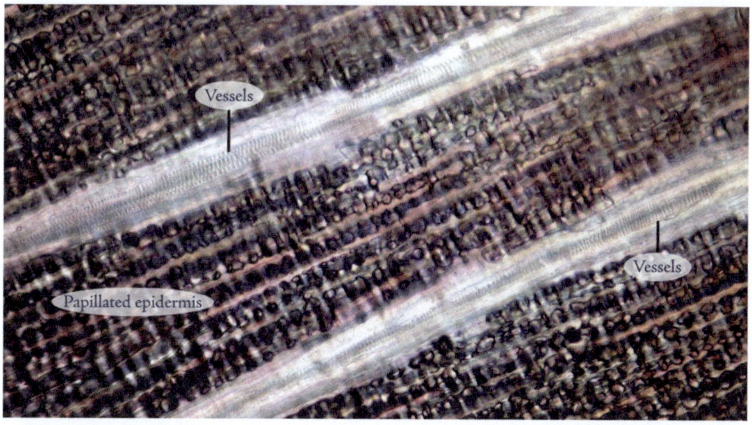

Fig. 14.19 Higher magnification light microscopic image of a tulip (*Tulipa gesneriana*) tepal

With this experiment, we can not only demonstrate the transportation itself, but with histological examinations, we can also determine which part of the vascular bundle transports the stained water upwards.

First, as a kind of control, let us make cross-sections from the unstained stem of the tulip and examine these under a light microscope!

Half an hour after placing in the coloured water, make thin cross-sections from the stem at different heights.

Where the dye appears in the stem cross-section, it can be clearly observed that it is concentrated in the xylem parts of the vascular bundles. This clearly proves that the upward flow of water occurs in the xylem of the vascular bundles (Figs. 14.21, 14.22, 14.23, and 14.24).

Fig. 14.20 White tulip (*Tulipa gesneriana*) bouquet and the dyed tepals (thumbnail)

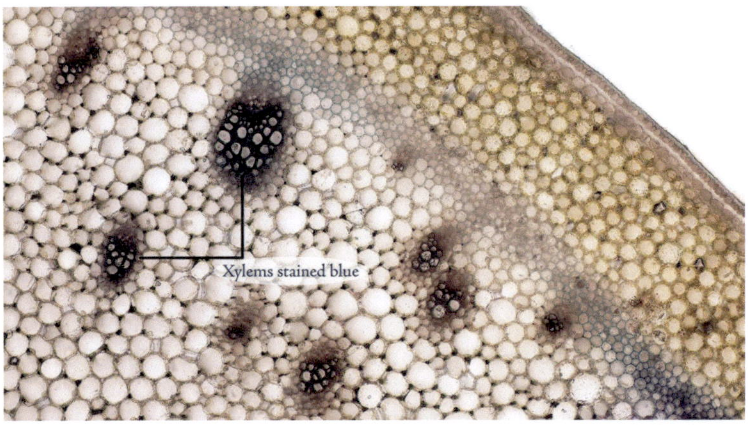

Fig. 14.21 Cross-section of a tulip (*Tulipa gesneriana*) stem placed in a solution of toluidine blue

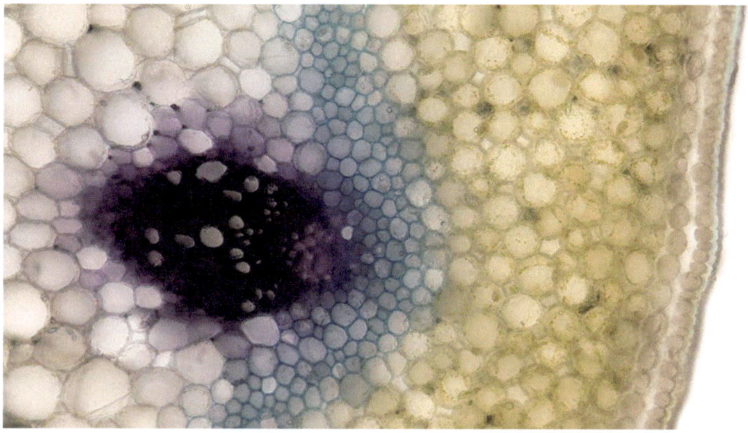

Fig. 14.22 Cross-section of a tulip (*Tulipa gesneriana*) vascular bundle stained with toluidine blue

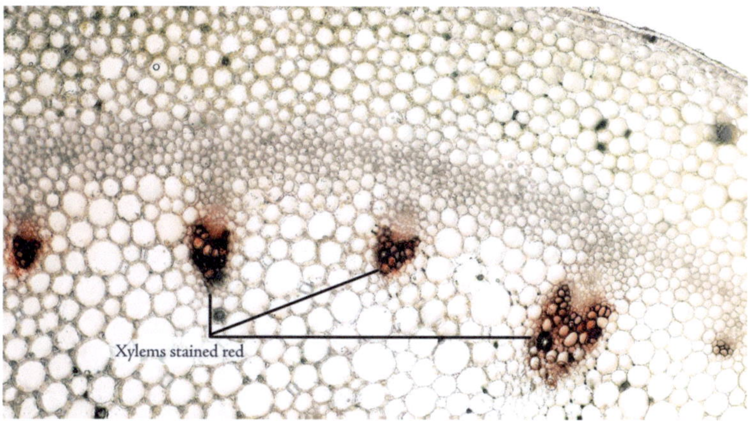

Fig. 14.23 Cross-section of a tulip (*Tulipa gesneriana*) vascular bundle stained with eosin

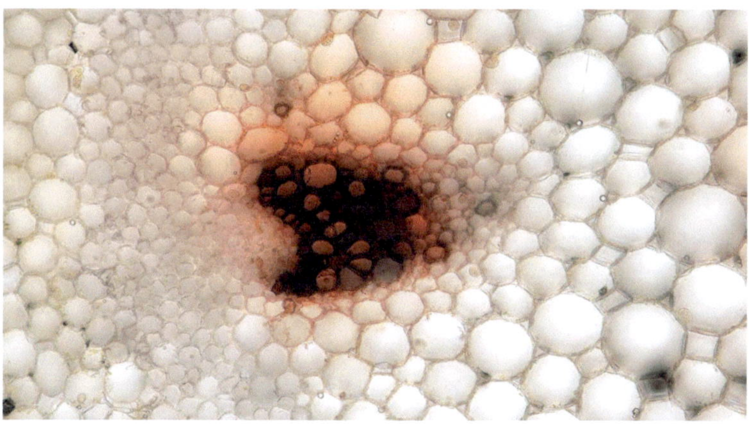

Fig. 14.24 Cross-section of a tulip (*Tulipa gesneriana*) vascular bundle stained with eosin

Digital Material 15

The digital annex of the book contains drop-down activity tests, which are based on the pictures in the book and can be solved using a mobile phone. This makes them suitable for recording and practicing knowledges.

Below are pictures on a mobile phone screen to illustrate the use of the programme. The task is to recognise the details of a picture and select the names of the parts marked with letter signs from a list. Clicking on the textblock next to the letter will display the selectable labels in large size (Fig. 15.1).

After completing the task, the programme will check your work by clicking on the Check button. Correct answers are displayed on a green background and incorrect answers on a red background. You can solve the task several times in a row to correct mistakes (Fig. 15.2).

15.1 Digital Tasks

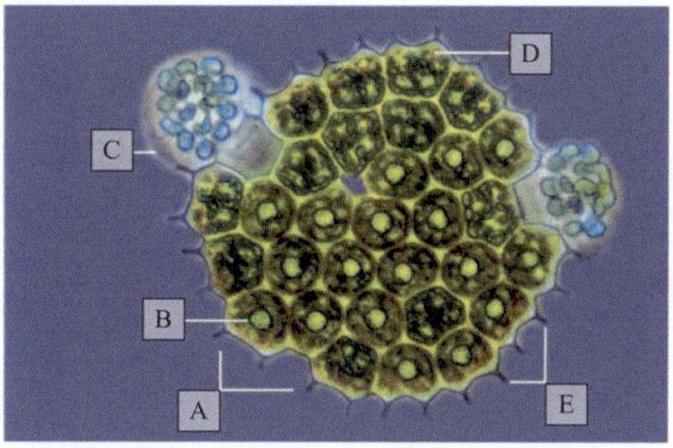

Algae, moss, lichens, ferns: 21 tasks

Supplementary Information The online version contains supplementary material available at https://doi.org/10.1007/978-3-031-84085-2_15.

G. Kriska, *Teaching Biology*, https://doi.org/10.1007/978-3-031-84085-2_15

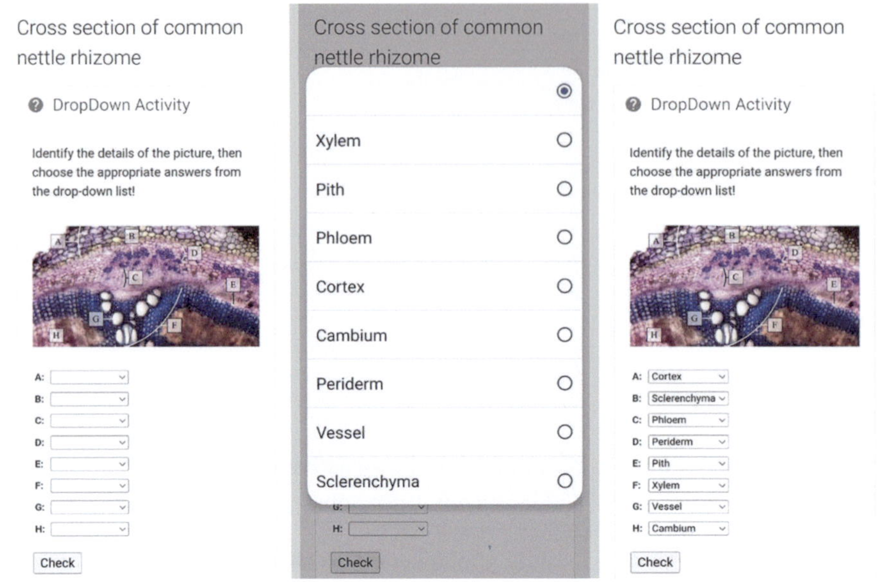

Fig. 15.1 Solve the task by mobile phone

Vegetative organs: 31 tasks

Flowers and inflorescence: 55 tasks

Fig. 15.2 Correcting the wrong answers

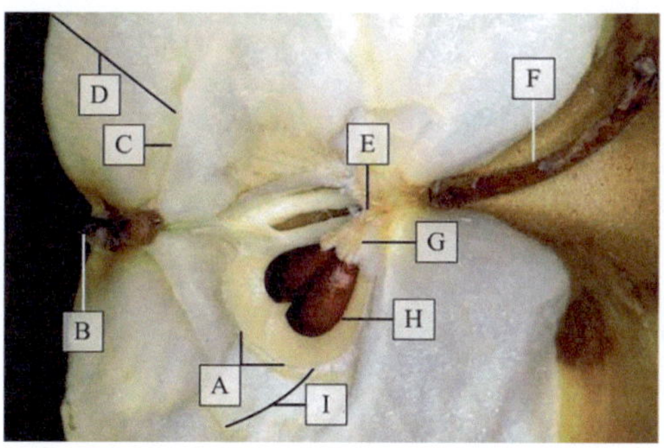

Fruits: 19 tasks

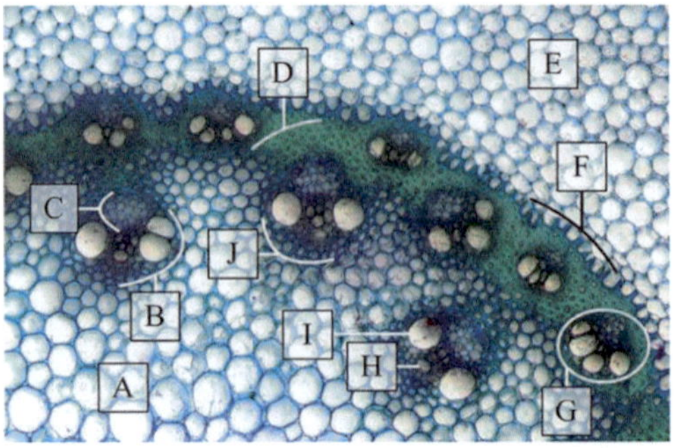

Root and rhizome tissues: 52 tasks

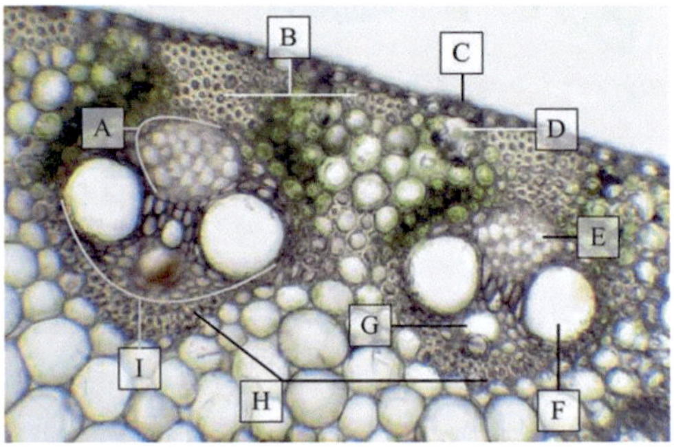

Horsetails and monocot stem, primary growth in dicot stem: 49 tasks

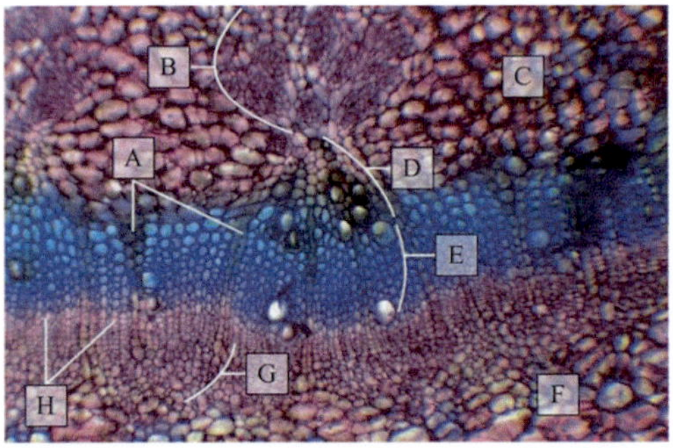

Secondary growth in dicot stem: 59 tasks

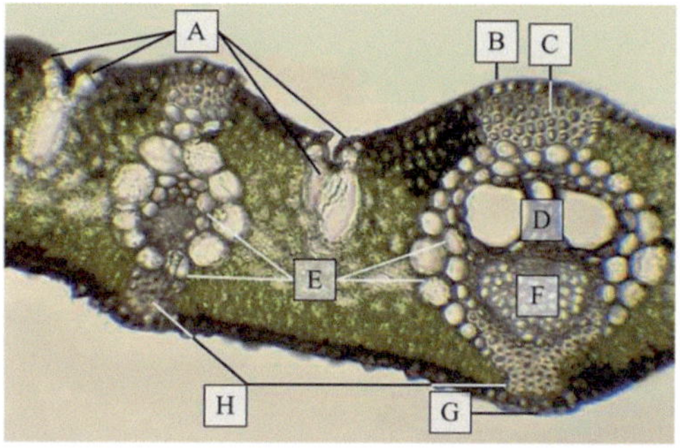

Leaf tissues: 26 tasks

Further Reading

Beck CB (2010) An introduction to plant structure and development: plant anatomy for the twenty-first century, 2nd edn. Cambridge University Press, Cambridge, p 459

Bresinsky A, Körner C, Kadereit JW, Neuhaus G, Sonnewald U (2013) Strasburger's plant sciences - including prokaryotes and fungi. Springer, Berlin, p 1302

Crang R, Lyons-Sobaski S, Wise R (2018) Plant anatomy-based approach to the structure of seed plants. Springer, Cham, p 725

Cutler DF, Botha T, Stevenson DW (2008) Plant anatomy: an applied approach. Wiley-Blackwell, Amsterdam p. 320

Haraszty Á, Fridvalszky L, Gracza P (1982) Microscopic plant anatomy. Textbook Publisher, Budapest, p 256 (in Hungarian)

Kriska Gy (2008) 3D photography and visualization. Flaccus Publisher, p 103 (in Hungarian)

Kriska Gy (2011) Preparation for graduation in biology. Photosynthesizing organisms 1. National Textbook Publisher, p 160 (in Hungarian)

Kriska Gy (2012) Preparation for graduation in biology. Photosynthesizing organisms 2. National Textbook Publisher, p. 203 (in Hungarian)

Kriska Gy (ed) (2013) Preparation for graduation in biology. Fungi, biochemistry, animal cell biology and histology, physiology. Knowledge of Generations Textbook Publisher, p 224 (in Hungarian)

Kristóf Z (ed) (2013) Structure of plants and fungi. Eötvös Loránd University, Budapest, p 101

Rudall PJ (2020) Anatomy of flowering plants. An introduction to plant structure and development, 4th edn. Cambridge University Press, Cambridge, p 148

Sárkány S, Szalai I (1966) Exercises in plant anatomy. Textbook Publisher, Budapest, p 708 (in Hungarian)

Tuba Z, Szerdahelyi T, Engloner A, Nagy J (2007) Botany I. Cell science, histology, anatomy. National Textbook Publisher, Budapest, p 378 (in Hungarian)

G. Kriska, *Teaching Biology*, https://doi.org/10.1007/978-3-031-84085-2

Index

G. Kriska, *Teaching Biology*, https://doi.org/10.1007/978-3-031-84085-2

MIX
Papier aus verantwortungsvollen Quellen
Paper from responsible sources
FSC® C105338

If you have any concerns about our products,
you can contact us on
ProductSafety@springernature.com

In case Publisher is established outside the EU,
the EU authorized representative is:
Springer Nature Customer Service Center GmbH
Europaplatz 3, 69115 Heidelberg, Germany

Printed by Libri Plureos GmbH
in Hamburg, Germany